Novel Wastewater Treatment Applications Using Polymeric Materials

Novel Wastewater Treatment Applications Using Polymeric Materials

Editors

Irene S. Fahim
Ahmed K. Badawi
Hossam E. Emam

MDPI • Basel • Beijing • Wuhan • Barcelona • Belgrade • Manchester • Tokyo • Cluj • Tianjin

Editors
Irene S. Fahim
The Nile University
Egypt

Ahmed K. Badawi
El-Madina Higher Institute for Engineering and Technology
Egypt

Hossam E. Emam
National Research Centre
Egypt

Editorial Office
MDPI
St. Alban-Anlage 66
4052 Basel, Switzerland

This is a reprint of articles from the Special Issue published online in the open access journal *Polymers* (ISSN 2073-4360) (available at: https://www.mdpi.com/journal/polymers/special_issues/Novel_Wastewater_Treatment_Applications_Using_Polymeric_Materials).

For citation purposes, cite each article independently as indicated on the article page online and as indicated below:

LastName, A.A.; LastName, B.B.; LastName, C.C. Article Title. *Journal Name* **Year**, *Volume Number*, Page Range.

ISBN 978-3-0365-6243-8 (Hbk)
ISBN 978-3-0365-6244-5 (PDF)

© 2023 by the authors. Articles in this book are Open Access and distributed under the Creative Commons Attribution (CC BY) license, which allows users to download, copy and build upon published articles, as long as the author and publisher are properly credited, which ensures maximum dissemination and a wider impact of our publications.

The book as a whole is distributed by MDPI under the terms and conditions of the Creative Commons license CC BY-NC-ND.

Contents

About the Editors vii

Stuti Jha, Rama Gaur, Syed Shahabuddin, Irfan Ahmad and Nanthini Sridewi
Kinetic and Isothermal Investigations on the Use of Low Cost Coconut Fiber-Polyaniline Composites for the Removal of Chromium from Wastewater
Reprinted from: *Polymers* **2022**, *14*, 4264, doi:10.3390/polym14204264 1

Hamid Najarzadekan, Muhammad Afzal Kamboh, Hassan Sereshti, Irfan Ahmad, Nanthini Sridewi, Syed Shahabuddin and Hamid Rashidi Nodeh
Headspace Extraction of Chlorobenzenes from Water Using Electrospun Nanofibers Fabricated with Calix[4]arene-Doped Polyurethane–Polysulfone
Reprinted from: *Polymers* **2022**, *14*, 3760, doi:10.3390/polym14183760 17

Hamid Najarzadekan, Hassan Sereshti, Irfan Ahmad, Syed Shahabuddin, Hamid Rashidi Nodeh and Nanthini Sridewi
Superhydrophobic Nanosilica Decorated Electrospun Polyethylene Terephthalate Nanofibers for Headspace Solid Phase Microextraction of 16 Organochlorine Pesticides in Environmental Water Samples
Reprinted from: *Polymers* **2022**, *14*, 3682, doi:10.3390/polym14173682 27

Haniyeh Najafvand Drikvand, Mitra Golgoli, Masoumeh Zargar, Mathias Ulbricht, Siamak Nejati and Yaghoub Mansourpanah
Thermo-Responsive Hydrophilic Support for Polyamide Thin-Film Composite Membranes with Competitive Nanofiltration Performance
Reprinted from: *Polymers* **2022**, *14*, 3376, doi:10.3390/polym14163376 41

Wafa Shamsan Al-Arjan
Zinc Oxide Nanoparticles and Their Application in Adsorption of Toxic Dye from Aqueous Solution
Reprinted from: *Polymers* **2022**, *14*, 3086, doi:10.3390/polym14153086 55

Noureddine Mahdhi, Norah Salem Alsaiari, Abdelfattah Amari and Mohamed Ali Chakhoum
Effect of TiO_2 Nanoparticles on Capillary-Driven Flow in Water Nanofilters Based on Chitosan Cellulose and Polyvinylidene Fluoride Nanocomposites: A Theoretical Study
Reprinted from: *Polymers* **2022**, *14*, 2908, doi:10.3390/polym14142908 81

Thanchanok Ratvijitvech
Fe-Immobilised Catechol-Based Hypercrosslinked Polymer as Heterogeneous Fenton Catalyst for Degradation of Methylene Blue in Water
Reprinted from: *Polymers* **2022**, *14*, 2749, doi:10.3390/polym14132749 101

Majed Al Anazi, Ismail Abdulazeez and Othman Charles S. Al Hamouz
Selective Removal of Iron, Lead, and Copper Metal Ions from Industrial Wastewater by a Novel Cross-Linked Carbazole-Piperazine Copolymer
Reprinted from: *Polymers* **2022**, *14*, 2486, doi:10.3390/polym14122486 119

Alejandro Onchi, Carlos Corona-García, Arlette A. Santiago, Mohamed Abatal, Tania E. Soto, Ismeli Alfonso and Joel Vargas
Synthesis and Characterization of Thiol-Functionalized Polynorbornene Dicarboximides for Heavy Metal Adsorption from Aqueous Solution
Reprinted from: *Polymers* **2022**, *14*, 2344, doi:10.3390/polym14122344 133

Abdallah Tageldein Mansour, Ahmed E. Alprol, Khamael M. Abualnaja, Hossam S. El-Beltagi, Khaled M. A. Ramadan and Mohamed Ashour
Dried Brown Seaweed's Phytoremediation Potential for Methylene Blue Dye Removal from Aquatic Environments
Reprinted from: *Polymers* **2022**, *14*, 1375, doi:10.3390/polym14071375 153

Iram Ayaz, Muhammad Rizwan, Jeffery Layton Ullman, Hajira Haroon, Abdul Qayyum, Naveed Ahmed, Basem H. Elesawy, Ahmad El Askary, Amal F. Gharib and Khadiga Ahmed Ismail
Lignocellulosic Based Biochar Adsorbents for the Removal of Fluoride and Arsenic from Aqueous Solution: Isotherm and Kinetic Modeling
Reprinted from: *Polymers* **2022**, *14*, 715, doi:10.3390/polym14040715 179

Yuxin Chen, Yujuan Chen, Dandan Lu and Yunren Qiu
Synthesis of a Novel Water-Soluble Polymer Complexant Phosphorylated Chitosan for Rare Earth Complexation
Reprinted from: *Polymers* **2022**, *14*, 419, doi:10.3390/polym14030419 195

Vairavel Parimelazhagan, Gautham Jeppu and Nakul Rampal
Continuous Fixed-Bed Column Studies on Congo Red Dye Adsorption-Desorption Using Free and Immobilized *Nelumbo nucifera* Leaf *Adsorbent*
Reprinted from: *Polymers* **2022**, *14*, 54, doi:10.3390/polym14010054 203

Nouf F. Al-Harby, Ebtehal F. Albahly and Nadia A. Mohamed
Kinetics, Isotherm and Thermodynamic Studies for Efficient Adsorption of Congo Red Dye from Aqueous Solution onto Novel Cyanoguanidine-Modified Chitosan Adsorbent
Reprinted from: *Polymers* **2021**, *13*, 4446, doi:10.3390/polym13244446 227

Adedapo Oluwasanu Adeola and Philiswa Nosizo Nomngongo
Advanced Polymeric Nanocomposites for Water Treatment Applications: A Holistic Perspective
Reprinted from: *Polymers* **2022**, *14*, 2462, doi:10.3390/polym14122462 259

Anton Manakhov, Maxim Orlov, Vyacheslav Grokhovsky, Fahd I. AlGhunaimi and Subhash Ayirala
Functionalized Nanomembranes and Plasma Technologies for Produced Water Treatment: A Review
Reprinted from: *Polymers* **2022**, *14*, 1785, doi:10.3390/polym14091785 283

Hamad Noori Hamad and Syazwani Idrus
Recent Developments in the Application of Bio-Waste-Derived Adsorbents for the Removal of Methylene Blue from Wastewater: A Review
Reprinted from: *Polymers* **2022**, *14*, 783, doi:10.3390/polym14040783 305

About the Editors

Irene S. Fahim

Irene S. Fahim is an associate Professor, Industrial and Service Engineering and Management department, Nile University Cairo, Egypt. She is the leader for the industrial and manufacturing track in the Smart Engineering systems research center, Nile University. She won the state encouragement award for women 2021. Irene participated in Fulbright Junior Faculty program for renewable energy, 2016 and participated in Entrepreneurship and leadership Program, 1000 Women, Goldman Sachs, 2016. She got Newton Mosharfa institutional link award for two years in collaboration with Nottingham University, UK. for manufacturing plastic bags from natural materials where she got acknowledged by Mr President El-SISI for her work at the 3rd national youth conference in Ismailia, April 2017. Irene is interested in studying the sustainability concept and she was a co-author of a book entitled "Sustainability and Innovation", AUC press, 2015. Irene is a volunteer in IEEE Smart Village committee and active member in the organizing committee for IEEE Conferences such as IEEE Power Africa 2017, where Irene plans to extend the IEEE smart village committee in Egypt to supply solar electricity in poor villages. She is also one of the members in the Events Committee of the IEEE Humanitarian Activities, 2018. She is also the technical chair for the first IEEE SIGHT Egypt ideation camp 2018.

Ahmed K. Badawi

Ahmed K. Badawi is an assistant professor in Environmental Engineering (E.E). He received his M.Sc. and Ph.D. degrees in E.E and was appointed as a university professor at 29 years old. Dr. Karam's research focuses on municipal and industrial wastewater treatment using innovative approaches. He has investigated the viability of applying algal–bacterial photo-bioreactors for wastewater treatment in large scale applications. He has also investigated several nano-sized advanced/hybrid materials for industrial wastewater treatment. Dr. Badawi is a specialist in pilot plant design, fabrication, and implementation. He has designed and constructed several pilot plants in different WWTPs and factories. He has published in several high-impact journals (Q1 & Q2) and international conferences and is an acting editor and reviewer at several reputable international journals. He has edited and reviewed over (120) manuscripts and was acknowledged as an editor/reviewer by several international journals. He has also served as an acting PI/Co-PI for many national and international funded grants.

Hossam E. Emam

Hossam E. Emam is a Professor in Department of Pretreatment and Finishing of Cellulosic Fibers, Textile Research Division, National Research Centre, Giza, Egypt.

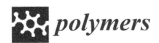

Article

Kinetic and Isothermal Investigations on the Use of Low Cost Coconut Fiber-Polyaniline Composites for the Removal of Chromium from Wastewater

Stuti Jha [1], Rama Gaur [1,*], Syed Shahabuddin [1,*], Irfan Ahmad [2] and Nanthini Sridewi [3,*]

[1] Department of Chemistry, School of Technology, Pandit Deendayal Energy University, Knowledge Corridor, Raysan, Gandhinagar 382426, Gujarat, India
[2] Department of Clinical Laboratory Sciences, College of Applied Medical Sciences, King Khalid University, Abha 61421, Saudi Arabia
[3] Department of Maritime Science and Technology, Faculty of Defence Science and Technology, National Defence University of Malaysia, Kuala Lumpur 57000, Malaysia
* Correspondence: rama.gaur@sot.pdpu.ac.in (R.G.); syedshahab.hyd@gmail.com or syed.shahabuddin@sot.pdpu.ac.in (S.S.); nanthini@upnm.edu.my (N.S.); Tel.: +91-8585932338 (S.S.); +60-124-675-320 (N.S.)

Citation: Jha, S.; Gaur, R.; Shahabuddin, S.; Ahmad, I.; Sridewi, N. Kinetic and Isothermal Investigations on the Use of Low Cost Coconut Fiber-Polyaniline Composites for the Removal of Chromium from Wastewater. *Polymers* **2022**, *14*, 4264. https://doi.org/10.3390/polym14204264

Academic Editors: Irene S. Fahim, Ahmed K. Badawi and Hossam E. Emam

Received: 5 September 2022
Accepted: 7 October 2022
Published: 11 October 2022

Publisher's Note: MDPI stays neutral with regard to jurisdictional claims in published maps and institutional affiliations.

Copyright: © 2022 by the authors. Licensee MDPI, Basel, Switzerland. This article is an open access article distributed under the terms and conditions of the Creative Commons Attribution (CC BY) license (https://creativecommons.org/licenses/by/4.0/).

Abstract: Pollution due to various heavy metals is increasing at an alarming rate. Removal of hexavalent chromium from the environment is a significant and challenging issue due to its toxic effects on the ecosystem. Development of a low-cost adsorbent with better adsorption efficiency is presently required. In this study, waste coconut fibers (CF) were used to prepare its composite with polyaniline (PANI) via in-situ oxidation. The obtained composites with varying loading of PANI (15, 25, 50, and 75% w/w) were characterized by FE-SEM, TGA, and FTIR spectroscopy. The prepared composites were evaluated for their adsorption performance for removal of Cr(VI). It was concluded that the composite with 50% w/w polyaniline loading on coconut fiber exhibited a maximum adsorption efficiency of 93.11% in 30 min. The effect of pH, dosage, and concentration of the aqueous solution of chromium on the Cr(VI) adsorption efficiency of the composite was also studied. From the optimization studies it was observed that the absorbents exhibited the best adsorption response for Cr(VI) removal with 0.25 mg/mL adsorbent at pH 4, in 30 min. The effect of pH, dosage, and concentration of the aqueous solution of chromium on the Cr(VI) adsorption efficiency of the composite was also studied. This study highlights the application of low-cost adsorbent as a potential candidate for the removal of hexavalent chromium. A detailed study on the adsorption kinetics and isothermal analysis was conducted for the removal of Cr(VI) from aqueous solution using coconut fiber-polyaniline composite. From the kinetic investigation, the adsorption was found to follow the pseudo second order model. The data obtained were best fitted to the Elovich model confirming the chemisorption of the Cr(VI) on coconut polymer composites. The analysis of the isothermal models indicated monolayer adsorption based on the Langmuir adsorption model.

Keywords: adsorption; heavy metals; environmental remediation; wastewater; polyaniline; coconut fiber

1. Introduction

Heavy metal pollution is a prime concern for the society due to their toxicity, persistent nature and bioaccumulation in the environment [1]. Heavy metals are metals with densities greater than 5 gm/cm^3 and atomic numbers greater than 20 [2]. Such metals pose a serious threat to human, plant, and animal health. Because of their toxicity, heavy metal removal should be considered. Heavy metals are omnipresent in the environment, the concentration of which is increasing due to modern day urbanization and industrialization [3]. Heavy metals include Cr, Hg, Pb, Co, Ni, Cu, Zn, Sn, and Cd, etc. Chromium is a naturally occurring element with valency ranging from II to VI [4]. The main oxidation state of chromium is III and VI. When chromium is released into the environment due to various activities, it is mainly in its hexavalent form [5]. The hexavalent state of chromium is more stable and mobile than its trivalent state. Cr(VI) is a common contaminant in many environmental systems as it is widely used in various processes such as in dyes and pigments, leather tanning, chrome plating, etc. [6–8]. Many methods are implemented for the removal of heavy metal like adsorption, electro dialysis, ion-exchange, reverse osmosis, and ultra-filtration, etc. [9–14]. Among all the methods used for heavy metal remediation, adsorption is the most widely adapted method [15]. The adsorption method involves a simple set up and has higher performance efficiency. It is a regenerative and a cost-effective method making it the most feasible method for heavy metal removal [16]. Many low-cost adsorbents have been used by researchers for the adsorption study of chromium. Several studies using agricultural wastes such as banana peels, citrus limetta peels, coconut husk, potato peels, palm pressed fibers, and sawdust, etc. as an adsorbent for the treatment of chromium have been reported [17–21]. However, their efficiency is limited and can be modified by combining them with other suitable materials. This awakens the necessity of development of new or modification of the already used adsorbents for effective Cr(VI) removal.

Recently, conducting polymers have attracted a lot of attention in pollutant adsorption due to their properties such as special morphologies, functional groups and simple synthetic procedure [22]. They have the ability to remove heavy metals through complexation and ion-exchange mechanism [23]. Polyaniline (PANI) is a polymer which has been explored in recent years for its potential as a heavy metal adsorbent. PANI, a conducting polymer with terminal amine (–NH_2) group has excellent properties such as high surface area, adjustable surface chemistry, desirable pore size distribution, rigidity, and economical regeneration [24]. Apart from PANI, PANI-based composites have also been studied for their application in heavy metal adsorption. PANI-based composites offer added advantages such as higher surface area, higher dispersibility, enhanced adsorption performance, and combined properties of the polymer and the substrate [22,25]. Dutta et al. (2021) synthesized polyaniline-polypyrrole copolymer coated green rice husk ash and investigated its potential for Cr(VI) removal [26]. PANI-jute fiber was synthesized by Kumar et al. (2008) for removal of hexavalent chromium from wastewater [27]. PANI-magnetic mesopores silica composite was used an adsorbent for chromium adsorption by Tang et al. (2014) [28]. Hexavalent chromium was adsorbed on the surface of PANI-rice husk nanocomposite by Ghorbani et al. (2011) [29]. Lei et al. has reported the use of PANI-magnetic chitosan composite for the removal of hexavalent chromium [30]. Rahmi et al. reported the use of using chitosan based composites chitosan for the removal of Cd(II) from its aqueous solution [31,32]. Cr(VI) was adsorbed using gelatine composites in a study reported by Marciano et al. [33]. From the literature review, it was inferred that all the similar studies have either reported the use of large amount of adsorbents (1 to 125 g/L) for the removal of contaminants or a more time consuming process (up to 5 to 6 h). In addition, the removal efficiency is also less in comparison to the present study. Thus, it was observed that the present study offers certain advantages such as using low-cost adsorbent, less adsorbent dosage, and high efficiency in less time. In this study, coconut fibers have been used as a substrate and PANI has been dispersed on its surface using in-situ polymerization. Agricultural waste such as coconut fiber has advantages over other substrates such as being easy collectable and available with less or no cost. Moreover, it involves simple processing steps (washing, drying, sieving) and thus reduces energy and production cost. This study aims to focus on the synthesis of PANI-coconut fiber and its application for Cr(VI) removal. It also reports the effect of various parameters such as pH, adsorbent dosage and concentration of Cr on the adsorption capacity of the composite. A detailed investigation on the kinetic aspects and adsorption isotherm has been performed.

2. Materials and Methods

2.1. Materials

In this study, coconut shells were collected from the local market of Gandhinagar, Gujarat, India. Potassium dichromate ($K_2Cr_2O_7$) (SRL (Ahemdabad, Gujarat, India) AR grade, extrapure, 99.9%) was used as the source of Cr(VI). Aniline used in this process was purified using distillation process prior use. All the chemicals used for the preparation of composites including aniline, ammonium persulfate (APS) (Sigma Aldrich (Ahemdabad, Gujarat, India), reagent grade, 98%), HCl (Finar (Ahemdabad, Gujarat, India), AR grade, 37% purity) were used as received. All the dilutions performed in this study were carried out using milli pore water.

2.1.1. Pre-Treatment of Coconut Fibers

The collected coconut shells were separated into coconut fibers (CF) and washed to remove dirt. The coconut fibers were then dried under shade. The dried coconut fibers were cut into pieces before grinding them to make a fine powder. The obtained coconut fiber powder was sieved to obtain particles of uniform size (≤ 75 microns).

2.1.2. Synthesis of Polyaniline (PANI) and Polyaniline-Coconut Fiber (PANI-CF) Composites

The composites were prepared with different $w/w\%$ loading of PANI on CF. PANI was prepared by in-situ oxidation method. For this process, a solution of aniline in 1 M HCl was prepared. Another solution of ammonium persulphate (APS) dissolved in 1 M HCl was added dropwise with constant stirring for 2–3 h. The reaction temperature was maintained between 0 to 5 °C. Subsequently, the reaction mixture was filtered and washed with 0.5 M HCl until the filtrate became colorless and then with deionized water until the filtrate became neutral. Then, the obtained PANI was dried in vacuum oven at 80°C overnight. The composites with varying loading of PANI were prepared using a similar approach. The schematic flow of both the processes is shown in Figure S1a,b (Supplementary Materials). The digital images of the prepared composites are shown in Figure S1c (Supplementary Materials).

Composites were prepared with different $w/w\%$ loading of PANI. The samples were coded as CFC15, CFC25, CFC50 and CFC75 for 15%, 25%, 50% and 75% PANI, respectively. For preparation of all the composites, the starting weight of CF was kept 0.5 gm and the weight of PANI was varied. Rest all steps were similar to the synthesis of PANI. Depending on the composition of PANI, the prepared samples were named CFC15, CFC25, CFC50, and CFC75 as listed in Table 1.

Table 1. Nomenclature and starting weight of the reagents for the preparation of composites.

Sample Code	CF (gm)	Weight of PANI (gm)	Weight of APS (gm)	Weight of Product (gm)
CFC15	0.5	0.075	0.231	0.34
CFC25	0.5	0.15	1.54	0.36
CFC50	0.5	0.5	1.54	0.81
CFC75	0.5	1.5	4.63	1.25

3. Characterization

All the samples in the present study were analyzed for their functional groups, morphology and thermal stability using different characterization techniques such as Fourier-transform infrared spectroscopy (FT-IR), Field Emission-Scanning Electron Microscopy (FE-SEM), and Thermal gravimetric analyzer (TGA). The Fourier transform- infrared spectra of the samples were recorded using FT-IR spectrometer Perkin Elmer (Mumbai, Maharashtra, India), spectrum 2 model in ATR mode in a scan range of 400 to 4000 cm^{-1}. The FE-SEM images of the samples were taken in Zeiss ultra 55 model (Bangalore, Karnataka, India) at acceleration voltage of 5.00 kV. For FE-SEM analysis the samples were sprinkled on clean aluminum stub over conducting carbon tape. The samples on aluminum stubs were coated with a thin gold layer using LEICA EM ACE200 (Wetzlar, Hesse, Germany) to make them conductive. The thermal stabilities of the samples were analyzed using Thermal gravimetric analyzer (Eltra sthermostep, (Hyderabad, Telangana, India) with a heating rate of 10 °C/min with temperature range 200 to 950 °C in O_2 atmosphere. The concentration of the aqueous solution of chromium during adsorption study was monitored using LABINDIA analytical model 3000$^+$ Ultraviolet-visible (UV-Vis) Spectrometer (Ahemdabad, Gujarat, India), in absorbance scan mode in the range of 200–800 nm.

4. Adsorption Studies

In the present study, adsorption of Cr(VI) in aqueous solution using CFC15, CFC25, CFC50, CFC75 and PANI was carried out at room temperature in batch mode. For the adsorption studies, a test solution of Cr(VI) (10 ppm) was prepared and used for adsorption. For the preparation of 10 ppm solution, 0.02828 gm of potassium dichromate was dissolved in 1 L of water. For the detailed study for the adsorption of chromium in aqueous solution, different sets of experiment were performed. The effect of different parameters such as dosage of adsorbent, concentration of chromium solution and pH were explored for adsorption of Cr(VI) in aqueous solution. In a typical adsorption experiment, 0.25 mg/mL of adsorbent was added to aqueous solution of chromium (10 ppm). The mixture was sonicated for uniform dispersion of the adsorbent. The suspension with the adsorbent was kept for constant stirring for 30 min for shaking on a mechanical shaker. After completion of 30 min, the solution was centrifuged to remove the adsorbent and the solution was analyzed using UV-Vis spectrometer. The analysis of the final concentration of chromium was done by monitoring the absorbance at a wavelength of 352 nm. The kinetics of the adsorption studies were monitored by taking aliquots at regular interval of time. The supernatant was analyzed using UV-Vis spectrometer. For investigating the effect of dosage, similar studies were conducted with amount of adsorbent from 0.05 mg/mL to 1 mg/mL in aqueous solution of chromium (10 ppm). To explore the effect of pH, the adsorption studies were conducted at different pH 2, 4, 7, and 9. The acidic pH was adjusted using 0.1 M HCl while the basic pH was adjusted using 0.1 M NaOH. The spectral data obtained were analyzed for each sample and fitted into different kinetic and isothermal models to determine the nature of the process. The % removal and adsorption at equilibrium were calculated using the following formula,

$$\% \text{ removal} = \frac{C_1 - C_2}{C_1} \times 100 \tag{1}$$

The adsorption at equilibrium will be calculated using,

$$Q_e = V(C_1 - C_2) \div M \tag{2}$$

Q_e = amount of adsorption at equilibrium (mmol/g);
V = volume of heavy metal solution taken (ml);
M = quantity of adsorbent added (mg);
C_1 and C_2 (mg/L) refer to the heavy metal concentration before and after adsorption respectively at the λ_{max}.

The kinetic data was fitted to different kinetic models such as first order, second order, pseudo first, pseudo second order, Elovich, and intra-particle diffusion. For adsorption isotherm analysis, experiments were performed using different concentration of chromium solution from 10 to 50 ppm. To each solution, 0.25 mg/mL of adsorbent was added. The adsorption process was carried out for 30 min following the same steps as mentioned above. The data collected was analyzed using Langmuir, Freundlich, and Temkin isotherm models. All the experimental data in terms of concentration and %removal were fitted to standard isotherm models using Origin pro 2021.

5. Results and Discussions

The synthesized samples were analyzed for structural, thermal, morphological analysis. The characterization and adsorption results are as follows.

5.1. Characterization of the Samples

5.1.1. Thermal Stability Analysis

Figure 1a shows the thermogram of CF, PANI and its composites. In general, a weight loss is observed with increase in the temperature for the raw materials and composites used in this study. From the TGA curve of CF, it was interpreted that the decomposition temperature was 343 °C where it suffered maximum weight loss (79.5%). On the other hand, the TGA curve of PANI showed that the weight loss was maximum (84.8%) at temperature 485 °C. The loading of PANI on CF lead to increased stability of composites as evident from TGA analysis. An increase in T_d is observed in composites with varying % of PANI as compared to CF. The decomposition temperature obtained from TGA has been listed in the Table 2. Thus, preparation of composites enables us to develop more thermally stable materials which make them a suitable candidate for adsorption process.

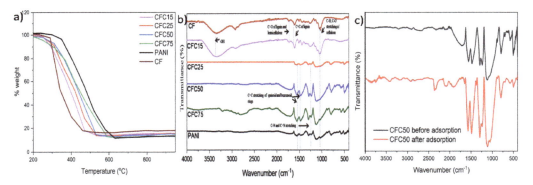

Figure 1. (a) TGA curve of CF, PANI, CFC15, CFC25, CFC50, and CFC75, (b) FTIR spectra of CF, PANI, and its composites, and (c) CFC50 before and after the adsorption of Cr(VI).

Table 2. Decomposition temperature and % weight loss with varying composition of PANI.

Sample Code	% Weight Loading of PANI (Theoretical)	Decomposition Temperature (T_d)	% Weight Loss from TGA
CF	0	343	79.5
CFC15	15	396	79.9
CFC25	25	424	77.6
CFC50	50	440	78
CFC75	75	455	77
PANI	100	485	84.8

5.1.2. Functional Group Analysis

Since adsorption is a surface phenomenon it becomes crucial to analyze the functional group present on the adsorbent surface. The understanding of functional groups helps to explore the adsorption mechanism and the nature of the process. Figure 1b shows the IR spectra of CF, PANI, and the prepared composites. In the IR spectrum of CF, the characteristic bands at around 3400 cm^{-1} is assigned to –OH stretching, the peaks at 1750 and 1240 cm^{-1} are attributed to C=O stretching of lignin and hemicellulose and C–H, C–O stretching of cellulose [26,34]. Other peaks at 1614 cm^{-1} is for C=C of lignin and 1440 cm^{-1}

corresponds to C–H vibration. Similarly in a study investigating thermally treated wood samples, Cheng et al. also observed C=C stretching vibrations at 1603 cm^{-1} [35]. IR peaks observed in the spectrum of PANI at 1568 and 1489 cm^{-1} correspond to C=C stretching of quinoid and benzenoid rings, respectively [36]. The peak at 1292 cm^{-1} correspond to C–N and C=N stretching. The peak for out-plane and in-plane C–H bonding is observed at 795 and 1106 cm^{-1} [37–40]. The oxygen and nitrogen containing functional groups offer potential binding sites for the adsorption of heavy metals. These functional groups tend to increase the cation-exchange capacity of the material by creating electron donor centers in the aqueous medium [41,42]. For the confirmation of the adsorption of Cr(VI) on the surface of CFC50, FTIR spectrum of CFC50 after chromium adsorption was also recorded (Figure 1c). A shift in IR peak of CFC50 at 1561 cm^{-1} corresponding to C=C stretching of quinoid to 1572 cm^{-1} was observed after the adsorption depicting the adsorption of chromium on the surface of CFC50. The intensity of peak at 1292 cm^{-1} corresponding to C–N and C=N stretching also changes before and after adsorption showing the involvement of nitrogen containing functional group in the adsorption process. In the IR spectrum of CFC50 after adsorption, a new peak at 1051 cm^{-1} can be observed. Similar results have been reported by Dula et al., Solgi et al., and Shooto et al. [43–45].

5.1.3. Morphological Analysis

Figure 2 shows the FE-SEM image of CF, PANI, and their composites. The FE-SEM image of CF clearly shows the presence of fibrous shape morphology (Figure 2a) while PANI exhibits a rod like shape with agglomerates as shown in Figure 2b. Similar fibrous morphology for CF and rod-like shape for PANI has been reported by Dutra et al. and Martina et al. respectively [36,46]. The FE-SEM images of composites shown in Figure 2c–f depict dispersion of rod like particles over CF confirming the formation of CF-PANI composites. Additionally, as the concentration of PANI increases in the composites, the amount of rod like particles dispersed on the CF increases which confirm the proper loading of PANI on CF.

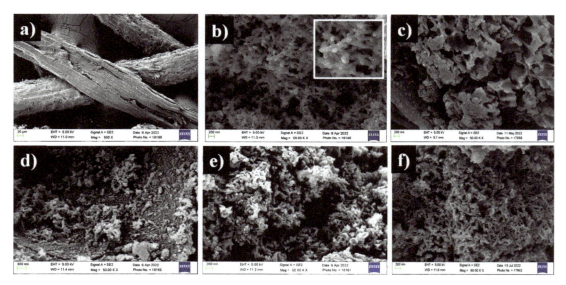

Figure 2. FE-SEM images of (**a**) CF, (**b**) PANI, (**c**) CFC15, (**d**) CFC25, (**e**) CFC50, and (**f**) CFC75. (inset of (**b**) shows the higher resolution image of PANI depicting the rod like shape with agglomerates).

5.2. Adsorption Study

The prepared composites namely CFC15, CFC25, CFC50, CFC75 and PANI were evaluated for Cr(VI) adsorption in aqueous solution. Figure 3a shows the UV spectra of the comparative performance/adsorption ability of CF, PANI, and composites. The results indicated a drastic reduction in the intensity of absorption peak at λ_{max} of 352 nm confirming the removal of Cr(VI) from aqueous solution. The adsorption efficiency of PANI, CFC15, CFC25, CFC50, and CFC75 was found to be 88.41%, 14.58%, 73.80%, 93.11%, and 82.60%, respectively as represented in Figure 3b. From the adsorption results, it was inferred that the preparation of CF-PANI composite (CFC50) showed enhanced adsorption efficiency as compared with PANI. This is attributed to the synergic effect of CF and PANI in the composites, for improved adsorption of Cr(VI) in aqueous solution. Thus, the preparation of composites of PANI with CF is a cost-effective and sustainable method for the removal of Cr(VI). An improved performance as compared with a pristine PANI sample was observed for smaller amount of PANI when dispersed over CF. Additionally, by further increasing the amount of PANI on CF (CFC75) the adsorption efficiency decreased from 93.11% to 82.60%. It is proposed that coconut fibers act as a support and promote uniform dispersion of PANI over its surface. The poor efficiency of CFC15 is attributed to smaller loading amounts of PANI and its non-uniform dispersion. An increase in the performance was observed with increased loading of PANI until 50%. The sudden decrease in efficiency for CFC75 is due to the agglomeration of particles and unavailability of more active surface sites. From the studies we can conclude that the development of CFC50 reduces the cost and results in improved performance as compared with PANI.

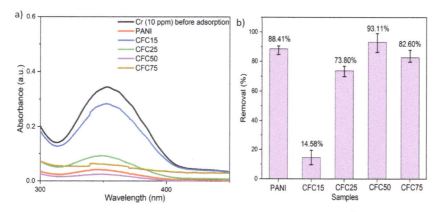

Figure 3. (**a**) UV-Vis spectra depicting the adsorption performance of the prepared samples (**b**) % adsorption of prepared samples for Chromium adsorption (pH = 6, adsorbent dosage = 0.25 mg/mL, Cr concentration = 10 ppm, contact time = 30 min).

5.3. Kinetic Studies

The kinetics of adsorption of Cr(VI)in aqueous solution were monitored at different time intervals. Figure S2a–e (Supplementary Materials) shows the UV-Vis spectrum for the kinetic studies for different adsorbents used. A continuous decrease in absorbance indicates the removal of Cr(VI) from aqueous solution. Figure S2f (Supplementary Materials) shows the absorbance vs. time graph for CF, PANI, and its composites. From the results, it was concluded that CFC15 and CFC75 showed desorption of Cr(VI) after 10 min, while no desorption could be seen in cases of PANI, CFC25, and CFC50.

The kinetic data obtained from the UV-Vis spectra was then analyzed using a different kinetics model to better understanding of the process. The kinetic data was fitted into Pseudo first order (PFO), first order (FO), pseudo second order (PSO), second order (SO), Elovich model, and intraparticle diffusion corresponding to the same concentration of

Cr(VI) aqueous solution (10 ppm). Equations of the kinetic models studied in the present work are as mentioned in Table S1 (Supplementary Materials).

Figure 4 shows the graph plotted to understand the kinetics for different models. The parameters for the linear fitting analysis such as R^2, rate constant (K), etc. are listed in Table 3. From the high R^2 values, the pseudo second order kinetic model was found to be the best fitted for the present adsorption study. The Elovich and Intraparticle diffusion models were also analyzed to understand the mechanism of adsorption. The intraparticle diffusion model states that the adsorption process is controlled by either film diffusion, pore diffusion, or surface diffusion or their combination [47]. Ofomaja et al. reported that the adsorption of chromium by magnetite coated biomass followed the intraparticle diffusion model [48]. The Elovich model is a widely adapted in adsorption kinetics, used to describe chemical adsorption [49]. The value of R^2 indicated that the Elovich model was better suited to understand the mechanism. Elovich models hints towards chemisorption and is more suited for the heterogeneous surface of the adsorbent [50]. Similar results were reported by Aworanti et al. for the adsorption of Cr(VI) by sawdust derived activated carbon [51].

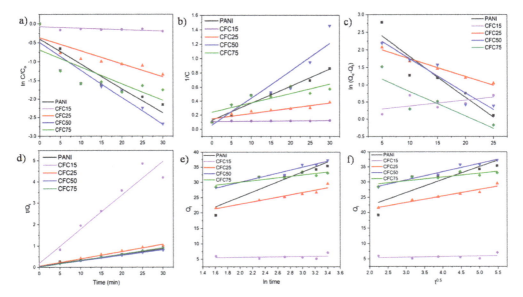

Figure 4. Kinetic models for the adsorption of Cr(VI) onto the prepared samples (**a**) First order, (**b**) Second order, (**c**) Pseudo first order, (**d**) Pseudo second order, (**e**) Elovich, (**f**) Intra-particle diffusion, (pH = 6, adsorbent dosage = 0.25 mg/mL, Cr concentration = 10 ppm, total contact time = 30 min).

5.4. Isotherm Study

The isothermal analysis was performed for the composite sample CFC50 as it was found to be the best adsorbent for the removal of Cr(VI) among all the samples. For isothermal analysis, different concentrations of chromium in its aqueous solution were used. The experimental data obtained were plotted in the form of Q_e versus C_e (concentration at equilibrium) to study the Langmuir, Freundlich, and Temkin adsorption isotherm [52,53]. The data were fitted with the non-linear form of all the isotherms shown in Figure S3 (Supplementary Materials). The linear and non-linear equations of the isotherms are listed in Table S2 (Supplementary Materials). The comparative analysis of R^2 value for all the isotherm models as shown in Table 4 indicated that the data were best fitted in Langmuir isotherm equation. Hence it can be concluded that the adsorption process of Cr(VI) using CFC50 follows the Langmuir model. From the analysis it is inferred that the adsorption occurs at a specific homogeneous site and is limited to one layer. The isotherm assumes that there is a formation of a monolayer adsorbate on the outer surface of adsorbent. After the formation of this layer no further adsorption takes place. Piccin et al. (2011) and Dada et al. (2012) reported similar results for the adsorption of food dye and Zn^{+2} by chitosan and rice husk, respectively. In both the cases the adsorption followed the Langmuir adsorption isotherm [53,54].

Table 3. The values of K and R^2 of different kinetic model fittings for the adsorption of Cr(VI) onto the prepared samples.

Sample Code	Q_e(mg/gm)	First Order		Second Order		Pseudo First Order		Pseudo Second Order		Elovich	Intra-Particle Diffusion	
		K	R^2	K	R^2	K	R^2	K	R^2	R^2	K_d	R^2
PANI	35.36	0.06592	0.8460	0.0242	0.9547	0.1172	0.8846	0.012	0.9836	0.8390	4.254	0.7478
CFC15	7.15	0.00419	0.473	0.00043	0.4727	0.0175	0.3347	0.134	0.9362	0.0549	0.198	0.1022
CFC25	29.52	0.03436	0.7558	0.007	0.8755	0.0503	0.9768	0.023	0.9884	0.9182	2.143	0.9427
CFC50	37.24	0.0733	0.8733	0.0383	0.8765	0.0981	0.9100	0.020	0.9937	0.9625	2.725	0.9630
CFC75	33.04	0.04476	0.5923	0.0134	0.7085	0.0718	0.7314	0.065	0.9984	0.7985	1.253	0.7239

K_d = intraparticle diffusion constant.

Table 4. Value of R^2 and different constants for the Freundlich, Langmuir, and Temkin isotherm models for the adsorption of different concentrations of Cr(VI) by CFC50.

Freundlich			Langmuir			Temkin	
R^2	K_f	n	R^2	B	q_{max}	R^2	K_T
0.9730	2.534	5.420	0.9888	36.630	2.4024	0.9867	1.586

Note: $K_f[(mg/g)(L/mg)^{1/n}]$ = Freundlich adsorption capacity constant; n = Freundlich intensity parameter; b (L/g) = constant indicating affinity between an adsorbent and adsorbate; q_{max}(mg/g) = maximum saturated monolayer adsorption capacity of the adsorbent; K_T = Temkin isotherm constant.

5.5. Effect of Adsorbent Dosage on Chromium Adsorption

To optimize the ideal dosage of adsorbent for the efficient removal of Cr from its aqueous solution, the effect of dosage was investigated. Different dosages of adsorbent (CFC50) that is 0.05, 0.1, 0.25, 0.5, and 1 mg/mL were explored for the removal of aqueous solution of Cr(VI) under similar conditions. Figure 5a shows that the adsorption efficiency of CFC50 increased as we increased its dosage from 0.05 to 0.5 mg/mL. The increase in %efficiency is due to the availability of more active sites as the adsorbent dosage is increased. The adsorption efficiency changed from 98.23% to 98.06% for an increase in adsorbent dosage from 0.5 to 1 mg/mL. This observation indicating saturation of adsorption by the sample is in agreement with the fact that the adsorption process follows the Langmuir isotherm model as inferred from isothermal analysis. After reaching the optimum dosage, the equilibrium was attained between the adsorbate and the adsorbent at a particular condition. Hence, the % efficiency also became saturated. Similar results were reported by Malhotra et al., (2018) where they found a decrease in adsorption with increase in dosage of adsorbent after attaining the optimum dosage condition [55]. It has been reported that increasing adsorbent dosage leads to the overcrowding of particles which led to a decrease in the adsorption performance. Keeping in mind the economic point of view, all the adsorption studies were performed taking 0.25 mg/mL as the optimum adsorbent dosage.

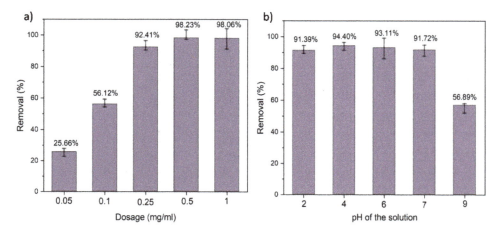

Figure 5. Removal (%) (**a**) at different dosage of adsorbent (pH = 6, adsorbent dosage = 0.25 mg/mL to 1 mg/mL, Cr concentration = 10 ppm, contact time = 30 min) and (**b**) at different pH values of the aqueous chromium solution (adsorbent dosage = 0.25 mg/mL, Cr concentration = 10 ppm, contact time = 30 min).

5.6. Effect of pH

The pH of the solution plays a vital role during the adsorption studies as it influences the adsorption mechanism. The effect of pH was analyzed by preparing Cr(VI) solutions with different pH values, i.e., 2, 4, 7, and 9. The adsorbent dosage was kept at 0.25 mg/mL in 10 ppm of chromium solution with a contact time of 30 min. The pH of the aqueous solution of Cr(VI) was found to be 6. It was observed that the adsorbent exhibited superior performance in acidic conditions as shown in Figure 5b. Removal efficiencies of 91.39%, 94.40%, and 93.11% were observed at pH 2, 4, and 6 respectively. On the contrary, a reduction in removal (%) was observed at higher pH values. With the increase in pH from 6 to 9, a reduction in removal (%) from 93.11% to 56.89% by CFC50 was observed. From the pH study performed, it is concluded that the optimum pH for the adsorption of Cr(VI) is acidic. At acidic pH, Cr(VI) species generally exist as $HCrO_4^-$ ions which increase their electrostatic attraction with the highly protonated polymer composite. Similar findings showing the protonation of PANI were reported by Sulimenko et al. and Stejskal et al. [56,57]. This allows additional removal of chromium in acidic media. At basic pH, $HCrO_4^-$ becomes converted into CrO_4^{2-}. This creates competition between OH^- and CrO_4^{2-} ions to become adsorbed on the surface of the adsorbent resulting in lower adsorption of chromium. The entire mechanism of the interaction of the adsorbent with the analyte (Cr(VI)) over the acidic and basic pH range is represented in Figure 6.

On comparing the results obtained from the present study with other similar research work as summarized in Table 5, we can conclude that the composites prepared in this study showed better results in terms of adsorption efficiency, adsorbent dosage, contact time, and utilizing low-cost adsorbent.

5.7. Adsorption Mechanism

Every adsorption process has a unique mechanism. The mechanism of adsorption depends on the interaction of adsorbent and adsorbate. The interaction is influenced by multiple factors such as surface charge, surface area, the nature of the analyte, and functional groups present on the surface of the adsorbent, etc. The understanding of the mechanism and its dependence on surface characterization are very important. Mechanism details enable us to modify the adsorbent and improve its performance. From the analysis of the adsorption process and characterization of the adsorbent, the proposed mechanism is as follows. The adsorption of chromium by CF-PANI composites followed the PSO kinetics, indicating that the adsorption by these materials occurs by chemisorption [69]. There is a strong attraction between positively charged amine functional groups (–NH–, –NH$_2$) present on the surface of the polymer (confirmed from the IR spectrum of PANI) and negatively charged HCrO$_4^-$ in acidic medium. Owing to this, it can be anticipated that electrostatic interaction could be a possible mechanism to explain the adsorption of Cr(VI) in the present study. As inferred from the IR results the presence of oxygen and nitrogen containing functional groups also help in the binding of Cr(VI) as mentioned in the results obtained from FTIR analysis [42]. Hence the preparation of composites provides more active sites for the enhanced removal of Cr(VI) due to dispersion of PANI on CF. Similar mechanisms have been reported by Deng et al. (2015) and Chigondo et al. (2019) for chromium adsorption using Polyethylenimine-modified fungal biomass and Magnetic arginine-functionalized polypyrrole, respectively [70,71].

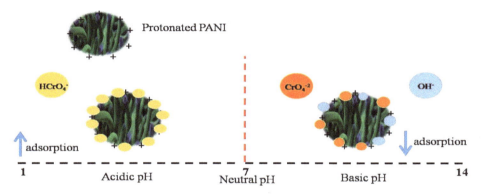

Figure 6. Schematic showing the mechanism of adsorption of chromium at acidic and basic pH ranges.

6. Conclusions

Composites of CF and PANI were prepared with different loadings of PANI (15, 25, 50, and 75 w/w%). The composites were characterized using FTIR, FE-SEM, and FTIR spectroscopy for their surface and functional analysis. Subsequently, the prepared composites were used as adsorbents (0.25 mg/mL) for the removal of hexavalent chromium (10 mg L^{-1}) from its aqueous solution. Our findings demonstrated that the CFC50 composite was most effective for the removal of Cr(VI) exhibiting 93.11% adsorption in 30 min. It showed an enhanced removal capacity as compared with pristine PANI (88.41%). The kinetics studies indicated towards the pseudo second order (R^2 = 0.9937) of the removal process. The effects of adsorbent dosage, pH of the chromium and concentration of the chromium solution on the adsorption performance of CFC50 were also studied. The adsorption efficiency increased with increasing adsorbent dosages of CFC50. Moreover, from the pH studies it was inferred that the acidic pH is more suitable for the adsorption of Cr(VI) in its aqueous solution. This is due to the existence of Cr(VI) as a HCrO$_4^-$ ion in acidic medium which increases its electrostatic attraction with the highly protonated polymer composite. The adsorption of Cr(VI) by CFC50 was well described by the Elovich kinetic (R^2 = 0.9625) and Langmuir isotherm models (R^2 = 0.9888). The nature of adsorption was found to be monolayer and occurred via chemisorption. Thus, our study effectively displayed the suitability of the prepared CF-PANI composite in the treatment of Cr(VI). Development of CF-PANI composites results in enhanced adsorption performance of PANI for Cr(VI) removal and also is feasible considering the economic aspect in mind. This will contribute towards the removal of harmful pollutants and environmental mitigation which will open new doors in the field of adsorption. Moreover, field studies using the CF-PANI composites as an adsorbent for wastewater treatment should be considered in future. The CF-PANI composites evaluated in this study can be further up scaled and used in water purification systems as a filtration medium owing to their excellent performance as adsorbents.

Table 5. Summary of research papers on the removal of chromium using PANI and bio-waste based adsorbents.

Adsorbent	Dosage of Adsorbent (g/L)	Time	Removal (%)	Q_e (mg/g)	Ref.
Rice husk ash—Ppy—PANI	0.8	300 min	98%		[26]
Polypyrole-calcium rectorite composite	1	-		714.29	[58]
Metal-organic framework-alginate beads	50	-	98%		[59]
PANI—jute	2	180 min		62.9	[27]
PANI—silica	0.8	430 min	193.85%		[28]
Calcinated wheat bran	1	24 h		29.3	[60]
Tea leaves	-	24 h	84.5%		[61]
Palm kernel	0.5	45 min		19	[62]
Eggshell powder	125	120 min	60.96%		[63]
PANI—sugarcane bagasse	1	100 min		35.2	[64]
CoFe(2)O$_4$_PANI	0.5	14 min		103.11	[65]
Arginine doped PANI—walnut shell	0.3	3 h	99%		[66]
Pomegranate peels—Ppy—PANI	10	90 min	95.35%		[67]
Sugarcane bagasse Oil cake Maize corn	20	60 min	92% 97% 62%		[68]
Coconut fiber-polyaniline composite	0.25	30 min	93.11%	37.24	Present study

Supplementary Materials: The following supporting information can be downloaded at: https://www.mdpi.com/article/10.3390/polym14204264/s1, Figure S1. Schematic flow for the synthesis of (a) PANI, (b) CF-PANI composites, and (c) the digital images of the prepared samples; Figure S2. UV-Vis absorbance spectra for the kinetic studies at different time intervals of time (a) PANI (b) CFC15 (c) CFC25 (d) CFC50 (e) CFC75(pH = 6, adsorbent dosage = 0.25 mg/mL, Cr concentration = 10 ppm, total contact time = 30 min) (f) Absorbance (at λmax = 352 nm) vs. time of prepared composites for Cr(VI); Figure S3. Freundlich, Langmuir, and Temkin isotherm models for different concentration of Cr(VI) by CFC50 (pH = 6, adsorbent dosage = 0.25 mg/mL, Cr concentration = 10 ppm to 50 ppm, contact time = 30 min); Table S1. Linear equation forms for different kinetic models; Table S2. Linear and non-linear equations for Freundlich, Langmuir, and Temkin isotherm models.

Author Contributions: Conceptualization, S.J., R.G. and S.S.; methodology, S.J., R.G. and S.S.; software, S.J.; validation, S.J., R.G. and S.S.; formal analysis, S.J.; resources, S.S., N.S. and I.A.; writing—original draft preparation, S.J. and R.G.; writing—review and editing, S.J., S.S., R.G., I.A. and N.S.; supervision, R.G and S.S.; funding acquisition, N.S., I.A. and S.S. All authors have read and agreed to the published version of the manuscript.

Funding: The authors would like to thank Pandit Deendayal Energy University, Scientific Research Deanship at King Khalid University, Abha, Saudi Arabia through the Large Research Group Project under grant number (RGP.02/219/43) and the Marine Pollution Special Interest Group, the National Defence University of Malaysia via SF0076-UPNM/2019/SF/ICT/6, for providing research facilities and funding.

Institutional Review Board Statement: Not applicable.

Informed Consent Statement: Not applicable.

Acknowledgments: The authors would like to thank Solar Research and development Center (SRDC) for FE-SEM characterization and Pandit Deendayal Energy University for providing institutional fellowship to Stuti Jha.

Conflicts of Interest: The authors have no conflict of interest.

References

1. Ali, H.; Khan, E.; Ilahi, I. Environmental Chemistry and Ecotoxicology of Hazardous Heavy Metals: Environmental Persistence, Toxicity, and Bioaccumulation. *J. Chem.* **2019**, *2019*, 6730305. [CrossRef]
2. Briffa, J.; Sinagra, E.; Blundell, R. Heavy Metal Pollution in the Environment and their Toxicological Effects on Humans. *Heliyon* **2020**, *6*, e04691. [CrossRef] [PubMed]
3. Tchounwou, P.B.; Yedjou, C.G.; Patlolla, A.K.; Sutton, D.J. Heavy metal toxicity and the environment. *Mol. Clin. Environ. Toxicol.* **2012**, *101*, 133–164. [CrossRef]
4. Jacobs, J.A.; Avakian, C.P. *Chromium (VI) Handbook*; Taylor & Francis: Oxfordshire, UK, 2005.
5. Wilbur, S.B. Toxicological profile for chromium: US Department of Health and Human Services, Public Health Service. *Agency Toxic Subst. Dis. Regist.* **2000**, *67*, 1054S–1060S.
6. Mondal, N.K.; Basu, S. Potentiality of Waste Human Hair Towards Removal of Chromium (VI) from Solution: Kinetic and Equilibrium Studies. *Appl. Water Sci.* **2019**, *9*, 49. [CrossRef]
7. Palaniappan, P.; Karthikeyan, S. Bioaccumulation and Depuration of Chromium in the Selected Organs and Whole Body Tissues of Freshwater fish *Cirrhinusmrigala* Individually and in Binary Solutions with Nickel. *J. Environ. Sci.* **2009**, *21*, 229–236. [CrossRef]
8. Norseth, T. The Carcinogenicity of Chromium and its Salts. *Occup. Environ. Med.* **1986**, *43*, 649–651. [CrossRef]
9. Ambaye, T.G.; Vaccari, M.; van Hullebusch, E.D.; Amrane, A.; Rtimi, S. Mechanisms and Adsorption Capacities of Biochar for the Removal of Organic and Inorganic Pollutants from Industrial Wastewater. *Int. J. Environ. Sci. Technol.* **2020**, *18*, 3273–3294. [CrossRef]
10. Möbius, C.H. Adsorption and Ion Exchange Processes for Treatment of White Water and Waste Water of Paper Mills. *Water Pollut. Res. Dev.* **1981**, *13*, 681–695.
11. Trishitman, D.; Cassano, A.; Basile, A.; Rastogi, N.K. Reverse osmosis for industrial wastewater treatment. In *Current Trends and Future Developments on (Bio-) Membranes*; Elsevier: Amsterdam, The Netherlands, 2020; pp. 207–228.
12. Fontanier, V.; Farines, V.; Albet, J.; Baig, S.; Molinier, J. Study of Catalyzed Ozonation for Advanced Treatment of Pulp and Paper Mill Effluents. *Water Res.* **2006**, *40*, 303–310. [CrossRef]
13. Park, J.H.; Choppala, G.K.; Bolan, N.S.; Chung, J.W.; Chuasavathi, T. Biochar Reduces the Bioavailability and Phytotoxicity of Heavy Metals. *Plant Soil* **2011**, *348*, 439–451. [CrossRef]
14. Brandl, F.; Bertrand, N.; Lima, E.; Langer, R. Nanoparticles with Photoinduced Precipitation for the Extraction of Pollutants from Water and Soil. *Nat. Commun.* **2015**, *6*, 7765. [CrossRef] [PubMed]
15. Rashed, M.N. Adsorption technique for the removal of organic pollutants from water and wastewater. *Org. Pollut. Monit. Risk Treat.* **2013**, *7*, 167–194.
16. Sadegh, H.; Ali, G.A. Potential applications of nanomaterials in wastewater treatment: Nanoadsorbents performance. In *Research Anthology on Synthesis, Characterization, and Applications of Nanomaterials*; IGI Global: Hershay, PA, USA, 2021; pp. 1230–1240.
17. Ali, A.; Saeed, K.; Mabood, F. Removal of Chromium (VI) from Aqueous Medium Using Chemically Modified Banana Peels as Efficient Low-Cost Adsorbent. *Alex. Eng. J.* **2016**, *55*, 2933–2942. [CrossRef]
18. Mondal, N.K.; Basu, S.; Sen, K.; Debnath, P. Potentiality of Mosambi (*Citrus limetta*) Peel Dust toward Removal of Cr (VI) from Aqueous Solution: An Optimization Study. *Appl. Water Sci.* **2019**, *9*, 116. [CrossRef]
19. Tan, W.; Ooi, S.; Lee, C. Removal of Chromium (VI) from Solution by Coconut Husk and Palm Pressed Fibres. *Environ. Technol.* **1993**, *14*, 277–282. [CrossRef]
20. Srivastava, H.; Mathur, R.; Mehrotra, I. Removal of Chromium from Industrial Effluents by Adsorption on Sawdust. *Environ. Technol. Lett.* **1986**, *7*, 55–63. [CrossRef]
21. Mutongo, F.; Kuipa, O.; Kuipa, P.K. Removal of Cr (VI) from aqueous solutions using powder of potato peelings as a low cost sorbent. *BioinorgChem Appl.* **2014**, *2014*, 973153. [CrossRef]
22. Maponya, T.C.; Hato, M.J.; Somo, T.R.; Ramohlola, K.E.; Makhafola, M.D.; Monama, G.R.; Maity, A.; Modibane, K.D.; Matata-Seru, L.M. Polyaniline-based nanocomposites for environmental remediation. In *Trace Metals in the Environment-New Approaches and Recent Advances*; IntechOpen: London, UK, 2019.
23. Samadi, A.; Xie, M.; Li, J.; Shon, H.; Zheng, C.; Zhao, S. Polyaniline-Based Adsorbents for Aqueous Pollutants Removal: A review. *Chem. Eng. J.* **2021**, *418*, 129425. [CrossRef]
24. Zhan, G.; Zeng, H.C. Integrated Nanocatalysts with Mesoporous Silica/Silicate and Microporous MOF Materials. *Co-ord. Chem. Rev.* **2016**, *320-321*, 181–192. [CrossRef]

25. Eskandari, E.; Kosari, M.; Farahani, M.H.D.A.; Khiavi, N.D.; Saeedikhani, M.; Katal, R.; Zarinejad, M. A Review on Polyaniline-Based Materials Applications in Heavy Metals Removal and Catalytic Processes. *Sep. Purif. Technol.* **2019**, *231*, 115901. [CrossRef]
26. Dutta, S.; Srivastava, S.K.; Gupta, A.K. Polypyrrole–Polyaniline Copolymer Coated Green Rice Husk Ash as an Effective Adsorbent for the Removal of Hexavalent Chromium from Contaminated Water. *Mater. Adv.* **2021**, *2*, 2431–2443. [CrossRef]
27. Kumar, P.A.; Chakraborty, S.; Ray, M. Removal and Recovery of Chromium from Wastewater Using Short Chain Polyaniline Synthesized on Jute Fiber. *Chem. Eng. J.* **2008**, *141*, 130–140. [CrossRef]
28. Tang, L.; Fang, Y.; Pang, Y.; Zeng, G.; Wang, J.; Zhou, Y.; Deng, Y.; Yang, G.; Cai, Y.; Chen, J. Synergistic Adsorption and Reduction of Hexavalent Chromium Using Highly Uniform Polyaniline–Magnetic Mesoporous Silica Composite. *Chem. Eng. J.* **2014**, *254*, 302–312. [CrossRef]
29. Ghorbani, M.; Lashkenari, M.S.; Eisazadeh, H. Application of Polyaniline Nanocomposite Coated on Rice Husk Ash for Removal of Hg (II) from Aqueous Media. *Synth. Met.* **2011**, *161*, 1430–1433. [CrossRef]
30. Lei, C.; Wang, C.; Chen, W.; He, M.; Huang, B. Polyaniline Magnetic Chitosan Nanomaterials for Highly Efficient Simultaneous Adsorption and In-Situ Chemical Reduction of Hexavalent Chromium: Removal Efficacy and Mechanisms. *Sci. Total Environ.* **2020**, *733*, 139316. [CrossRef]
31. Rahmi; Julinawati; Nina, M.; Fathana, H.; Iqhrammullah, M. Preparation and Characterization of New Magnetic Chitosan-Glycine-PEGDE (Fe_3O_4/Ch-G-P) Beads for Aqueous Cd (II) Removal. *J. Water Process Eng.* **2022**, *45*, 102493. [CrossRef]
32. Rahmi, R.; Lelifajri, L.; Iqbal, M.; Fathurrahmi, F.; Jalaluddin, J.; Sembiring, R.; Farida, M.; Iqhrammullah, M. Preparation, Characterization and Adsorption Study of PEDGE-Cross-linked Magnetic Chitosan (PEDGE-MCh) Microspheres for Cd2+ Removal. *Arab. J. Sci. Eng.* **2022**, 1–9. [CrossRef]
33. Marciano, J.S.; Ferreira, R.R.; de Souza, A.G.; Barbosa, R.F.S.; de Moura Junior, A.J.; Rosa, D.S. Biodegradable Gelatin Composite Hydrogels Filled with Cellulose for Chromium (VI) Adsorption from Contaminated Water. *Int. J. Biol. Macromol.* **2021**, *181*, 112–124. [CrossRef]
34. Andrade, R.T.; da Silva, R.C.S.; Pereira, A.C.; Borges, K.B. Self-Assembly Pipette Tip-Based Cigarette Filters for Micro-Solid Phase Extraction of Ketoconazole Cis-Enantiomers in Urine Samples Followed by High-Performance Liquid Chromatography/Diode Array Detection. *Anal. Methods* **2015**, *7*, 7270–7279. [CrossRef]
35. Cheng, S.; Huang, A.; Wang, S.; Zhang, Q. Effect of Different Heat Treatment Temperatures on the Chemical Composition and Structure of Chinese Fir Wood. *BioResources* **2016**, *11*, 4006–4016. [CrossRef]
36. Dutra, F.V.A.; Pires, B.C.; Nascimento, T.A.; Mano, V.; Borges, K.B. Polyaniline-Deposited Cellulose Fiber Composite Prepared Via in Situ Polymerization: Enhancing Adsorption Properties for Removal of Meloxicam from Aqueous Media. *RSC Adv.* **2017**, *7*, 12639–12649. [CrossRef]
37. Bhadra, S.; Singha, N.K.; Khastgir, D. Polyaniline by New Miniemulsion Polymerization and the Effect of Reducing Agent on Conductivity. *Synth. Met.* **2006**, *156*, 1148–1154. [CrossRef]
38. Palaniappan, S. Chemical and Electrochemical Polymerization of Aniline Using Tartaric Acid. *Eur. Polym. J.* **2001**, *37*, 975–981. [CrossRef]
39. Wei, Y.; Hsueh, K.F.; Jang, G.W. A Study of Leucoemeraldine and Effect of Redox Reactions on Molecular Weight of Chemically Prepared Polyaniline. *Macromolecules* **1994**, *27*, 518–525. [CrossRef]
40. Shao, W.; Jamal, R.; Xu, F.; Ubul, A.; Abdiryim, T. The Effect of a Small Amount of Water on the Structure and Electrochemical Properties of Solid-State Synthesized Polyaniline. *Materials* **2012**, *5*, 1811–1825. [CrossRef]
41. Xie, R.; Wang, H.; Chen, Y.; Jiang, W. Walnut Shell-Based Activated Carbon with Excellent Copper (II) Adsorption and Lower Chromium (VI) Removal Prepared by Acid-Base Modification. *Environ. Prog. Sustain. Energy* **2012**, *32*, 688–696. [CrossRef]
42. Iqhrammullah, M.; Suyanto, H.; Rahmi; Pardede, M.; Karnadi, I.; Kurniawan, K.H.; Chiari, W.; Abdulmadjid, S.N. Cellulose Acetate-Polyurethane Film Adsorbent with Analyte Enrichment for in-Situ Detection and Analysis of Aqueous Pb Using Laser-Induced Breakdown Spectroscopy (LIBS). *Environ. Nanotechnol. Monit. Manag.* **2021**, *16*, 100516. [CrossRef]
43. Dula, T.; Siraj, K.; Kitte, S.A. Adsorption of Hexavalent Chromium from Aqueous Solution Using Chemically Activated Carbon Prepared from Locally Available Waste of Bamboo (*Oxytenantheraabyssinica*). *ISRN Environ. Chem.* **2014**, *2014*, 1–9. [CrossRef]
44. Solgi, M.; Najib, T.; Ahmadnejad, S.; Nasernejad, B. Synthesis and Characterization of Novel Activated Carbon from Medlar Seed for Chromium Removal: Experimental Analysis and Modeling with Artificial Neural Network and Support Vector Regression. *Resour. Technol.* **2017**, *3*, 236–248. [CrossRef]
45. Shooto, N.D. Removal of Toxic Hexavalent Chromium (Cr (VI)) and Divalent Lead (Pb (II)) Ions from Aqueous Solution by Modified Rhizomes of *Acorus calamus*. *Surf. Interfaces* **2020**, *20*, 100624. [CrossRef]
46. Martina, V.; De Riccardis, M.F.; Carbone, D.; Rotolo, P.; Bozzini, B.; Mele, C. Electrodeposition of Polyaniline–Carbon Nanotubes Composite Films and Investigation on Their Role in Corrosion Protection of Austenitic Stainless Steel by SNIFTIR Analysis. *J. Nanopart. Res.* **2011**, *13*, 6035–6047. [CrossRef]
47. Sheha, R.R.; El-Zahhar, A.A. Synthesis of Some Ferromagnetic Composite Resins and Their Metal Removal Characteristics in Aqueous Solutions. *J. Hazard. Mater.* **2008**, *150*, 795–803. [CrossRef] [PubMed]
48. Pholosi, A.; Naidoo, E.B.; Ofomaja, A.E. Intraparticle Diffusion of Cr (VI) Through Biomass and Magnetite Coated Biomass: A Comparative Kinetic and Diffusion Study. *South Afr. J. Chem. Eng.* **2020**, *32*, 39–55. [CrossRef]
49. Wu, F.-C.; Tseng, R.-L.; Juang, R.-S. Characteristics of Elovich Equation Used for the Analysis of Adsorption Kinetics In Dye-Chitosan Systems. *Chem. Eng. J.* **2009**, *150*, 366–373. [CrossRef]

50. López-Luna, J.; Ramírez-Montes, L.E.; Martinez-Vargas, S.; Martínez, A.I.; Mijangos-Ricardez, O.F.; González-Chávez, M.D.C.A.; Carrillo-González, R.; Solís-Domínguez, F.A.; Cuevas-Díaz, M.D.C.; Vázquez-Hipólito, V. Linear and Nonlinear Kinetic and Isotherm Adsorption Models for Arsenic Removal by Manganese Ferrite Nanoparticles. *SN Appl. Sci.* **2019**, *1*, 1–19. [CrossRef]
51. Aworanti, O.; Agarry, S.E. Kinetics, Isothermal and Thermodynamic Modelling Studies of Hexavalent Chromium Ions Adsorption from Simulated Wastewater onto ParkiaBiglobosa-Sawdust Derived Acid-Steam Activated Carbon. *Methods* **2017**, *10*, 11.
52. Li, A.; Deng, H.; Jiang, Y.; Ye, C. High-Efficiency Removal of Cr (VI) from Wastewater by Mg-Loaded Biochars: Adsorption Process and Removal Mechanism. *Materials* **2020**, *13*, 947. [CrossRef]
53. Piccin, J.S.; Dotto, G.L.; Pinto, L.A.A. Adsorption Isotherms and Thermochemical Data of FD&C Red n 40 binding by Chitosan. *Braz. J. Chem. Eng.* **2011**, *28*, 295–304. [CrossRef]
54. Dada, A.; Olalekan, A.; Olatunya, A.; Dada, O. Langmuir, Freundlich, Temkin and Dubinin-RadushkevichIsotherms Studies of Equilibrium Sorption of Zn2+ Unto Phosphoric Acid Modified Rice Husk. *J. Appl. Chem.* **2012**, *3*, 38–45.
55. Malhotra, M.; Suresh, S.; Garg, A. Tea Waste Derived Activated Carbon for the Adsorption of Sodium Diclofenac from Wastewater: Adsorbent Characteristics, Adsorption Isotherms, Kinetics, and Thermodynamics. *Environ. Sci. Pollut. Res.* **2018**, *25*, 32210–32220. [CrossRef] [PubMed]
56. Sulimenko, T.; Stejskal, J.; Prokeš, J. Poly (Phenylenediamine) Dispersions. *J. Colloid Interface Sci.* **2001**, *236*, 328–334. [CrossRef] [PubMed]
57. Stejskal, J.; Hlavatá, D.; Holler, P.; Trchová, M.; Prokeš, J.; Sapurina, I. Polyaniline Prepared in the Presence of Various Acids: A Conductivity Study. *Polym. Int.* **2004**, *53*, 294–300. [CrossRef]
58. Xu, Y.; Chen, J.; Chen, R.; Yu, P.; Guo, S.; Wang, X. Adsorption and Reduction of Chromium (VI) from Aqueous Solution Using Polypyrrole/Calcium Rectorite Composite Adsorbent. *Water Res.* **2019**, *160*, 148–157. [CrossRef]
59. Daradmare, S.; Xia, M.; Le, V.N.; Kim, J.; Park, B.J. Metal–Organic Frameworks/Alginate Composite Beads as Effective Adsorbents for the Removal of Hexavalent Chromium from Aqueous Solution. *Chemosphere* **2020**, *270*, 129487. [CrossRef] [PubMed]
60. Ogata, F.; Nagai, N.; Itami, R.; Nakamura, T.; Kawasaki, N. Potential of Virgin and Calcined Wheat Bran Biomass for the Removal of Chromium (VI) Ion from a Synthetic Aqueous Solution. *J. Environ. Chem. Eng.* **2020**, *8*, 103710. [CrossRef]
61. Das, S.H.; Saha, J.; Saha, A.; Rao, A.K.; Chakraborty, B.; Dey, S. Adsorption Study of Chromium (VI) by Dried Biomass of Tea Leaves. *J. Indian Chem. Soc.* **2019**, *96*, 447–454.
62. Hanafiah, M.M.; Hashim, N.A.; Ahmed, S.; Ashraf, M.A. Removal of Chromium from Aqueous Solutions Using a Palm Kernel Shell Adsorbent. *Desalination Water Treat.* **2018**, *118*, 172–180. [CrossRef]
63. Abatan, O.G.; Alaba, P.A.; Oni, B.A.; Akpojevwe, K.; Efeovbokhan, V.; Abnisa, F. Performance of Eggshells Powder as an Adsorbent for Adsorption of Hexavalent Chromium and Cadmium from Wastewater. *SN Appl. Sci.* **2020**, *2*, 1–13. [CrossRef]
64. Kumari, B.; Tiwary, R.K.; Yadav, M.; Singh, K.M.P. Nonlinear Regression Analysis and Response Surface Modeling of Cr (VI) Removal from Synthetic Wastewater by an Agro-Waste *Cocos Nucifera:* Box-Behnken Design (BBD). *Int. J. Phytoremediation* **2020**, *23*, 791–808. [CrossRef]
65. Mohammadi, H.; Ghaedi, M.; Fazeli, M.; Sabzehmeidani, M.M. Removal of Hexavalent Chromium Ions and Acid Red 18 By Superparamagnetic Cofe2o4/Polyaniline Nanocomposites under External Ultrasonic Fields. *Microporous Mesoporous Mater.* **2021**, *324*, 111275. [CrossRef]
66. Hsini, A.; Naciri, Y.; Laabd, M.; El Ouardi, M.; Ajmal, Z.; Lakhmiri, R.; Boukherroub, R.; Albourine, A. Synthesis and Characterization of Arginine-Doped Polyaniline/Walnut Shell Hybrid Composite with Superior Clean-Up Ability for Chromium (VI) from Aqueous Media: Equilibrium, Reusability and Process Optimization. *J. Mol. Liq.* **2020**, *316*, 113832. [CrossRef]
67. Rafiaee, S.; Samani, M.R.; Toghraie, D. Removal of Hexavalent Chromium from Aqueous Media Using Pomegranate Peels Modified by Polymeric Coatings: Effects of Various Composite Synthesis Parameters. *Synth. Met.* **2020**, *265*, 116416. [CrossRef]
68. Garg, U.K.; Kaur, M.; Garg, V.; Sud, D. Removal of Hexavalent Chromium from Aqueous Solution by Agricultural Waste Biomass. *J. Hazard. Mater.* **2007**, *140*, 60–68. [CrossRef]
69. Guo, X.; Liu, A.; Lu, J.; Niu, X.; Jiang, M.; Ma, Y.; Liu, X.; Li, M. Adsorption Mechanism of Hexavalent Chromium on Biochar: Kinetic, Thermodynamic, and Characterization Studies. *ACS Omega* **2020**, *5*, 27323–27331. [CrossRef] [PubMed]
70. Deng, S.; Ting, Y.P. Polyethylenimine-Modified Fungal Biomass as a High-Capacity Biosorbent for Cr (VI) Anions: Sorption Capacity and Uptake Mechanisms. *Environ. Sci. Technol.* **2005**, *39*, 8490–8496. [CrossRef] [PubMed]
71. Chigondo, M.; Paumo, H.K.; Bhaumik, M.; Pillay, K.; Maity, A. Magnetic arginine-functionalized polypyrrole with improved and selective chromium (VI) ions removal from water. *J. Mol. Liq.* **2018**, *275*, 778–791. [CrossRef]

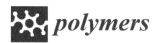

Article

Headspace Extraction of Chlorobenzenes from Water Using Electrospun Nanofibers Fabricated with Calix[4]arene-Doped Polyurethane–Polysulfone

Hamid Najarzadekan [1], Muhammad Afzal Kamboh [2], Hassan Sereshti [1,*], Irfan Ahmad [3], Nanthini Sridewi [4,*], Syed Shahabuddin [5,*] and Hamid Rashidi Nodeh [6]

1. School of Chemistry, College of Science, University of Tehran, Tehran 1417614411, Iran
2. Department of Chemistry, Shaheed Benazir Bhutto University, Shaheed Benazirabad, Sindh 67450, Pakistan
3. Department of Clinical Laboratory Sciences, College of Applied Medical Sciences, King Khalid University, Abha 61421, Saudi Arabia
4. Department of Maritime Science and Technology, Faculty of Defence Science and Technology, National Defence University of Malaysia, Kuala Lumpur 57000, Malaysia
5. Department of Chemistry, School of Technology, Pandit Deendayal Energy University, Raisan, Gandhinagar 382426, India
6. Food Technology and Agricultural Products Research Center, Standard Research Institute, Karaj 3174734463, Iran
* Correspondence: sereshti@ut.ac.ir (H.S.); nanthini@upnm.edu.my (N.S.); syedshahab.hyd@gmail.com or syed.shahabuddin@sot.pdpu.ac.in (S.S.); Tel.: +98-21-6113-735 (H.S.); +60-124-675-320 (N.S.); +91-858-593-2338 (S.S.)

Abstract: Chlorobenzenes (CBs) are persistent and potentially have a carcinogenic effect on mammals. Thus, the determination of CBs is essential for human health. Hence, in this study, novel polyurethane–polysulfone/calix[4]arene (PU-PSU/calix[4]arene) nanofibers were synthesized using an electrospinning approach over in-situ coating on a stainless-steel wire. The nanosorbent was comprehensively characterized using scanning electron microscopy (SEM) and Fourier transform infrared spectroscopy (FT-IR) techniques. The SEM analysis depicted the nanofiber's unique morphology and size distribution in the range of 50–200 nm. To determine the levels of 1,2,4-trichlorobenzene, 1,2,3-trichlorobenzene, and 1,2,3,4-tetrachlorobenzene in water samples, freshly prepared nanosorbent was employed using headspace-solid phase microextraction (HS-SPME) in combination with gas chromatography micro electron capture detector (GC-µECD). Other calixarenes, such as sulfonated calix[4]arene, p-tert-calixarene, and calix[6]arene were also examined, and among the fabricated sorbents, the PU–PSU/calix[4]arene showed the highest efficiency. The key variables of the procedure, including ionic strength, extraction temperature, extraction duration, and desorption conditions were examined. Under optimal conditions, the LOD (0.1–1.0 pg mL^{-1}), the LDR (0.4–1000 pg mL^{-1}), and the R^2 > 0.990 were determined. Additionally, the repeatability from fiber to fiber and the intra-day and inter-day reproducibility were determined to be 1.4–6.0, 4.7–10.1, and 0.9–9.7%, respectively. The nanofiber adsorption capacity was found to be 670–720 pg/g for CBs at an initial concentration of 400 pg mL^{-1}. A satisfactory recovery of 80–106% was attained when the suggested method's application for detecting chlorobenzenes (CBs) in tap water, river water, sewage water, and industrial water was assessed.

Keywords: electrospun-nanofibers; Calix[4]arene; polyurethane; polysulfone; extraction; chlorobenzenes

1. Introduction

Over the past few years, chlorinated volatile organic compounds have attracted substantial attention because they can cause environmental water contamination, which seriously affects the environment and health of human beings [1]. The chlorobenzenes (CBs) are renowned persistent environmental pollutants that widely exist in the discharged industrial effluent of various industrial units, such as the petrochemical, pharmaceutical, textile, and painting industries [1,2]. It has been determined that it is detrimental to human health for individuals to be exposed to effluents that include CBs, which can potentially produce histopathological changes, genotoxicity, mutagenicity, and carcinogenicity in humans [3]. CBs contamination has recently been identified as one of the worst health problems by the United States Environmental Protection Agency (US-EPA) and the World Health Organization (WHO) [4,5]. The maximum level of contamination goals (MLCG) was determined to be less than 0.1 µg mL^{-1}. Consequently, introducing an effective analytical technique for precisely monitoring CBs in aqueous media is of prime importance and the most challenging issue of the modern era.

A review of the relevant literature revealed the importance of sample preparation techniques, such as the extraction and preconcentration of CBs, for their subsequent determination, alongside advances in modern analytical tools. For extraction, scientists have turned to many different methods, including liquid–liquid extraction, solid-phase extraction, liquid-phase microextraction, dispersive liquid–liquid microextraction, vortex-assisted liquid–liquid microextraction, single drop microextraction, and magnetic headspace adsorptive extraction. Solid-phase microextraction (SPME) [6,7] is one of these techniques, and it is both a powerful sample preparation approach and a non-exhaustive extraction method, without the need for solvents [8]. When this method is combined with GC, the fiber coating is first exposed to the sample matrix through either direct immersion or headspace mode. Next, the target analyte or analytes are thermally desorbed in the GC injection port, and are simultaneously released into the column. The commercially available sorbents for extraction of various CBs are usually made up of polymers, including polydimethylsiloxane (PDMS) and polyacrylate (P.A.), and they are coated on fused silica [9]. The conventional fibers are easily broken and are relatively expensive. Consequently, the fabrication of innovative fiber-based sorbents enhances their high capacity, imparting diverse functionalities, polarities, and stability. Additionally, wire metals have been employed in place of brittle fused silica [10].

In alkaline conditions, formaldehyde and para-substituted phenols can be cyclically condensed to generate calixarenes, which are macrocyclic ligands. The calixarene framework as a host–guest platform provides space for accepting various analytes [11]. They provide well-defined cavities, which simultaneously supply non-polar and polar features with modification of the upper-rim and lower-rim, respectively. Because of the cyclic structure of different cavity sizes and functionalities of calixarenes, they are considered suitable carriers for cationic, anionic, and nonionic species [12].

A suitable platform is required to separate, extract, and enrich target analytes. Nanofibers and nanostructured materials consist of all these properties owing to their high surface-to-volume ratio and many active sites for adsorption [13]. Recently, electrospun nanofibers have been utilized as adsorbents for SPE, micro-SPE, microextraction in the packed syringe, membrane extraction, filtration/removal, and SPME [14]. Calixarenes-electrospun nanofibers have been successfully employed for catalytic activity [15], studying toxic anion binding [16] and binding efficiency towards chromium and uranium ions [17] with polyacrylonitrile nanofibers.

This research was conducted to fabricate a novel electrospun PU–PSU/C4A nanofiber to be used as a coating for HS-SPME to extract CB compounds from water samples. Accordingly, various calix[4]arene derivatives were electrospun on the surface of a stainless steel wire and examined for headspace preconcentration of CBs under the best experimental conditions. The presence of functionalized calixarenes in the structure of nanofiber provides an appropriate force interactive with CBs, including electrostatic and π-π interactions. To our knowledge, PU–PSU/calix[4]arene has never been synthesized and used for extracting the chosen CBs as model compounds from water media.

2. Experimental Section
2.1. Chemicals and Reagents

Polyurethane (PU) and polysulfone (PSU) were purchased from Bayer Company (Leverkusen, Germany). The Merck Company (Darmstadt, Germany) provided the methanol, 1,2,4-

trichlorobenzene (1,2,4-TCB), 1,2,3-trichlorobenzene (1,2,3-TCB), 1,2,3,4-tetrachlorobenzene (1,2,3,4-TCB), and N, N-dimethylformamide (DMF). The standard stock solutions of CBs were produced in methanol at a concentration of 2000 mg L^{-1} and stored at a temperature of 4 °C until their subsequent usage. Stock standard solutions of CBs were prepared, and working standard solutions were prepared daily before the extraction process.

2.2. Instrumentation

A gas chromatograph manufactured by Agilent (6890N, Santa Clara, CA, USA) was utilized to determine the composition of the extracted analytes. This particular model was equipped with a μ-ECD detector, an HP-5 fused silica capillary column (30 m × 0.32 mm × 0.25 μm), and a split/splitless injection port. Helium with a purity of 99.999% and a flow rate of 1 mL min^{-1}, and nitrogen with a purity of 99.999% and a flow rate of 30 mL min^{-1} were employed as a carrier gas and makeup gas, respectively. 3 min of splitless injection were used to inject the sample. Temperature settings for the injector and detector were 170 °C and 290 °C, respectively. Initially, the oven was set to preheat to 70 °C for 2 min, then heated to 170 °C at a rate of 40 °C per minute for 2 min, and finally heated to 200 °C at a rate of 20 °C per minute for 5 min.

Electrospinning was carried out using a spinal needle attached to a rotating motor (to act as a collector), a syringe pump, and a direct current high voltage power source. The SPME syringe consisted of two spinal needles, an internal needle, G27 as an SPME coated needle, an external needle, and G22 as an SPME barrel.

Micrographs of nanofibers were recorded using a Zeiss DSM-960 scanning electron microscope (SEM) (Oberkochen, Germany). The functional groups of produced nanosorbent were monitored using Fourier transform infrared spectroscopy, with an Equinox 55 FTIR–A.T.R. spectrometer (Bruker, Bremen, Germany) in the 400–4000 cm^{-1} range.

2.3. Electrospinning

The four calixarene derivatives (Figure 1), including calix[4]arene, sulfonated calix[4]arene, p-tert-calixarene, and calix[6]arene were synthesized according to the literature [18]. After 90 min stirring 210 mg of PU, 60 mg of PSU, and 4 mg of calixarenes in 2 mL of DMF, the mixture was sufficiently homogeneous for electrospinning.

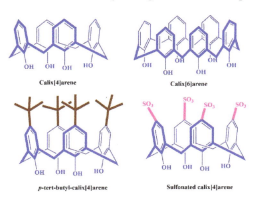

Figure 1. Chemical structure of calixarene derivatives.

After that, the solution was drawn into a syringe with a capacity of 2 mL and mounted on the syringe pump set. At a distance of 10 cm from the tip of the syringe needle, electrospun nanofibers were collected for 8 min. The flow rate of 0.15 mL h^{-1} and the electrospinning voltage of 15.5 kV was used. The SPME fiber was then placed in the GC inlet and heated to 150 °C for 1 h as the final step in the procedure.

2.4. The Procedure

At first, the PU–PSU/calix[4]arene was placed in the GC inlet and heated to 150 °C for 1 h. After that, a vial with a capacity of 10 mL was filled with 5 mL of the sample solution, which contained 100 ng mL^{-1} of each CP, and 1 g of sodium chloride. The mixture was stirred for 5 min. Afterward, the solution was sealed using a polytetrafluoroethylene septum and an open-top aluminum cap. Next, the sample solution's headspace was allowed to interact with the fiber for a total of 5.5 min. At the end of the process, the fiber was removed, then placed into the inlet of the GC to undergo thermal desorption of the adsorbed analytes at a temperature of 170 °C for 3 min.

3. Results and Discussion

3.1. Effect of Calixarene Type

PU–PSU nanofibers were treated with calix[4]arene, calix[6]arene, sulfonated calix[4]arene, and p-tert-butyl-calix[4]arene to see how each type of calixarene affected the extraction efficiency of the fiber. The process described in Section 2.4 was used to test the manufactured nanofibers, and the results were compared to those obtained using electrospun PU and PU–PSU nanofibers. As shown in Figure 2, utilizing nanofibers that included calixarenes improved the extraction efficiency. However, the maximum efficiency was obtained with PU–PSU/calix[4]arene nanofibers. Thus, it was selected as the adsorbent for subsequent studies.

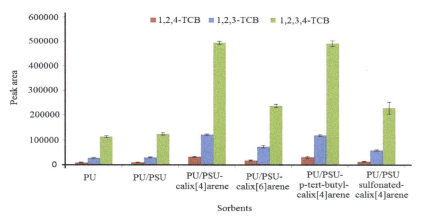

Figure 2. The extraction efficiency of different fabricated nanofibers.

3.2. Characterization

Figure 3 shows the FT-IR spectra of the PSU, PU–PSU, and PU–PSU/calix[4]arene nanofibers. The prominent peaks at 2955 cm^{-1} (C-H), 1580 and 1500 cm^{-1} (aromatic ring), 1160 cm^{-1} (S=O of sulfone), 1250 and 1090 cm^{-1} (C-O), and 850 cm^{-1} (C-S) in Figure 3A can be ascribed to the PSU [19]. Changes that correlate to PU could be seen in the spectrum of PU–PSU (Figure 3B), namely at 3391 cm^{-1} (O-H and N-H), 2955 cm^{-1} (C-H), 1726 cm^{-1} (C=O), 1582 and 1563 cm^{-1} (C-N stretching and N-H bending), 1125 cm^{-1} (C-O), and 1068 cm^{-1} (C-N) [20]. The major calix[4]arene moiety peaks (Figure 3C) have been attributed to the stretching vibrations of the O-H, C-H, C=C, and C-O functional groups, respectively, at 3340, 2950/2966, 1485, and 1281 cm^{-1} [21]. Some of these peaks overlapped with strong vibration bands of PU–PSU.

Figure 4 shows the SEM micrographs and diameter histograms of PU–PSU and PU–PSU/calix[4]arene nanofibers. The nanofibers are composed of fibers that are randomly aligned and have a smooth surface form, and they have a consistent three-dimensional porosity structure (Figure 4A,B). The fiber diameter distribution (Figure 4C) also indicates almost uniform diameters (50–450 nm) for PU–PSU/calix[4]arene as compared to PU–PSU

(50–1000 nm). This is most likely because of the characteristics of the polymer solution, such as its viscosity and the effect of the high voltage electric field on the creation of nanofibers [22].

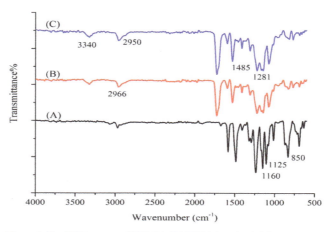

Figure 3. The FT-IR spectra of PSU (**A**), PU–PSU (**B**), and calix[4]arene modified PU–PSU nanofiber (**C**).

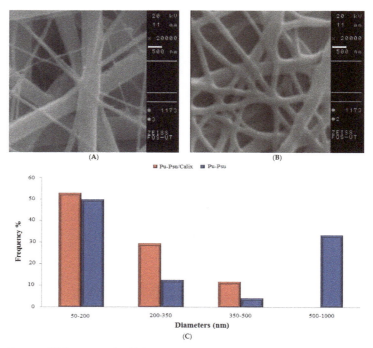

Figure 4. SEM micrograph of (**A**) PU–PSU nanofiber, (**B**) calix[4]arene modified PU–PSU nanofiber, and (**C**) fibers diameter histogram.

3.3. Impact of Ionic Strength

The effectiveness of the extraction was correlated with the sample solution's ionic strength [23]. By adding 0–20% (w/v) salt (NaCl) to the sample solution, it was possible to determine how this parameter affected the extraction efficiency. As seen in Figure 5A, increasing NaCl caused the extraction efficiency to rise steadily. The analytes may migrate to the sample headspace and be absorbed by the fiber due to the salting-out phenomenon [6]. Salt concentrations above 20% (w/v) need more time for solving and preparation. For this reason, a 20% (w/v) concentration of NaCl was decided upon as the optimum condition in the extraction process.

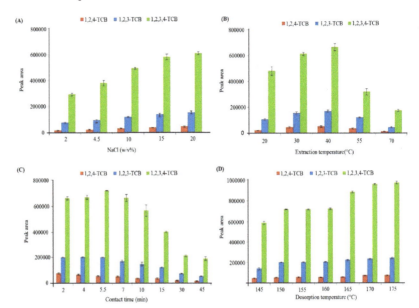

Figure 5. Influence of NaCl concentration (**A**), extraction temperature (**B**), extraction time (**C**), and desorption temperature (**D**) on the HS-SPME performance.

3.4. Impact of Extraction Temperature

The sample solution's temperature is a critical factor in the HS-SPME method since it impacts the extraction rate and equilibrium [24]. The analytes must be equilibrated between the headspace and both the sample solution and fiber. The increase in temperature accelerates the transfer of analytes between the solution and the headspace; thus, the partition coefficient of the headspace and sample solution (Khs/s) increases while the partition coefficient of the headspace and fiber (Khs/f) decreases [23]. The temperature range of 20–70 °C was studied to determine how temperature affected the technique's effectiveness. As seen in Figure 5B, the efficiency improves as the temperature rises to 40 °C, then drops. Therefore, 40 °C was chosen as the best extraction temperature for subsequent tests.

3.5. Impact of Extraction Time

The time required for extraction is also a significant factor that affects extraction yield [25]. The reaction duration was studied by varying the contact time between the adsorbent and the sample solution from 2–45 min. It can be shown in Figure 5C that the analytical signal rose to 5.5 min, then decreased. This reduction of extraction duration after equilibrium may be due to the stainless steel wire, which gets warmer with passing time, and this phenomenon can desorb analytes of nanofiber sorbent. The rapid equilibrium period may result from the adsorbent's high porosity and large surface area. As a result, the ideal amount of time for the extraction was determined to be 5.5 min.

3.6. Impact of Desorption Time and Temperature

The complete migration of analytes to the column was ensured by employing the optimum time and temperature conditions and carrying out the desorption process in the split mode. Figure 5D displays the findings from a study on desorption temperature in the split mode between 145 and 175 °C. As can be seen, raising the temperature to 170 °C improved the extraction efficiency. As can also be seen, the incorporation of PSU and calixarene into nanofibers increases the sorbent's thermal stability. At a temperature of 170 °C, the desorption time (1–4 min) was also examined. It was determined that a desorption time of 3 min was optimal to maximize fiber lifetime, minimize the risk of coating damage, and ensure that there was no carryover impact detected.

3.7. Method Validation

Matrix match calibration was used to quantitatively evaluate the nanofiber's performance under optimum conditions (salt concentration of 20% (w/v), extraction temperature of 40 °C, extraction time of 5.5 min, desorption temperature of 170 °C, and desorption time of 3 min). The limit of quantification (LOQ) was determined based on a calculation using 10 Sd/m (where Sd is the standard deviation of the blank and m represents the slope of the calibration graph), which is equal to 0.4–4 pg mL^{-1}. With an acceptable R^2 of more than 0.991, the linear dynamic range (LDR) was determined to be in the region of 0.4–1000 pg mL^{-1}. The limit of detection (LOD) based on 3Sd/m was in the range of 0.1–1 pg mL^{-1}. The relative standard deviation (RSD%) values (40 and 400 pg mL^{-1}, n = 3) were in the range of 0.9–6.0%. The within-day precision was calculated on 3 different days with 3 replicates each day equal to 0.9–9.7% (C = 40 and 400 pg mL^{-1}). The fiber manufactured can be utilized at least 60 times, which was determined through the sequential analysis of distilled water samples spiked with standard CB solutions at a concentration of 40 pg mL^{-1} (Table 1).

Table 1. Analytical figures of merit obtained for the selected CB mixtures using the HS-SPME method based on PU–PSU/calix[4]arene coupled with GC-ECD.

Compound	LOD [a]	LOQ [b]	LDR [c]	R^2 [d]
1,2,4-TCB	1.0	4.0	4–800	0.9909
1,2,3-TCB	0.1	0.4	0.4–1000	0.9911
1,2,3,4-TCB	0.1	0.4	0.4–1000	0.9917

Compound	RSD% [e]	RSD% [f]	RSD% [g]	RSD% [h]	RSD% [i]	RSD% [j]
1,2,4-TCB	1.7	10.1	9.7	3.6	8.8	3.4
1,2,3-TCB	6.0	8.9	6.7	0.9	4.7	3.6
1,2,3,4-TCB	3.8	5.5	3.5	1.4	5.2	0.9

[a] Limit of detection (S/N = 3, pg mL^{-1}); [b] Limit of quantification (S/N = 10, pg mL^{-1}); [c] Linear dynamic range (pg mL^{-1}); [d] Determination coefficient; [e] Inter-day (40 pg mL^{-1}); [f] Intra-day (40 pg mL^{-1}); [g] Fiber to fiber (400 pg mL^{-1}); [h] Intra-day (400 pg mL^{-1}); [i] Inter-day (400 pg mL^{-1}); [j] Fiber to fiber (400 pg mL^{-1}).

3.8. Analysis of Real Samples

The applicability of the developed SPME method was evaluated for the determination of the selected CBs in real samples such as tap water, sewage water (collected from the university campus), an industrial water sample (collected from an industrial park), and a river water sample (collected from Sepahsalar, Chalous, north of Iran). All the sample solutions (non-spiked and spiked) were tested in accordance with the proposed procedure (Section 2.4. The relative recoveries (RR%) were calculated using Equation (1) [6] as follows:

$$RR\,(\%) = \frac{C_{found} - C_{real}}{C_{added}} \times 100 \tag{1}$$

$$qe\ (\%) = \frac{V \times \left(C_{real} - C_{equal}\right)}{W} \qquad (2)$$

where C_{found}, C_{real}, and C_{added} represent the concentrations of the analytes in the real water samples that have been spiked, the concentrations of analytes in the real sample, and the concentration of the standard solution that has been added to the water samples, respectively; qe is the adsorption capacity and C_{equal} represents the residual concentrations of analysts in solution after the adsorption process. In addition, V is the sample volume 10 mL and W is the fiber mass 5 mg. The results for the non-spiked samples showed no detection of the selected CBs. Table 2 displays the RR% values for the spiked samples with the target analytes' 40 and 400 pg mL^{-1} standard solutions. The adsorption capacity of the nanofiber is calculated based on 400 pg mL^{-1} with Equation (2) [26], which is 720 pg/g.

Table 2. Results of HS-SPME analysis of CB mixtures using the PU–PSU/Calix[4]arene in different real water samples.

Compound	RR [a]% (RSD%) [b]			
40 pg mL^{-1}	Industrial Water [c]	Sewage Water [d]	Tap Water [d]	River Water [e]
1,2,4-TCB	105 (6.1)	89 (5.6)	101 (3.2)	91 (7.5)
1,2,3-TCB	101 (2.8)	94 (3.3)	106 (3.9)	95 (2.4)
1,2,3,4-TCB	90 (3.4)	95 (2.3)	104 (2.3)	103 (1.7)
400 pg mL^{-1}				
1,2,4-TCB	102 (3.9)	82 (5.1)	82 (3.7)	81 (2.6)
1,2,3-TCB	89 (1.8)	89 (2.8)	89 (2.2)	97 (2.0)
1,2,3,4-TCB	95 (1.0)	80 (1.8)	80 (1.8)	100 (2.5)

[a] Relative recovery; [b] Relative standard deviation (n = 3); [c] Collected from an industrial park near Tehran (Iran); [d] Collected from our university campus; [e] Collected from a river in the north of Iran (Chalous city, Mazandaran province).

A review of the relevant literature was carried out to evaluate the current method in comparison to other methods for calculating CBs that have been reported. Table 3 presents the findings of this evaluation. As is shown, the method used in this study has several advantages over previous studies in this area, including (i) a significantly shorter extraction time, (ii) a lower LOD, and (iii) a higher linear dynamic range (LDR).

Table 3. Comparison of HS-SPME/GC-ECD analysis with other methods for determination of CBs.

Sorbent	Method	Sample	LOD (pg mL^{-1})			LDR (pg mL^{-1})	Extraction Time (min)	Recovery%	RSD%	Ref.
			1,2,4-TCB	1,2,3-TCB	1,2,3,4-TCB					
PU–PSU/calix nanofibers	HS-SPME-GC-ECD	water	1	0.1	0.1	0.4–1000	5.5	80–106	4.7–10.1	This work
Polyacrylate [a]-SiO$_2$ nanofibers	HS-SPME-GC-FID	water	5	-	-	5–1000	15	94–103	4–12	[27]
PU nanofibers	HS-SPME-GC-MS	water	10	10	10	50–1000	10	94–102	3–8	[28]
PDMS [b]	HS-SPME-GC-MS	water	4	4	3	20–2000	30	91–107	1.8–6.7	[29]
PDMS	HS-SPME-GC-MS	soil	2.35	4.48	0.92	13.3–1333	15	-	-	[30]
Diglycidyloxycalix[4]areene	HS-SPME-GC-ECD	soil	0.2	0.34	0.18	5.33–533	15	-	-	[30]
Diglycidyloxycalix[4]areene	HS-SPME-GC-ECD	soil	0.14	0.16	0.16	0.267–26.7	15	76–100	2.9–13.4	[30]
PDMS [b]	SPME-GC-IT-MS	soil	46	30	40	-	50	-	2–15	[31]

[a] Polyamide, [b] Polydimethylsiloxane.

4. Conclusions

In this study, the PU–PSU/calix[4]arene was fabricated and used as an effective SPME fiber coating in the headspace extraction of CBs in aquatic samples. The large surface area and porous structure of the nanofibrous PU–PSU/calix[4]arene mat provided fast adsorption of the analytes (5.5 min). The π-π stacking interactions are possibly the main reason for adsorbing CB molecules that diffused to the nanofibers/calixarene mat. The method is eco-friendly since it requires no organic solvents in the extraction and analysis steps. Moreover, the low LOD (0.1–1 pg mL^{-1}) and short extraction and analysis times are characteristics that provide an excellent solution for the detection of the analytes.

Author Contributions: Conceptualization, H.N. and H.S.; methodology, H.N., H.S. and H.R.N.; formal analysis, H.N., H.S. and S.S.; investigation, H.N., H.S. and H.R.N.; writing—original draft preparation, H.N., M.A.K., H.S., S.S. and H.R.N.; writing—review and editing, H.N., H.S., M.A.K., S.S., N.S., I.A. and H.R.N.; supervision, H.S. and H.R.N.; funding acquisition, H.S., I.A. and N.S. All authors have read and agreed to the published version of the manuscript.

Funding: This research was supported by the Marine Pollution Special Interest Group, and the National Defence University of Malaysia via SF0076-UPNM/2019/SF/ICT/6. Syed Shahabuddin and Irfan Ahmad are grateful to the Scientific Research Deanship at King Khalid University, Abha, Saudi Arabia for their support through the Small Research Group Project under grant number (RGP.01-115-43).

Institutional Review Board Statement: Not applicable.

Informed Consent Statement: Not applicable.

Data Availability Statement: The data presented in this study are available on request from the corresponding author.

Conflicts of Interest: The authors declare no conflict of interest.

References

1. Li, G.; Wang, J.; Zhu, P.; Han, Y.; Yu, A.; Li, J.; Sun, Z.; Row, K.H. The Separation of Chlorobenzene Compounds from Environmental Water Using a Magnetic Molecularly Imprinted Chitosan Membrane. *Polymers* **2022**, *14*, 3221. [CrossRef] [PubMed]
2. Sharma, J.; Ahuja, S.; Arya, R.K. Experimental designing of polymer-polymer-solvent coatings: Poly (styrene)-poly (ethylene glycol)-chlorobenzene coating. *Prog. Org. Coat.* **2019**, *128*, 181–195. [CrossRef]
3. Zhang, S.; Lin, D.; Wu, F. The effect of natural organic matter on bioaccumulation and toxicity of chlorobenzenes to green algae. *J. Hazard. Mater.* **2016**, *311*, 186–193. [CrossRef] [PubMed]
4. UNEP (United Nations Environment Program). *Stockholm Convention on Persistent Organic Pollutants*; UNEP: Geneva, Switzerland, 2011.
5. WHO (World Health Organization). DDT use in disease vector control under the Stockholm Convention on Persistent Organic Pollutants. In *WHO: Guidelines for Drinking-Water Quality*; Regional Office for Americas/Pan American Sanitary Bureau (AMRO/PAHO) 525: Washington, DC, USA, 2004.
6. Shirani, M.; Parandi, E.; Nodeh, H.R.; Akbari-Adergani, B.; Shahdadi, F. Development of a rapid efficient solid-phase microextraction: An overhead rotating flat surface sorbent based 3-D graphene oxide/lanthanum nanoparticles@ Ni foam for separation and determination of sulfonamides in animal-based food products. *Food Chem.* **2022**, *373*, 131421. [CrossRef]
7. Shirani, M.; Aslani, A.; Sepahi, S.; Parandi, E.; Motamedi, A.; Jahanmard, E.; Nodeh, H.R.; Akbari-Adergani, B. An efficient 3D adsorbent foam based on graphene oxide/AgO nanoparticles for rapid vortex-assisted floating solid phase extraction of bisphenol A in canned food products. *Anal. Methods* **2022**, *14*, 2623–2630. [CrossRef] [PubMed]
8. Shah, H.U.R.; Ahmad, K.; Bashir, M.S.; Shah, S.S.A.; Najam, T.; Ashfaq, M. Metal organic frameworks for efficient catalytic conversion of CO_2 and CO into applied products. *Mol. Catal.* **2022**, *517*, 112055. [CrossRef]
9. Cheng, H.; Wang, F.; Bian, Y.; Ji, R.; Song, Y.; Jiang, X. Co-and self-activated synthesis of tailored multimodal porous carbons for solid-phase microextraction of chlorobenzenes and polychlorinated biphenyls. *J. Chromatogr. A* **2019**, *1585*, 1–9. [CrossRef]
10. Hussain, D.; Naqvi, S.T.R.; Ashiq, M.N.; Najam-ul-Haq, M. Analytical sample preparation by electrospun solid phase microextraction sorbents. *Talanta* **2020**, *208*, 120413. [CrossRef]
11. Kumar, R.; Sharma, A.; Singh, H.; Suating, P.; Kim, H.S.; Sunwoo, K.; Shim, I.; Gibb, B.C.; Kim, J.S. Revisiting fluorescent calixarenes: From molecular sensors to smart materials. *Chem. Rev.* **2019**, *119*, 9657–9721.
12. Chung, T.-S.; Lai, J.-Y. The potential of calixarenes for membrane separation. *Chem. Eng. Res. Des.* **2022**, *183*, 538–545.
13. Zhu, F.; Zheng, Y.-M.; Zhang, B.-G.; Dai, Y.-R. A critical review on the electrospun nanofibrous membranes for the adsorption of heavy metals in water treatment. *J. Hazard. Mater.* **2021**, *401*, 123608. [CrossRef] [PubMed]
14. Háková, M.; Havlíková, L.C.; Solich, P.; Švec, F.; Šatínský, D. Electrospun nanofiber polymers as extraction phases in analytical chemistry—The advances of the last decade. *TrAC Trends Anal. Chem.* **2019**, *110*, 81–96. [CrossRef]
15. Chen, M.; Wang, C.; Fang, W.; Wang, J.; Zhang, W.; Jin, G.; Diao, G. Electrospinning of calixarene-functionalized polyacrylonitrile nanofiber membranes and application as an adsorbent and catalyst support. *Langmuir* **2013**, *29*, 11858–11867. [CrossRef] [PubMed]
16. Bayrakcı, M.; Özcan, F.; Ertul, Ş. Synthesis of calixamide nanofibers by electrospinning and toxic anion binding to the fiber structures. *Tetrahedron* **2015**, *71*, 3404–3410. [CrossRef]

17. Özcan, F.; Bayrakcı, M.; Ertul, Ş. Synthesis and characterization of novel nanofiber based calixarene and its binding efficiency towards chromium and uranium ions. *J. Incl. Phenom. Macrocycl. Chem.* **2016**, *85*, 49–58. [CrossRef]
18. Belardi, R.P.; Pawliszyn, J.B. The application of chemically modified fused silica fibers in the extraction of organics from water matrix samples and their rapid transfer to capillary columns. *Water Qual. Res. J.* **1989**, *24*, 179–191. [CrossRef]
19. Shah, H.U.R.; Ahmad, K.; Naseem, H.A.; Parveen, S.; Ashfaq, M.; Aziz, T.; Shaheen, S.; Babras, A.; Shahzad, A. Synthetic routes of azo derivatives: A brief overview. *J. Mol. Struct.* **2021**, *1244*, 131181. [CrossRef]
20. Ahmad, K.; Khan, M.S.; Iqbal, A.; Potrich, E.; Amaral, L.S.; Rasheed, S.; Ashfaq, M. Lead In drinking water: Adsorption method and role of zeolitic imidazolate frameworks for its remediation: A review. *J. Clean. Prod.* **2022**, *368*, 133010. [CrossRef]
21. Solangi, I.B.; Bhatti, A.A.; Kamboh, M.A.; Memon, S.; Bhanger, M. Comparative fluoride sorption study of new calix [4] arene-based resins. *Desalination* **2011**, *272*, 98–106. [CrossRef]
22. Parandi, E.; Pero, M.; Kiani, H. Phase change and crystallization behavior of water in biological systems and innovative freezing processes and methods for evaluating crystallization. *Discov. Food* **2022**, *2*, 6. [CrossRef]
23. Mosleh, N.; Ahranjani, P.J.; Parandi, E.; Nodeh, H.R.; Nawrot, N.; Rezania, S.; Sathishkumar, P. Titanium lanthanum three oxides decorated magnetic graphene oxide for adsorption of lead ions from aqueous media. *Environ. Res.* **2022**, *214*, 113831. [CrossRef]
24. Xu, S.; Dong, P.; Liu, H.; Li, H.; Chen, C.; Feng, S.; Fan, J. Lotus-like Ni@ NiO nanoparticles embedded porous carbon derived from MOF-74/cellulose nanocrystal hybrids as solid phase microextraction coating for ultrasensitive determination of chlorobenzenes from water. *J. Hazard. Mater.* **2022**, *429*, 128384. [CrossRef] [PubMed]
25. Mosleh, N.; Najmi, M.; Parandi, E.; Nodeh, H.R.; Vasseghian, Y.; Rezania, S. Magnetic sporopollenin supported polyaniline developed for removal of lead ions from wastewater: Kinetic, isotherm and thermodynamic studies. *Chemosphere* **2022**, *300*, 134461. [CrossRef]
26. Ahmad, K.; Ashfaq, M.; Nawaz, H. Removal of decidedly lethal metal arsenic from water using metal organic frameworks: A critical review. *Rev. Inorg. Chem.* **2022**, *42*, 197–227. [CrossRef]
27. Bagheri, H.; Roostaie, A. Electrospun modified silica-polyamide nanocomposite as a novel fiber coating. *J. Chromatogr. A* **2014**, *1324*, 11–20. [CrossRef] [PubMed]
28. Bagheri, H.; Aghakhani, A. Novel nanofiber coatings prepared by electrospinning technique for headspace solid-phase microextraction of chlorobenzenes from environmental samples. *Anal. Methods* **2011**, *3*, 1284–1289. [CrossRef]
29. He, Y.; Wang, Y.; Lee, H.K. Trace analysis of ten chlorinated benzenes in water by headspace solid-phase microextraction. *J. Chromatogr. A* **2000**, *874*, 149–154. [CrossRef]
30. Li, X.; Zeng, Z.; Xu, Y. A solid-phase microextraction fiber coated with diglycidyloxycalix [4] arene yields very high extraction selectivity and sensitivity during the analysis of chlorobenzenes in soil. *Anal. Bioanal. Chem.* **2006**, *384*, 1428–1437. [CrossRef]
31. Sarrion, M.; Santos, F.; Galceran, M. Strategies for the analysis of chlorobenzenes in soils using solid-phase microextraction coupled with gas chromatography–ion trap mass spectrometry. *J. Chromatogr. A* **1998**, *819*, 197–209. [CrossRef]

Article

Superhydrophobic Nanosilica Decorated Electrospun Polyethylene Terephthalate Nanofibers for Headspace Solid Phase Microextraction of 16 Organochlorine Pesticides in Environmental Water Samples

Hamid Najarzadekan [1], Hassan Sereshti [1,*], Irfan Ahmad [2], Syed Shahabuddin [3,*], Hamid Rashidi Nodeh [4] and Nanthini Sridewi [5,*]

1 School of Chemistry, College of Science, University of Tehran, Tehran 1417614411, Iran
2 Department of Clinical Laboratory Sciences, College of Applied Medical Sciences, King Khalid University, Abha 61421, Saudi Arabia
3 Department of Chemistry, School of Technology, Pandit Deendayal Energy University, Raisan 382426, Gujarat, India
4 Food Technology and Agricultural Products Research Center, Standard Research Institute, Karaj 3174734563, Iran
5 Department of Maritime Science and Technology, Faculty of Defence Science and Technology, National Defence University of Malaysia, Kuala Lumpur 57000, Malaysia
* Correspondence: sereshti@ut.ac.ir (H.S.); syedshahab.hyd@gmail.com or syed.shahabuddin@sot.pdpu.ac.in (S.S.); nanthini@upnm.edu.my (N.S.); Tel.: +98-21-6113735 (H.S.); +91-8585932338 (S.S.); +60-124-675-320 (N.S.)

Citation: Najarzadekan, H.; Sereshti, H.; Ahmad, I.; Shahabuddin, S.; Rashidi Nodeh, H.; Sridewi, N. Superhydrophobic Nanosilica Decorated Electrospun Polyethylene Terephthalate Nanofibers for Headspace Solid Phase Microextraction of 16 Organochlorine Pesticides in Environmental Water Samples. *Polymers* **2022**, *14*, 3682. https://doi.org/10.3390/polym14173682

Academic Editors: Ahmed K. Badawi, Irene S. Fahim and Hossam E. Emam

Received: 8 August 2022
Accepted: 23 August 2022
Published: 5 September 2022

Publisher's Note: MDPI stays neutral with regard to jurisdictional claims in published maps and institutional affiliations.

Copyright: © 2022 by the authors. Licensee MDPI, Basel, Switzerland. This article is an open access article distributed under the terms and conditions of the Creative Commons Attribution (CC BY) license (https://creativecommons.org/licenses/by/4.0/).

Abstract: A new solid phase micro extraction (SPME) fiber coating composed of electrospun polyethylene terephthalate (PET) nanofibrous mat doped with superhydrophobic nanosilica (SiO$_2$) was coated on a stainless-steel wire without the need of a binder. The coating was characterized by scanning electron microscopy (SEM) and Fourier transform infrared spectrometer (FTIR) techniques and it was used in headspace-SPME of 16 organochlorine pesticides in water samples prior to gas chromatography micro electron capture detector (GC-µECD) analysis. The effects of main factors such as adsorption composition, electrospinning flow rate, salt concentration, extraction temperature, extraction time, and desorption conditions were investigated. Under the optimum conditions, the linear dynamic range (8–1000 ng L^{-1}, R^2 > 0.9907), limits of detection (3–80 ng L^{-1}), limits of quantification (8–200 ng L^{-1}), intra-day and inter-day precisions (at 400 and 1000 ng L^{-1}, 1.7–13.8%), and fiber-to-fiber reproducibility (2.4–13.4%) were evaluated. The analysis of spiked tap, sewage, industrial, and mineral water samples for the determination of the analytes resulted in satisfactory relative recoveries (78–120%).

Keywords: electrospun nanofibers; polyethylene terephthalate; superhydrophobic nanosilica; solid-phase microextraction; organochlorine pesticides

1. Introduction

Organochlorine pesticides (OCPs) are persistent lipophilic organic pollutants that are highly resistant to biodegradation in the environment [1]. These compounds exhibit high toxicity and bioaccumulation and have been demonstrated to be carcinogenic in animals and humans. For these reasons, they have been included in the list of priority pollutants compiled by The United States-Environmental Protection Agency (US-EPA) [2]. Although the application of OCPs has been forbidden for a considerable period in many countries, the residues continue to induce a significant impact on the environment and its ecosystems [3]. The maximum possible limits are 0.1 µg L^{-1} for each OCP and 0.5 µg L^{-1} for the total concentration of all pesticides based on toxicological considerations [4,5]. Therefore, it is necessary to develop fast, simple, and valid methods for their determination in different matrices [2]. A variety of sample preparation techniques, such as pressurized liquid extraction [6], in-cell accelerated solvent extraction [7], microsolid-phase extraction [8], magnetic solid-phase extraction (MSPE) [9], vortex-assisted MSPE [10], and SPME [11] coupled with gas chromatography, have been used for the extraction and determination of OCPs in different matrices. Among these methods, SPME was developed in the 1990s by Arthur and Pawliszyn et al. [12], which satisfies the requirements of green analytical chemistry. Hence, this method has widely been applied for sampling a broad spectrum of analytes from gaseous, liquid, and solid media with diverse matrix compositions [13]. Polydimethylsiloxane (PDMS), divinylbenzene (DVB), carboxen (CAR), polyethylene glycol (PEG), carbowax (CW), polyacrylate, and their combinations, such as PDMS/DVB, PDMS/CAR, CW/DVB, and DVB/CAR/PDMS, are the commercially-available SPME fiber coatings [14,15]. However, the ordinary SPME fibers suffer from poor mechanical strength, low recommended operating temperature, fragility, restricted lifetime, and limited applicability [16]. To overcome these drawbacks, researchers have tried to synthesize new sorption materials for SPME purposes. In recent years, the use of nanoscale materials with diverse functionalities and polarities for the fabrication of new SPME fiber coatings with enhanced selectivity, sorption capacity, and stability has attracted the attention of researchers [17,18]. Moreover, nanoscale materials with a high specific surface area can improve the sample loading capacity and extraction efficiency [17,19–21]. Therefore, due to the unique features of the nanoscale materials, they have emerged and become important in the analytical detection and remediation of environmental pollutants [22]. Furthermore, in some SPME applications, low-cost metal wires have been replaced with fragile fused silica fibers [23,24]. The electrospinning method, which is a popular technique for the production of polymeric nanofibers [25], has been widely used as an SPME fiber coating [26–29]. The polymers can be embedded with different materials to enhance the performance of the coating by increasing the surface area-to-volume ratio and the functionality of the produced nanofibers [30,31].

In the present study, an effective head space solid phase microextraction (HS-SPME) adsorbent using electrospun composite nanofibers was used for the extraction of OCPs from aqueous solutions. Accordingly, different PET-based nanofibers, such as PET/nanoclay, PET/nano-SiO$_2$, and PET/calixarene, were electrospun on the surface of a stainless-steel wire and tested under the same experimental conditions. To the best of our knowledge, this is the first time that these PET-based composite nanofibers have been fabricated and applied for the extraction of OCPs.

2. Experimental

2.1. Chemicals and Reagents

The 16 OCP mixture (EPA 608 pesticide mix, a stock standard solution of 20 µg mL^{-1} OCPs in toluene:n-hexane, (1:1)) was purchased from Sigma-Aldrich (St Louis, MO, USA). The mixture consisted of α-hexachlorocyclohexane (HCH), β-HCH, γ-HCH, δ-HCH, aldrin, dieldrin, endosulfan I, endosulfan II, endosulfan sulfate, endrin, endrin aldehyde, heptachlor, heptachlor epoxide-isomer B, p,p'-DDE, p,p'-DDD, and p,p'-DDT. All the standard solutions were diluted with distilled water and acetone and later stored at −4 °C in the dark. Trifluoroacetic acid (TFA, 99%) was supplied from Samchun Pure Chemical (Seoul, South Korea). Polyethylene terephthalate (PET) was bought from Merck Chemicals (Darmstadt, Germany). NaCl (ACS reagent, ≥99.0%, Sigma-Aldrich, St Louis, MO, USA) was used as received. The superhydrophobic nano-SiO$_2$ and nano-clay were obtained from Nanosav Company (Tehran, Iran).

2.2. Gas Chromatography

An Agilent Technologies gas chromatograph (6890N, Santa Clara, CA, USA) equipped with a micro electron capture detector (µ-ECD) and an HP-5 capillary fused silica column (30 m × 0.32 mm × 0.25 µm) was used for the separation of the extracted OCPs. Helium (99.999%) at a flow rate of 1 mL min^{-1} and nitrogen (99.999%) at a flow rate of 30 mL min^{-1} were used as the carrier gas and makeup gas, respectively. The injector and detector temperatures were set at 200 and 300 °C, respectively. The sample introduction was performed in the splitless mode for 3 min. The column temperature program was initiated at 60 °C (held for 1 min), increased at 30 °C min^{-1} to 180 °C, and then raised to 250 °C at 7 °C min^{-1} (fixed for 5 min).

2.3. Instrumentation

A Zeiss DSM-960 SEM (Oberkochen, Germany) at the accelerating voltage of 20 kV was used for morphology characterization. An Equinox 55 FT-IR spectrometer (Bruker, Bremen, Germany) was utilized for recording the infrared spectra with KBr pallet. The electrospinning device (Fanavaran Nano-Meghyas Co., Tehran, Iran) consisted of a DC high-voltage power supply and a syringe pump, which was applied for the electrospinning of nanofibers. A homemade SPME syringe with two spinal needles—the internal G 27as SPME-coated needle and the external needle of G 22 as an SPME barrel—was used in headspace–SPME.

2.4. Electrospinning of Nanofibers

The PET/nano-SiO$_2$ nanofibers were fabricated as follows. First, 180 mg of polyethylene terephthalate (PET) was dissolved in 1 mL of trifluoroacetic acid (TFA). Then, 5 mg of nano-SiO$_2$ was added to the solution and stirred for 100 min to obtain a homogeneous solution. Next, the mixture was transferred into a 2 mL syringe, placed in the syringe pump, and pumped at the rate of 0.15 mL h^{-1}. The electrospun nanofibers were collected on the stainless-steel wire attached to a rotating electric motor at a distance of 10 cm from the tip of the syringe's needle for 8 min (Figure 1A). The electrospinning voltage was 16 kV.

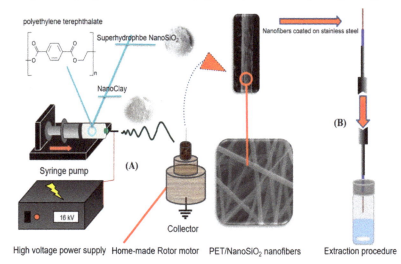

Figure 1. (**A**) The setup design for electrospinning process, and (**B**) headspace–SPME procedure.

2.5. The Procedure

The fabricated fiber coating was conditioned prior to use by inserting it in the GC injection port at 200 °C for 15 min. Then, 5 mL of the sample solution was placed in a vial, and 1.25 g of NaCl was added to it and stirred for 5 min. After that, the solution was spiked with the standard mixture of the OCPs (100 ng mL^{-1}) and sealed by a Polytetrafluoroethylene (PTFE) (CNW, Beijing, China) septum. Next, the needle coated with fibrous PET/nano-SiO$_2$ was exposed to the headspace of the sample solution at 40 °C (Figure 1B). Finally, the fiber was withdrawn and immediately inserted into the GC injection port for thermal desorption of the analytes at 200 °C for 3 min.

3. Results

3.1. Effect of Adsorbent Composition

Polyethylene terephthalate was doped with different materials, such as nanoclay, nano-SiO_2, and calixarene, and electrospun. The nanofibers were tested with the procedure in Section 2.5 and the data was presented in Figure 2A. Clearly, the PET/nano-SiO_2 had the highest extraction efficiency. Thus, it was selected as the fiber coating for further analyses. In addition, the effect of the nano-SiO_2 dose in the electrospun nanofibers was also investigated by adding various amounts of nano-SiO_2 (1–9 mg) into the PET/TFA polymer solution before electrospinning. Figure 2B shows that with an increasing nano-SiO_2 dose, until 5 mg, the efficiency was increased and remained almost constant after that. Thus, 5 mg was chosen as the optimum amount of nano-SiO_2.

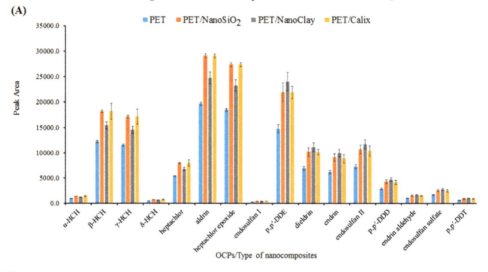

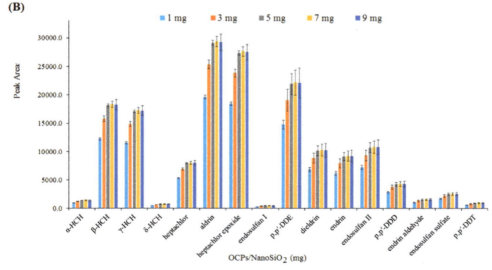

Figure 2. *Cont.*

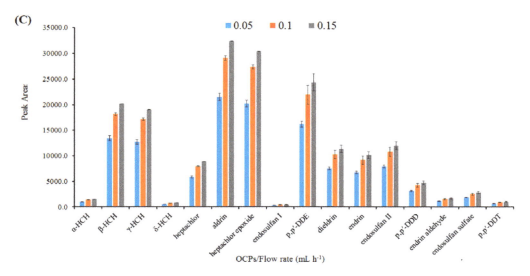

Figure 2. (**A**) Performance of different types of PET nanofibers, (**B**) dosage of nano-SiO$_2$ (mg) in PET nanofibers, and (**C**) the effect of polymer solution flow rate (mL h^{-1}).

3.2. Effect of Electrospinning Flow Rate

The influence of the electrospinning flow rate on the efficiency of the PET/nano-SiO$_2$ nanofibers was studied in the range of 0.05–0.15 mL h^{-1}. The higher flow rates were not tested because the thickness of the adsorbent layer was limited by the internal diameter of the SPME needle. As shown in Figure 2C, by increasing the flow rate, the extraction efficiency was also increased. This increase could be due to the higher surface area and porosity of the coating. Therefore, 0.15 mL h^{-1} was selected as the optimum flow rate.

3.3. Characterization

The nano-SiO$_2$ and PET/nano-SiO$_2$ surface functional groups were studied using FT-IR spectroscopy. The spectrum of nano-SiO$_2$ in Figure 3 shows the peaks at 2925, 1631, and 1073 cm^{-1} that can be assigned to the alkane C–H stretching vibrations, O–H bending vibrations, and Si–O stretching vibrations, respectively. Figure 3 presents the FTIR spectrum of PET/nano-SiO$_2$, in which the bands at 1728 (C=O), 1243 (C–C–O), 1034 (O–C–C), and 1069 cm^{-1} (Si–O) indicate the successful deposition of nano-SiO$_2$ particles on PET [32].

Figure 4A–D shows the SEM micrograph of the PET and PET/nano-SiO$_2$ nanofibers in two scales with a histogram of the diameter distribution for nanofibers. These images indicate the three-dimensional porous structure of randomly-oriented fibers with approximately uniform diameters in the ranges of 300–890 nm and 190–615 nm for PET and PET/nano-SiO$_2$, respectively. Figure 4E depict the histogram for size distribution of fabricated nanofiber, which is average size obtained 300 to 500 nm.

3.4. Effect of Parameters on Extraction Efficiency
3.4.1. Effect of Salt Concentration

The effect of NaCl concentration on the headspace extraction of OCPs was investigated in the range of 5–30% (w/v). As shown in Figure 5A, the extraction efficiency increased from the addition of salt by up to 25% w/v due to the salting-out effect. At higher concentrations, the analytical signals remained almost constant, thus 25% was selected as the optimum salt concentration.

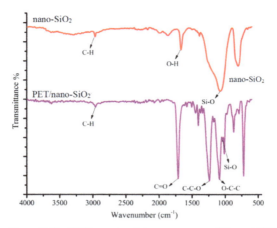

Figure 3. The FT-IR spectra of electrospun nano-SiO$_2$ and PET/nano-SiO$_2$.

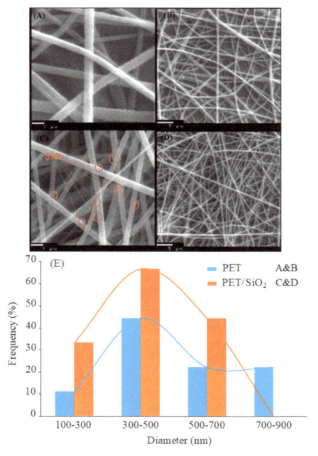

Figure 4. The SEM images of (**A**,**B**) electrospun PET and (**C**,**D**) PET/nano-SiO$_2$. (**E**) histogram for size distribution of fabricated nanofiber.

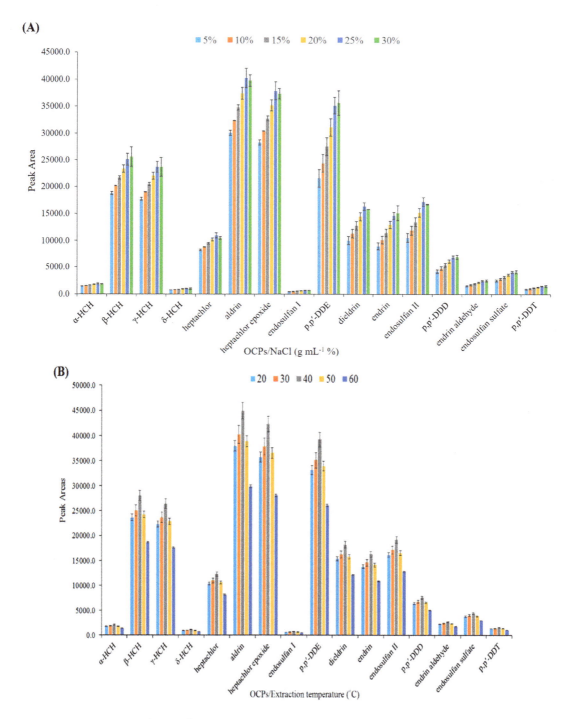

Figure 5. *Cont.*

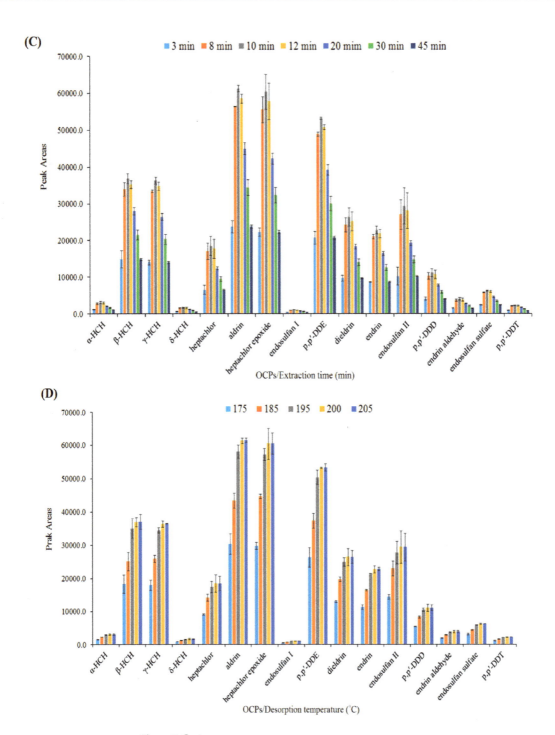

Figure 5. Cont.

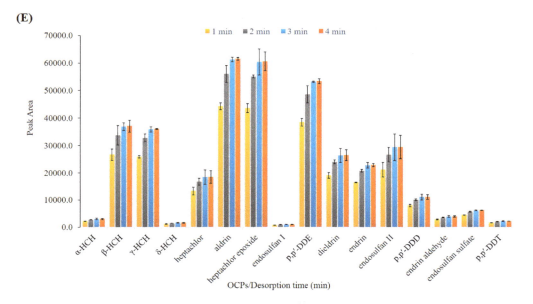

Figure 5. Optimization of effective parameters on extraction efficiency of OCPs. (**A**) Effect of salt (w/v %), (**B**) extraction temperature (°C), (**C**) extraction time (min), (**D**) desorption temperature (°C), and (**E**) desorption time (min).

3.4.2. Effect of Extraction Temperature

In the HS-SPME technique, the temperature of the sample solution can affect the extraction rate and equilibrium. The increase in temperature accelerates the transfer of the analyte between phases and affects the extraction efficiency through partition coefficients, i.e., $K_{hs/s}$ and $K_{f/hs}$ (Equations (1) and (2)), in such a way that improves the former and worsens the latter.

$$K_{hs/s} = \frac{C_{hs}}{C_{os}} \quad (1)$$

$$K_{f/hs} = \frac{C_f}{C_{hs}} \quad (2)$$

where $K_{hs/s}$, $K_{f/hs}$ are the partition coefficients of the analyte in sample/headspace, and headspace/fiber, respectively; and C_{hs}, C_s, and C_f show the concentrations of analytes in the headspace, sample solution, and nanofibers, respectively. The influence of temperature on the extraction efficiency was studied in the range of 20–60 °C. Figure 5B shows that the extraction efficiency increased from 20–40 °C and declined at higher temperatures. The increase in efficiency in the first region is due to the increase in the concentration of analytes in the headspace, but the decrease in efficiency in the second region can be attributed to the decrease of $K_{f/hs}$ [13]. Thus, 40 °C was selected as the optimum temperature for the subsequent experiments.

3.4.3. Effect of Extraction Time

The performance of HS-SPME is based on the equilibrium between the adsorbent, headspace, and sample solution. Therefore, the diffusion of the analytes through this triple-phase system is essential. The time taken to reach equilibrium is usually long and the extraction is often involved in non-equilibrium conditions. The influence of extraction time was studied by varying the exposure time of the fiber to the headspace of the sample solution in the range of 3–45 min. Figure 5C shows that the extraction efficiency increased by increasing the extraction time to 10 min, and was thereafter reduced. This short equilibrium time could be attributed to the high surface area and porosity of the nanofibers. Therefore, 10 min was selected as the optimum extraction time.

3.4.4. Effect of Desorption Temperature and Time

The temperature and time of desorption are important factors that influence the efficiency of the SPME process. To ensure the complete transfer of analytes from fiber coating to the GC column, the GC inlet system was operated in the splitless mode. The effect of the desorption temperature (GC inlet) was studied in the range of 175–205 °C. As shown in Figure 5D, the extraction efficiency increased from 175–200 °C and remained almost constant afterward. Thus, 200 °C was chosen as the optimum desorption temperature. Then, the effect of desorption time was investigated in the range of 2–4 min at 200 °C. Therefore, regarding the obtained results in Figure 5E, 3 min was selected as the optimum desorption time to achieve the total desorption of all analytes with no carryover effect.

3.5. Method Validation

Under the optimum conditions (salt concentration, 25% (w/v); extraction temperature, 40 °C; extraction time, 10 min; and desorption temperature, 200 °C; and desorption time, 3 min), the quantitative performance of the developed method was assessed, and the results were given in Table 1. The calibration curves were prepared using the mixed standard solutions of the Ops at nine concentration levels. The curves were linear in the range of 8–1000 ng L^{-1}, with a satisfactory determination coefficient (R^2) of >0.9971. The limits of detection based on the signal-to-noise ratio (S/N) of three replicates that were 3–80 ng L^{-1}. The limits of quantification (S/N, 10) were calculated as 8–200 ng L^{-1}. The intra-day relative standard deviations (RSD%, n = 3) at the concentration levels of 400 and 1000 ng L^{-1} were within the 1.7–13.8% range. The inter-day RSD% was calculated using fiber in three different days with three replicates on each day equal to 1.7–11.8% (C = 400 and 1000 ng mL^{-1}). The reusability of the PET/nano-SiO$_2$ fiber was also evaluated by assessment of its extraction performance in successive adsorption/desorption cycles under the same experimental conditions. After completing each cycle, the adsorbent was washed three times with MeOH and water sequentially. Then, the dried adsorbent was reused for the next run. The results indicated that the fiber can be reused at least 75 times without a significant reduction in efficiency (<5%). Therefore, the PET/nano-SiO$_2$ fiber qualified for frequent use in solid-phase extraction-based methods.

3.6. Analysis of Real Samples

Four real water samples including tap water, sewage water, industrial wastewater, and mineral water were selected to assess the applicability of the proposed method for the determination of the selected OCPs. The real samples were spiked with the mixed standard solutions of the target analytes at two concentration levels (400 and 1000 pg mL^{-1}). The unspiked and spiked sample solutions were analyzed with the proposed procedure in Section 2.5. The relative recoveries for the spiked samples were calculated by Equation (3), and the results were given in Table 2.

$$RR(\%) = \frac{C_{found} - C_{real}}{C_{added}} \times 100 \qquad (3)$$

where C_{found} is the concentration of analytes after spiking the real sample with a standard solution, C_{real} is the concentration of analytes in the real sample, and C_{added} is the concentration of standard solution added to the real sample.

A comparison study was conducted based on the literature survey for the previously reported works for the determination of OCPs, and the data is shown in Table 3. The results indicated that the extraction time of the proposed method is shorter than that of the other methods. In addition, the recovery, linear dynamic range (LDR), and limit of detection (LOD) of the developed method are better than most of the other methods.

Table 1. Analytical data obtained by HS-SPME of OCPs mixture using the PET/nano-SiO$_2$ adsorbent.

Compound	LOD [a]	LOQ [b]	LDR [c]	R^2	RSD [d]			RSD [e]		
					Intraday	Interday	Fiber to Fiber	Intraday	Interday	Fiber to Fiber
α-HCH	30	80	80–10,000	0.9967	6.3	8.6	7.5	2.2	6.4	5.4
β-HCH	20	50	50–5000	0.9914	5.3	1.7	3.5	3.1	2.4	2.4
γ-HCH	20	50	50–5000	0.9914	4.8	9.5	7.1	3.2	7.0	6.3
δ-HCH	80	200	200–10,000	0.9983	8.1	10.6	4.8	8.2	8.8	11.9
Heptachlor	20	50	50–5000	0.9907	10.5	7.3	8.9	5.5	6.0	6.4
Aldrin	3	8	8–2000	0.9977	6.2	4.5	5.4	4.0	3.9	4.2
Heptachlor epoxide	3	8	8–2000	0.9930	5.5	2.6	4.0	5.5	2.7	4.0
Endosulfan I	80	200	200–10,000	0.9987	10.9	10.3	11.6	5.7	10.6	12.0
p,p'-DDE	3	8	8–2000	0.9989	7.2	6.5	6.8	2.6	4.2	4.5
Dieldrin	30	80	80–5000	0.9909	11.3	7.6	11.5	4.2	6.0	5.9
Endrin	30	80	80–5000	0.9987	9.8	3.1	6.5	1.7	3.4	2.4
Endosulfan II	30	80	80–5000	0.9932	12.3	8.1	10.2	5.6	4.3	6.9
p,p'-DDD	30	80	80–2000	0.9912	12.4	8.5	5.4	10.0	8.8	9.3
Endrin aldehyde	30	80	80–2000	0.9942	13.8	8.0	5.4	9.5	11.0	12.3
Endosulfan sulfate	30	80	80–5000	0.9921	11.4	9.3	7.3	8.5	6.6	11.4
p,p'-DDT	30	80	80–5000	0.9965	11.3	10.5	11.4	11.3	11.8	13.4

[a] Limit of detection (ng L^{-1}). [b] Limit of quantification (ng L^{-1}). [c] Linear dynamic range (ng L^{-1}). [d] C = 400 ng L^{-1}. [e] C = 1000 ng L^{-1}.

Table 2. Analytical data obtained after HS-SPME of OCPs mixture using the PET/Superhydrophobe NanoSiO$_2$.

Compound	Industrial Water [a] (RR%) [b]				Sewage Water [c] (RR%)				Tap Water (RR%)				River Water [d] (RR%)			
	S1 [e]	RSD%	S2 [f]	RSD%	S1	RSD%	S2	RSD%	S1	RSD%	S2	RSD%	S1	RSD%	S2	RSD%
α-HCH	100.3	6	99.6	6.6	93.7	7.8	99.6	6.6	90.8	6.3	99.4	2.2	96.5	11.3	101	4.9
β-HCH	114.9	2.9	97.7	2.4	98	1.5	97.7	2.4	89.8	5.3	99.8	3.1	105.1	2.5	99.3	3.9
γ-HCH	88.8	3.5	93.3	7.7	103.3	1.5	93.3	7.7	91.2	4.8	100.1	3.2	90.3	1.7	96.4	5.0
δ-HCH	95.4	18	95.1	13.6	120.7	7.3	95.1	13.6	103.1	48.1	100.4	8.2	113.2	17.0	98.1	9.8
Heptachlor	98.9	11	97.1	6.9	98.4	10.1	97.1	6.9	100.8	10.5	99.6	5.5	103.1	6.9	100.9	6.3
Aldrin	96.0	1.3	98	4.6	98.2	5.2	98	4.6	102.3	6.2	100	4.0	101.5	4.9	100.5	4.3
Heptachlorepoxide	98.9	5.9	96.7	1.7	101.3	4.4	96.7	1.7	100.1	5.5	99.7	5.5	102.0	3.1	96.0	1.7
Endosulfan I	100	10.4	96.2	13.1	116.1	10.7	96.2	13.1	95.3	10.9	100.5	5.7	102.5	9.2	101.1	5.2
p,p'-DDE	95.2	1.3	99.6	4.2	98.9	7.0	99.6	4.2	96.0	7.2	100.3	2.6	101.5	6.1	102.2	2.2
Dieldrin	87.3	7.0	97.1	6.4	90.9	4.8	97.1	6.4	92.8	21.3	99.4	4.2	105.8	14.7	96.5	5.5
Endrin	103	7.3	100.8	3.7	99.9	4.1	100.8	3.7	101.4	9.8	100.5	1.7	100.3	9.8	99.6	3.0
Endosulfan II	97.9	13.7	96.0	5.1	95.7	10.1	96	5.1	99.7	12.3	100.2	5.6	103.2	8.7	95.8	5.2
p,p'-DDD	88.3	7.0	96.1	11.0	92.6	13.8	96.1	11	90.0	22.4	99.6	10.0	106.1	15.5	96.4	11.5
Endrin aldehyde	81.3	10.1	100	13.2	94.1	12	100	13.2	97.8	23.8	101.3	9.5	104.7	16.6	99.4	12.2
Endosulfan sulfate	82.4	17	98.3	8.0	79.1	14.5	98.3	8.0	98.5	14.4	100.8	8.5	102.3	11.2	99.2	9.6
p,p'-DDT	78.7	7.0	100.6	18	88.4	14.1	100.6	18	93.1	21.3	100.1	11.3	99.9	16.2	97.3	12.8

[a] Collected from Khoramdasht industrial park. [b] Relative recovery. [c] Collected from our university campus. [d] A river in the north of Iran. [e] Spiked with 400 ng mL^{-1}. [f] Spiked with 1000 ng mL^{-1}.

Table 3. Comparison of HS-SPME/GC-ECD with other published method for determination of OCPs.

Methods	Sorbent	LOD [a]	LDR [b]	Extraction Time (min)	RR% [c]	Ref
SPME	PET/NanoSiO$_2$	3–80	8–10,000	10	78–120	Current study
SPE-GC/ECD	Florisil	400–2000	5–1000	-	77–105	[33]
SMPE	PDMS [d] /PA [e]	20–80	50–1000	20	91.4 (average)	[34]
SB-μ-SPE	Hydroxide/graphene	300–1400	1000–200,000	20	84.2–100.2	[8]
ASE [f] & SPME	PDMS/PA	0.2–4.9 (ng m^{-3})	50–3000 (ng m^{-3})	40	-	[35]

[a] Limit of detection (ng L^{-1}). [b] Linear dynamic range (ng L^{-1}). [c] Relative Recovery. [d] Polydimethylsiloxane. [e] Polyacrylate. [f] Accelerated solvent extraction.

4. Conclusions

The novel electrospun composite nanofibers of PET superhydrophobic nano-SiO$_2$ were fabricated and used as effective fiber coating in HS-SPME. The novel nono fiber was applied for SME extraction of 16 organochlorine pesticides from water samples and thermally desorbed and analysed with GC-μECD. After characterization, the large specific surface area, the porous structure of the adsorbent were obtained. Based on this, the adsorbent and the analytes led to a fast equilibrium (10 min) and efficient extraction with recovery > 90%. Moreover, the low limit of detection (3–80 ng L^{-1}) and good linearity (8–1000 ng L^{-1}) are the characteristics that provide an excellent sensitivity of the OCPs analytes. Hence, the high efficiency, low LOD and appropriate repeatability are probably due to the large pores structure and effective π-π interactions between the OCPs and PET/SiO$_2$. In addition, the method is eco-friendly since it requires no organic solvent in the extraction and analysis steps.

Author Contributions: Conceptualization, H.N. and H.S.; methodology, H.N., H.S. and H.R.N.; formal analysis, H.N. and H.S. and S.S.; investigation, H.N., H.S. and H.R.N.; writing—original draft preparation, H.N., H.S., S.S. and H.R.N.; writing—review and editing, H.N., H.S., S.S., N.S., I.A. and H.R.N.; supervision, H.S. and H.R.N.; funding acquisition, H.S., I.A. and N.S. All authors have read and agreed to the published version of the manuscript.

Funding: The authors would like to thank the University of Tehran for the facilitation and financial support through the Research Grants. This research was supported by the Marine Pollution Special Interest Group, the National Defence University of Malaysia via SF0076-UPNM/2019/SF/ICT/6, and Dr. Syed Shahabuddin and Dr. Irfan Ahmad are grateful to Scientific Research Deanship at King Khalid University, Abha, Saudi Arabia through the Small Research Group Project under grant number (RGP.01-115-43).

Conflicts of Interest: The authors declare no conflict of interest.

References

1. Moreno, D.V.; Ferrera, Z.S.; Santana Rodríguez, J.J. Microwave assisted micellar extraction coupled with solid phase microextraction for the determination of organochlorine pesticides in soil samples. *Anal. Chim. Acta.* **2006**, *571*, 51–57. [CrossRef] [PubMed]
2. Zhao, R.; Wang, X.; Yuan, J.; Jiang, T.; Fu, S.; Xu, X. A novel headspace solid-phase microextraction method for the exact determination of organochlorine pesticides in environmental soil samples. *Anal. Bioanal. Chem.* **2006**, *384*, 1584–1589. [CrossRef] [PubMed]
3. El-Shahawi, M.S.; Hamza, A.; Bashammakh, A.S.; Al-Saggaf, W.T. An overview on the accumulation, distribution, transformations, toxicity and analytical methods for the monitoring of persistent organic pollutants. *Talanta* **2010**, *80*, 1587–1597. [CrossRef]
4. Lallas, P.L. The Stockholm Convention on persistent organic pollutants. *Am. J. Int. Law* **2001**, *95*, 692–708. [CrossRef]
5. WHO (Water Health Organization). The Use of DDT in Malaria Vector Control. WHO Position Statement. Available online: http://whqlibdoc.who.int/hq/2011/WHO_HTM_GMP_2011_eng.pdf (accessed on 17 July 2017).
6. Choi, M.; Lee, I.-S.; Jung, R.-H. Rapid determination of organochlorine pesticides in fish using selective pressurized liquid extraction and gas chromatography–mass spectrometry. *Food Chem.* **2016**, *205*, 1–8. [CrossRef] [PubMed]
7. Duodu, G.O.; Goonetilleke, A.; Ayoko, G.A. Optimization of in-cell accelerated solvent extraction technique for the determination of organochlorine pesticides in river sediments. *Talanta* **2016**, *150*, 278–285. [CrossRef]
8. Sajid, M.; Basheer, C.; Daudc, M.; Alsharaa, A. Evaluation of layered double hydroxide/graphene hybrid as a sorbent in membrane-protected stir-bar supported micro-solid-phase extraction for determination of organochlorine pesticides in urine samples. *J. Chromatogr. A* **2017**, *1489*, 1–8. [CrossRef]
9. Liu, Y.; Gao, Z.; Wu, R.; Wang, Z.; Chen, X.; Chan, T.-W.D. Magnetic porous carbon derived from a bimetallic metal–organic framework for magnetic solid-phase extraction of organochlorine pesticides from drinking and environmental water samples. *J. Chromatogr. A* **2017**, *1479*, 55–61. [CrossRef]
10. Mahpishanian, S.; Sereshti, H. One-step green synthesis of β-cyclodextrin/iron oxide-reduced graphene oxide nanocomposite with high supramolecular recognition capability: Application for vortex-assisted magnetic solid phase extraction of organochlorine pesticides residue from honey samples. *J. Chromatogr. A* **2017**, *1485*, 32–43.
11. Xie, L.; Liu, S.; Han, Z.; Jiang, R.; Zhu, F.; Xu, W.; Su, C.; Ouyang, G. Amine-functionalized MIL-53 (Al)-coated stainless steel fiber for efficient solid-phase microextraction of synthetic musks and organochlorine pesticides in water samples. *Anal. Bioanal. Chem.* **2017**, *409*, 5239–5247. [CrossRef]
12. Arthur, C.L.; Pawliszyn, J. Solid phase microextraction with thermal desorption using fused silica optical fibers. *Anal. Chem.* **1990**, *62*, 2145–2148. [CrossRef]
13. Spietelun, A.; Kloskowski, A.; Chrzanowski, W.; Namieśnik, J. Understanding solid-phase microextraction: Key factors influencing the extraction process and trends in improving the technique. *Chem. Rev.* **2012**, *113*, 1667–1685. [CrossRef] [PubMed]

14. Spietelun, A.; Pilarczyk, M.; Kloskowski, A.; Namieśnik, J. Current trends in solid-phase microextraction (SPME) fibre coatings. *Chem. Soc. Rev.* **2010**, *39*, 4524–4537. [CrossRef] [PubMed]
15. Fumes, B.H.; Silva, M.R.; Andrade, F.N.; Nazario, C.E.D.; Lanças, F.M. Recent advances and future trends in new materials for sample preparation. *TrAC Trends Anal. Chem.* **2015**, *71*, 9–25. [CrossRef]
16. Kabir, A.; Furton, K.G.; Malik, A. Innovations in sol-gel microextraction phases for solvent-free sample preparation in analytical chemistry. *TrAC Trends Anal. Chem.* **2013**, *45*, 197–218. [CrossRef]
17. Aziz-zanjani, M.O.; Mehdinia, A. A review on procedures for the preparation of coatings for solid phase microextraction. *Microchim. Acta* **2014**, *181*, 1169–1190. [CrossRef]
18. Mehdinia, A.; Aziz-zanjani, M.O.; Mwcnts, P.E.G. Recent advances in nanomaterials utilized in fiber coatings for solid-phase microextraction. *TrAC Trends Anal. Chem.* **2013**, *42*, 205–215. [CrossRef]
19. Mehdinia, A.; Mousavi, M.F. Enhancing extraction rate in solid-phase microextraction by using nanostructured polyaniline coating. *J. Sep. Sci.* **2008**, *31*, 3565–3572. [CrossRef]
20. Wang, J.-X.; Jiang, D.-Q.; Gu, Z.-Y.; Yan, X.-P. Multiwalled carbon nanotubes coated fibers for solid-phase microextraction of polybrominated diphenyl ethers in water and milk samples before gas chromatography with electron-capture detection. *J. Chromatogr. A* **2006**, *1137*, 8–14. [CrossRef]
21. Tian, J.; Xu, J.; Zhu, F.; Lu, T.; Su, C.; Ouyang, G. Application of nanomaterials in sample preparation. *J. Chromatogr. A* **2013**, *1300*, 2–16. [CrossRef]
22. Liu, Y.; Su, G.; Zhang, B.; Jiang, G.; Yan, B. Nanoparticle-based strategies for detection and remediation of environmental pollutants. *Anal.* **2011**, *136*, 872–877. [CrossRef] [PubMed]
23. Li, Q.; Wang, X.; Yuan, D. Preparation of solid-phase microextraction fiber coated with single-walled carbon nanotubes by electrophoretic deposition and its application in extracting phenols from aqueous samples. *J. Chromatogr. A* **2009**, *1216*, 1305–1311. [CrossRef] [PubMed]
24. Bagheri, H.; Aghakhani, A. Novel nanofiber coatings prepared by electrospinning technique for headspace solid-phase microextraction of chlorobenzenes from environmental samples. *Anal. Methods* **2011**, *6*, 1284–1289. [CrossRef]
25. Zewe, J.W.; Steach, J.K.; Olesik, S.V. Electrospun fibers for solid-phase microextraction, Anal. Chem. 82 (2010) 5341–5348.
26. Sereshti, H.; Amini, F.; Najarzadekan, H. Electrospun polyethylene terephthalate (PET) nanofibers as a new adsorbent for micro-solid phase extraction of chromium (VI) in environmental water samples. *RSC Adv.* **2015**, *5*, 89195–89203. [CrossRef]
27. Eskandarpour, N.; Sereshti, H.; Najarzadekan, H.; Gaikani, H. Polyurethane/polystyrene-silica electrospun nanofibrous composite for the headspace solid-phase microextraction of chlorophenols coupled with gas chromatography. *J. Sep. Sci.* **2016**, *39*, 4637–4644. [CrossRef]
28. Sereshti, H.; Bakhtiari, S.; Najarzadekan, H.; Samadi, S. Electrospun polyethylene terephthalate/graphene oxide nanofibrous membrane followed by HPLC for the separation and determination of tamoxifen in human blood plasma. *J. Sep. Sci.* **2017**, *40*, 3383–3391. [CrossRef]
29. Eskandarpour, N.; Sereshti, H. Electrospun polycaprolactam-manganese oxide fiber for headspace solid phase microextraction of phthalate esters in water samples. *Chemosphere* **2018**, *191*, 36–430. [CrossRef]
30. Reyes-Gallardo, E.M.; Lucena, R.; Cárdenas, S. Electrospun nanofibers as sorptive phases in microextraction. *TrAC Trends Anal. Chem.* **2016**, *84*, 3–11. [CrossRef]
31. Rutledge, G.C.; Fridrikh, S.V. Formation of fibers by electrospinning. *Adv. Drug Deliv. Rev.* **2007**, *59*, 1384–1391. [CrossRef]
32. Kamboh, M.A.; Akoz, E.; Memon, S.; Yilmaz, M. Synthesis of amino-substituted p-tert-butylcalix [4] arene for the removal of chicago sky blue and tropaeolin 000 azo dyes from aqueous environment. *Water Air Soil Pollut.* **2013**, *224*, 1424. [CrossRef]
33. Yavuz, H.; Guler, G.O.; Aktumsek, A.; Cakmak, Y.S.; Ozparlak, H. Determination of some organochlorine pesticide residues in honeys from Konya, Turkey. *Environ. Monit. Assess.* **2010**, *168*, 277–283. [CrossRef] [PubMed]
34. Campillo, N.; Penalver, R.; Aguinaga, N.; Hernández-Córdoba, M. Solid-phase microextraction and gas chromatography with atomic emission detection for multiresidue determination of pesticides in honey. *Anal. Chim. Acta* **2006**, *562*, 9–15. [CrossRef]
35. Mokbel, H.; Al Dine, E.J.; Elmoll, A.; Liaud, C.; Millet, M. Simultaneous analysis of organochlorine pesticides and polychlorinated biphenyls in air samples by using accelerated solvent extraction (ASE) and solid-phase micro-extraction (SPME) coupled to gas chromatography dual electron capture detection. *Environ. Sci. Pollut. Res.* **2016**, *23*, 8053–8063. [CrossRef] [PubMed]

Article

Thermo-Responsive Hydrophilic Support for Polyamide Thin-Film Composite Membranes with Competitive Nanofiltration Performance

Haniyeh Najafvand Drikvand [1], Mitra Golgoli [2], Masoumeh Zargar [2], Mathias Ulbricht [3,*], Siamak Nejati [4] and Yaghoub Mansourpanah [1,3,*]

[1] Membrane Research Laboratory, Lorestan University, Khorramabad 68151-44316, Iran
[2] School of Engineering, Edith Cowan University, Joondalup, WA 6027, Australia
[3] Lehrstuhl für Technische Chemie II and Center for Water and Environmental Research (ZWU), Universität Duisburg-Essen, 45117 Essen, Germany
[4] Department of Chemical and Biomolecular Engineering, University of Nebraska-Lincoln, Lincoln, NE 68588, USA
* Correspondence: mathias.ulbricht@uni-essen.de (M.U.); mansourpanah.y@lu.ac.ir (Y.M.)

Citation: Drikvand, H.N.; Golgoli, M.; Zargar, M.; Ulbricht, M.; Nejati, S.; Mansourpanah, Y. Thermo-Responsive Hydrophilic Support for Polyamide Thin-Film Composite Membranes with Competitive Nanofiltration Performance. *Polymers* 2022, 14, 3376. https://doi.org/10.3390/polym14163376

Academic Editors: Irene S. Fahim, Ahmed K. Badawi and Hossam E. Emam

Received: 20 July 2022
Accepted: 13 August 2022
Published: 18 August 2022

Publisher's Note: MDPI stays neutral with regard to jurisdictional claims in published maps and institutional affiliations.

Copyright: © 2022 by the authors. Licensee MDPI, Basel, Switzerland. This article is an open access article distributed under the terms and conditions of the Creative Commons Attribution (CC BY) license (https://creativecommons.org/licenses/by/4.0/).

Abstract: Poly(N-isopropylacrylamide) (PNIPAAm) was introduced into a polyethylene terephthalate (PET) nonwoven fabric to develop novel support for polyamide (PA) thin-film composite (TFC) membranes without using a microporous support layer. First, temperature-responsive PNIPAAm hydrogel was prepared by reactive pore-filling to adjust the pore size of non-woven fabric, creating hydrophilic support. The developed PET-based support was then used to fabricate PA TFC membranes via interfacial polymerization. SEM–EDX and AFM results confirmed the successful fabrication of hydrogel-integrated non-woven fabric and PA TFC membranes. The newly developed PA TFC membrane demonstrated an average water permeability of 1 L/m^2 h bar, and an NaCl rejection of 47.0% at a low operating pressure of 1 bar. The thermo-responsive property of the prepared membrane was studied by measuring the water contact angle (WCA) below and above the lower critical solution temperature (LCST) of the PNIPAAm hydrogel. Results proved the thermo-responsive behavior of the prepared hydrogel-filled PET-supported PA TFC membrane and the ability to tune the membrane flux by changing the operating temperature was confirmed. Overall, this study provides a novel method to fabricate TFC membranes and helps to better understand the influence of the support layer on the separation performance of TFC membranes.

Keywords: poly(N-isopropylacrylamide) (PNIPAAm); thermo-responsive membrane; hydrophilic hydrogel support; interfacial polymerization

1. Introduction

With the rapid growth of the world's population and water contamination, the universal need for freshwater has increased more rapidly than in the past [1]. Along these lines, membrane technology is becoming one of the most effective strategies to purify water and wastewater and produce freshwater [2–5]. Depending on the membranes' pore size and rejection mechanism in pressure-driven processes, they are classified into microfiltration, ultrafiltration, nanofiltration, and reverse osmosis membranes [6]. Among different membranes used in water purification, thin-film composite (TFC) membranes have been widely used in the nanofiltration and reverse osmosis process due to their high selectivity, tunable structure, and chemical and thermal stability [7]. The TFC membranes are composed of a dense polyamide (PA) active layer supported by porous polymeric layers. The microporous layers, immediately in contact with the PA active layer, are normally prepared via the phase inversion technique, and interfacial polymerization is the dominant method to prepare the active PA layer on the surface of the support [8]. This method enables independent optimization of the support layer and the active layer to achieve significant selectivity and permeability [9]. Although researchers have mostly focused on PA layer modification, there are a few studies that investigated the impact of the support layer properties on the TFC membranes' performance [10–12]. The porous support layer mainly provides the required mechanical strength; however, its properties such as hydrophobicity, porosity, pore size, and roughness influence the formation of the PA layer [11,12]. Polysulfone (PSF) and polyethersulfone (PES) are mainly used as TFC support layers owing to their relatively high thermal resistance, chemical stability, and easy fabrication [13–16]. Usually, the PSF or PES support membrane is cast on a PET non-woven support to provide the desired mechanical strength. The main disadvantage of PSF and PES membranes is that they are hydrophobic whereas hydrophilic support layers facilitate the interfacial polymerization reaction and PA selective film formation and can ultimately result in higher permeability and performance of the developed TFC membranes [17–19]. Many studies have explored the hydrophilic modification of the conventional support layers to enhance the performance of TFC membranes [20,21]. However, modification methods commonly require a complicated or harsh additional step to achieve the desired properties [9,22]. Hence, developing new hydrophilic support layers could make a breakthrough in designing highly efficient TFC membranes [23].

Recently, stimuli-responsive hydrogels have gained increasing attention in membrane fabrication owing to their ability to control their properties in response to environmental changes (e.g., pH, temperature, light) [6,24–26]. Hydrogels are three-dimensional network structures composed of polymeric chains that can absorb water, undergo significant volume expansion (swelling), and form a hydration layer on their surfaces [27,28]. These characteristics make them promising candidates for integration into membranes for enhanced anti-fouling properties [6,28]. For instance, Zhang et al. [29] prepared a modified PES membrane by grafting a novel polyampholyte hydrogel onto the membrane surface and achieved low fouling and high flux recovery of the modified membrane due to high hydration of the grafted hydrogel. Additionally, introducing hydrogel into the architecture of membranes has two important features. First, the three-dimensional network of hydrogels can be considered microscopically as a porous structure [30–32]. Second, the sieving coefficient as a function of hydrogel mesh size may be tuned by an external stimulus such as pH or temperature. For instance, membranes with grafted thermo-responsive hydrogels are able to alter their pore size and surface properties by changing temperature; this attribute makes separation efficiency tunable [33,34].

PNIPAAm is a common thermo-responsive polymer that demonstrates a lower critical solution temperature (LCST) at 32 °C [22,35,36]. A porous membrane that is grafted with PNIPAAm reduces the membrane pore size at temperatures below 32 °C (LCST) due to swelling, and hydrogel dehydration and collapse lead to pore opening above 32 °C [22,37]; this enables control over the water permeation efficiency of the membranes by temperature alteration. Pressure-driven mass transfer through swollen hydrogels is only possible when the gel is stabilized in a porous support matrix to maintain its integrity [22,38,39]. For example, Adrus and Ulbricht [40] used a polyethylene terephthalate (PET) support for developing a hydrogel pore-filled microfiltration membrane. The PNIPAAm was grafted into the PET structure altering the accessible pore volume while not changing the overall membrane thickness. The developed microfiltration membrane showed size-selective barrier properties in response to the temperature due to swelling and deswelling of PNIPAAm, confirming its thermo-responsive behavior. To the best of the authors' knowledge, the incorporation of thermo-responsive hydrogels into the PET support layer of TFC membranes has never been explored. PNIPAAm can be a promising hydrogel to conduct a proof of concept study due to its hydrophilicity and thermo-responsive characteristics. A PET non-woven that is pore-filled with PNIPAAm can replace the microporous support layer of TFC membranes, thereby reducing the overall thickness of the TFC structure, which can also rectify the high energy demand associated with thick support layers [12].

Here, we report on a thermo-responsive TFC membrane prepared by interfacial polymerization of PA active layer on a hydrogel-filled PET support scaffold without the application of a microporous membrane interlayer. To integrate PNIPAAm within non-woven PET and create a hydrophilic substrate, we used a reactive pore filling approach. The fabricated hydrogel support and PA TFC membrane (with no microporous support layer) were analyzed using different characterization techniques to confirm the successful formation of the PA layer on the developed novel support. Water permeability and salt rejection of the developed membranes were explored. Finally, the separation performance was measured below and above the LCST of PNIPAAm to investigate the thermo-responsive behavior of the membranes.

2. Materials and Methods

2.1. Materials

N-isopropylacrylamide (NIPAAm) (97%) was purchased from Sigma (Taufkirchen, Germany). Tetramethylethylenediamine (TEMED), ammonium persulfate (APS), N,N'-methylenebisacrylamide (MBA), 1,3,5-benzenetricarbonyltrichloride (TMC, 98%), n-hexane, p-phenylenediamine (PPD, 99%), triethylamine (TEA, 99.5%), polyethylene glycol (PEG-600), benzophenone, sodium chloride (NaCl,) propanol, and ethanol were obtained from Merck (Steinheim am Albuch, Germany).

2.2. Membrane Fabrication

2.2.1. Fabrication of a Novel Hydrogel Support

First, the voids of the non-woven PET scaffold were filled with hydrogel. For this purpose, the non-woven PET fabric was immersed in a solution of 0.1 M benzophenone in an ethanol–water mixture with a ratio of 10:1 ($v/v\%$) and allowed to rest for 2 h. After adsorption of the photo-initiator, the non-woven fabric was placed in a Petri dish containing 0.2 M aqueous solution of NIPAAm and was irradiated with a UV source (160 W) for 15 min. The non-woven PET was then placed in distilled water for 24 h to remove the unreacted materials and was dried at room temperature [40,41]. For the reactive pore-filling, a solution containing 0.9 g of NIPAAm, 0.018 g of MBA as the cross-linker, and 270 µL of APS (10 wt.%) as the initiator in 10.8 mL of water was prepared. The PET scaffolds were inserted into a filter holder and the solution was continuously circulated for 1 h (Figure 1). The membrane was then removed and placed in a beaker containing the same solution and stirred for 1 h at 250 rpm at room temperature. Thereafter, 20 µL of TEMED as a promoter of the initiator APS for hydrogel formation was added to the solution, which was stirred for another 30 s. The membrane was then placed between two glass plates for 24 h for the cross-linking polymerization reaction to occur. Finally, the prepared membrane (noted as PET-PNIPAAm membrane) was placed in distilled water for 24 h to remove unreacted materials [39,41,42].

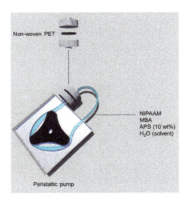

Figure 1. Circulation of reactive monomer solution through non-woven PET.

2.2.2. Fabrication of PA TFC Membrane

A PA thin layer was fabricated by the interfacial polymerization (IP) method, in which active amine monomers are allowed to react with organic monomers at the interface between the aqueous and organic phases to form a network structure [8]. To do so, the prepared support was placed in a frame with a 2.2 cm diameter and 1.0 cm depth and an aqueous solution containing 0.4 wt.% PPD and 0.8 wt.% TEA was poured on the membrane. The solution was allowed to remain for 5 min within the frame before being drained at room temperature. The membrane surface was pressed by a soft rubber roller to remove the extra solution. The organic solution (TMC, 0.2 wt.% in n-hexane) was poured on the membrane and kept for 3 min. The excess organic solution was then removed from the frame, and the prepared membrane was placed in an oven at 70 °C for 30 min to dry and cure. The resulting TFC membrane was finally washed with a hexane-propanol mixture with a ratio of 3:1. Finally, the as-prepared composite (noted as PET-PNIPAAm-PA membrane) was kept in DI water at 4 °C until further characterization and performance evaluations.

2.2.3. Modified Hydrogel Support

PEG as a pore-forming agent was used to improve the membrane water permeability. The method for the preparation of hydrogels modified with PEG was the same as the plain membrane fabrication noted in Section 2.2.1; the only difference was the addition of 1.8 g PEG (as a pore-forming agent in the hydrogel network) to the same composition of the aqueous solution. The membrane prepared in this step is noted as a PET-PEG-PNIPAAm membrane. The PA layer was fabricated on the modified membrane according to the method noted in Section 2.2.2 and the final membrane is referred to as the PET-PEG-PNIPAAm-PA membrane.

2.3. Membrane Characterization

2.3.1. Scanning Electron Microscopy (SEM) and Energy Dispersive X-ray Spectroscopy (EDX)

To investigate the hydrogel loading of pore-filled polyester and the formation of PA, an FE-SEM (TESCAN, Brno, Czech Republic) coupled with energy-dispersive X-ray spectroscopy (EDX) was used. Membrane samples were freeze-fractured in liquid nitrogen, sputter coated with gold, and mounted on SEM stubs. The non-woven hydrogel membrane sample was dried using a freeze dryer for better observation of the structure of hydrogels. The surface and cross-section of membranes were observed by SEM at 15.0 kV. The elemental composition information of the samples was obtained by EDX accordingly.

2.3.2. Atomic Force Microscopy (AFM)

AFM device (DME model C-21, Copenhagen, Denmark) was used to determine the surface roughness and morphology of the membranes. Small pieces of prepared membrane samples (1 cm^2) were cut and attached to a glass plate using double-sided tape; all samples were air-dried overnight before AFM analysis. The membrane samples were scanned and observed in a non-contact mode in the air at room temperature with a silicon probe in a scan size of 2 µm × 2 µm. SPM-DME software was used to measure roughness values. Three measurements were done to calculate the roughness for each sample, and average values are reported.

2.3.3. Water Contact Angle (WCA)

The hydrophilicity of the membranes was determined by the contact angle device (G10, KRÜSS Co., Nürnberg, Germany). In all measurements, a water droplet of 1 µL was placed on the membrane surface. Furthermore, the water contact angle was also determined at both room temperature and at 45 °C to investigate the hydrophilicity–hydrophobicity behavior of membranes upon temperature change. A heater was placed under the sample to evaluate the membrane hydrophilicity at different temperatures while the temperature was equilibrated for 10 min before measuring the contact angles. For all samples, at least five measurements at different locations of the sample were performed and the average is reported.

2.4. Membrane Separation Performance Evaluation

A dead-end filtration unit was used to evaluate the performance of membranes. The permeate flux of all membrane samples was measured for 90 min at 1 bar. The permeate flux (J) and water permeability (A) were calculated by measuring the changes in the volume of the permeate over time using Equations (1) and (2), respectively:

$$J = \Delta V/(Am\ \Delta t)\ (L/m^2\ h) \quad (1)$$

$$A = J/\Delta P\ (L/m^2\ h\ bar) \quad (2)$$

where Am is the specific surface area of the membrane, Δt is the measurement time and ΔP is the applied pressure.

To perform the membrane rejection tests, a 1000 ppm solution of NaCl was prepared as a feed solution. The salt concentration of permeate was evaluated with a conductivity meter using a calibration curve established for NaCl. Finally, the rejection of NaCl was calculated using Equation (3):

$$R\ (\%) = 1 - C_p/C_0 \quad (3)$$

where C_p is the concentration of the permeate and C_0 is the feed concentration. At least three tests were done to investigate the salt rejection performance for each salt and the average values are reported.

2.5. Effect of Temperature on the Performance of the Novel PA TFC Membrane

To investigate the effect of temperature, the hydrogel support was prepared according to Section 2.2.1 and then dried at 70 °C for 15 min to reach a hydrophobic state. The new non-woven hydrogel membrane was then used for PA layer fabrication according to Section 2.2.2 and is noted as PET-hydrophobic PNIPAAm-PA. Finally, to investigate the effect of temperature on membrane performance, the water permeability and salt rejection tests were performed at both room temperature and at 45 °C.

3. Results and Discussions

3.1. Membrane Characterization

3.1.1. SEM, AFM, and EDX

The surfaces of PET, PET-PNIPAAm, and PET-PNIPAAm-PA membranes were analyzed by SEM and AFM to evaluate their surface topography and surface morphology. Surface and cross-sectional SEM images as well as their AFM images are presented in Figure 2. The results show that the PET scaffold consists of fibers that provide enough voids for hydrogel loading between them (Figure 2a). Figure 2b corresponds to the PET-PNIPAAm membrane and confirms PNIPAAm hydrogel integration inside the non-woven fabric, filling the empty space between the fibers. Finally, Figure 2c shows a thin layer of PA covering the PET-PNIPAAm membrane and confirms the formation of PA ridge and valley structure on the hydrogel pore-filled non-woven membrane in both surface and cross-section SEM images. The results affirm the suitability of the proposed technique to develop a hydrogel-reinforced support layer for TFC membranes and its suitability for PA formation.

The average roughness (R_a), root mean square roughness (R_q), and maximum roughness (R_z) values are reported in Table 1. PET, PET-PNIPAAm, and PET-PNIPAAm-PA membranes showed roughness (R_q) of 24, 16, and 16 nm, respectively. The distinctive topography of the pristine non-woven PET membrane is shown in the AFM image (Figure 2a) with the light regions representing the highest points and the dark regions being the depth of the valleys. The results confirm that the PET backbone has a relatively rough surface compared to the PET-PNIPAAm membrane due to its non-uniform non-woven fiber structure. The topography of the PET textiles after the hydrogel grafting is shown in Figure 2b, implying the smooth surface of the PET-PNIPAAm membrane that is in accordance with the SEM result. The surface displayed a less pronounced peak to valley topography leading to a decreased roughness (cf. Table 1). This trend is consistent with the literature [43,44]. For instance, Kurşun et al. [45] developed a thermo-responsive membrane by grafting PNIPAAm on a poly(vinyl alcohol) membrane. They reported a smoother surface after PNIPAAm integration on a polymeric surface and correlated that to the PNIPAAm's presence. Hence, the surface morphological changes of the PET-PNIPAAm membrane suggest the successful integration of the hydrogel layer with the PET fabric. After the interfacial polymerization and PA formation on the hydrogel membrane (PET-PNIPAAm-PA membrane), the roughness was slightly increased compared to the PET-PNIPAAm membrane surface (cf. Table 1), which can be attributed to the formation of PA ridge and valley structure [46]. Therefore, the roughness data confirmed the successful formation of non-woven hydrogel and PA layer on top of it.

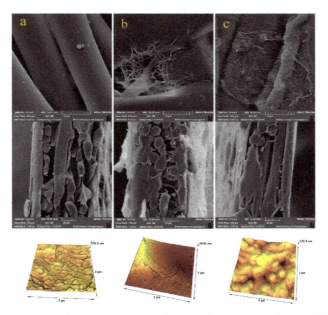

Figure 2. Surface and cross-sectional SEM and AFM images of (**a**) PET (**b**) PET-PNIPAAm (**c**) PET-PNIPAAm-PA membranes.

Table 1. Membrane surface roughness parameters from AFM.

Membrane	Ra (nm)	Rz (nm)	Rq (nm)
PET	3.5	23.3	24.2
PET-PNIPAAm	2.3	9.6	15.9
PET-PNIPAAm-PA	2.7	11.8	16.4

The EDX detector on the SEM instrument was used to evaluate the fraction of elements (i.e., carbon, oxygen, and nitrogen here). The results are presented in Table 2. The PET scaffold showed an oxygen content of 34.8 wt.% and carbon content of 65.2 wt.% which is consistent with the value expected for the structure of the material [47]. The nitrogen element emerged in the PET-PNIPAAm membrane (13.8 wt.%), which is correlated to the presence of the amide group after pore-filling of PNIPAAm hydrogel in PET non-woven [48].

The fraction of nitrogen was further increased from 13.8 to over 24 wt.% upon PA formation in the PET-PNIPAAm-PA membrane which is correlated to the nitrogen-rich structure of the polyamide active layer. The results confirm the successful hydrogel integration in the PET non-woven and the PA formation on top of the PET-PNIPAAm membrane.

Table 2. Elements fraction (in atom and weight percent) from EDX for the different membranes.

Element	PET		PET-PNIPAAm		PET-PEG-PNIPAAm-PA	
	A %	W %	A %	W %	A %	W %
C	71.3	65.2	68.5	63.2	67.2	61.8
O	28.7	34.8	18.7	23	19.7	24.1
N	0	0	12.8	13.8	13.1	24.1

The structural difference between the conventional PES-supported PA thin layer compared with the PET-PNIPAAm-based thin PA layer is clearly shown in Figure 3. The second one clearly depicts a relatively rougher skin layer structure. The main reason for such a difference can be found in the impact of the high hydrophilic properties of PET-PNIPAAm support. PNIPAAm effectively enhanced the hydrophilicity of the backbone support which impacted the diffusion rate of the amine PPD toward the organic phase during the interfacial polymerization reaction. Hence, a very thin PA layer with a denser structure in conjunction with less surface defect has been fabricated [49,50].

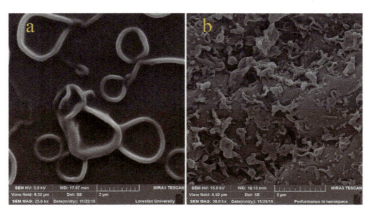

Figure 3. SEM micrographs of the surface of PA thin layer upon (**a**) conventional PES support and (**b**) PET-PNIPAAm support.

3.1.2. Water Contact Angle

The surface hydrophilicity of the membranes as well as their potential ability to change between more hydrophobic and more hydrophilic characteristics due to the temperature change was determined through water contact angle measurement [48,51]. These were done by measuring the contact angle of a static water drop at room temperature (23 °C) and 45 °C on membrane samples using a sessile drop technique with a tensiometer (Figure 4). This temperature range covers values that are higher and lower compared to the LCST, to account for thermo-responsive characteristics of the hydrogel membrane. WCA results at 23 °C show that the integration of the PNIPAAm hydrogel within the fabric as a pore filler decreased the contact angle of the PET scaffold from 57° to 23°, indicating that the membranes are more hydrophilic which is attributed to the incorporation of the hydrophilic functional groups of the hydrogel PNIPAAm [52,53]. After the thin PA layer formation on the PET-PNIPAAm membrane, the contact angle value increased compared to the PET-PNIPAAm and reached 41°. This can be attributed to the aromatic characteristics of the dense PA layer leading to reduced hydrophilicity of the PET-PNIPAAm-PA membrane compared to its PET-PNIPAAm precursor [54]. WCA results are also in accordance with the roughness values of membranes as increased roughness correlates with increased hydrophobicity [55,56].

WCA results at 45 °C indicate that by increasing the temperature, no significant change in the contact angle of the PET scaffold was observed. However, the contact angle value for the PET-PNIPAAm membrane increased from 23° to 72°, showing the pronounced thermo-responsive behavior of the fabricated membrane [6,43,57–60]. PNIPAAm hydrogel has an insoluble three-dimensional network structure that has reversible swelling properties in water. An increase in temperature causes the alteration of PNIPAAm chains' structure, leading to a pronounced deswelling, i.e., the release of water, and exposure of hydrophobic isopropyl groups on the membrane surface. This results in an increase in the WCA value [6]. Modigunta et al. [48] also investigated the WCA of a porous honeycomb-patterned polystyrene film integrated with PNIPAAm below and above the LCST and reported higher WCA above the LCST. In addition, the WCA value of the PA membrane at 45 °C was higher than its WCA at 23 °C. The PA layer itself is not responsive to temperature; therefore, this change is due to the effect of the hydrogel used as a support. This indicates that the PET-PNIPAAm membrane can preserve to some extent its thermo-responsive characteristic upon PA formation on its surface. This is a surprising finding because the SEM and AFM results indicate that the PA film should cover the support completely. More insights into the barrier function of the PA film will be obtained from the nanofiltration studies (see Sections 3.2 and 3.3).

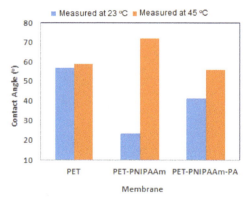

Figure 4. Water contact angle (WCA) values of membranes.

3.2. Membrane Separation Performance Evaluation

The flux and salt rejection of PET, PET-PNIPAAm, and PET-PNIPAAm-PA membranes were investigated to determine their separation ability and flux performance. As shown in Table 3, the water permeability of the prepared membranes was sharply reduced compared to the PET non-woven. Water permeability decreased from 16,920 (L/m^2 h bar) for PET scaffold to 1.5 (L/m^2 h bar) for PET-PNIPAAm membrane because the hydrogel was able to fill the empty spaces of the non-woven well, whereas the fabrication of the PA layer in the next step caused a further water permeability reduction to 1.0 (L/m^2 h bar) due to the compact structure of the PA layer [46]. The reactive pore filling has been performed by a cross-linking copolymerization of the PNIPAAm hydrogel. The tight entanglement ensures that the composite is stable in water. This had also been demonstrated in earlier work by Adrus and Ulbricht [40], where almost the same methodology had been used for reactive pore-filling of PET track-etched membranes that had thereafter been proven to be stable ultrafiltration membranes.

Although the hydrogel has sieving properties due to its polymer network mesh structure, the PET-PNIPAAm membrane did not show any rejection of NaCl. This is reasonable considering the mesh size in the range of several nm [40] and the uncharged structure of the polymer. However, after the interfacial polymerization, the PET-PNIPAAm-PA membrane on this support yielded 33.8% rejection against NaCl. This demonstrates a good salt selectivity compared to reported NaCl rejections of 10–20% for two industrial PA TFC nanofiltration membranes [61]. After the modification of the PET-PNIPAAm membrane by the integration of PEG as a pore-forming agent during the cross-linking polymerization, the performance of the modified hydrogel membrane (PET-PEG-PNIPAAm) and PA TFC membrane fabricated on the modified support (PET-PEG-PNIPAAm-PA) was investigated to evaluate the impact of the modified hydrogel structure. The water flux of the fabricated membranes and their NaCl rejection are presented in Table 3. Resulting from the application of PEG as a pore-forming agent in the structure of the hydrogel, the membrane flux changed from 1.5 L/m^2 h (for PET-PNIPAAm membrane) to 37.4 L/m^2 h (for PET-PEG-PNIPAAm- membrane). It is known that the addition of PEG into the polymerization mixture leads to the formation of a porous, phase-separated PNIPAAm hydrogel compared to a homogenous one without PEG [62]. This is the reason for the much larger permeability. However, the permeate flux of the PET-PEG-PNIPAAm-PA membrane did not show much difference compared with the PET-PNIPAAm-PA membrane. This is not very surprising when considering that the PA barrier film has the largest contribution to the resistance of a PA TFC membrane. A similar result was reported by Jimenez-Solomon et al. [63] that used PEG for modification of PA TFC support where the modified support membrane did not impact the flux of PA TFC prepared on it. Despite that, the NaCl rejection increased from 34% for the PET-PNIPAAm-PA membrane to 47% for the PET-PEG-PNIPAAm-PA membrane. This could be due to the higher chance of PA formation within the pores of the support layer due to the PPD monomers penetration in the larger pores generated through PEG influence onto the formation of a macroporous instead of a homogeneous hydrogel [12,64].

In addition, considering the low operating pressure of 1 bar used in this study, the developed novel TFC membrane showed a good balance between salt rejection and water permeability in comparison with different TFC nanofiltration membranes reported in the literature (Table 4). Overall, this study was able to achieve a suitable efficiency by reducing energy consumption by using low operating pressure.

Table 3. Overview of membrane water permeability and salt rejection.

Membrane	Water Permeability (L/m^2 h bar)	NaCl Rejection (%)
PET	16,920	0
PET-PNIPAAm	1.5	0
PET-PEG-PNIPAAm	37.4	0
PET-PNIPAAm-PA	1.0	33.8
PET-PEG-PNIPAAm-PA	1.0	47.0

Table 4. The performance comparison of PA TFC membranes reported in the literature with the ones established in this work.

Membrane	Water Permeability (L/m^2 h bar)	NaCl Rejection (%) (bar)	Ref.
PA TFC/modified polyacrylonitrile	0.84	37.8 (5)	[65]
Commercial TFC-SR3	2.1	38 (10)	[66]

Table 4. Cont.

Membrane	Water Permeability (L/m² h bar)	NaCl Rejection (%) (bar)	Ref.
Commercial TFC-SR2	7.5	24 (10)	[66]
Modified PA TFC/PES	7.8	25.6 (6)	[67]
PA TFC/modified PES	11.4	31 (2)	[9]
PET-PNIPAAm-PA	1	33.8 (1)	This work
PET-PEG-PNIPAAm-PA	1	47.0 (1)	This work

3.3. The Effect of Temperature on Membrane Performance

The fabrication of a thermo-responsive TFC membrane may lead to the development of controllable and more efficient processes. For example, the efficiency of thermo-responsive TFC membrane in desalination and wastewater treatment can be designed according to the seasonal changes in temperature or using external waste heat in cases where the increase in temperature can improve membrane performance [68]. The effect of temperature on the developed hydrogel pore-filled non-woven was first investigated by drying hydrogel after fabrication which caused the membrane to be in a hydrophobic state [6]. This caused lower water permeability after PA fabrication (0.3 L/m² h bar) compared to the PA TFC fabricated on the hydrophilic hydrogel (1 L/m² h bar). This is due to the resistance of hydrophobic support to the penetration of PPD aqueous solution limiting the penetration of PPD into the pores that result in the formation of a thicker PA layer with lower permeability [12]. In addition, the effect of temperature on the performance of the fabricated thermo-responsive membranes was also studied by comparing the water permeability and rejection to NaCl at room temperature and beyond LCST of PNIPAAm hydrogel (45 °C) at a constant pressure of 1 bar. According to Table 5, by increasing the temperature, the water permeability increased by 6 times, whereas a slight decrease in the NaCl rejection was observed. At room temperature, water permeability and rejection were 0.3 L/m² h bar and 32.4%, respectively, which changed to 1.8 L/m² h bar and 27.9%, respectively, by increasing the temperature. At a temperature below LCST, hydration of the polymer network by water leads to the absorption of water and swelling of the hydrogel; hence, the resistance to water flow through the swollen pore-filling PNIPAAm hydrogel is high. However, at temperatures higher than the LCST of PNIPAAm, the hydrogel deswelling process occurs due to the breaking of hydrogen bonds; this leads to a phase separation that will open up channels through the hydrogel, which results in lower resistance to water flow [6,43,57–60,69]. Such temperature-dependent behavior of water flux has been studied and discussed in detail for PNIPAAm pore-filled PET track-etched membranes [40], which had been prepared by the same methodology used here. Wang et al. [69] also reported a higher flux for their membrane that was incorporated with PNIPAAm at a temperature above its LCST due to the swelling and deswelling of the membrane by temperature change. Guo et al. [70] developed a PNIPAAm-gelatin hydrogel membrane and observed the linear increase in water flux with the increasing temperature (from 25 to 45 °C) that was correlated to the change in the pore size of the membranes due to swelling and deswelling. The results of the present study indicate that the resistance of the support has a significant influence on the overall resistance of the TFC membrane. The surprising effect of the temperature-responsive wetting properties of the PNIPAAm in the support onto WCA of the PA TFC membrane (cf. Section 3.1.2) may be related to the fact that the barrier layer has nanofiltration characteristics as indicated by the only modest NaCl rejection. The slightly decreased NaCl rejection at the higher temperature would then be directly caused by the higher water permeability at the same pressure. Overall, the thermo-responsive behavior of developed TFC can be used for tuning and controlling membrane flux by temperature.

Table 5. The effect of temperature on TFC PA membrane performance.

Membrane	Temperature	Water Permeability (L/m^2 h bar)	NaCl Rejection (%)
PET-PNIPAAm-PA	Room temperature	1	33.8
PET-hydrophobic PNIPAAm-PA	Room temperature	0.3	32.4
PET-hydrophobic PNIPAAm-PA	45 °C	1.8	27.9

4. Conclusions

A novel thermo-responsive membrane with switchable hydrophilicity/-phobicity was successfully developed and employed as the support of a PA TFC membrane, replacing the conventional microporous membrane support. The characterizations of the fabricated membranes clearly indicated the attachment of PNIPAAm hydrogel into PET nonwoven and the successful PA fabrication on the PET-PNIPAAm membrane. The effect of temperature on the hydrophobic–hydrophilic behavior of the fabricated membranes was studied using water contact angle measurement above and below the LCST of PNIPAAm. The results showed thermo-responsive behavior of both the PET-PNIPAAm membrane and the PA TFC membrane, fabricated on top of the pore-filling PNIPAAm PET scaffold, due to the thermal sensitivity of integrated PNIPAAm hydrogel. Two versions of the newly fabricated PET-PNIPAAm-PA membrane, based on supports obtained by reactive pore-filling with or without PEG as a porogen, respectively, showed a rejection of 34% and 47% for NaCl, respectively, and flux of 1 L/m^2 h at a low operating pressure of 1 bar. Finally, the impact of increasing temperature above LCST on the membrane performance was evaluated, which indicated a much higher flux with a slight decrease in NaCl rejection. The developed hydrogel-filled PET membranes in this study showed good potential to substitute the microporous membrane support layer (on a PET non-woven) used for conventional TFC membranes and yield smart thermo-responsive PA TFC membranes with improved efficiency and pronounced switchability.

Author Contributions: H.N.D.: Investigation, data curation; M.G.: Data curation, writing—review and editing; M.Z.: Data curation, writing—review and editing; M.U.: Supervision, project administration, writing—review and editing; S.N.: writing—review and editing; Y.M.: Conceptualization, supervision, project administration, writing—original draft, funding acquisition. All authors have read and agreed to the published version of the manuscript.

Funding: Y.M. greatly appreciates the Water and Wastewater Company of Lorestan, Iran, for financial support of the research through the grant number 17/12173.

Institutional Review Board Statement: Not applicable.

Informed Consent Statement: Not applicable.

Data Availability Statement: The data presented in this study are available on request from the corresponding author.

Conflicts of Interest: The authors declare no conflict of interest.

References

1. Elimelech, M. The Global Challenge for Adequate and Safe Water. *J. Water Supply Res. Technol.* **2006**, *55*, 3–10. [CrossRef]
2. Zuo, K.; Wang, K.; DuChanois, R.M.; Fang, Q.; Deemer, E.M.; Huang, X.; Xin, R.; Said, I.A.; He, Z.; Feng, Y.; et al. Selective Membranes in Water and Wastewater Treatment: Role of Advanced Materials. *Mater. Today* **2021**, *50*, 516–532. [CrossRef]
3. Golgoli, M.; Khiadani, M.; Shafieian, A.; Sen, T.K.; Hartanto, Y.; Johns, M.L.; Zargar, M. Microplastics Fouling and Interaction with Polymeric Membranes: A Review. *Chemosphere* **2021**, *283*, 131185. [CrossRef] [PubMed]
4. Mansourpanah, Y.; Madaeni, S.S.; Rahimpour, A.; Farhadian, A. The Effect of Non-Contact Heating (Microwave Irradiation) and Contact Heating (Annealing Process) on Properties and Performance of Polyethersulfone Nanofiltration Membranes. *Appl. Surf. Sci.* **2009**, *255*, 8395–8402. [CrossRef]

5. Masjoudi, M.; Golgoli, M.; Ghobadi Nejad, Z.; Sadeghzadeh, S.; Borghei, S.M. Pharmaceuticals Removal by Immobilized Laccase on Polyvinylidene Fluoride Nanocomposite with Multi-Walled Carbon Nanotubes. *Chemosphere* **2021**, *263*, 128043. [CrossRef]
6. Xu, D.; Zheng, J.; Zhang, X.; Lin, D.; Gao, Q.; Luo, X.; Zhu, X.; Li, G.; Liang, H.; Van der Bruggen, B. Mechanistic Insights of a Thermoresponsive Interface for Fouling Control of Thin-Film Composite Nanofiltration Membranes. *Environ. Sci. Technol.* **2022**, *56*, 1927–1937. [CrossRef]
7. Mansourpanah, Y.; Rahimpour, A.; Tabatabaei, M.; Bennett, L. Self-Antifouling Properties of Magnetic Fe_2O_3/SiO_2-Modified Poly (Piperazine Amide) Active Layer for Desalting of Water: Characterization and Performance. *Desalination* **2017**, *419*, 79–87. [CrossRef]
8. Zargar, M.; Jin, B.; Dai, S. An Integrated Statistic and Systematic Approach to Study Correlation of Synthesis Condition and Desalination Performance of Thin Film Composite Membranes. *Desalination* **2016**, *394*, 138–147. [CrossRef]
9. Li, Y.; Su, Y.; Li, J.; Zhao, X.; Zhang, R.; Fan, X.; Zhu, J.; Ma, Y.; Liu, Y.; Jiang, Z. Preparation of Thin Film Composite Nanofiltration Membrane with Improved Structural Stability through the Mediation of Polydopamine. *J. Memb. Sci.* **2015**, *476*, 10–19. [CrossRef]
10. Tiraferri, A.; Yip, N.Y.; Phillip, W.A.; Schiffman, J.D.; Elimelech, M. Relating Performance of Thin-Film Composite Forward Osmosis Membranes to Support Layer Formation and Structure. *J. Memb. Sci.* **2011**, *367*, 340–352. [CrossRef]
11. Liu, F.; Wang, L.; Li, D.; Liu, Q.; Deng, B. A Review: The Effect of the Microporous Support during Interfacial Polymerization on the Morphology and Performances of a Thin Film Composite Membrane for Liquid Purification. *RSC Adv.* **2019**, *9*, 35417–35428. [CrossRef]
12. Huang, L.; McCutcheon, J.R. Impact of Support Layer Pore Size on Performance of Thin Film Composite Membranes for Forward Osmosis. *J. Memb. Sci.* **2015**, *483*, 25–33. [CrossRef]
13. Ghosh, A.K.; Hoek, E.M.V. Impacts of Support Membrane Structure and Chemistry on Polyamide–Polysulfone Interfacial Composite Membranes. *J. Memb. Sci.* **2009**, *336*, 140–148. [CrossRef]
14. Yu, C.; Li, H.; Zhang, X.; Lü, Z.; Yu, S.; Liu, M.; Gao, C. Polyamide Thin-Film Composite Membrane Fabricated through Interfacial Polymerization Coupled with Surface Amidation for Improved Reverse Osmosis Performance. *J. Memb. Sci.* **2018**, *566*, 87–95. [CrossRef]
15. Choi, W.; Jeon, S.; Kwon, S.J.; Park, H.; Park, Y.I.; Nam, S.E.; Lee, P.S.; Lee, J.S.; Choi, J.; Hong, S.; et al. Thin Film Composite Reverse Osmosis Membranes Prepared via Layered Interfacial Polymerization. *J. Memb. Sci.* **2017**, *527*, 121–128. [CrossRef]
16. Li, D.; Wang, H. Recent Developments in Reverse Osmosis Desalination Membranes. *J. Mater. Chem.* **2010**, *20*, 4551–4566. [CrossRef]
17. Wang, L.; Kahrizi, M.; Lu, P.; Wei, Y.; Yang, H.; Yu, Y.; Wang, L.; Li, Y.; Zhao, S. Enhancing Water Permeability and Antifouling Performance of Thin–Film Composite Membrane by Tailoring the Support Layer. *Desalination* **2021**, *516*, 115193. [CrossRef]
18. Zhao, S.; Liao, Z.; Fane, A.; Li, J.; Tang, C.; Zheng, C.; Lin, J.; Kong, L. Engineering Antifouling Reverse Osmosis Membranes: A Review. *Desalination* **2021**, *499*, 114857. [CrossRef]
19. Lian, Y.; Zhang, G.; Wang, X.; Yang, J. Impacts of Surface Hydrophilicity of Carboxylated Polyethersulfone Supports on the Characteristics and Permselectivity of PA-TFC Nanofiltration Membranes. *Nanomaterials* **2021**, *11*, 2470. [CrossRef]
20. Kim, H.I.; Kim, S.S. Plasma Treatment of Polypropylene and Polysulfone Supports for Thin Film Composite Reverse Osmosis Membrane. *J. Memb. Sci.* **2006**, *286*, 193–201. [CrossRef]
21. Xie, Q.; Zhang, S.; Hong, Z.; Ma, H.; Zeng, B.; Gong, X.; Shao, W.; Wang, Q. A Novel Double-Modified Strategy to Enhance the Performance of Thin-Film Nanocomposite Nanofiltration Membranes: Incorporating Functionalized Graphenes into Supporting and Selective Layers. *Chem. Eng. J.* **2019**, *368*, 186–201. [CrossRef]
22. Frost, S.; Ulbricht, M. Thermoresponsive Ultrafiltration Membranes for the Switchable Permeation and Fractionation of Nanoparticles. *J. Memb. Sci.* **2013**, *448*, 1–11. [CrossRef]
23. Mansourpanah, Y.; Samimi, M. Preparation and Characterization of a Low-Pressure Efficient Polyamide Multi-Layer Membrane for Water Treatment and Dye Removal. *J. Ind. Eng. Chem.* **2017**, *53*, 93–104. [CrossRef]
24. Koetting, M.C.; Peters, J.T.; Steichen, S.D.; Peppas, N.A. Stimulus-Responsive Hydrogels: Theory, Modern Advances, and Applications. *Mater. Sci. Eng. R Rep.* **2015**, *93*, 1–49. [CrossRef] [PubMed]
25. Yang, Z.; Chen, L.; McClements, D.J.; Qiu, C.; Li, C.; Zhang, Z.; Miao, M.; Tian, Y.; Zhu, K.; Jin, Z. Stimulus-Responsive Hydrogels in Food Science: A Review. *Food Hydrocoll.* **2022**, *124*, 107218. [CrossRef]
26. Hartanto, Y.; Zargar, M.; Cui, X.; Jin, B.; Dai, S. Non-Ionic Copolymer Microgels as High-Performance Draw Materials for Forward Osmosis Desalination. *J. Memb. Sci.* **2019**, *572*, 480–488. [CrossRef]
27. Ahmed, E.M. Hydrogel: Preparation, Characterization, and Applications: A Review. *J. Adv. Res.* **2015**, *6*, 105–121. [CrossRef] [PubMed]
28. Fu, W.; Pei, T.; Mao, Y.; Li, G.; Zhao, Y.; Chen, L. Highly Hydrophilic Poly(Vinylidene Fluoride) Ultrafiltration Membranes Modified by Poly(N-Acryloyl Glycinamide) Hydrogel Based on Multi-Hydrogen Bond Self-Assembly for Reducing Protein Fouling. *J. Memb. Sci.* **2019**, *572*, 453–463. [CrossRef]
29. Zhang, W.; Yang, Z.; Kaufman, Y.; Bernstein, R. Surface and Anti-Fouling Properties of a Polyampholyte Hydrogel Grafted onto a Polyethersulfone Membrane. *J. Colloid Interface Sci.* **2018**, *517*, 155–165. [CrossRef]
30. Ebara, M.; Kotsuchibashi, Y.; Uto, K.; Aoyagi, T.; Kim, Y.-J.; Narain, R.; Idota, N.; Hoffman, J.M. *Smart Hydrogels*; Springer: Berlin/Heidelberg, Germany, 2014; pp. 9–65. [CrossRef]

31. Yang, Q.; Adrus, N.; Tomicki, F.; Ulbricht, M. Composites of Functional Polymeric Hydrogels and Porous Membranes. *J. Mater. Chem.* **2011**, *21*, 2783–2811. [CrossRef]
32. Marchetti, M.; Cussler, E.L. Hydrogels As Ultrafiltration Devices. *Sep. Purif. Rev.* **1989**, *18*, 177–192. [CrossRef]
33. Li, J.J.; Zhou, Y.N.; Luo, Z.H. Smart Fiber Membrane for PH-Induced Oil/Water Separation. *ACS Appl. Mater. Interfaces* **2015**, *7*, 19643–19650. [CrossRef]
34. Weidman, J.L.; Mulvenna, R.A.; Boudouris, B.W.; Phillip, W.A. Unusually Stable Hysteresis in the PH-Response of Poly(Acrylic Acid) Brushes Confined within Nanoporous Block Polymer Thin Films. *ACS Publ.* **2016**, *138*, 7030–7039. [CrossRef]
35. Gisbert Quilis, N.; van Dongen, M.; Venugopalan, P.; Kotlarek, D.; Petri, C.; Moreno Cencerrado, A.; Stanescu, S.; Toca Herrera, J.L.; Jonas, U.; Möller, M.; et al. Actively Tunable Collective Localized Surface Plasmons by Responsive Hydrogel Membrane. *Adv. Opt. Mater.* **2019**, *7*, 1900342. [CrossRef]
36. Tiwari, A.; Sancaktar, E. Poly(N-Isopropylacrylamide) Grafted Temperature Responsive PET Membranes: An Ultrafast Method for Membrane Processing Using KrF Excimer Laser at 248 nm. *J. Memb. Sci.* **2018**, *552*, 357–366. [CrossRef]
37. Darvishmanesh, S.; Qian, X.; Wickramasinghe, S.R. Responsive Membranes for Advanced Separations. *Curr. Opin. Chem. Eng.* **2015**, *8*, 98–104. [CrossRef]
38. Kapur, V.; Charkoudian, J.; Anderson, J.L. Transport of Proteins through Gel-Filled Porous Membranes. *J. Memb. Sci.* **1997**, *131*, 143–153. [CrossRef]
39. Lin, X.; Huang, R.; Ulbricht, M. Novel Magneto-Responsive Membrane for Remote Control Switchable Molecular Sieving. *J. Mater. Chem. B* **2016**, *4*, 867–879. [CrossRef]
40. Adrus, N.; Ulbricht, M. Novel Hydrogel Pore-Filled Composite Membranes with Tunable and Temperature-Responsive Size-Selectivity. *J. Mater. Chem.* **2012**, *22*, 3088–3098. [CrossRef]
41. Geismann, C.; Yaroshchuk, A.; Ulbricht, M. Permeability and Electrokinetic Characterization of Poly(Ethylene Terephthalate) Capillary Pore Membranes with Grafted Temperature-Responsive Polymers. *Langmuir* **2006**, *23*, 76–83. [CrossRef]
42. Zhang, X.Z.; Yang, Y.Y.; Chung, T.S.; Ma, K.X. Preparation and Characterization of Fast Response Macroporous Poly(N-Isopropylacrylamide) Hydrogels. *Langmuir* **2001**, *17*, 6094–6099. [CrossRef]
43. Suradi, S.S.; Naemuddin, N.H.; Hashim, S.; Adrus, N. Impact of Carboxylation and Hydrolysis Functionalisations on the Anti-Oil Staining Behaviour of Textiles Grafted with Poly(N-Isopropylacrylamide) Hydrogel. *RSC Adv.* **2018**, *8*, 13423–13432. [CrossRef] [PubMed]
44. Wang, X.; McCord, M.G. Grafting of Poly(N-Isopropylacrylamide) onto Nylon and Polystyrene Surfaces by Atmospheric Plasma Treatment Followed with Free Radical Graft Copolymerization. *J. Appl. Polym. Sci.* **2007**, *104*, 3614–3621. [CrossRef]
45. Kurşun, F.; Işıklan, N. Development of Thermo-Responsive Poly(Vinyl Alcohol)-g-Poly(N-Isopropylacrylamide) Copolymeric Membranes for Separation of Isopropyl Alcohol/Water Mixtures via Pervaporation. *J. Ind. Eng. Chem.* **2016**, *41*, 91–104. [CrossRef]
46. Zargar, M.; Hartanto, Y.; Jin, B.; Dai, S. Understanding Functionalized Silica Nanoparticles Incorporation in Thin Film Composite Membranes: Interactions and Desalination Performance. *J. Memb. Sci.* **2017**, *521*, 53–64. [CrossRef]
47. Zhang, S.; Li, R.; Huang, D.; Ren, X.; Huang, T.S. Antibacterial Modification of PET with Quaternary Ammonium Salt and Silver Particles via Electron-Beam Irradiation. *Mater. Sci. Eng. C* **2018**, *85*, 123–129. [CrossRef]
48. Modigunta, J.K.R.; Kim, J.M.; Cao, T.T.; Yabu, H.; Huh, D.S. Pore-Selective Modification of the Honeycomb-Patterned Porous Polystyrene Film with Poly(N-Isopropylacrylamide) and Application for Thermo-Responsive Smart Material. *Polymer* **2020**, *201*, 122630. [CrossRef]
49. Karan, S.; Jiang, Z.; Livingston, A.G. Sub-10 Nm Polyamide Nanofilms with Ultrafast Solvent Transport for Molecular Separation. *Science* **2015**, *348*, 1347–1351. [CrossRef]
50. Mansourpanah, Y. MXenes and Other 2D Nanosheets for Modification of Polyamide Thin Film Nanocomposite Membranes for Desalination. *Sep. Purif. Technol.* **2022**, *289*, 120777. [CrossRef]
51. Zhang, W.; Liu, N.; Zhang, Q.; Qu, R.; Liu, Y.; Li, X.; Wei, Y.; Feng, L.; Jiang, L. Thermo-Driven Controllable Emulsion Separation by a Polymer-Decorated Membrane with Switchable Wettability. *Angew. Chemie* **2018**, *130*, 5842–5847. [CrossRef]
52. Zhao, G.; Chen, W.N. Design of Poly(Vinylidene Fluoride)-g-p(Hydroxyethyl Methacrylate-Co-N-Isopropylacrylamide) Membrane via Surface Modification for Enhanced Fouling Resistance and Release Property. *Appl. Surf. Sci.* **2017**, *398*, 103–115. [CrossRef]
53. Hartanto, Y.; Zargar, M.; Wang, H.; Jin, B.; Dai, S. Thermoresponsive Acidic Microgels as Functional Draw Agents for forward Osmosis Desalination. *Environ. Sci. Technol.* **2016**, *50*, 4221–4228. [CrossRef]
54. Yang, Y.; Song, C.; Wang, P.; Fan, X.; Xu, Y.; Dong, G.; Liu, Z.; Pan, Z.; Song, Y.; Song, C. Insights into the Impact of Polydopamine Modification on Permeability and Anti-Fouling Performance of Forward Osmosis Membrane. *Chemosphere* **2022**, *291*, 132744. [CrossRef]
55. Balcıoğlu, G.; Yılmaz, G.; Gönder, Z.B. Evaluation of Anaerobic Membrane Bioreactor (AnMBR) Treating Confectionery Wastewater at Long-Term Operation under Different Organic Loading Rates: Performance and Membrane Fouling. *Chem. Eng. J.* **2021**, *404*, 126161. [CrossRef]
56. McGaughey, A.L.; Gustafson, R.D.; Childress, A.E. Effect of Long-Term Operation on Membrane Surface Characteristics and Performance in Membrane Distillation. *J. Memb. Sci.* **2017**, *543*, 143–150. [CrossRef]
57. Feil, H.; Bae, Y.H.; Feijen, J.; Kim, S.W. Effect of Comonomer Hydrophilicity and Ionization on the Lower Critical Solution Temperature of N-Isopropylacrylamide Copolymers. *Macromolecules* **1993**, *26*, 2496–2500. [CrossRef]

58. Taylor, L.D.; Cerankowski, L.D. Preparation of Films Exhibiting a Balanced Temperature Dependence to Permeation by Aqueous Solutions—A Study of Lower Consolute Behavior. *J. Polym. Sci. Polym. Chem. Ed.* **1975**, *13*, 2551–2570. [CrossRef]
59. Tanaka, T.; Ohmine, I.; Hirokawa, Y.; Fujishige, S.; Kubota, K.; Ando, I.; Otake, K.; Inomata, H.; Konno, M.; Saito, S.; et al. *Principles of Polymer Chemistry*; Cornell University Press: Ithaca, NY, USA, 1990; Volume 23, p. 761.
60. Bokias, G.; Hourdet, D.; Iliopoulos, I.; Staikos, G.; Audebert, R. Hydrophobic Interactions of Poly(N-Isopropylacrylamide) with Hydrophobically Modified Poly(Sodium Acrylate) in Aqueous Solution. *Macromolecules* **1997**, *30*, 8293–8297. [CrossRef]
61. Mohammad, A.W.; Teow, Y.H.; Ang, W.L.; Chung, Y.T.; Oatley-Radcliffe, D.L.; Hilal, N. Nanofiltration Membranes Review: Recent Advances and Future Prospects. *Desalination* **2015**, *356*, 226–254. [CrossRef]
62. Fänger, C.; Wack, H.; Ulbricht, M. Macroporous Poly(N-Isopropylacrylamide) Hydrogels with Adjustable Size "Cut-off" for the Efficient and Reversible Immobilization of Biomacromolecules. *Macromol. Biosci.* **2006**, *6*, 393–402. [CrossRef]
63. Jimenez-Solomon, M.F.; Gorgojo, P.; Munoz-Ibanez, M.; Livingston, A.G. Beneath the Surface: Influence of Supports on Thin Film Composite Membranes by Interfacial Polymerization for Organic Solvent Nanofiltration. *J. Memb. Sci.* **2013**, *448*, 102–113. [CrossRef]
64. Mokarizadeh, H.; Moayedfard, S.; Maleh, M.S.; Mohamed, S.I.G.P.; Nejati, S.; Esfahani, M.R. The Role of Support Layer Properties on the Fabrication and Performance of Thin-Film Composite Membranes: The Significance of Selective Layer-Support Layer Connectivity. *Sep. Purif. Technol.* **2021**, *278*, 119451. [CrossRef]
65. Wang, N.; Ji, S.; Zhang, G.; Li, J.; Wang, L. Self-Assembly of Graphene Oxide and Polyelectrolyte Complex Nanohybrid Membranes for Nanofiltration and Pervaporation. *Chem. Eng. J.* **2012**, *213*, 318–329. [CrossRef]
66. Zazouli, M.A.; Ulbricht, M.; Naseri, S.; Susanto, H. Effect of Hydrophilic and Hydrophobic Organic Matter on Amoxicillin and Cephalexin Residuals Rejection from Water by Nanofiltration. *J. Environ. Health Sci. Eng.* **2010**, *7*, 15–24.
67. Hu, D.; Xu, Z.L.; Chen, C. Polypiperazine-Amide Nanofiltration Membrane Containing Silica Nanoparticles Prepared by Interfacial Polymerization. *Desalination* **2012**, *301*, 75–81. [CrossRef]
68. Sun, Z.; Dong, F.; Wu, Q.; Tang, Y.; Zhu, Y.; Gao, C.; Xue, L. High Water Permeating Thin Film Composite Polyamide Nanofiltration Membranes Showing Thermal Responsive Gating Properties. *J. Water Process. Eng.* **2020**, *36*, 101355. [CrossRef]
69. Wang, W.; Tian, X.; Feng, Y.; Cao, B.; Yang, W.; Zhang, L. Thermally On–Off Switching Membranes Prepared by Pore-Filling Poly(N-Isopropylacrylamide) Hydrogels. *Ind. Eng. Chem. Res.* **2009**, *49*, 1684–1690. [CrossRef]
70. Guo, J.W.; Wang, C.F.; Lai, J.Y.; Lu, C.H.; Chen, J.K. Poly(N-Isopropylacrylamide)-Gelatin Hydrogel Membranes with Thermo-Tunable Pores for Water Flux Gating and Protein Separation. *J. Memb. Sci.* **2021**, *618*, 118732. [CrossRef]

Article

Zinc Oxide Nanoparticles and Their Application in Adsorption of Toxic Dye from Aqueous Solution

Wafa Shamsan Al-Arjan

Department of Chemistry, College of Science, King Faisal University, P.O. Box 400, Al-Ahsa 31982, Saudi Arabia; walarjan@kfu.edu.sa

Abstract: Dye waste is one of the most serious types of pollution in natural water bodies, since its presence can be easily detected by the naked eye, and it is not easily biodegradable. In this study, zinc oxide nanoparticles (ZnO-NPs) were generated using a chemical reduction approach involving the zinc nitrate procedure. Fourier transform infrared (FTIR), scanning electron microscopy (SEM), Brunauer-Emmett-Teller (BET), and UV-vis techniques were used to analyse the surface of ZnO-NPs. The results indicate the creation of ZnO-NPs with a surface area of 95.83 m² g^{-1} and a pore volume of 0.058 cm³ g^{-1}, as well as an average pore size of 1.22 nm. In addition, the ZnO-NPs were used as an adsorbent for the removal of Ismate violet 2R (IV2R) dye from aqueous solutions under various conditions (dye concentration, pH, contact time, temperature, and adsorbent dosage) using a batch adsorption technique. Furthermore, FTIR and SEM examinations performed before and after the adsorption process indicated that the surface functionalisation and shape of the ZnO-NP nanocomposites changed significantly. A batch adsorption analysis was used to examine the extent to which operating parameters, the equilibrium isotherm, adsorption kinetics, and thermodynamics affected the results. The results of the batch technique revealed that the best results were obtained in the treatment with 0.04 g of ZnO-NP nanoparticles at 30 °C and pH 2 with an initial dye concentration of 10 mg L^{-1}, which removed 91.5% and 65.6% of dye from synthetic and textile industry effluents, respectively. Additionally, six adsorption isotherm models were investigated by mathematical modelling and were validated for the adsorption process, and error function equations were applied to the isotherm model results in order to find the best-fit isotherm model. Likewise, the pseudo-second-order kinetic model fit well. A thermodynamic study revealed that IV2R adsorption on ZnO-NPs is a spontaneous, endothermic, and feasible sorption process. Finally, the synthesised nanocomposites prove to be excellent candidates for IV2R removal from water and real wastewater systems.

Keywords: Ismate violet 2R dye; IV2R; ZnO-NP; BET; kinetics; isotherm modelling; thermodynamic; adsorption

Citation: Al-Arjan, W.S. Zinc Oxide Nanoparticles and Their Application in Adsorption of Toxic Dye from Aqueous Solution. *Polymers* **2022**, *14*, 3086. https://doi.org/10.3390/polym14153086

Academic Editors: Irene S. Fahim, Ahmed K. Badawi and Hossam E. Emam

Received: 24 June 2022
Accepted: 26 July 2022
Published: 29 July 2022

Publisher's Note: MDPI stays neutral with regard to jurisdictional claims in published maps and institutional affiliations.

Copyright: © 2022 by the author. Licensee MDPI, Basel, Switzerland. This article is an open access article distributed under the terms and conditions of the Creative Commons Attribution (CC BY) license (https://creativecommons.org/licenses/by/4.0/).

1. Introduction

In terms of industrial pollution, textile industries are a major source of pollution since they use a lot of water and chemicals and discharge high levels of toxic and non-biodegradable dye effluents [1,2]. Dyes are widely applied in a variety of textile-based industries due to their advantageous properties [3,4], such as bright colours, water resistance, and ease of application [5]. Several sectors, including textile dyeing (60%), paper (10%), and plastic materials (10%), utilise large amounts of synthetic dyes (10%). According to some reports, there are approximately 100,000 commercially accessible dyes with a production capacity of more than 7 × 10⁵ metric tons per year, with the textile industry using about 10,000 compounds [6]. Dye industry effluents, in particular, necessitate not only the treatment of problematic wastewater with high chemical and biological oxygen demands, suspended particles, and hazardous chemicals but also the treatment of dyes that are perceived by human eyes at very low concentrations [5]. When dyes are released into receiving water bodies, they form hazardous amines through reductive cleavage of azo linkages, which can harm essential organs such as the brain, liver, kidney, central nervous system, and reproductive system. Furthermore, synthetic dyes may have an unfavourable effect on some aquatic life's photosynthetic activity due to the presence of aromatics, metals, chlorides, and other contaminants. Therefore, as a consequence, their removal from aquatic environments is important and the target of numerous scientific studies [7–9]. Chemical precipitation [10], flocculation/coagulation [11,12], membrane technology [13], oxidation technology [14], electrolytic reduction [15], ion exchange [16], and biological adsorption [17–20] are developed methods for dyes and pollutant removal from water/wastewater. Recently, the adsorption process has been one of the most commonly used procedures for dye removal due to its simplicity and high efficacy, as well as the suitability of the use of a wide range of adsorbents [21–23]. Moreover, various nanoparticles have been investigated for dye adsorption due to the ease of changing their surface functionality and their high surface-to-volume ratio for increased adsorption capacity and efficiency [24]. Nanosized metal oxides, such as nanosized manganese oxides, ferric oxides, aluminium oxides, magnesium oxides, and cerium oxides, are thought to be capable of adsorbing dyes from aqueous solutions [25]. Additionally, these nanoparticles have been intensively investigated as extremely efficient absorbents for the removal of heavy metal ions from water and wastewater. They have a number of advantages, including high ability, unsaturated surfaces, ease of use, quick kinetics, and favourable dye sorption in water and wastewater [25].

Zinc oxide (ZnO) has a wide range of uses as a low-toxicity material, including in the catalyst industry [26], gas sensors [27], solar cells [28], and medicine [29]. In addition, zinc oxide (ZnO) belongs to the class of metal oxides that are commercially very important due to their remarkable applications in various industrial fields, such as catalysis, solar cells, paints, UV light-emitting devices, electronic devices, biomedicine, and cosmetics. Likewise, ZnO NPs as semiconductors have gained attention for their wide range of applications, including optoelectronics, optics, electronics, and dye removal employing environmentally benign synthesis components, including fungi, bacteria, and marine macroalgae [30]. Additionally, ZnO has been discovered to be more effective, possibly beneficial, than other metals for the bio-synthesis of nanoparticles (NPs) for medical applications [31].

Zinc oxide nanoparticles (ZnO-NPs) are the most important of the metal oxide nanoparticles (MO-NPs) because of their unique chemical and physical features, which increase their applicability [32]. The elimination of several pollutants from the environment is a challenge, and adsorption methods are generally thought to be more facile and effective. Bearing a large theoretical specific surface area [33], the practical use of ZnO in water cleaning, including decontamination and reuse, has attracted excessive attention in recent years. As it contains additional functional groups, ZnO has shown higher adsorption performance. Therefore, ZnO might have more potential in adsorption technologies. In addition, ZnO was found to be more effective as an adsorbent than other absorbents, such as phosphate, iron oxide, and activated carbon, for the removal of sulphur compounds and H_2S [34].

ZnO-NPs have recently been reported to efficiently absorb dyes from aqueous environments [35]. In addition, Ismate violet 2R was chosen as a model compound in the current study because of its broad application range, which includes colouring paper, dyeing cotton and wool, coating paper stock, and medicinal applications, as well as its potential harm.

In this work, a composite of zinc oxide nanoparticles was produced by chemical reduction using zinc nitrate as an adsorbent for the removal of Ismate violet 2R (IV2R) dye from aqueous solutions under various conditions by employing the batch adsorption technique. The initial dye concentration, pH, contact time, temperature, and adsorbent dosage were the main parameters assessed. Moreover, SEM, FTIR, UV, and BET methods were used to reveal the surface functionalisation, morphology, and pore size of the composite. In addition, the experimental equilibrium was applied using several different adsorption isotherm models to assess the adsorption mechanism, as well as thermodynamic and kinetic analyses. Furthermore, error functions were applied to reveal the most suitable model.

2. Materials and Methods

2.1. Materials

Polyvinyl pyrrolidone (PVP), H_2O_2, zinc nitrate, and sodium hydrosulphite were provided by M/s. Himedia Laboratories Pvt. Ltd., Mumbai, India. Table 1 shows the chemical structure of IV2R dye (molecular formula, molecular weight, and λ max). All of the used compounds were of analytical grade.

Table 1. The chemical and physical properties of the ISMATE 2R dye [2].

Features	Data
Dye name	ISMATE violet 2R
Mol. wt.	700
Molecular formula	C22H14N4O11S3CuCl
Wavelength (λ max)	550 nm
Molecular structure	(structure image)

2.2. Synthesis of ZnO Nanoparticles

An improved approach was used to create ZnO-NPs, as described previously [26]. Briefly, 0.1 M Zn $(NO_3)_2$ was hydrolyzed with 250 mL of 0.2 M NaOH, and 1 % PVP was added and stirred continuously for 2 h using 100 mL of deionised water (Millipore, Milli-Q, Buenos Aires, Argentina). The pellet was produced by centrifuging the suspension at 3000 rpm for 5 min at 4 °C. At 75 °C, 1 M H_2O_2 was added and agitated for 1 h. The sample was also dried for 3 h at 65 °C in an oven before being maintained at 350 °C for 6 h.

2.3. Preparation of Dye Solution

The IV2R dye stock solution was made by weighing 1.00 g of powdered dye. The dye was transferred quantitatively into a 1 L measuring flask, which was then filled with distilled water to achieve a dye concentration of 1000 mg L^{-1} in the solution. The stock dye solutions (1000 mg L^{-1}) were prepared separately and stored at 4 °C in distilled water. Dilution of the stock solutions with distilled water generated the working solutions.

2.4. ZnO-NP Characterisation

A scanning electron microscope was used to examine the morphologies of the adsorbents (JEOL JSM 6360 LA, Austin, TX, USA). Brunauer-Emmett-Teller (BET) desorption-adsorption studies were also performed using a sorptiometer (Quantachrome TouchWinTM v1.2, Downers Grove, IL, USA) and N_2 adsorption isotherms. Using N_2 desorption-adsorption measurements in a N_2 solution with a saturated vapour pressure of 33.5 atm and an adsorption temperature of 77.3 K, the average particle radius, mean pore diameter, and pore volume of the ZnO-NPs were estimated. Furthermore, a Shimadzu FTIR-8400 S was employed to analyse the functional groups of the pre-prepared ZnO-NP materials using FTIR analysis spectrophotometry (Japan), and a UV-vis spectrophotometer was used to take spectrophotometric readings (UV 4000, MRI, Germany).

2.5. Adsorption Experiments

2.5.1. Batch Adsorption Experiments

In a 100 mL airtight Erlenmeyer flask, batch studies for IV2R elimination were carried out with 50 mL of 10 mg L^{-1} dye solution. The temperature of the system was kept constant at 25 °C. During the experiment, a weighed amount of adsorbent was placed in the flask and mixed with various concentrations of IV2R at constant and moderate mixing rates. At predetermined contact times, samples were removed and filtered to separate the adsorbent from the dye solution. The concentration of the dye solution was assessed using spectrometric techniques, and all experiments were carried out in triplicate, with the mean result reported. Furthermore, the effects of contact time (15 to 180 min) and adsorbent dosage (0.005 to 0.08 g) were investigated at a constant initial dye concentration of 10 mg L^{-1}. Experiments were conducted by altering the initial dye concentrations from 10 to 80 mg L^{-1}. The temperature (25 to 55 °C) was evaluated at a constant dose of 0.02 g of ZnO-NP adsorbent to determine the effects of the initial dye concentration on dye uptake. Moreover, the effects of final pH on IV2R adsorption were investigated by altering the dye's initial pH from 2 to 10. NaOH and HCl solutions were used to adjust the dye solutions' initial pH. A UV-visible spectrophotometer was used to measure the quantity of IV2R dye in the clear supernatant at 550 nm at any time after shaking.

2.5.2. Analytical Techniques

The dye concentrations were measured using a Spectronic Genesy 2PC UV-vis spectrophotometer at the wavelength of its maximum absorbance, λ max. Using the Beer-Lambert equation, the final dye concentration was measured spectrometrically to correspond to the dye's maximum concentration:

$$\text{Absorbance} = \varepsilon CSl \tag{1}$$

where ε is the molar absorptivity, CS is the sample concentration, and l is the thickness of the absorbing material (1 cm). A pH/ion meter (WTW Inolab pH/ION Level 2, Germany) was used to determine the pH of the dye solution. Equations (2) and (3) were used to calculate the adsorption capacity (q_e) and dye elimination percentages [4]:

$$q_e = \frac{(C_i - C_e) \times V}{W} \tag{2}$$

$$\text{Removal Percentage (\%)} = \left(\frac{(C_i - C_e)}{C_i} \times 100 \right) \tag{3}$$

where C_i and C_e (mg L^{-1}) indicate the initial and final concentrations of IV2R at a given time, respectively, while V represents the volume of the dye mixture (L), and W represents the weight of the dry adsorbent (g).

2.6. Study of the Adsorption Isotherm

2.6.1. Experiment with Isotherms

Experiments were carried out to develop adsorption isotherms for the ZnO-NP adsorbent by adjusting the adsorbent dose to 0.02 g at an initial dye concentration of 10 to 80 mg L^{-1} and an ambient pH of 2 for 3 h at 150 rpm, which was mixed with 50 mL of dye solution [36,37]. The following isotherm models were used to determine the most suitable one and calculate the data.

2.6.2. Freundlich Model

The ability of the Freundlich model to fit the experimental data was used to calculate the slope of n and the intercept value of K_f by displaying a curve of log q_e with respect to log C_e. By visualizing the Freundlich model in logarithmic form [38], it is simple to linearise it:

$$\text{Log } q_e = \text{Log } K_f + 1/n \text{ Log } C_e \quad (4)$$

The isotherm was used to determine the Freundlich constants K_f and n.

2.6.3. The Langmuir Model

The Langmuir model is represented by the following mathematical formula [39]:

$$q_e = q_{max} bC_e/(1 + bC_e) \quad (5)$$

where q_{max} (mg g^{-1}) is the maximum sorption capacity corresponding to the saturation capacity, and b (L mg^{-1}) is a coefficient relating to the affinity between ZnO-NP and IV2R dye ions. The linear relationship obtained by graphing the curve $(1/q_e)$ vs. $(1/C_e)$ is given in Equation (6).

$$1/q_e = 1/(bq_{max} C_e) + 1/q_{max} \quad (6)$$

The slope and intercept, respectively, are used to calculate b and q_{max}.

2.6.4. Henderson and Halsey Models

These models perform well with heteroporous substances and multilayer sorption. Using the equation below, the Halsey model [40] was calculated, as given in Equation (7).

$$\text{Ln } q_e = \frac{1}{n} \text{Ln } K + \frac{1}{n} \text{Ln } C_e \quad (7)$$

where n and K are Halsey constants. Meanwhile, the Henderson model was obtained from the following equation:

$$\ln[-\ln(1 - C_e)] = \ln K + \left(\frac{1}{n}\right) \ln q_e \quad (8)$$

where the Henderson constants are nh and Kh.

2.6.5. The Harkins-Jura Model

This model describes multilayer adsorption, as well as the presence of heterogeneous pore scattering in an adsorbent [41]. This model was obtained from Equation (9).

$$\frac{1}{q_e^2} = \left(\frac{B_2}{A}\right) - \left(\frac{1}{A}\right) \log C_e \quad (9)$$

where the isotherm constants are A and B.

2.7. Error Function Test

Various error functions were investigated in order to find the best and most appropriate model for investigating the equilibrium data. The error function tests were employed as follows:

2.7.1. Hybrid Fractional Error (HYBRID)

Because it solves for low concentrations by balancing absolute deviation against fractional error and is more dependable than other error functions, the hybrid fractional error function was used. Equation (10) gives the hybrid error:

$$HYBRID = \frac{100}{N-P} \sum \left| \frac{q_{e,exp} - q_{e,calc}}{q_{e,exp}} \right|_i \quad (10)$$

where N is the total number of data points, and $q_{e,exp}$ and $q_{e,cal}$ (mg g^{-1}) are the respective experimental and calculated adsorption capacity values. In addition, P is the number of isotherm factors.

2.7.2. Error Percentage Average (APE)

The APE model exhibits the suitability or trend between the predicted and observed values of the sorption capacity used to create model curves (APE) and can be designed using Equation (11).

$$\text{APE}(\%) = \frac{100}{N}\sum_{i=1}^{N}\left|\frac{q_{e,isotherm} - q_{e,calc}}{q_{e,isotherm}}\right|_i \tag{11}$$

where N is the number of data points under investigation.

2.7.3. Nonlinear Chi-Square Test (Nonlinear Chi-Square Test)

The nonlinear chi-square test is a statistical method for determining which treatment system is most suitable. The following Equation (12) defines the approach to determining the chi-square error:

$$X^2 = \frac{(q_{e,isotherm} - q_{e,calc})^2}{q_{e,isotherm}} \tag{12}$$

2.7.4. Sum of Absolute Errors (EABS)

An increase in errors improves the fit, resulting in a bias toward high-concentration data. The following Equation (13) can be used to evaluate EABS tests:

$$\text{EABS} = \sum_{i=1}^{N}\left|q_{e,calc} - q_{e,isotherm}\right|_i \tag{13}$$

2.8. Adsorption Kinetics

Adsorption kinetic experiments were carried out using 0.05 g of adsorbent mixed separately with 50 mL of IV2R solution containing 10 mg L^{-1} concentrations in 100 mL conical flasks at a solution pH of 2, and the mixture was agitated at room temperature for 15, 30, 60, 120, and 180 min. The clear solutions were examined for any remaining IV2R concentrations.

2.8.1. Pseudo-First-Order Kinetics

The following equation gives the linear version of the generalised pseudo-first-order kinetics of dyes adsorbed at time t (mg g^{-1}).

$$dq/dt = K_1 (q_e - q_t) \tag{14}$$

The linear formula of the pseudo-first-order kinetics is expressed in Equation (15).

$$\text{Log}(q_e - q_t) = \log(q_e) - k_1 t/2.303 \tag{15}$$

where q_e is the amount of dye adsorbed at equilibrium (mg g^{-1}), q_t denotes the amount of time t, and K_1 denotes the pseudo-first-order rate constant (min^{-1}).

2.8.2. Pseudo-Second-Order Kinetics

The pseudo-second-order Equation (16) is as follows:

$$T/q_t = 1/K_2 q_e^2 + t/q_e \tag{16}$$

where K_2 denotes the second-order rate constant (g mg^{-1} min). Plotting (t/q_t) versus (t) yields a linear relationship, and the slope and intercept can be used to derive the q_e and K_2 parameters, respectively.

2.8.3. The Intraparticle Diffusion Model

The following is the intraparticle diffusion equation:

$$K_{dif} \, t^{1/2} + C = q_t \tag{17}$$

where C is the intercept, and K_{dif} (mg g^{-1} min$^{0.5}$) is the intraparticle diffusion rate constant, which is calculated from the regression line's slope.

2.9. Thermodynamics of Adsorption

The value of Gibbs free energy change ($\Delta G°$) is a fundamental principle of non-spontaneity, and a thermodynamic analysis is required to determine whether the nature of the sorption process is spontaneous or not. The following nonlinear forms (18)–(21) can be used to compute the Gibbs free energy change ($\Delta G°$ kJ mol^{-1}), enthalpy change ($\Delta H°$ kJ mol^{-1}), and entropy change ($\Delta S°$, J mol^{-1} K^{-1}) parameters for the sorption process at various temperatures (e.g., 25, 30, 45, and 55 °C):

$$K_d = q_e / C_e \tag{18}$$

$$\Delta G° = -RT \ln K_d \tag{19}$$

$$\Delta G° = \Delta H° - T \Delta S° \tag{20}$$

$$\Delta G° = T(\Delta S°) + \Delta H° \tag{21}$$

In addition to $\Delta G°$, which can be obtained from Equation (20) or (21), the values of $\Delta H°$ and $\Delta S°$ were computed using the intercept and slope of the plotted curve of T vs. $\Delta G°$ from Equation (20) or (21).

3. Results and Discussion

3.1. ZnO-NP Characterisation

3.1.1. Infrared Spectroscopy (FTIR)

The produced ZnO-NPs were identified using FTIR measurements, as depicted in Figure 1. Figure 1A shows the existence of IR signals before adsorption in the range of 3600–3000 cm^{-1} in dye samples, which indicate the presence of –OH and –NH$_2$ groups. Similarly, CH$_2$ stretching vibration is indicated by the peak at 2935 cm^{-1}. The stretching vibration of the C–H bond in –CH$_2$ groups is attributed to the band at 2860 cm^{-1}. According to the data, the peak at 1546 cm^{-1} is O–H bending vibration. Correspondingly, the peaks around 1394 and 1508 cm^{-1} indicate stretching of the C–H bond, which is attributed to CH$_3$ vibration. Likewise, the absorption bands in the range of 1000–1300 cm^{-1} indicate the presence of the C–O group. In addition, the spectra of ZnO-NPs reveal a strong peak at 426 cm^{-1}, which corresponds to the zinc-oxygen stretching mode [42]. The peaks at 949, 834, and 700 cm^{-1} generally correlate to C–H bending vibration.

On the other hand, Figure 1B shows that the bands at 3737, 3294, 2924, 2860, 1546, 1508, 1394, 1044, and 949 cm^{-1} were shifted to 3343, 2929, 2637, 1545, 1408, 1037, and 948 cm^{-1} after IV2R adsorption. According to this hypothesis, OH, C–H, C=C, and C=O groups may play a role in the adsorption of IV2R onto ZnO-NPs, while the C=O bond may be seen in the peaks between 1400 and 1500 cm^{-1} [43]. Additionally, the new peak at 1633 cm^{-1} is compatible with the C=O stretching of proteins [44]. C–H bending, C–O, or C–C stretching vibrations are represented by absorption bands that are located between 1100 and 1000 cm^{-1}. The presence of an aromatic ring on the dye compound is indicated by the band seen at 885 cm^{-1}, which is caused by aromatic C–H out-of-plane vibration [45]. Free O–H and N–H stretching vibrations are responsible for the robust and wide absorption band located at 3371 cm^{-1}. While the observed band at 1408 cm^{-h} is related to the C–N stretching bond of amino acids, the band observed at a wavelength of 1037 cm^{-h} can be ascribed to C–O–C. Moreover, C–H bending vibration is responsible for the weak absorption band with a centre radius of 667 cm^{-1}. Finally, the stretching vibrations of Zn-O are responsible for the peak range of 400–600 cm^{-1} [46].

It is well known that the production of ZnO particles from the hydrolysis of Zn^{2+} ions in aqueous medium is a complicated process. Many polyvalent cationic species can form when Zn^{2+} ions interact with OH ions, and their formation is significantly influenced by the pH of the solution. However, based on the pH, temperature, and synthesis processes, the precipitation of ZnO particles is usually defined by a development unit that can be either Zn (OH)$_2$ or Zn(OH)$^2{}_4$ ions. The dissolution-reprecipitation mechanism has been proposed as a mechanism for the formation of ZnO from Zn(OH)$_2$ [47]. ZnO will be produced by the chemical reaction [48]:

$$Zn(OH)_2 \longrightarrow ZnO + H_2O.$$

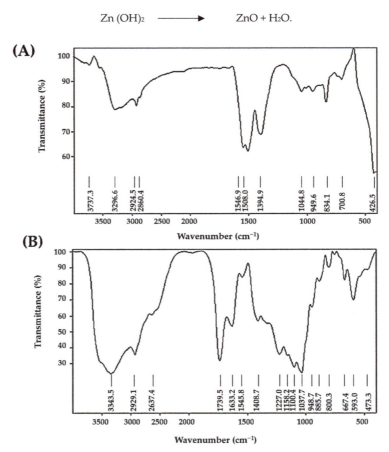

Figure 1. FTIR pattern of ZnO-NP before (**A**) and after adsorption of IV2R dye (**B**).

3.1.2. BET Surface Analysis

The pore diameter, pore size distribution, and specific surface area of the ZnO-NPs were measured using BET and the nitrogen adsorption-desorption isotherm technique, as shown in Figure 2. It was revealed that an isotherm existed when the BET analysis was conducted at high P/P0 levels. Table 2 shows the BET results. The total pore volume and specific surface area of the ZnO-NPs were 95.83 m^2 g^{-1} and 0.058 cm^3 g^{-1}, respectively. The specific surface area calculated using the Langmuir technique was 140.692 m^2 g^{-1}. This superior property could provide ZnO-NPs with a larger surface area, allowing more active sorption sites to occur. The micropore volume of ZnO-NPs was determined to be 0.015 cm^3 g^{-1} using a cumulative BJH adsorption experiment. Furthermore, the average particle radius was 1.42 nm.

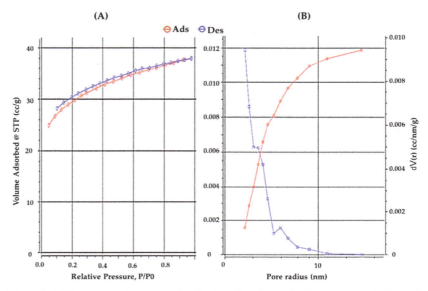

Figure 2. BET specific surface area (**A**) and adsorption-desorption isotherm examination (**B**) of ZnO-NPs.

Table 2. Physicochemical properties of ZnO-NPs.

Characteristics	Value/Unit
Density	2.2 (g cm^{-3})
Langmuir method	140.692 m^2 g^{-1}
BJH adsorption	10.682 m^2 g^{-1}
BJH desorption	8.847 m^2 g^{-1}
BET surface area	95.838 m^2 g^{-1}
Average pore size	1.228 nm
Total pore volume	0.058 C2 g^{-1}
BJH adsorption cumulative micropore volume	0.014 C2 g^{-1}
Average particle radius	1.422 nm

3.1.3. SEM Examination

SEM measurements are helpful in determining the surface morphology of the ZnO-NP structure. SEM pictures of ZnO-NP powders can be seen in Figure 3, which shows the particle morphologies. The nanostructure, which resembles nanoflowers and has agglomerated nanoparticles with an average pore size of approximately 1.22 nm, refers to ZnO-NP powders that result in formations that resemble flowers in uniformly sized nanoporous channels. The length and thickness of these formations are in nanometres. SEM micrographs of pure ZnO-NPs reveal that they have a porous character and a large surface area. These flowers and swollen structures are preferable for the absorption of dye contaminants.

Figure 3. SEM examination at 1 µm for the prepared ZnO-NP at magnifications of 10,000× (**A**) and 25,000× (**B**).

3.1.4. UV–Visible Spectra

The wavelength range of metal nanoparticles was determined by UV, and the results are shown in Table 3. The maximum absorbance is shown by the highest peak. The production of zinc oxide nanoparticles is confirmed by UV spectrometer absorption peaks in the 300–400 nm (390 nm) range. These peaks also indicate that the particles are nanosized and have a narrow particle size distribution. The absorption peak at 243 nm, which is associated with $\pi \rightarrow \pi^*$ transitions in the sesquiterpene system, can hardly be seen in the UV spectrum of ZnO-NPs, which could be owing to a change in the sesquiterpene structure and the absence of these $\pi \rightarrow \pi^*$ transitions [49].

Table 3. UV-vis spectra of chemical ZnO-NPs.

Wavelength (nm)	Abs.
390.50	0.100
243.000	0.528
380.51	0.350
500.86	0.537

3.2. Adsorption Experiments

3.2.1. Influence of pH

The shape and chemistry of the target dye ions, as well as the binding sites on the adsorbent, can be affected by the pH of a solution. In addition, the speciation of IV2R dye affects leads to an alteration in the reaction kinetics in addition to the equilibrium features of the sorption process [50]. Experiments were investigated by using different initial pH values of 2, 4, 6, 8, and 10 for the adsorbent to explore the effect of pH on the adsorption of IV2R. The percentage of adsorption tended to increase as the pH increased from 2 to 6. Furthermore, when the pH of the original dye solution was 2 and 6, the percentage of dye removal was at its maximum (90.7% and 89.17%, respectively) for the ZnO-NP adsorbent (Figure 4). The maximum adsorption capacity was observed at pH 2 and 6 with 4.31 and 4.24 mg g^{-1}, respectively. These results indicate that solutions that are acidic in nature are efficient for the adsorption of IV2R dye when the adsorption is below pH 6. In addition, these results prove that IV2R removal was slightly decreased at pH 4 with a percentage removal of 86.28% and an adsorption capacity of 4.10 mg g^{-1}. This result implies that there is a considerable electrostatic attraction between the adsorbent surface and the dye; this is likely owing to an increase in positively charged sites on the adsorbent surface when pH decreases, as reported by Netpradist et al. [51].

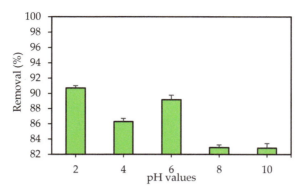

Figure 4. Influence of the pH value on the sorption of IV2R dye.

This study also revealed that anionic dye IV2R sorption is best in an acidic solution (positively charged adsorbent surface). Increased solution pH caused deprotonation and, as a result, a negatively charged surface of the adsorbent; consequently, this event may cause IV2R adsorption to decrease. Moreover, the nature of the surface charge of the adsorbent is affected by the pH of the solution in any adsorption system. In an acidic solution, the adsorbent's oxide surface acquires a net positive charge. Therefore, anionic dyes have a stronger electrostatic attraction in acidic solutions than in basic media [52]. Other probable reasons for such observations include dye-adsorbent interactions owing to hydrogen bonding and hydrophobic-hydrophobic interaction mechanisms. Furthermore, the surface area and pore size, which remain unaffected by pH changes, play a significant role during the process [53,54].

Ghoneim et al. [55] found that at a higher pH, elimination is reduced in comparison to the maximum condition. This can be explained by the binding site's ability to activate under normal conditions.

3.2.2. Influence of ZnO-NP Dose

The adsorbent dose is one of the most important factors for examining the impacts of the adsorption process to achieve the maximum adsorption capacity of the adsorbent by measuring the amount of ZnO-NP adsorbent. Experiments were performed with different adsorbent doses from 0.005 g to 0.08 g at a constant initial dye concentration of 10 mg L^{-1} to investigate the impact of the ZnO-NP dose on the adsorption system. At their respective equilibrium contact times, the percent adsorption increased from roughly 83.5 to 91.4 % as the adsorbent dose increased from 0.005 g to 0.04 g (Figure 5). It is self-evident that as the adsorbent dose increases, the dye uptake is enhanced. It is well established that as the adsorbent dosage increases, the % adsorption increases. The adsorbed amount per unit mass decreases. The amount adsorbed per unit mass was only 3.4 mg g^{-1} adsorbent when the dose was 0.08 g. The removal percentage was roughly 89%. The adsorption density decreases as the adsorbent dose increases due to unsaturated adsorption sites during the adsorption process [56]. Another cause could be intraparticle interactions, such as aggregation, as a result of a high adsorbent dose. The total surface area of the adsorbent is reduced because of this aggregation, while the diffusion path length rises [57].

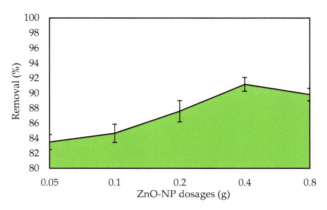

Figure 5. Influence of the ZnO-NP dose on the removal of IV2R.

3.2.3. Influence of Contact Time

The equilibrium time is essential when considering economical water and wastewater applications, and contact time is an important component in all transfer phenomena for the adsorption process. The adsorption process was studied to establish the best contact time between 15 and 180 min. Figure 6 shows the adsorption removal of IV2R dye by the ZnO-NP adsorbent. Equilibrium was reached, and the optimum contact time for IV2R dye was chosen to be 120 min for the sorbate-sorbent contact. The uptake of IV2R was observed to occur in two phases as a function of time. The first phase involved fast dye uptake during the first 10 min of sorbate-sorbent interactions, followed by a slow dye removal phase that lasted significantly longer (>120 min) until equilibrium was reached. The higher sorption value at the start of the process could be attributable to the abundance of active sites on the sorbent at this time. The sorption process becomes less efficient during the slower phase as these sites are gradually occupied [58]. The IV2R removal effectiveness by ZnO-NPs was found to be 88.9% (4.23 mg g^{-1}) at 180 min. Two-stage sorption has been extensively documented in the literature, with the first being quick and quantitatively dominant and the second being slower and quantitatively insignificant [59].

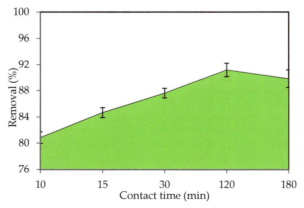

Figure 6. Effect of contact time on the removal of IV2R by ZnO-NPs.

Furthermore, Ananta et al. [60] reported that adsorption was rapid at the first contact time, which happens in the early stage of adsorption after a few minutes, since most of the binding sites are free and the adsorbent sites are empty, allowing the dye ions to bind quickly to the adsorbent. David and Joseph [61] mentioned that adsorption occurs quickly and is often controlled by diffusion from the majority of the solution on the surface. Moreover, Kumar and Gayathri [62] stated that adsorption increases as the contact time increases, which is possible due to a larger surface area of ZnO-NPs being available at the start and the exhaustion of the conversion of external adsorption sites, where the adsorbate (dye particles) is transported from the external to the internal sites of ZnO-NP adsorbent molecules.

3.2.4. Influence of the Initial Dye Concentration

The amount of dye removed from an aqueous mixture is highly dependent on the dye concentration. At a constant temperature, the adsorption process for IV2R was examined at concentrations ranging from 10 to 80 mg L^{-1}. The effect of IV2R dye concentrations on adsorption is seen in Figure 7. As can be seen, raising the initial dye concentration from 10 to 80 mg L^{-1} improved the ZnO-NP adsorption capacity from 4.33 to 20.58 mg g^{-1}. This could be owing to the strong driving force that occurs when the initial concentration of the adsorbates is increased, overcoming the mass transfer resistance between the aqueous and solid phases [20]. Furthermore, the increased dye clearance at higher concentrations is most likely due to greater diffusion and decreased dye absorption resistance [63]. The availability of active sites on the adsorbent and the final occupancy of these sites are attributed to the rapid first stage of dye removal with this adsorbent; the sorption then becomes less efficient. In addition, the greatest uptake of IV2R dye was found at a concentration of 10 mg L^{-1} (91.14%). The equilibrium loading capacity and initial dye concentration have a strong linear relationship with correlation coefficients greater than 0.99 for the ZnO-NP adsorbent. A dye concentration of 10 mg L^{-1} was used for further experiments. In general, the decreasing percentage of dye removal with increasing dye concentration could be due to the increase in sorption sites on the adsorbent surface [64]. Furthermore, the high probability of dye ions colliding with the adsorbent surface and the high rate of dye ion diffusion onto the adsorbent surface could be linked to the large amount of dye adsorbed at a high dye concentration [65]. This finding could point to the potential of treating textile effluent with a higher dye concentration.

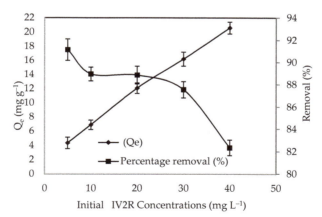

Figure 7. Influence of the initial IV2R dye concentration.

3.2.5. Influence of Temperature

Figure 8 illustrates the percentage removal of IV2R dye ions by ZnO-NPs. The maximum percentage of dye removal was obtained at 45 °C with 94.9%, which rose when the temperature was increased from 25 to 45 °C for the IV2R dye on ZnO-NPs. Because of the decreased solution viscosity as the temperature rose from 25 to 45 °C, the diffusion rate of the adsorbate molecules within the pores changed, as did the equilibrium capacity of the ZnO-NPs for a specific adsorbate. Temperature increases (above 45 °C) resulted in a decrease in the percentage removed. This is attributed to a reduction in surface activity [66]. The decrease in adsorption efficiency can be attributed to a variety of factors, including deactivation of the adsorbent surface, a growing tendency for dyes to migrate from the solid to the bulk stage, and the destruction of specific active sites on the adsorbent surface due to bond ruptures. According to Sivaprakash et al. [67], this could be due to an increase in the mobility of the large dye ion as temperature rises. A growing number of molecules may be able to obtain enough energy to interact with active areas on the surface. Furthermore, rising temperatures may cause a swelling effect within the adsorbent's internal structure, allowing large dyes to penetrate further. Additionally, there are two possible explanations for this outcome. At high temperatures, the pore diameters of adsorbent particles will expand [68]. Second, due to the breakage of some internal bonds, such as hydrogen bonds between the dye ion and the hydroxyl groups on the surface of the adsorbent's active surface sites, the number of adsorption sites will increase [69,70].

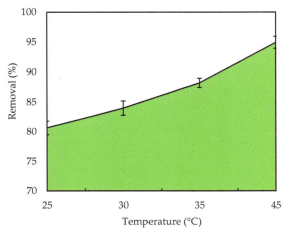

Figure 8. Effect of temperature on the removal of IV2R.

3.3. Isothermal Analysis

The equilibrium isotherm is a plot of the quantity of sorbate extracted per unit sorbent (q_e) as the sorbent's solid phase concentration against the sorbate's liquid phase concentration (C_e). For the design and optimisation of an adsorption system for the removal of a dye from an aqueous solution, equilibrium isotherm data are necessary [71]. As a consequence, the most appropriate correlation for the equilibrium curve must be determined. To test the validity of the experimental data, a number of isotherm models were applied [22]. Therefore, in the present study, the most commonly used models, namely, the Langmuir, Harkins-Jura, Freundlich, Halsey, Henderson, and Tempkin isotherms, were used to describe the adsorption equilibrium.

3.3.1. Freundlich Isotherm

On the basis of the assumption of energy surface heterogeneity, the Freundlich isotherm model is the earliest known relationship describing non-ideal and reversible adsorption, which can be extended to multilayer adsorption. The obtained results were fit to the Freundlich isotherm model's experimental data, which was supported by a strong correlation coefficient of $R^2 = 0.994$ for the ZnO-NPs, showing that this model is beneficial for the adsorption process, as shown in Figure 9. The IV2R dye and the adsorbents formed a strong bond, as indicated by the value of $1/n$, also known as the heterogeneity factor, which describes the divergence from sorption linearity as follows: When $1/n$ equals 1, the adsorption is linear, and the concentration of dye particles has no effect on the two stages. When 1/n is less than 1, chemical adsorption occurs; when 1/n is greater than 1, cooperative adsorption occurs, which is more physically advantageous and contains strong contacts among the adsorbate particles [72]. The value of the factor "1/n" was smaller than 1 in this study; the results indicate that a chemical sorption method on an external surface is preferable with this isotherm equation [22].

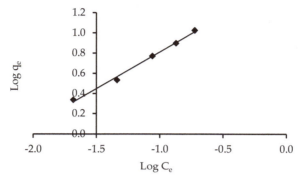

Figure 9. Freundlich isotherm of IV2R adsorption onto ZnO-NPs.

3.3.2. Langmuir Isotherm

The most popular model for quantifying the amount of adsorbate on an adsorbent as a function of partial pressure or concentration at a particular temperature is the Langmuir adsorption model. The Langmuir isotherm is based on the assumption of monolayer adsorption on a structurally homogeneous adsorbent, in which all adsorption sites are similar and energetically equivalent, adsorption occurs at specific homogeneous sites on the adsorbent, and once a dye molecule occupies a site, no further adsorption can occur. Table 4 shows the estimated parameters. The results obtained using the Langmuir isotherm coincided with data obtained throughout the experiment, with a strong correlation coefficient of $R^2 = 0.974$ for the ZnO-NPs. In addition, the ZnO-NP dye has a high maximum absorption capacity (q_{max}) of 119.05 mg g^{-1} (Figure 10). This is in line with the creation of a full monolayer on the adsorbent surface. The Langmuir constant (b), which is related to the heat of adsorption, was found to be 0.119. The dimensionless separation factor (R_L), which is described below, can also be used to predict the affinity between the sorbate and the sorbent using Langmuir parameters (22):

$$RL = \frac{1}{1 + (b * \text{initial concentration})} \quad (22)$$

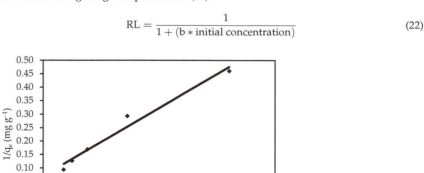

Figure 10. Langmuir isotherm for the sorption of IV2R.

Table 4. Parameters and error function of the isotherm models for IV2R removal by ZnO-NPs.

Isotherm Model	Isotherm Parameters	Value	EABS	X^2	APE (%)	Hybrid
Freundlich	$1/n$ n K_F (mg$^{1-1/n}$ L$^{1/n}$ g^{-1}) R^2	0.725 1.37 34.30 0.994	1.87	0.116	0.1173	0.343
Langmuir	Q_{max} (mg g^{-1}) b R_L R^2	119.05 0.119 0.597 0.974	36.21	71.84	7.934	312.359
Harkins-Jura	A_{HJ} B_{HJ} R^2	0.0047 0.80 0.905	762.50	754.6	3.95	3281.2
Halsey	n K_H R^2	1.379 131 0.994	740.50	711.75	3.845	3094.5
Henderson	$1/n_h$ K_h R^2	1.426 0.007 0.995	0.178	0.000	0.024	0.005
Tempkin	A_T B_T b_T R^2	68.03 3.72 255.5 0.928	0.104	0.000	0.014	0.002

According to the parameters in Table 3, the value of R_L can be used to determine whether a sorption system is "favourable" or "unfavourable." The sorption of IV2R onto ZnO-NPs has an R_L value of 0.597, which shows that adsorption of IV2R on ZnO-NPs was "favourable".

3.3.3. Harkins-Jura Isotherm

Multilayer adsorption is accounted for by the Harkins-Jura model, which can be explained by the existence of a heterogeneous pore distribution [41]. Equation (8) can be used to solve the Harkins-Jura adsorption isotherm, which can be seen in Figure 11 as a plot of $1/q_e$ vs. log C_e. The presence of a heterogeneous pore distribution and multilayer adsorption can be explained by the Harkins-Jura model. The isotherm constants are presented in Table 4, and the correlation coefficient was found to be R^2 = 0.905. This could mean that the Harkins-Jura model is useful for adsorption data.

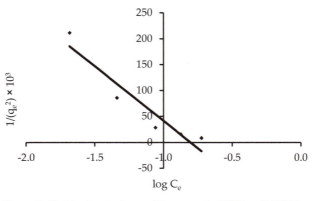

Figure 11. Harkins-Jura isotherm of the removal of IV2R on ZnO-NPs.

3.3.4. Isotherm Models of Halsey and Henderson

The Halsey and Henderson isotherm models are suitable for multilayer adsorption, and the fitting of the Halsey equation can be applied to heteroporous solids [73]. Figures 12 and 13 show plots of Ln q_e versus Ln C_e for Halsey and ln [ln(1C_e)] versus Ln q_e for Henderson isotherms, respectively. Table 4 summarises the isotherm constants and correlation coefficients. The correlation coefficient for Halsey was $R^2 = 0.994$, while Henderson's was $R^2 = 0.995$. The results of Halsey and Henderson suggest that both models can be used to predict IV2R adsorption on ZnO-NPs.

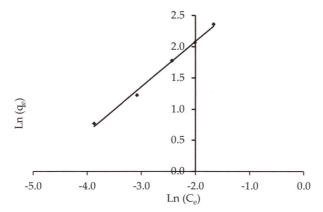

Figure 12. Halsey isotherm of the removal of IV2R by ZnO-NPs.

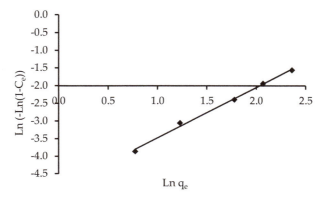

Figure 13. Linear Henderson isotherm of the removal of IV2R by ZnO-NPs.

3.3.5. Tempkin Isotherm

The Tempkin isotherm equation includes a component that describes the interactions between the adsorbing species and adsorbate [74]. It is assumed that due to adsorbate-adsorbate repulsions, the heat of adsorption of all molecules in the layer decreases linearly with coverage, and that adsorption is a uniform distribution of maximum binding energy [75]. The Tempkin adsorption isotherm model was used to evaluate the adsorption potentials of the ZnO-NPs for IV2R. The derivation of the Tempkin isotherm assumes that the fall in the heat of adsorption is linear rather than logarithmic, as implied in the Freundlich equation. The data indicate that the Tempkin isotherm model applies to the adsorption of IV2R dye onto ZnO-NPs, as shown by the high linear regression correlation coefficient ($R^2 = 0.928$), as presented in Figure 14, which indicates that the Tempkin isotherm is appropriate for the equilibrium data attained for the adsorption of IV2R on ZnO-NPs.

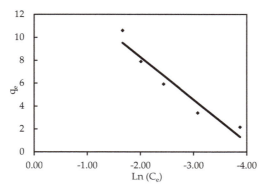

Figure 14. Tempkin isotherm of the removal of IV2R on ZnO-NP.

3.4. Examining Error Functions to Find the Most Appropriate Isotherm Model

Several error functions, such as the hybrid fractional error, average percentage error (APE) equation, chi-square error (X^2) equation, and the sum of absolute errors, were used to determine the best-fit model for the investigational data (EABS). Table 4 summarises the data gathered from the various error functions. For each isotherm model, the analysed error functions produced varying findings, and the comparison across isotherm models should focus on each error function independently [2]. Tempkin > Henderson > Freundlich > Langmuir > Halsey > Harkins-Jura are the most appropriate isotherm models based on the observed data. Nonetheless, the error function test offered variable data for all models, and the model evaluation focused on each error function individually.

3.5. Adsorption Kinetics

3.5.1. Model of Pseudo-First-Order Kinetics

The pseudo-first-order equation explains the adsorption rate based on the adsorption capacity. According to this model [76], the ratio of occupied to empty adsorption sites is proportional to the number of vacant sites. At a dye concentration of 10 mg L^{-1}, the linear figure of log (q_e−q_t) versus t is shown in Figure 15. The intercept and the slope of the linear plots for the removal of the IV2R dye from ZnO-NPs were used to compute the q_e and K_1 values. The values of K_1, the experimental and calculated values of q_e, and the correlation coefficients for the pseudo-first-order kinetic plots are provided in Table 5. The obtained experimental data do not agree with the theoretical values of q_e. This implies that the kinetic data and the pseudo-first-order model do not match well. Moreover, because the correlation coefficients (R^2) for the current experimental results for IV2R dye on ZnO-NPs were small, the pseudo-first-order equation was ruled out.

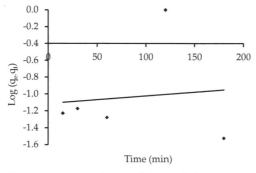

Figure 15. Adsorption kinetics of the pseudo-first-order kinetics of IV2R adsorption onto the ZnO-NPs.

Table 5. Parameters of the different adsorption kinetic models.

Kinetic Models	Parameters	Value
First-order	q_e (calc.) (mg g^{-1}) $k_1 \times 10^3$ (min^{-1}) R^2	12.94 2.07 0.011
Second-order	q_e (calc.) (mg g^{-1}) $k_2 \times 10^3$ (mg g^{-1} min^{-1}) R^2	1.06 818.54 0.999
Intraparticle diffusion	K_{dif} (mg g^{-1} min$^{-0.5}$) C cal (mg g^{-1}) R^2	0.0028 1.01 0.596

3.5.2. Model of Pseudo-Second-Order Kinetics

The rate-limiting step is assumed to be due to chemical adsorption containing valence forces through the exchange and/or sharing of electrons between dye ions and the adsorbent in the pseudo-second-order equation (adsorbent) [76]. As shown in Figure 16 and Table 5, plotting (t/q_t) against (t) should yield a linear correlation from which the data for parameters q_e and k_2 may be determined from the slope and intercept, respectively. The second-order kinetic model's correlation coefficient was more than 0.999, pointing to the fact that the pseudo-second-order kinetic model resulted in a good correlation for IV2R adsorption onto ZnO-NPs. Apart from that, it was clear that the k_2 parameter value was higher than the corresponding k_1 parameter value. This is because, according to David and Joseph [61], the adsorption rate is proportional to the square of the number of empty sites in the pseudo-second-order model.

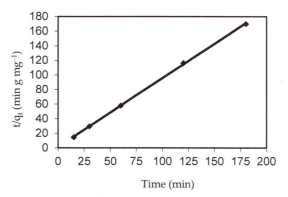

Figure 16. Adsorption kinetics of the pseudo-first-order kinetics of IV2R adsorption onto the ZnO-NPs.

3.5.3. The Intraparticle Diffusion Equation

The adsorption technique requires a number of steps, including the transport of solute particles from the aqueous part to the exterior of the solid molecules, followed by the diffusion of solute molecules into the cavities' interior decoration, which is likely to be a slow process and a rate-determining step [77]. The figures of q_t inverse $t^{0.5}$ may indicate a multilinear correlation, indicating that the adsorption process occurs in two or more stages (Figure 17 and Table 5). The slope directly estimates the rate constant K_{dif}, and the intercept is C, as shown in Table 4. Because the barrier to exterior mass transfer increases as the intercept increases, the value of the C factor provides information about the thickness of the border layer.

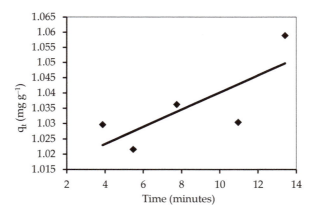

Figure 17. Adsorption kinetics of the intraparticle diffusion equation of IV2R adsorption onto ZnO-NPs.

3.6. Adsorption Thermodynamics

Table 6 shows the thermodynamic parameters of IV2R adsorption onto ZnO-NPs. The large negative value of ΔG° demonstrates that dye sorption was spontaneous and feasible. Nonetheless, the free energy values in Table 6 increase with increasing temperature, demonstrating that the adsorption method is endothermic [78,79] and suggesting that the adsorption method's spontaneity decreases at lower temperatures. For physisorption, the range of ΔG° values is between −20 and 0 kJ mol^{-1}, whereas, for chemisorption, the range is between −80 and −400 kJ mol^{-1} [3]. The activation energy measurements in this work indicate that IV2R sorption onto ZnO-NPs occurs via a physisorption process.

Table 6. Thermodynamic factors of the sorption of IV2R onto ZnO-NPs.

Temperature (°C)	ΔG° (kJ mol^{-1})	ΔH° (kJ mol^{-1})	ΔS° (kJ mol^{-1})
25	−12.98629244		
30	−12.8458117		
35	−12.81290783	47.92	−0.202
45	−20.94418521		
55	−16.5176115		

3.7. Application to Real-Life Wastewater

To test the validity of using ZnO-NPs as adsorbents, real wastewater was mixed with simulated dye samples to see if the adsorbent could remove IV2R under optimum conditions. The results revealed that changing the kind of water had a significant impact on dye removal, with deionised water having the least impact on the adsorption process, with 91.51% of dye removed at an acidic pH after 180 min. On the other hand, real wastewater contains very high quantities of interfering ions from a variety of contaminants, which had a major impact on the IV2R dye removal effectiveness, with 62.6% of dye removed after 180 min. Regardless, the results show that ZnO-NP adsorbents may be employed successfully to remove IV2R dye from aqueous mixes and wastewater at a reasonable cost.

3.8. Comparative Studies of ZnO-NP Sorption Capacity

To demonstrate the efficacy of ZnO-NPs, the results of this experiment were compared to those of other studies on the adsorption capabilities of various dyes. This adsorption capacity of ZnO-NPs for IV2R was also found to be significantly higher in comparison with some other recently reported adsorbents, as reported in Table 7.

Table 7. Summary of the elimination of several dyes from aquatic mixtures by various ZnO-NPs.

ZnO-NP.	Dye Adsorbed	q_e (mg g^{-1})	Ref.
ZnO-NPs-AC	Acid yellow 119	116.29	[80]
AC-ZnO	Acid orange 7	32.13	[81]
AC-ZnO	Methylene blue (MB)	32.22	[81]
ZnO-NR-AC	Bromophenol red	200	[82]
ZnO-NP-AC	Malachite green	322.58	[83]
ZnO	Malachite green	310.50	[84]
ZnO-NRs-AC		113.64	[85]
Chloroacetic Acid-Modified Ferula	Basic dye	354.89	[86]
ZnO-NP	IV2R	119.05	Present study

4. Conclusions

SEM, FTIR, UV, and BET surface analyses were used to investigate ZnO-NPs. The presence of functional groups such as N–H, O–H, CH_2, C–O, and Zn–O stretching was revealed by FTIR, which increased their ion-exchange capabilities for the selective adsorption of oppositely charged molecules. The total pore volume and specific surface area of the ZnO-NPs were 95.83 m^2 g^{-1} and 0.058 cm^3 g^{-1}, respectively. The specific surface area calculated using the Langmuir technique was 140.692 m^2 g^{-1}. Likewise, ZnO-NPs are agglomerated nanoparticles with a size of 1.22 nm, similar to nanoflowers. The optimal operational parameters were found to be 10 mg L^{-1} IV2R with 0.04 g of ZnO-NPs and 60 min contact time at pH 2 and 45 °C. Moreover, the Halsey, Freundlich, Langmuir, Harkins-Jura, Henderson, and Tempkin models, on the other hand, were used to investigate the equilibrium isotherm adsorption results. The Langmuir model produced a greater adsorption capacity (q_{max}) of 119.05 mg g^{-1}. Different error function models were used to find the best-fitting isotherm model in this study. In addition, the intraparticle diffusion, pseudo-first order, and pseudo-second-order models were used in a kinetic investigation. The pseudo-second-order model effectively represented experimental data pertaining to the system under study with $R^2 = 0.999$. Furthermore, the percentage of dye removal by ZnO-NPs from real wastewater was 65 %. Additionally, the thermodynamic parameters ($\Delta G°$, $\Delta H°$, and $\Delta S°$) of the sorption processes were estimated. IV2R dye adsorption was endothermic and spontaneous, and the adsorption reaction was a physisorption reaction. This technique takes advantage of the ability to remove a large quantity of dye in a short amount of time with a small amount of adsorbent.

Funding: This work was supported by the Deanship of Scientific Research, Vice Presidency for Graduate Studies and Scientific Research, King Faisal University, Saudi Arabia (Grant No. 884).

Institutional Review Board Statement: Not applicable.

Informed Consent Statement: Not applicable.

Data Availability Statement: The data are available from the corresponding author upon reasonable request.

Acknowledgments: The author acknowledges the Deanship of Scientific Research, Vice Presidency for Graduate Studies and Scientific Research, King Faisal University, Saudi Arabia (Grant No. 884).

Conflicts of Interest: The author declares no conflict of interest.

References

1. Yogendra, K.; Mahadevan, K.; Naik, S.; Madhusudhana, N. Photocatalytic activity of synthetic ZnO composite against Coralene red F3BS dye in presence of solar light. *Int. J. Environ. Sci.* **2011**, *1*, 839–846.
2. Abualnaja, K.M.; Alprol, A.E.; Ashour, M.; Mansour, A.T. Influencing Multi-Walled Carbon Nanotubes for the Removal of Ismate Violet 2R Dye from Wastewater: Isotherm, Kinetics, and Thermodynamic Studies. *Appl. Sci.* **2021**, *11*, 4786. [CrossRef]
3. Abualnaja, K.M.; Alprol, A.E.; Abu-Saied, M.A.; Ashour, M.; Mansour, A.T. Removing of Anionic Dye from Aqueous Solutions by Adsorption Using of Multiwalled Carbon Nanotubes and Poly (Acrylonitrile-styrene) Impregnated with Activated Carbon. *Sustainability* **2021**, *13*, 7077. [CrossRef]
4. Abualnaja, K.M.; Alprol, A.E.; Abu-Saied, M.A.; Mansour, A.T.; Ashour, M. Studying the Adsorptive Behavior of Poly(Acrylonitrile-co-Styrene) and Carbon Nanotubes (Nanocomposites) Impregnated with Adsorbent Materials towards Methyl Orange Dye. *Nanomaterials* **2021**, *11*, 1144. [CrossRef] [PubMed]
5. Aksu, Z. Application of biosorption for the removal of organic pollutants: A review. *Process Biochem.* **2005**, *40*, 997–1026. [CrossRef]
6. Guivarch, E.; Trevin, S.; Lahitte, C.; Oturan, M.A. Degradation of azo dyes in water by electro-Fenton process. *Environ. Chem. Lett.* **2003**, *1*, 38–44. [CrossRef]
7. Alprol, A.E.; Ashour, M.; Mansour, A.T.; Alzahrani, O.M.; Mahmoud, S.F.; Gharib, S.M. Assessment of Water Quality and Phytoplankton Structure of Eight Alexandria Beaches, Southeastern Mediterranean Sea, Egypt. *J. Mar. Sci. Eng.* **2021**, *9*, 1328. [CrossRef]
8. Alprol, A.E.; Heneash, A.M.M.; Soliman, A.M.; Ashour, M.; Alsanie, W.F.; Gaber, A.; Mansour, A.T. Assessment of Water Quality, Eutrophication, and Zooplankton Community in Lake Burullus, Egypt. *Diversity* **2021**, *13*, 268. [CrossRef]
9. Bayramoğlu, G.; Arıca, M.Y. Biosorption of benzidine based textile dyes "Direct Blue 1 and Direct Red 128" using native and heat-treated biomass of Trametes versicolor. *J. Hazard. Mater.* **2007**, *143*, 135–143. [CrossRef]
10. Harper, T.R.; Kingham, N.W. Removal of arsenic from wastewater using chemical precipitation methods. *Water Environ. Res.* **1992**, *64*, 200–203. [CrossRef]
11. Song, Z.; Williams, C.; Edyvean, R. Treatment of tannery wastewater by chemical coagulation. *Desalination* **2004**, *164*, 249–259. [CrossRef]
12. Semerjian, L.; Ayoub, G. High-pH–magnesium coagulation–flocculation in wastewater treatment. *Adv. Environ. Res.* **2003**, *7*, 389–403. [CrossRef]
13. Schwermer, C.U.; Krzeminski, P.; Wennberg, A.C.; Vogelsang, C.; Uhl, W. Removal of antibiotic resistant *E. coli* in two Norwegian wastewater treatment plants and by nano-and ultra-filtration processes. *Water Sci. Technol.* **2018**, *77*, 1115–1126. [CrossRef]
14. Deng, Y.; Zhao, R. Advanced oxidation processes (AOPs) in wastewater treatment. *Curr. Pollut. Rep.* **2015**, *1*, 167–176. [CrossRef]
15. Golder, A.K.; Chanda, A.K.; Samanta, A.N.; Ray, S. Removal of hexavalent chromium by electrochemical reduction–precipitation: Investigation of process performance and reaction stoichiometry. *Sep. Purif. Technol.* **2011**, *76*, 345–350. [CrossRef]
16. Guida, S.; Van Peteghem, L.; Luqmani, B.; Sakarika, M.; McLeod, A.; McAdam, E.J.; Jefferson, B.; Rabaey, K.; Soares, A. Ammonia recovery from brines originating from a municipal wastewater ion exchange process and valorization of recovered nitrogen into microbial protein. *Chem. Eng. J.* **2022**, *427*, 130896. [CrossRef]
17. Abo-Taleb, H.A.; Ashour, M.; Elokaby, M.A.; Mabrouk, M.M.; El-feky, M.M.M.; Abdelzaher, O.F.; Gaber, A.; Alsanie, W.F.; Mansour, A.T. Effect of a New Feed Daphniamagna (Straus, 1820), as a Fish Meal Substitute on Growth, Feed Utilization, Histological Status, and Economic Revenue of Grey Mullet, Mugil cephalus (Linnaeus 1758). *Sustainability* **2021**, *13*, 7093. [CrossRef]
18. Mansour, A.T.; Ashour, M.; Alprol, A.E.; Alsaqufi, A.S. Aquatic Plants and Aquatic Animals in the Context of Sustainability: Cultivation Techniques, Integration, and Blue Revolution. *Sustainability* **2022**, *14*, 3257. [CrossRef]
19. Zaki, M.A.; Ashour, M.; Heneash, A.M.M.; Mabrouk, M.M.; Alprol, A.E.; Khairy, H.M.; Nour, A.M.; Mansour, A.T.; Hassanien, H.A.; Gaber, A.; et al. Potential Applications of Native Cyanobacterium Isolate (*Arthrospira platensis* NIOF17/003) for Biodiesel Production and Utilization of Its Byproduct in Marine Rotifer (*Brachionus plicatilis*) Production. *Sustainability* **2021**, *13*, 1769. [CrossRef]
20. Mansour, A.T.; Alprol, A.E.; Ashour, M.; Ramadan, K.M.; Alhajji, A.H.; Abualnaja, K.M. Do Red Seaweed Nanoparticles Enhance Bioremediation Capacity of Toxic Dyes from Aqueous Solution? *Gels* **2022**, *8*, 310. [CrossRef]
21. Ashour, M.; Alprol, A.E.; Heneash, A.M.M.; Saleh, H.; Abualnaja, K.M.; Alhashmialameer, D.; Mansour, A.T. Ammonia Bioremediation from Aquaculture Wastewater Effluents Using *Arthrospira platensis* NIOF17/003: Impact of Biodiesel Residue and Potential of Ammonia-Loaded Biomass as Rotifer Feed. *Materials* **2021**, *14*, 5460. [CrossRef]
22. Mansour, A.T.; Alprol, A.E.; Abualnaja, K.M.; El-Beltagi, H.S.; Ramadan, K.M.A.; Ashour, M. Dried Brown Seaweed's Phytoremediation Potential for Methylene Blue Dye Removal from Aquatic Environments. *Polymers* **2022**, *14*, 1375. [CrossRef]
23. Mansour, A.T.; Alprol, A.E.; Abualnaja, K.M.; El-Beltagi, H.S.; Ramadan, K.M.A.; Ashour, M. The Using of Nanoparticles of Microalgae in Remediation of Toxic Dye from Industrial Wastewater: Kinetic and Isotherm Studies. *Materials* **2022**, *15*, 3922. [CrossRef]
24. Ghiloufi, I.; El Ghoul, J.; Modwi, A.; El Mir, L. Ga-doped ZnO for adsorption of heavy metals from aqueous solution. *Mater. Sci. Semicond. Process.* **2016**, *42*, 102–106. [CrossRef]

25. Agrawal, A.; Sahu, K. Kinetic and isotherm studies of cadmium adsorption on manganese nodule residue. *J. Hazard. Mater.* **2006**, *137*, 915–924. [CrossRef]
26. Zeng, H.; Cai, W.; Liu, P.; Xu, X.; Zhou, H.; Klingshirn, C.; Kalt, H. ZnO-based hollow nanoparticles by selective etching: Elimination and reconstruction of metal− semiconductor interface, improvement of blue emission and photocatalysis. *ACS Nano* **2008**, *2*, 1661–1670. [CrossRef]
27. Jing, Z.; Zhan, J. Fabrication and gas-sensing properties of porous ZnO nanoplates. *Adv. Mater.* **2008**, *20*, 4547–4551. [CrossRef]
28. Chou, T.P.; Zhang, Q.; Fryxell, G.E.; Cao, G. Hierarchically Structured ZnO Film for Dye-Sensitized Solar Cells with Enhanced Energy Conversion Efficiency. *Adv. Mater.* **2007**, *19*, 2588–2592. [CrossRef]
29. Azizi, S.; Mohamad, R.; Bahadoran, A.; Bayat, S.; Rahim, R.A.; Ariff, A.; Saad, W.Z. Effect of annealing temperature on antimicrobial and structural properties of bio-synthesized zinc oxide nanoparticles using flower extract of Anchusa italica. *J. Photochem. Photobiol. B Biol.* **2016**, *161*, 441–449. [CrossRef]
30. Azizi, S.; Ahmad, M.B.; Namvar, F.; Mohamad, R. Green biosynthesis and characterization of zinc oxide nanoparticles using brown marine macroalga Sargassum muticum aqueous extract. *Mater. Lett.* **2014**, *116*, 275–277. [CrossRef]
31. Naseer, M.; Aslam, U.; Khalid, B.; Chen, B. Green route to synthesize Zinc Oxide Nanoparticles using leaf extracts of Cassia fistula and Melia azadarach and their antibacterial potential. *Sci. Rep.* **2020**, *10*, 1–10. [CrossRef] [PubMed]
32. Elia, P.; Zach, R.; Hazan, S.; Kolusheva, S.; Porat, Z.e.; Zeiri, Y. Green synthesis of gold nanoparticles using plant extracts as reducing agents. *Int. J. Nanomed.* **2014**, *9*, 4007.
33. Yu, J.-G.; Yu, L.-Y.; Yang, H.; Liu, Q.; Chen, X.-H.; Jiang, X.-Y.; Chen, X.-Q.; Jiao, F.-P. Graphene nanosheets as novel adsorbents in adsorption, preconcentration and removal of gases, organic compounds and metal ions. *Sci. Total Environ.* **2015**, *502*, 70–79. [CrossRef] [PubMed]
34. Hassan, K.H.; Khammas, Z.A.; Rahman, A.M. Zinc oxide hydrogen sulfide removal catalyst/preparation, activity test and kinetic study. *Al-Khwarizmi Eng. J.* **2008**, *4*, 74–84.
35. Wang, X.; Cai, W.; Lin, Y.; Wang, G.; Liang, C. Mass production of micro/nanostructured porous ZnO plates and their strong structurally enhanced and selective adsorption performance for environmental remediation. *J. Mater. Chem.* **2010**, *20*, 8582–8590. [CrossRef]
36. Abd El-Mohdy, H.; Mostafa, T.B. Synthesis of Polyvinyl Alcohol/Maleic Acid Hydrogels by Electron Beam Irradiation for Dye Uptake. *J. Macromol. Sci. Part A* **2013**, *50*, 6–17. [CrossRef]
37. Dada, A.; Olalekan, A.; Olatunya, A.; Dada, O. Langmuir, Freundlich, Temkin and Dubinin–Radushkevich isotherms studies of equilibrium sorption of Zn2+ unto phosphoric acid modified rice husk. *IOSR J. Appl. Chem.* **2012**, *3*, 38–45.
38. Freundlich, H. *Über die Adsorption in Lösungen. Habilitationsschrift durch Welche... zu Haltenden Probevorlesung" Kapillarchemie und Physiologie" einladet Dr. Herbert Freundlich*; W. Engelmann: Berlin, German, 1906.
39. Langmuir, I. The constitution and fundamental properties of solids and liquids. Part I. Solids. *J. Am. Chem. Soc.* **1916**, *38*, 2221–2295.
40. Halsey Jr, G. The Rate of Adsorption on a Nonuniform Surface. *J. Phys. Chem.* **1951**, *55*, 21–26. [CrossRef]
41. Harkins, W.D.; Jura, G. An adsorption method for the determination of the area of a solid without the assumption of a molecular area, and the area occupied by nitrogen molecules on the surfaces of solids. *J. Chem. Phys.* **1943**, *11*, 431–432. [CrossRef]
42. Pandimurugan, R.; Thambidurai, S. Novel seaweed capped ZnO nanoparticles for effective dye photodegradation and antibacterial activity. *Adv. Powder Technol.* **2016**, *27*, 1062–1072. [CrossRef]
43. Abul, A.; Samad, S.; Huq, D.; Moniruzzaman, M.; Masum, M. Textile dye removal from wastewater effluents using chitosan-ZnO nanocomposite. *J. Text. Sci. Eng.* **2015**, *5*, 3–7.
44. Jabs, A. Determination of Secondary Structure in Proteins by Fourier Transform Infrared Spectroscopy (FTIR). Jena Library of Biologica Macromolecules. 2005. Available online: http://www.imb-jena.de/ImgLibDoc/ftir/IMAGEpFTIR.html (accessed on 30 December 2020).
45. Figueira, M.; Volesky, B.; Mathieu, H. Instrumental analysis study of iron species biosorption by Sargassum biomass. *Environ. Sci. Technol.* **1999**, *33*, 1840–1846. [CrossRef]
46. Sharma, D.; Sabela, M.I.; Kanchi, S.; Bisetty, K.; Skelton, A.A.; Honarparvar, B. Green synthesis, characterization and electrochemical sensing of silymarin by ZnO nanoparticles: Experimental and DFT studies. *J. Electroanal. Chem.* **2018**, *808*, 160–172.
47. Samaele, N.; Amornpitoksuk, P.; Suwanboon, S. Effect of pH on the morphology and optical properties of modified ZnO particles by SDS via a precipitation method. *Powder Technol.* **2010**, *203*, 243–247. [CrossRef]
48. Sagar, P.; Shishodia, P.; Mehra, R. Influence of pH value on the quality of sol–gel derived ZnO films. *Appl. Surf. Sci.* **2007**, *253*, 5419–5424. [CrossRef]
49. Belda, I.; Ruiz, J.; Esteban-Fernández, A.; Navascués, E.; Marquina, D.; Santos, A.; Moreno-Arribas, M.V. Microbial contribution to wine aroma and its intended use for wine quality improvement. *Molecules* **2017**, *22*, 189. [CrossRef] [PubMed]
50. Boukoussa, B.; Hamacha, R.; Morsli, A.; Bengueddach, A. Adsorption of yellow dye on calcined or uncalcined Al-MCM-41 mesoporous materials. *Arab. J. Chem.* **2017**, *10*, S2160–S2169. [CrossRef]
51. Netpradit, S.; Thiravetyan, P.; Towprayoon, S. Application of 'waste'metal hydroxide sludge for adsorption of azo reactive dyes. *Water Res.* **2003**, *37*, 763–772. [CrossRef]
52. O'mahony, T.; Guibal, E.; Tobin, J. Reactive dye biosorption by Rhizopus arrhizus biomass. *Enzym. Microb. Technol.* **2002**, *31*, 456–463. [CrossRef]

53. Blaga, A.C.; Tanasă, A.M.; Cimpoesu, R.; Tataru-Farmus, R.-E.; Suteu, D. Biosorbents Based on Biopolymers from Natural Sources and Food Waste to Retain the Methylene Blue Dye from the Aqueous Medium. *Polymers* **2022**, *14*, 2728. [CrossRef]
54. Karaca, S.; Gürses, A.; Açıkyıldız, M.; Ejder, M. Adsorption of cationic dye from aqueous solutions by activated carbon. *Microporous Mesoporous Mater.* **2008**, *115*, 376–382. [CrossRef]
55. Ghoneim, M.M.; El-Desoky, H.S.; El-Moselhy, K.M.; Amer, A.; Abou El-Naga, E.H.; Mohamedein, L.I.; Al-Prol, A.E. Removal of cadmium from aqueous solution using marine green algae, Ulva lactuca. *Egypt. J. Aquat. Res.* **2014**, *40*, 235–242. [CrossRef]
56. Malik, R.; Ramteke, D.; Wate, S. Adsorption of malachite green on groundnut shell waste based powdered activated carbon. *Waste Manag.* **2007**, *27*, 1129–1138. [CrossRef]
57. Reddy, S.S. The removal of composite reactive dye from dyeing unit effluent using sewage sludge derived activated carbon. *Turk. J. Eng. Environ. Sci.* **2006**, *30*, 367–373.
58. Iqbal, M.; Saeed, A. Biosorption of reactive dye by loofa sponge-immobilized fungal biomass of Phanerochaete chrysosporium. *Process Biochem.* **2007**, *42*, 1160–1164. [CrossRef]
59. Zafar, S.I.; Bisma, M.; Saeed, A.; Iqbal, M. FTIR spectrophotometry, kinetics and adsorption isotherms modelling, and SEM-EDX analysis for describing mechanism of biosorption of the cationic basic dye Methylene blue by a new biosorbent (Sawdust of Silver Fir; Abies Pindrow). *Fresenius Environ. Bull.* **2008**, *17*, 2109–2121.
60. Ananta, S.; Saumen, B.; Vijay, V. Adsorption isotherm, thermodynamic and kinetic study of arsenic (III) on iron oxide coated granular activated charcoal. *Int. Res. J. Environ. Sci.* **2015**, *4*, 64–77.
61. David, A.; Joseph, L. The Effect of Ph and Biomass Concentration on Lead (Pb) Adsorption by Aspergillus Niger from Simulated Waste Water. Bachelor's Thesis, University Malaysia Pahang, Pahang, Malaysia, 2008.
62. Kumar, P.S.; Gayathri, R. Adsorption of Pb2+ ions from aqueous solutions onto bael tree leaf powder: Isotherms, kinetics and thermodynamics study. *J. Eng. Sci. Technol.* **2009**, *4*, 381–399.
63. Bao, C.; Chen, M.; Jin, X.; Hu, D.; Huang, Q. Efficient and stable photocatalytic reduction of aqueous hexavalent chromium ions by polyaniline surface-hybridized ZnO nanosheets. *J. Mol. Liq.* **2019**, *279*, 133–145. [CrossRef]
64. Daija, L.; Selberg, A.; Rikmann, E.; Zekker, I.; Tenno, T.; Tenno, T. The influence of lower temperature, influent fluctuations and long retention time on the performance of an upflow mode laboratory-scale septic tank. *Desalination Water Treat.* **2016**, *57*, 18679–18687. [CrossRef]
65. Putra, W.P.; Kamari, A.; Yusoff, S.N.M.; Ishak, C.F.; Mohamed, A.; Hashim, N.; Isa, I.M. Biosorption of Cu (II), Pb (II) and Zn (II) ions from aqueous solutions using selected waste materials: Adsorption and characterisation studies. *J. Encapsulation Adsorpt. Sci.* **2014**, *4*, 43532.
66. Malakootian, M.; Toolabi, A.; Moussavi, S.G.; Ahmadian, M. Equilibrium and kinetic modeling of heavy metals biosorption from three different Real industrial wastewaters onto Ulothrix Zonata algae. *Aust. J. Basic Appl. Sci.* **2011**, *5*, 1030–1037.
67. Sivaprakasha, S.; Kumarb, P.S.; Krishnac, S. Adsorption study of various dyes on Activated Carbon Fe3O4 Magnetic Nano Composite. *Int. J. Appl. Chem.* **2017**, *13*, e266.
68. Tang, C.; Huang, X.; Wang, H.; Shi, H.; Zhao, G. Mechanism investigation on the enhanced photocatalytic oxidation of nonylphenol on hydrophobic TiO2 nanotubes. *J. Hazard. Mater.* **2020**, *382*, 121017. [CrossRef] [PubMed]
69. Acharya, J.; Sahu, J.; Mohanty, C.; Meikap, B. Removal of lead (II) from wastewater by activated carbon developed from Tamarind wood by zinc chloride activation. *Chem. Eng. J.* **2009**, *149*, 249–262. [CrossRef]
70. Sun, D.; Zhang, Z.; Wang, M.; Wu, Y. Adsorption of reactive dyes on activated carbon developed from Enteromorpha prolifera. *Am. J. Anal. Chem.* **2013**, *4*, 33867.
71. Saeed, A.; Sharif, M.; Iqbal, M. Application potential of grapefruit peel as dye sorbent: Kinetics, equilibrium and mechanism of crystal violet adsorption. *J. Hazard. Mater.* **2010**, *179*, 564–572. [CrossRef]
72. Al Prol, A. Study of environmental concerns of dyes and recent textile effluents treatment technology: A review. *Asian J. Fish. Aquat. Res.* **2019**, *3*, 1–18. [CrossRef]
73. Halsey, G. Physical adsorption on non-uniform surfaces. *J. Chem. Phys.* **1948**, *16*, 931–937. [CrossRef]
74. Hossain, M.; Ngo, H.; Guo, W.; Nguyen, T. Palm oil fruit shells as biosorbent for copper removal from water and wastewater: Experiments and sorption models. *Bioresour. Technol.* **2012**, *113*, 97–101. [CrossRef]
75. Kavitha, D.; Namasivayam, C. Experimental and kinetic studies on methylene blue adsorption by coir pith carbon. *Bioresour. Technol.* **2007**, *98*, 14–21. [CrossRef] [PubMed]
76. Kooh, M.R.R.; Dahri, M.K.; Lim, L.B. The removal of rhodamine B dye from aqueous solution using Casuarina equisetifolia needles as adsorbent. *Cogent Environ. Sci.* **2016**, *2*, 1140553. [CrossRef]
77. E Al Prol, A.; EA El-Metwally, M.; Amer, A. *Sargassum latifolium* as eco-friendly materials for treatment of toxic nickel (II) and lead (II) ions from aqueous solution. *Egypt. J. Aquat. Biol. Fish.* **2019**, *23*, 285–299. [CrossRef]
78. Babarinde, N.A.; Oyesiku, O.; Dairo, O.F. Isotherm and thermodynamic studies of the biosorption of copper (II) ions by Erythrodontium barteri. *Int. J. Phys. Sci.* **2007**, *2*, 300–304.
79. Potgieter, J.; Pearson, S.; Pardesi, C. Kinetic and thermodynamic parameters for the adsorption of methylene blue using fly ash under batch, column, and heap leaching configurations. *Coal Combust. Gasif. Prod.* **2018**, *10*, 23–33.
80. Jamshidi, H.; Ghaedi, M.; Sabzehmeidani, M.M.; Bagheri, A.R. Comparative study of acid yellow 119 adsorption onto activated carbon prepared from lemon wood and ZnO nanoparticles loaded on activated carbon. *Appl. Organomet. Chem.* **2018**, *32*, e4080. [CrossRef]

81. Nourmoradi, H.; Ghiasvand, A.; Noorimotlagh, Z. Removal of methylene blue and acid orange 7 from aqueous solutions by activated carbon coated with zinc oxide (ZnO) nanoparticles: Equilibrium, kinetic, and thermodynamic study. *Desalination Water Treat.* **2015**, *55*, 252–262. [CrossRef]
82. Ghaedi, M.; Ghayedi, M.; Kokhdan, S.N.; Sahraei, R.; Daneshfar, A. Palladium, silver, and zinc oxide nanoparticles loaded on activated carbon as adsorbent for removal of bromophenol red from aqueous solution. *J. Ind. Eng. Chemistry* **2013**, *19*, 1209–1217. [CrossRef]
83. Ghaedi, M.; Mosallanejad, N. Study of competitive adsorption of malachite green and sunset yellow dyes on cadmium hydroxide nanowires loaded on activated carbon. *J. Ind. Eng. Chem.* **2014**, *20*, 1085–1096. [CrossRef]
84. Lammi, S.; Barakat, A.; Mayer-Laigle, C.; Djenane, D.; Gontard, N.; Angellier-Coussy, H. Dry fractionation of olive pomace as a sustainable process to produce fillers for biocomposites. *Powder Technol.* **2018**, *326*, 44–53. [CrossRef]
85. Sarabadan, M.; Bashiri, H.; Mousavi, S.M. Removal of crystal violet dye by an efficient and low cost adsorbent: Modeling, kinetic, equilibrium and thermodynamic studies. *Korean J. Chem. Eng.* **2019**, *36*, 1575–1586. [CrossRef]
86. Salih, S.J. *Removal of Basic Dyes from Aqueous Solution by Chloroacetic Acid Modified Ferula Communis Based Adsorbent: Thermodynamic and Kinetic Studies*; Eastern Mediterranean University (EMU)-Doğu Akdeniz Üniversitesi (DAÜ): Gazimağusa, Cyprus, 2014.

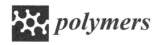

Article

Effect of TiO₂ Nanoparticles on Capillary-Driven Flow in Water Nanofilters Based on Chitosan Cellulose and Polyvinylidene Fluoride Nanocomposites: A Theoretical Study

Noureddine Mahdhi [1,*], Norah Salem Alsaiari [2], Abdelfattah Amari [3,4,*] and Mohamed Ali Chakhoum [5]

[1] Laboratory Materials Organizations and Properties, Tunis El Manar University, Tunis 2092, Tunisia
[2] Department of Chemistry, College of Science, Princess Nourah Bint Abdulrahman University, Riyadh 11671, Saudi Arabia; nsalsaiari@pnu.edu.sa
[3] Department of Chemical Engineering, College of Engineering, King Khalid University, Abha 61411, Saudi Arabia
[4] Research Laboratory of Processes, Energetics, Environment and Electrical Systems, National School of Engineers, Gabes University, Gabes 6072, Tunisia
[5] Laboratoire des Sciences de la Matière Condensée (LSMC), Université Oran 1 Ahmed Ben Bella, Oran 31100, Algeria; chakhoum.mohammed@edu.univ-oran1.dz
* Correspondence: noureddine.maadhi@fst.rnu.tn (N.M.); abdelfattah.amari@enig.rnu.tn (A.A.)

Citation: Mahdhi, N.; Alsaiari, N.S.; Amari, A.; Chakhoum, M.A. Effect of TiO₂ Nanoparticles on Capillary-Driven Flow in Water Nanofilters Based on Chitosan Cellulose and Polyvinylidene Fluoride Nanocomposites: A Theoretical Study. *Polymers* **2022**, *14*, 2908. https://doi.org/10.3390/polym14142908

Academic Editors: Irene S. Fahim, Ahmed K. Badawi and Hossam E. Emam

Received: 8 June 2022
Accepted: 12 July 2022
Published: 17 July 2022

Publisher's Note: MDPI stays neutral with regard to jurisdictional claims in published maps and institutional affiliations.

Copyright: © 2022 by the authors. Licensee MDPI, Basel, Switzerland. This article is an open access article distributed under the terms and conditions of the Creative Commons Attribution (CC BY) license (https://creativecommons.org/licenses/by/4.0/).

Abstract: In this study, a novel concept of nanofiltration process of drinking water based on capillary-driven nanofiltration is demonstrated using a bio-based nanocomposites' nanofilter as free power: a green and sustainable solution. Based on Lifshitz and Young–Laplace theories, we show that the chitosan (CS), cellulose acetate (CLA), and Polyvinylidene fluoride (PVDF) polymer matrixes demonstrate hydrophobic behavior, which leads to the draining of water from nanopores when negative capillary pressure is applied and consequently prevents the capillary-driven nanofiltration process. By incorporating 10%, 20%, and 30% volume fraction of titanium dioxide (TiO₂) nanoparticles (NPs) to the polymers' matrixes, we demonstrate a wetting conversion from hydrophobic to hydrophilic behavior of these polymer nanocomposites. Subsequently, the threshold volume fraction of the TiO₂ NPs for the conversion from draining (hydrophobic) to filling (hydrophilic) by capillary pressure were found to be equal to 5.1%, 10.9%, and 13.9%, respectively, for CS/TiO₂, CLA/TiO₂, and PVDF/TiO₂ nanocomposites. Then, we demonstrated the negligible effect of the gravity force on capillary rise as well as the capillary-driven flow for nanoscale pore size. For nanofilters with the same effective nanopore radius, porosity, pore shape factor, and tortuosity, results from the modified Lucas–Washburn model show that the capillary rise as well as the capillary-driven water volume increase with increased volume fraction of the TiO₂ NPs for all nanocomposite nanofilter. Interestingly, the capillary-driven water volume was in range (5.26–6.39) L/h·m² with 30% volume fraction of TiO₂ NPs, which support our idea for capillary-driven nanofiltration as zero energy consumption nano-filtration process. Correspondingly, the biodegradable CS/TiO₂ and CLA/TiO₂ nanocomposites nanofilter demonstrate capillary-driven water volume higher, ~1.5 and ~1.2 times, respectively, more than the synthetic PVDF/TiO₂ nanocomposite.

Keywords: nanofilter; nanocomposite; water; purification; capillary-driven flow; biocompatible; biodegradable

1. Introduction

The occurrence of micro-sized particle wastes such as heavy metals, microplastics, and pathogens in drinking water presents great risks for human health and biodiversity. Recently, Darren et al. performed water chemistry analysis and size fractionation sampling of drinking water at four houses in the city of Newark, New Jersey [1]. They found the existence of pyromorphite (Pb₅(PO4)₃Cl) microparticles with size < 100 nm in drinking water samples that passed through the tap or pitcher filtration units. Barbara et al. investigated 32 water bottle samples from 21 different brands of mineral water from Bavarian food stores [2]. They detected variable amounts of microsized plastics and pigments in both reusable and glass bottles. They ascribe their occurrence and possible contamination to various sources, such as from the washing machinery, the bottle cap, and other steps during the filling process. Additionally, other sources of drinking water (e.g., surface water, groundwater) are likely to be polluted by microparticle wastes such as heavy metals, pathogens, and pesticides [3–5].

Regarding this state-of-the-art, a supplementary domestic filtration process is an adequate solution for purification of the tap and bottled drinking water. Lately, a considerable literature has developed around the theme of the purification of drinking water from the nano- and micro-sized particles using reverse osmosis (RO), nanofiltration (NF), chemical coagulation, adsorption, and magnetic nanoparticles processes [6–12]. Typically, the Nanofiltration (NF) process is the suitable domestic purification process for tap and bottled drinking water. In fact, NF has greater flux and less energy use rates, operational under normal conditions, and its unique advantages of retaining at an optimum the essential multivalent and all monovalent minerals ions required for the human body [6,7,13,14]. However, the main common disadvantages of the previous processes are their energy consumption, removing all minerals essential ions required for human body (RO), and more difficult installation as a domestic purification solution. Moreover, the NF process commonly introduces supplementary pressure to drive water inside the nanopores and to increase the permeation rate of the nanofilter membranes, which leads to an increase in their cost-effectiveness [15,16].

Very recently, emerging studies focused on exploring the so-called capillary-driven nanofiltration process as a natural, free energy, and sustainable process for water purification [17–20]. This ubiquitous nanofiltration natural process allows water transport throughout nanoporous materials only using capillary pressure and without the help of external forces [19,20]. In fact, at the nanoscale, the flow of the water inside the nanoporous structure, known as imbibition, is driven by the capillary pressure. For hydrophobic materials, the capillary pressure tends to pull water from the nanopores. The reason for that is the nanofilter based on hydrophobic materials requires supplementary pressure driven to overcome the capillary pressure and to drive water inside the nanopores [21]. This fact may cause membrane destruction and then increase the cost-effectiveness of the NF process and loss in purification efficiency of the nanofilter membranes [15,16]. For hydrophilic materials, the water flows spontaneously throughout the nanostructure due to the capillary pressure and then generates a natural flow through the nanofilter without any additional pressure [17–20].

For instance, there are poor attempts to implant the capillary-driven nanofiltration using bio-based materials as a free energy, biocompatible, and sustainable solution for the purification of drinking water.

Therefore, the purpose of this study is to demonstrate how to develop a capillary-driven NF on bio-based porous nanocomposites as a newly sustainable domestic purification process of tap and bottled drinking water with zero energy consumption. To do so, it is essential to study quantitatively and qualitatively the effect nanofilter materials composition have on their wettability as well as their water permeability in terms of capillary-driven flow.

Actually, an emerging trend is underway to develop novel synthetic and biodegradable nanocomposite polymers for use in the NF process [22–26]. They suggest the incorporation of NPs-based metal oxides materials for increasing water permeability, mechanical strength, separation efficiency, and reducing fouling of the membrane [27–29].

In this paper, we aim to give a detailed theoretical investigation of the effect of NPs filling rates on the physical properties that involves the spontaneous capillary-driven NF: the surface energy, wettability, capillary pressure, and capillary-driven water volume for some biodegradable and synthetic nanocomposites commonly used as nanofilters for drinking water.

Titanium dioxide (TiO_2) NPs was chosen as filler for the nanocomposites because of their prominent properties such as abundance in nature, biocompatibility, low-cost preparation with different sized and shaped particles (including nanowires, nanotubes, nanofibers, core-shell structures, and hollow nanostructures), and antiseptic and antibacterial properties [30]. Chitosan (CS) and cellulose acetate (CLA) were chosen as bio-based and biodegradable polymer matrix, whereas polyvinylidene fluoride (PVDF) was chosen as synthetic non-biodegradable polymer [31–33].

The following part of this study is dedicated to the presentation of the theoretical methods used for the evaluation of the surface energy, water contact angle, capillary pressure, and the capillary-driven water volume as function of the TiO_2 NPs volume fraction. In the next section, we demonstrate that the pure polymer matrix exhibits a hydrophobic behavior, which leads to negative capillary pressure around the nanopores, that acts as draining force of water from the nanopores of the nanofilter and, therefore, prevents the capillary-driven NF process. Afterwards, the threshold filler rate of TiO_2 NPs required to activate capillary-driven flow is determined for each polymer's nanocomposite. Later, the enhancement of water flow throughout the nanocomposite's nanofilters in terms of capillary rise and capillary-driven water volume will be demonstrated by increasing the volume fraction of the TiO_2 NPs. At the end, using a recapitulative comparison study, we demonstrate that the nanofilter based on biodegradable CS/TiO_2 and CLA/TiO_2 nanocomposites provides better capillary-driven NF performance than the synthetic nanocomposites $PVDF/TiO_2$ one.

2. Materials and Methods
2.1. Capillary Nanofiltration
2.1.1. Nanofiltration and Capillary Pressure

NF is purification process based on the flow of water through nanopores/nanochannel sizes from 1 to 10 nm for the removal of waste microparticles that have a diameter larger than the nanopores of the membrane [34]. Figure 1 shows a brief description of the principle of capillary-driven NF proposed in this study. The flow of water on both sides, superior and inferior, of the membrane is governed by gravity (from top to bottom). However, the flow of water throughout the NF membrane is driven by action of capillary pressure which depends on the wettability [17,35].

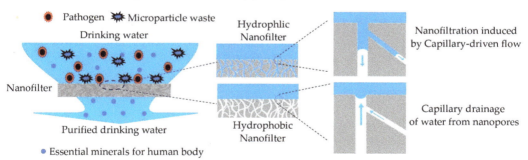

Figure 1. Schematic illustration of capillary-driven NF principle for $r_p \ll \lambda_c$.

In fact, at normal conditions of temperature and pressure, the spontaneous flow of the water across porous structure is determined by an interplay of surface and gravity forces. The influence of the gravity force on the flow of water inside nanopore can be found by calculation of the so-called capillary length of water [36]:

$$\lambda_c = \sqrt{\frac{\gamma}{\rho g}} \quad (1)$$

where γ and ρ are surface tension and the density of water and g is the acceleration of the gravity. For water (With γ = 72 10^{-3} N·m^{-1} and ρ = 1000 kg·m^{-3}). Equation (1) yields a capillary length of water equal to 2.7 mm.

Usually, for surfaces with characteristic lengths smaller than λ_c, the gravitational force can be neglected. Therefore, for nanofilter membrane with pore radius ranges from 1 nm to 10 nm, we have $r_p \ll \lambda_c$ and then the influence of the gravity force in water flow is ignored. Consequently, the flow of water throughout the nanofilter membrane is a function of surface forces i.e., wettability and capillary pressure.

As presented in Figure 1, for the hydrophilic nanofilter membrane, the transport phenomenon of water throughout the nanopores is governed by so-called capillary-driven flow, whereas for a hydrophobic nanofilter membrane, the capillary forces pull water from the nanopore and this results in the drainage of water.

To evaluate the strength of the capillary pressure, we use the Young–Laplace equation, which relies on the capillary pressure to the contact angle of water θ, the surface tension of water γ_l, and the capillary pore radius r_p by [37]:

$$P_c = P_{nw} - P_w = \frac{2\gamma_l \cdot \cos\theta}{r_p} \quad (2)$$

The capillary pressure (P_c) acts as the difference between pressure non-wetting (P_{nw}) and wetting (P_w) phases.

As shown in Figure 2a, at the interface water–membrane–air, when the adhesion force (i.e., force with which water molecules adhere to a membrane surface) is less than the cohesion force of water molecules, the water does not penetrate spontaneously into nanopores and it forms a convex meniscus with contact angle above $\theta > 90°$.

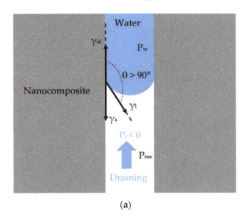 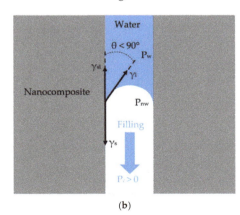

(a) (b)

Figure 2. Capillary pressure action in nanopore $r_p \ll \lambda_c$ (a) capillary filling with hydrophilic nanocomposite (b) capillary draining with hydrophobic nanocomposite.

However, when the force of adhesion is greater than the cohesion force of water molecules and it forms a concave meniscus with contact angle less than 90° (Figure 2b), water molecules tend to penetrate spontaneously into nanopores [38]. This latter situation is defined as the capillary-driven flow known as spontaneous imbibition [17–19,35].

2.1.2. Capillary Rise in Nanopore

Capillary rise is defined as the movement of water within porous materials due to the forces of adhesion, cohesion, and surface tension. In this study, we consider a straight nanopore of radius $r_p \ll \lambda_c$ in contact with water of viscosity μ. For $\theta < 90°$, the water is pulled inside the nanopore by the net interfacial force $2\pi r_p \gamma \cos\theta$. At the equilibrium, the final capillary rise y_f is determined by the force balance between the capillary force $f_c = 2\pi r_p \gamma \cos\theta$ and the gravity force ($f_g = \rho g \pi r_p^2 y_f$) known as Jurin's law [38]:

$$y_f = \frac{2\gamma_l \cdot \cos\theta}{\rho g r_p} \qquad (3)$$

When y is much smaller than y_f, the capillary force f_c and viscous friction force ($f_f = \mu \cdot V \cdot y$) become the two drived forces which govern the flow of the water. This is the so-called viscous regime in which the capillary rise y as function of time t is described by Washburn's law [39]:

$$y(t) = \sqrt{\frac{r_p \gamma_l \cos\theta}{2\mu} t} \qquad (4)$$

2.1.3. Modified Lucas–Washburn for Predicting Capillary-Driven Water Volume

To take account of the morphological characteristics of nanoporous medium on capillary filling kinetics, Benavente et al. [40] developed the Lucas–Washburn (LW) model for predicting the capillary-driven liquid volume V over time t throughout nanoporous material, which is related to the capillary rise as in the following equation:

$$V(t) = S \cdot \varphi \sqrt{\frac{\delta}{\tau}} \cdot y(t) = S \cdot \varphi \sqrt{\frac{\delta r_p \gamma_l \cos\theta}{2\mu\tau} t} \qquad (5)$$

where S is the cross section of the nanofilter surface, φ is the porosity of the nanofilter membrane, δ is the pore shape factor ($\delta = 1$ when the cross section of the pore is a perfect circle), τ is the correction factor of tortuosity (equal 1 to the straight pore channel and $1 < \tau < 3$ for random pore channels).

The tortuosity is directly related to the porosity within the classical fluid flow approach [41] as in the following equation:

$$\tau = 1 - 0.77\ln(\varphi) \qquad (6)$$

2.1.4. Wettability: Contact Angle and Surface Energy

Wettability plays a crucial role in determining water permeability of porous membrane. It is well-known that hydrophilic porous materials exhibit a higher permeability than hydrophobic ones [17]. According to relation 2, this evidence can be understood in terms of capillary pressure. In fact, the capillary pressure is positive (filling action) for hydrophilic materials and negative (draining action) for hydrophobic.

- Contact angle

According to Young's equation, the contact angle θ (Figure 2) results from the thermodynamic equilibrium between the surface tension of water γ_l, the interfacial tension water/nanocomposite γ_{sl}, and the surface energy of the nanocomposites γ_s as [42]:

$$\cos\theta = \frac{\gamma_s - \gamma_{sl}}{\gamma_l} \qquad (7)$$

where γ_{sl} is determined by [43] (p. 417):

$$\gamma_{sl} = \gamma_s + \gamma_l - 2(\sqrt{\gamma_l^D \gamma_s^D} + \sqrt{\gamma_l^P \gamma_s^P}) \qquad (8)$$

where γ^D and γ^P denote the dispersive and polar component of surface energy.

- Surface energy

To evaluate the surface energy of nanocomposites as function of the TiO_2 NPs volume fraction, we used a theoretical method that allowed calculating of the surface energy from the evaluation of the Hamaker constant as in the following equation [43] (p. 415):

$$\gamma = \frac{H}{24 \cdot \pi \cdot x_0^2} \quad (9)$$

where x_0 the interatomic distance between two surfaces in contact called the cut-off distance typically equals to 0.165 nm and H is the Hamaker constant.

2.2. Method for Calculation of the Hamaker Constant

The Hamaker constant H is a fundamental constant for describing qualitatively and quantitatively the van der Waals (vdW) intermolecular interaction that governs the surface forces. On the basis of the Lifshitz theory, the Hamaker constant is function of the bulk properties of the interacting mediums: their dielectric constants ε, and their refractive indexes n [43] (p. 260):

$$H = H^P + H^D \quad (10)$$

where H^P is the polar part of the Hamaker constant. It regroups the Keesom interaction that arises from permanent molecular dipoles and the Debye interaction that arises from permanent dipoles and induced dipoles [44,45]. It is expressed as a function as dielectric permittivity as in the following equation [43] (p. 260):

$$H^P = \frac{3}{4} k_B T \left(\frac{\varepsilon_1 - \varepsilon_2}{\varepsilon_1 + \varepsilon_2} \right)^2 \quad (11)$$

where k_B is the Boltzmann constant, T is the temperature, (ε_1, ε_2) are, respectively, the dielectric permittivity of the nanocomposite and air.

However, H^D is the dispersive part of the Hamaker constant H that regroups the London van der Waals forces resulting from the fluctuations in the charge densities of the electron clouds surrounding the nuclei of the atoms [46]. It is expressed as function of the refractive index of the three interacting mediums and it is given by the following equation [43] (p. 260):

$$H^D = \frac{3h\nu_e}{16\sqrt{2}} \frac{(n_1^2 - n_2^2)^2}{(n_1^2 + n_2^2)\sqrt{(n_1^2 + n_2^2)}} \quad (12)$$

ν_e is the main electronic absorption frequency in the UV typically around 3×10^{15} s^{-1} [43] (p. 260), h is the Planck constant, and (n_1, n_2) are, respectively, the refractive index of nanocomposite and air in the visible spectrum (at 600 nm).

2.3. Model for Calculation the Dielectric Constant of Nanocomposites

The power-law model is used to calculate the dielectric constant as function of the volume fraction (ϕ) of the NPs filler added to the nanocomposites. Hypothetically, we consider that the NPs filler have a spherical form and they are uniformly dispersed in the continuous polymer matrix [47]:

$$\varepsilon_c^{1/3} = \Phi \varepsilon_m^{1/3} + (1 - \Phi) \varepsilon_f^{1/3} \quad (13)$$

2.4. Models for Calculation of the Refractive Index

The refractive index of the nanocomposites is calculated using the popular mixing theory of Maxwell garnet [48]:

$$n_c^2 = n_m^2 \frac{(n_f^2 + 2n_m^2) + 2\Phi(n_f^2 - n_m^2)}{(n_f^2 + 2n_m^2) - \Phi(n_f^2 - n_m^2)} \quad (14)$$

where Φ is the volume fraction of NPs filler, n_c, n_m and n_f are the refractive indexes, respectively, of nanocomposites, polymer matrix, and TiO_2 NPs filler.

The method and models used in this study for evaluation of the surface energy have been validated by many pertinent research studies [43] (p. 278), [49–52].

2.5. Materials

The material chosen in this study are from two categories. Chitosan (CS) and cellulose acetate (CLA) were chosen as the bio-based and biodegradable polymer matrix. Whereas, polyvinylidene fluoride (PVDF) was chosen as synthetic polymer. These polymers have facilitated a great interest in drinking water and wastewater treatment using many processes, such as adsorption, NF, and reverse osmosis [22–29]. Their main advantageous properties are abundance, chemical stability in aqueous medium, mechanical strength, and cost-effectiveness [53–55].

However, TiO_2 anatase NPs were chosen as filler to these polymer matrixes. Moreover, their common advantages with the polymer matrix, TiO_2 NPs, provide many prominent properties for water treatment, such as biocompatibility, antibacterial. and good photostability [56,57].

Actually, there are many easy and ecofriendly techniques for manufacturing nanocomposites nanofilter membrane for water purification, such as the phase inversion process, thermally induced phase separation, chemical bond connection, and interfacial polymerization [58–60].

Hereafter, the volume fraction rates of TiO_2 NPs added to CS, CLA and PVDF polymer matrix are fixed to 0%, 10%, 20%, and 30%, and they are designated in the study as presented in Table 1.

Table 1. Affected names of CS, CLA, and PVDF filled TiO_2 NPs nanocomposites.

Φ% of TiO_2 NPs	Nanocomposite		
	CS/TiO_2	CLA/TiO_2	$PVDF/TiO_2$
0	CS	CLA	PVDF
10	CS/TiO_2-10	CLA/TiO_2-10	$PVDF/TiO_2$-10
20	CS/TiO_2-20	CLA/TiO_2-20	$PVDF/TiO_2$-20
30	CS/TiO_2-30	CLA/TiO_2-30	$PVDF/TiO_2$-30

In Table 2, we summarized the dielectric constant and the refractive index of the materials for the calculation of the Hamaker constants.

Table 2. Dielectric constants and refractive indexes of polymers matrixes, TiO_2 NPs filler, water, and air at room temperature (298.15 K).

Material		ε at 1 MHz	n at 600 nm
Bio-based polymer	CS	5.5 [61]	1.53 [62]
	CLA	5 [63]	1.47 [64]
Synthetic polymer	PVDF	8.5 [65]	1.42 [66]
Filler	TiO_2 (Anatase)	86 [67]	2.60 [67]
	Air	1 [43]	1 [43]
	Water	78.4 [68]	1.33 [68]

3. Results and Discussion

All the calculations and the discussions of the results were made at standard temperature (T = 25 °C = 298.15 K) and pressure (P = 1 atm = 101.325 kPa).

3.1. Effect of TiO_2 NPs Filling on Surface Energy

We can deduce from relation 2 the capillary pressure is function of the contact angle, the surface tension of water, and pore radius. Hence, we report the variation in the contact angle by varying the surface energy of nanocomposites during filling with TiO_2 NPs and nanopore radius (1 to 10 nm), while the surface tension of water is always constant ($\gamma_l = 72 \cdot 10^{-3}$ J·m^{-2}).

Table 3 shows an increase of the refractive index and the dielectric constant of all nanocomposites after incorporation TiO_2 NPs filler. This raise is attributed to the elevated refractive index and dielectric constant of TiO_2 NPs compared to those for the pure CS, CLA, and PVDF polymer matrices.

Table 3. Refractive index and dielectric constant of CS/TiO_2, CLA/TiO_2 and $PVDF/TiO_2$ nanocomposites as function of volume fraction of TiO_2 NPs.

Φ% (TiO_2)	n			ε		
	CS/TiO_2	CLA/TiO_2	$PVDF/TiO_2$	CS/TiO_2	CLA/TiO_2	$PVDF/TiO_2$
0	1.53	1.47	1.42	5.5	5	8.5
10	1.61	1.56	1.51	8.36	7.76	11.82
20	1.70	1.65	1.61	12.08	11.40	15.91
30	1.80	1.75	1.70	16.77	16.02	20.86

Table 4 reports the Hamaker constants calculated from relations (10-11-12). It is apparent from this table that the Hamaker constants for the system nanocomposite–air are still positive and in the range of $\times 10^{-20}$ J [43] (p. 261). Interestingly, this suggests the preciseness of our used model for computing of the vdW intermolecular interactions that governs surface forces [43] (pp. 253–289).

Table 4. Hamaker constants of CS/TiO_2, CLA/TiO_2 and $PVDF/TiO_2$ nanocomposites as function of volume fraction of TiO_2 NPs.

Φ% (TiO_2)	CS/TiO_2			CLA/TiO_2			$PVDF/TiO_2$		
	H^P (10^{-21} J)	H^D (10^{-20} J)	H (10^{-20} J)	H^P (10^{-21} J)	H^D (10^{-20} J)	H (10^{-20} J)	H^P (10^{-21} J)	H^D (10^{-20} J)	H (10^{-20} J)
0	2.08	5.11	5.325	2.00	4.62	4.82	2.37	4.19	4.42
10	2.36	5.81	6.049	2.32	5.36	5.59	2.53	4.98	5.23
20	2.54	6.48	6.738	2.52	6.09	6.34	2.65	5.75	6.01
30	2.670	7.13	7.399	2.65	6.78	7.04	2.73	6.48	6.75

After providing the Hamaker constant, the surface energy is then calculated using Equation (9). As it can be observed in Tables 5 and 6, for purely CS, CLA, and PVDF polymer matrices, the surface energy range values (20–43·10^{-3} J·m^{-2}) are in good agreement with those reported by many studies (Table 6), which confirms the accuracy of our used model for the evaluation of the surface energy.

Table 5. Surface energy of CS/TiO_2, CLA/TiO_2 and $PVDF/TiO_2$ nanocomposites as function of volume fraction of TiO_2 NPs.

Φ% (TiO_2)	CS/TiO_2			CLA/TiO_2			$PVDF/TiO_2$		
	γ_s^P (10^{-3} J·m^{-2})	γ_s^D (10^{-3} J·m^{-2})	γ_s (10^{-3} J·m^{-2})	γ_s^P (10^{-3} J·m^{-2})	γ_s^D (10^{-3} J·m^{-2})	γ_s (10^{-3} J·m^{-2})	γ_s^P (10^{-3} J·m^{-2})	γ_s^D (10^{-3} J·m^{-2})	γ_s (10^{-3} J·m^{-2})
0	0.72	37.82	38.54	0.66	30.80	31.47	0.93	25.33	26.26
10	0.93	48.91	49.84	0.89	41.68	42.57	1.07	35.87	36.94
20	1.08	61.04	62.12	1.05	53.78	54.84	1.17	47.83	49.00
30	1.18	74.15	75.33	1.17	67.05	68.22	1.24	61.16	62.40

However, in accordance with our earlier observations on the Hamaker constant, the surface energy was, remarkably, increasing with filler rates (Φ = 10%, 20% and 30%) of TiO_2 NPs for all nanocomposites. This increase was significantly important in the order $PVDF/TiO_2 < CLA/TiO_2 < CS/TiO_2$. Therefore, it is well-known that the incorporation of TiO2 NPs give rise to an increase of the dispersion and polar vdW intermolecular interactions that governs surface forces of the nanocomposites.

Table 6. Surface energy of CS, CLA and PVDF polymer matrixes provided by this study and from literature.

Polymer	γ_s (10^{-3} J·m^{-2})	
	This Study	Literature
CS	38.54	34–43 [69,70]
CLA	31.47	30–34 [71,72]
PVDF	26.26	22–29 [73,74]

To evaluate the changes in the intermolecular interactions at the interface nanocomposites-water after incorporation of the TiO$_2$ NPs, the interfacial tension γ_{sl} was calculated using Equation (8) and summarized in Table 7. According to the observations on the surface energy, there was also an increase of the γ_{sl} with filler rates for all nanocomposites. In particular, for the same rate of Φ, we find that the γ_{sl} was more important in the order PVDF/TiO$_2$-Water < CLA/TiO$_2$-Water < CS/TiO$_2$-Water.

Table 7. Interfacial tension of CS/TiO$_2$, CLA/TiO$_2$ and PVDF/TiO$_2$ with water as function Φ of TiO$_2$ NPs.

Φ% (TiO$_2$)	γ_{sl} (10^{-3} J·m^{-2})		
	CS/TiO$_2$	CLA/TiO$_2$	PVDF/TiO$_2$
0	43.16	41.92	39.18
10	45.23	43.04	40.32
20	49.09	46.25	43.40
30	54.43	51.20	48.22

Interestingly, during the incorporation of TiO$_2$ NPs, the bio-based CS/TiO$_2$ and CLA/TiO$_2$ nanocomposites exhibit an improved γ_s as well as and γ_{sl} than for PVDF/TiO$_2$ synthetic one.

3.2. Contact Angle and Capillary Pressure

As mentioned in Section 2.1.1, for nanopore with radius r_p (1–10 nm) < λ_c (water) = 2.7 mm, the effect gravitational force on the contact angle configuration is then neglected and only the intermolecular vdW surface forces contribute to the contact angle [36] (p. 36).

To study the capillary pressure behavior of the nanocomposites, we first determined the contact angle at the three-phase boundary between nanocomposite–water–air using Young's relation (relation 7). Figure 3 shows a decrease of the contact angle for all nanocomposites with increasing the volume fraction of TiO$_2$ NPs. The CS, CLA and PVDF polymer matrix (Φ = 0%) exhibits a hydrophobic behavior with the water contact angle > 90°. Simultaneously, we can see from Tables 5 and 7 for Φ = 0%, the interfacial tension γ_{sl} is greater than surface energy γ_s of polymer matrix which leads to the formation of a convex meniscus ($P_c = P_{nw} - P_w < 0$) at the interface water-air that prevent water to flow inside of the nanopore.

After incorporation of TiO$_2$ NPs, a conversion from hydrophobic to hydrophilic character for CS/TiO$_2$ nanocomposite is identified at Φ < 10%, while the hydrophobic character persists by filling with 10% < Φ < 20% for CLA/TiO$_2$ and PVDF/TiO$_2$ nanocomposites. Therefore, the CS/TiO$_2$ nanocomposite provides enhanced hydrophilic behavior with low filling with TiO$_2$ NPs than CLA/TiO$_2$ and PVDF/TiO$_2$ nanocomposites.

For depicting the changes in the contact angle on the capillary pressure, we calculated the evolution of the capillary pressure as function of the nanopore radius (1–10 nm) using relation 2. Then, we normalized the capillary pressure P_c to the atmospheric pressure (Pa = 101.325 kPa) to show the rate of changes caused by the incorporation of TiO$_2$ NPs.

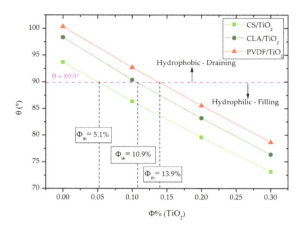

Figure 3. Water Contact angle of CS/TiO$_2$, CLA/TiO$_2$ and PVDF/TiO$_2$ nanocomposites as function of volume fraction of TiO$_2$ NPs.

For a pure polymer matrix (Figure 4a), the normalized capillary pressure was negative along the nanopore radius r$_p$. As can be interpreted from relation 2, the negative value of P$_c$ originates hydrophobic behavior of the CS, CLA and PVDF polymer matrix which leads to negative value of the term cosθ (Figure 3). Consequently, the meniscus is convex, and the capillary pressure tends to pull water from unwetted to wetted regions of the nanopore (Figure 2a), resulting in the drainage of water from the nanopore. In the other hand, such negative capillary pressure describes the necessary opposite pressure to provide rising water inside the nanopore in the case of the use of a nanofilter based on a purely polymer matrix. Therefore, for the CS, CLA, and PVDF polymeric nanofilter membranes, the NF process requires supplementary pressure with opposite and higher strength than the capillary pressure to squeeze water throughout the nanopore.

We now turn to study the capillary pressure after incorporation of the TiO$_2$ NPs. As noted Figure 4b, the normalized capillary pressure for CS/TiO$_2$-10 becomes positive which reveals a changing from draining to filling of water inside the nanopores. However, for CLA/TiO$_2$-10 and PVDF/TiO$_2$-10, the normalized capillary pressure becomes less important than for CLA and PVDF purely matrix but it still drains water from the nanopores. However, as demonstrated in Figure 4c,d for Φ = (20%, 30%), the capillary pressure becomes positive for all nanocomposites and still increases with increasing the volume fractions of TiO$_2$ NPs.

Figure 5 depicts the relative evolution of the capillary pressure for Φ = (20%, 30%) at the same effective nanopore radius r$_{pe}$ = 5 nm. For all nanocomposites, the P$_c$/P$_a$ (Φ = 30%) ≈ 2. P$_c$/P$_a$ (Φ = 20%). Interestingly, this demonstrates that the capillary pressure increases almost twofold with the addition rate of 10% (from 20% to 30%) of TiO$_2$ NPs. In addition, it should be noted that the capillary pressure was more important in the order PVDF/TiO$_2$ < CLA/TiO$_2$ < CS/TiO$_2$ nanocomposites.

Therefore, for future experimental valorizations of these finding, it is crucial to determine the threshold volume fraction that corresponds to conversion from draining to filling by capillary pressure. To do so, we determined the threshold of the volume fraction Φ$_{th}$ of TiO$_2$ NPs for what the contact angle equal 89.9°. Correspondingly, Φ$_{th}$ at which there is a transition from P$_c$ < 0 that is in favor of draining, to P$_c$ > 0 that is in favor of capillary filling. As depicted in Figure 4, the Φ$_{th}$ for CS/TiO$_2$, CLA/TiO$_2$ and PVDF/TiO$_2$ were, respectively, 5.1%, 10.90%, and 13.90%. In Table 8, we summarized the corresponding threshold dielectric constant ε$_{th}$, refractive n$_{th}$, Hamaker constant H, the surface energy γ$_{th}$ for CS/TiO$_2$, CLA/TiO$_2$, and PVDF/TiO$_2$ nanocomposites.

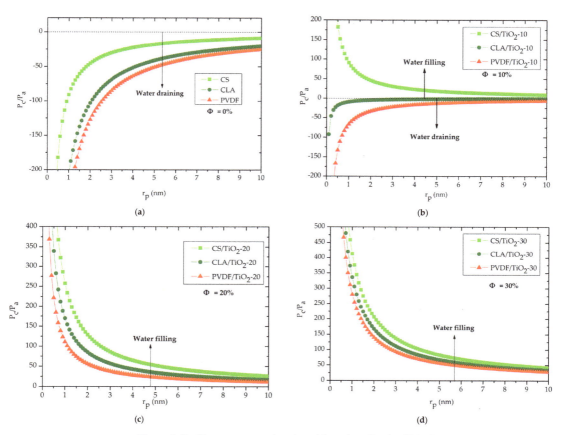

Figure 4. Capillary pressure as function of the pore radius for CS/TiO_2, CLA/TiO_2 and $PVDF/TiO_2$: (**a**) $\phi = 0\%$, (**b**) $\phi = 10\%$, (**c**) $\phi = 20\%$, (**d**) $\phi = 30\%$.

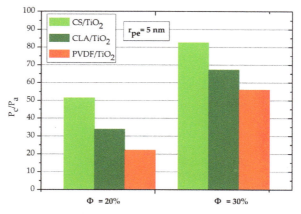

Figure 5. Normalized capillary pressure P_c/P_a at $r_{pe} = 5$ nm and $\Phi = (20\%$ and $30\%)$ for CS/TiO_2, CLA/TiO_2 and $PVDF/TiO_2$ nanocomposites.

Table 8. Corresponding threshold ε_{th}, n_{th}, H and γ_{th} for $\theta = 89.9°$ of CS/TiO$_2$, CLA/TiO$_2$ and PVDF/TiO$_2$ nanocomposites.

Nanocomposite	CS/TiO$_2$	CLA/TiO$_2$	PVDF/TiO$_2$
ϕ_{th} (%)	5.10	10.90	13.90
n_{th}	1.57	1.57	1.55
ε_{th}	6.83	7.92	13.24
H (10^{-20} J)	9.08	8.97	8.55
γ_{th} (10^{-3} J·m^{-2})	43.83	42.62	41.51

3.3. Capillary Rise

We highlighted the effect of TiO$_2$ NPs on surface energy, capillary pressure, and contact angle in the previous section. Now, we will explore the dynamics aspects of the filling of water by evaluation of capillary rise and volume uptake of water throughout the NF membrane for $\phi > \phi_{th}$, i.e., when water flows inside the nanopore of the membrane under action of capillary pressure. Therefore, the capillary rise (y(t)) and water filling dynamics (V(t)) of water in nanocomposites were restrictively studied for CS/TiO$_2$ at ϕ = (10%, 20% and 30%) > ϕ_{th} (CS/TiO$_2$) = 5.1%, and for CLA/TiO$_2$ and PVDF/TiO$_2$, at ϕ = (20% and 30%) > ϕ_{th} (PVDF/TiO$_2$) > ϕ_{th} (CLA/TiO$_2$). For comparative purposes, they will juxtaposed with those and the y_{th}(t) and V_{th}(t) threshold ones.

To demonstrate the negligible effect of gravity in capillary filling kinetics, we show the gravity and capillary forces for each nanocomposite with effective nanopore radius r_{pe} = 5 nm. As it can be observed in Figure 6a, the capillary force balance with gravity force only for y_f that is over y > 10 m that is very higher to the ordinary dimension of the nanofilter. Incidentally, for a y < 10 m, the gravity force was less than capillary force by one order of magnitude $f_g < 10^{-1} \cdot f_c$ (Figure 6b), which is in good agreement with previous studies [19] (p. 1624), [36] (p. 47). Consequently, for the here examined rise levels restricted by the maximum nanofilter height to less than 1 m, the contribution of the gravity force on the capillary filling kinetic of the water inside the nanopores is neglected ($f_g << f_c$).

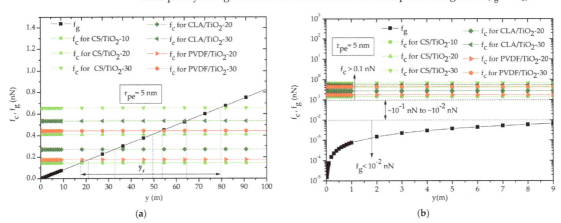

Figure 6. Gravity and capillary forces as function of distance rise in one nanopore: (a) the gravity force balance capillary force for at y_f, (b) the strength of gravity force is negligible compared to the capillary force for distance (1 nm to 9 m).

Therefore, the capillary rise is governed only by the balance between the viscous friction force (f_f) and capillary force (f_c). It is the so-called viscous regime, during which the capillary filling is well described by Washburn's law (Section 2.1.2). Hereafter, to carry out the effect of the volume fraction on the capillary rise, we consider single capillary nanopore of uniform internal circular cross section with effective radius r_{pe} = 5 nm.

Figure 7 shows that the capillary rise y(t) increase with increasing volume fraction of TiO_2 NPs. Obviously, the water rises spontaneously in nanopores under action of capillary pressure and without any other additional force for all nanocomposites. Interestingly, this result supports our idea for capillary-driven NF using nanoporous CS/TiO_2, CLA/TiO_2 and $PVDF/TiO_2$ nanocomposites as nanofiltration process with zero energy consumption. Typically, the rise times in our study are in agreement with recent experimental and simulation studies [20,75]. In fact, for capillary nanopore, water rises slowly over time because the viscous friction forces become more important for nanoporous narrowed capillary, which reduces the overall capillary dynamics significantly.

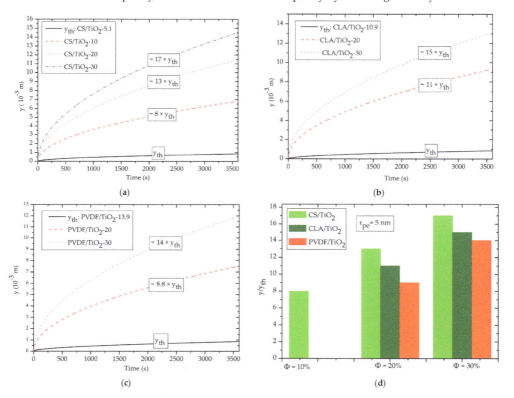

Figure 7. Effect of the volume fraction of TiO_2 NPs on capillary rise y(t) with r_{pe} = 5 nm: (**a**) CS/TiO_2, (**b**) CLA/TiO_2, (**c**) $PVDF/TiO_2$, (**d**) y/y_{th} as function of ϕ of TiO_2 NPs.

In accordance with Section 3.2, for ϕ = 10%, only the CS/TiO_2-10 nanocomposite that exhibits a capillary rise (Figure 7d). Contrarily, for the CLA/TiO_2-10 and $PVDF/TiO_2$-10, the capillary pressure remains negative, which prevents the rise of water inside the nanopores. However, for ϕ = (20% and 30%) (Figure 7a–c), all the nanocomposites exhibit spontaneous capillary rise. As expected from the study of capillary pressure, the capillary rise increases with volume fraction of the TiO_2 NPs.

Figure 7d depicts the increases rates of the normalized capillary rise (y/y_{th}) with volume fraction filler of TiO_2 NPs. In particular, at ϕ = 30%, the y/y_{th} were about ~17, ~15 and ~14, respectively, for CS/TiO_2, CLA/TiO_2 and $PVDF/TiO_2$ nanofilter. Interestingly, this finding demonstrates that the biodegradable CS/TiO_2 and CLA/TiO_2 nanofilter exhibit higher capillary rise than the synthetic $PVDF/TiO2$ nanofilter. From a review of Tables 3 and 5, we can thus reveal that the enhancement of the capillary rise can be made by increasing the refractive index of the based nanofilter materials.

3.4. Effect of Volume Fraction of TiO$_2$ NPs on Capillary-Driven Water Volume

As detailed in Section 2.1.3, the modified LW model allows the calculation of the capillary-driven water volume throughout nanoporous membrane as a function of the morphological characteristics of the nanofilter, (φ, r_{pe}, δ and τ), the physical properties of the water, (ρ, η, and γ_l), and the contact angle θ which depends on the volume fraction of TiO$_2$ NPs.

Therefore, to explore the effect of the incorporation of TiO$_2$ NPs on the capillary-driven water volume, we consider that the morphological characteristics the nanocomposite nanofilter are the same for all nanocomposites. Hereafter, we consider the porosity of nanofilter $\varphi = 0.7$ with corresponding tortuosity $\tau = 1.27$ (Relation 6) and non-circular cross section of nanopores with roundness $\delta = 0.5$.

The capillary-driven water volume was calculated for surface nanofilter membrane S = 1 m^2 and effective nanopore radius r_{pe} = 5 nm using relation 5. As shown in Figure 8a–c, for $\phi = \phi_{th}$ (CS/TiO$_2$-5.1 $\cong$ CLA/TiO$_2$-10.9 $\cong$ PVDF/TiO$_2$-13.9), the threshold capillary-driven water volume (V$_{th}$) after one hour of filling was ~30 mL. In fact, at θ_{th} = 89.9°, the term cos(θ_{th}) in Equation (5) is equal 1.7·10^{-3}, which leads to a weak value of V$_{th}$.

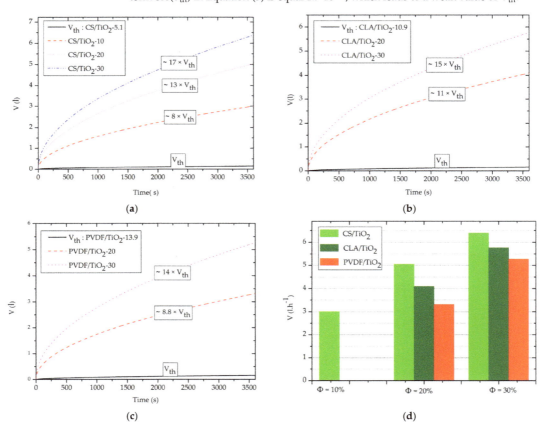

Figure 8. Evolution of the capillary-driven water volume V(t) for with S = 1 m^2 and r_{pe} = 5 nm. (**a**) CS/TiO$_2$, (**b**) CLA/TiO$_2$, (**c**) PVDF/TiO$_2$, (**d**) capillary-driven water flow rate (L·h^{-1}) as function of the ϕ of TiO$_2$ NPs.

However, for ϕ = 10% (Figure 8a–d), only the CS/TiO$_2$-10 nanocomposites' nanofilter exhibits a spontaneous capillary-driven nanofiltration with V $\approx$ 3 L·h^{-1}. In agreement with capillary pressure (Section 3.2) and capillary rise (Section 3.3), this finding indicates the enhanced capillary-driven NF of CS/TiO$_2$-10 nanocomposite in comparison with other CLA/TiO$_2$-10 and PVDF/TiO$_2$-10 nanocomposites, which do not demonstrate a capillary-driven flow at ϕ = 10%.

While, for ϕ = (20% and 30%), we clearly observe in Figure 8b-c-d a great increase of the capillary-driven water volume. In particular, CS/TiO$_2$-30 has the highest capillary-driven nanofiltration with V = 6.39 L·h^{-1}, followed by CLA/TiO$_2$-30 with V = 5.76 L·h^{-1} and PVDF/TiO$_2$-30 with V = 5.26 L·h^{-1}. Accordingly, it is important to note that these orders of magnitude of capillary-driven water volume are directly in line with previous findings (2–25 L·h^{-1}·m^{-2}) [17,40,76]. This validates our theorical findings on the one hand, and motivate future experimental valorizations of this study on the other.

Interestingly, for the same incorporation rates ϕ = (20% or 30%) of the TiO$_2$ NPs, the increased V(t) of the CS/TiO$_2$, in comparison to CLA/TiO$_2$ and PVDF/TiO$_2$ nanocomposites nanofilter, originate from the elevated surface energy of the CS polymer matrix (γ_s (CS) = 38.54 J·m^{-2}) > (γ_s (CLA) = 31.47 J·m^{-2}) > (γ_s (PVDF) = 26.26 J·m^{-2}), which strongly proportional to the refractive index of the polymers matrixes (n (CS) = 1.53) > (n (CLA) = 1.47) > (n (PVDF) = 1.42). In addition, from study of the capillary-driven water volume, we demonstrate that all nanocomposites are prominent candidates for use as domestic supplementary purification of drinking water for removal of microparticle waste with zero energy consumption.

Crucially, the biodegradable CS/TiO$_2$-30 and CLA/TiO$_2$-30 nanocomposites nanofilter demonstrate capillary-driven water volume higher ~1.5 and ~1.2 times, respectively, more than the synthetic PVDF/TiO$_2$-30 nanocomposite. In particular, CS/TiO$_2$ nanocomposite offer the most improved capillary-driven NF behavior during the incorporation of the TiO$_2$ NPs.

It is important to note that for the experimental validation, the nanocomposite materials do not have the same morphological characteristics (tortuosity, porosity, pore size, etc.). Therefore, the capillary-driven water volume calculated by the modified L-W model (relations 5 and 6) is directly proportional to the pore radius, shape factor, porosity, and cross section nanofilter, and inversely proportional to the tortuosity of the nanofilter materials.

The methodologies used along this study constitute simple and complete theoretical methods for prediction the wettability as well as the spontaneous capillary pressure and the capillary-driven flow for any material with known refractive index dielectric permittivity and physical properties of the fluids.

Overall, these findings constitute a quantitative and qualitative background for further experimental elaboration of capillary-driven nanofilter based on green and biodegradable CS/TiO$_2$ and CLA/TiO$_2$ nanocomposites.

4. Conclusions

In this study, from Lifshitz and Young-Laplace theories, it is shown that the polymer matrixes CS, CLA and PVDF provide hydrophobic contact angle (θ > 90°) and draining capillary pressure. Consequently, the pure polymers matrix does not perform a capillary-driven flow and then cannot be used as capillary-driven nanofilter for water purification.

However, after incorporation of TiO$_2$ NPs, a wetting conversion from hydrophobic to hydrophilic were depicted for all CS/TiO$_2$, CLA/TiO$_2$ and PVDF/TiO$_2$ with, respectively, threshold volume fraction TiO$_2$ NPs for wetting conversion equal to 5.1%, 10.9%, and 13.9%.

For incorporation with $\phi > \phi_{th}$, the contribution of gravity forces on capillary rise as well as the capillary-driven flow is neglected for the considered nanofilter with effective nanopore radius r_{pe} (~10 nm) < λ_c (water) = 2.7 mm and with macroscopic size of nanofilter < 10 m.

For a single nanopore, an increase of capillary rise with increased volume fraction was demonstrated for all CS/TiO$_2$, CLA/TiO$_2$ and PVDF/TiO$_2$ nanocomposites. However, for the same incorporation rates of TiO$_2$, the biodegradable CS/TiO$_2$ and CLA/TiO$_2$ exhibits a higher capillary rise than PVDF/TiO$_2$. This increase was attributed to the higher surface energy proportional to the refractive index of the dielectric permittivity. Consequently, this fact reveled that water rise sin nanopores without the help of any external force for all nanoporous nanocomposites. This important finding supported our proposed idea for capillary-driven nanofiltration as zero energy consumption nanofiltration process.

Based on the modified L-W model, we demonstrated that capillary-driven water volume for CS/TiO_2, CLA/TiO_2, and $PVDF/TiO_2$ nanocomposites increases with increasing TiO_2 NPs volume fraction. Interestingly, at 30% volume fraction of TiO_2 NPs, the biodegradable CS/TiO_2 and CLA/TiO_2 nanocomposites nanofilter demonstrate capillary-driven water volume higher by ~1.5 and ~1.2 times, respectively; more than the synthetic $PVDF/TiO_2$ nanocomposite.

Collectively, CS/TiO_2 and CLA/TiO_2 biodegradable nanocomposites were the better candidates for purification of drinking water by capillary-driven nanofiltration process as sustainable, environmentally safe, and zero energy consumption.

Author Contributions: Conceptualization, N.M. and A.A.; methodology, N.M.; validation, N.M., N.S.A. and M.A.C.; formal analysis, N.M.; investigation, N.M. and N.S.A.; resources, N.M., M.A.C. and N.S.A.; data curation, N.M., N.S.A. and A.A.; writing—original draft preparation, N.M., A.A. and N.S.A.; writing—review and editing, N.M. and A.A.; visualization, N.M.; supervision, N.M. and N.S.A.; project administration, N.M., N.S.A. and A.A.; funding acquisition, A.A. and N.S.A. All authors have read and agreed to the published version of the manuscript.

Funding: This research was funded by the Deanship of Scientific Research at King Khalid University under grant number RGP. 2/57/43. Additionally, this research was funded by Princess Nourah bint Abdulrahman University Researchers Supporting Project number (PNURSP2022R19), Princess Nourah bint Abdulrahman University, Riyadh, Saudi Arabia.

Institutional Review Board Statement: Not applicable.

Informed Consent Statement: Not applicable.

Data Availability Statement: Not applicable.

Acknowledgments: The authors extend their appreciation to the Deanship of Scientific Research at King Khalid University for funding this work through the Research Groups Program under grant number RGP. 2/57/43. Also, this research was funded by Princess Nourah bint Abdulrahman University Researchers Supporting Project number (PNURSP2022R19), Princess Nourah bint Abdulrahman University, Riyadh, Saudi Arabia.

Conflicts of Interest: The authors declare no conflict of interest.

References

1. Lytle, D.A.; Schock, M.R.; Formal, C.; Bennett-Stamper, C.; Harmon, S.; Nadagouda, M.N.; Williams, D.; DeSantis, M.K.; Tully, J.; Pham, M. Lead Particle Size Fractionation and Identification in Newark, New Jersey's Drinking Water. *Environ. Sci. Technol.* **2020**, *54*, 13672–13679. [CrossRef] [PubMed]
2. Oßmann, B.E.; Sarau, G.; Holtmannspötter, H.; Pischetsrieder, M.; Christiansen, S.H.; Dicke, W. Small-sized microplastics and pigmented particles in bottled mineral water. *Water Res.* **2018**, *141*, 307–316. [CrossRef] [PubMed]
3. Chowdhury, S.; Mazumder, M.J.; Al-Attas, O.; Husain, T. Heavy metals in drinking water: Occurrences, implications, and future needs in developing countries. *Sci. Total Environ.* **2016**, *569–570*, 476–488. [CrossRef] [PubMed]
4. Mraz, A.L.; Tumwebaze, I.K.; McLoughlin, S.R.; McCarthy, M.E.; Verbyla, M.E.; Hofstra, N.; Rose, J.B.; Murphy, H.M. Why pathogens matter for meeting the united nations' sustainable development goal 6 on safely managed water and sanitation. *Water Res.* **2021**, *189*, 116591. [CrossRef]
5. Danopoulos, E.; Twiddy, M.; Rotchell, J.M. Microplastic contamination of drinking water: A systematic review. *PLoS ONE* **2020**, *15*, e0236838. [CrossRef] [PubMed]
6. Liu, M.; Lü, Z.; Chen, Z.; Yu, S.; Gao, C. Comparison of reverse osmosis and nanofiltration membranes in the treatment of biologically treated textile effluent for water reuse. *Desalination* **2011**, *281*, 372–378. [CrossRef]
7. Naidu, L.D.; Saravanan, S.; Chidambaram, M.; Goel, M.; DAS, A.; Babu, J.S.C. Nanofiltration in Transforming Surface Water into Healthy Water: Comparison with Reverse Osmosis. *J. Chem.* **2015**, *2015*, 326869. [CrossRef]
8. Bouchareb, R.; Derbal, K.; Özay, Y.; Bilici, Z.; Dizge, N. Combined natural/chemical coagulation and membrane filtration for wood processing wastewater treatment. *J. Water Process Eng.* **2020**, *37*, 101521. [CrossRef]

9. Fanourakis, S.K.; Peña-Bahamonde, J.; Bandara, P.C.; Rodrigues, D.F. Nano-based adsorbent and photocatalyst use for pharmaceutical contaminant removal during indirect potable water reuse. *NPJ Clean Water* **2020**, *3*, 1. [CrossRef]
10. Liosis, C.; Karvelas, E.G.; Karakasidis, T.; Sarris, I.E. Numerical study of magnetic particles mixing in waste water under an external magnetic field. *J. Water Supply: Res. Technol.* **2020**, *69*, 266–275. [CrossRef]
11. Karvelas, E.G.; Lampropoulos, N.K.; Karakasidis, T.E.; Sarris, I.E. A computational tool for the estimation of the optimum gradient magnetic field for the magnetic driving of the spherical particles in the process of cleaning water. *Desalination Water Treat.* **2017**, *99*, 27–33. [CrossRef]
12. Karvelas, E.; Liosis, C.; Benos, L.; Karakasidis, T.; Sarris, I. Micromixing Efficiency of Particles in Heavy Metal Removal Processes under Various Inlet Conditions. *Water* **2019**, *11*, 1135. [CrossRef]
13. Hafiz, M.; Hawari, A.H.; Alfahel, R.; Hassan, M.K.; Altaee, A. Comparison of Nanofiltration with Reverse Osmosis in Reclaiming Tertiary Treated Municipal Wastewater for Irrigation Purposes. *Membranes* **2021**, *11*, 32. [CrossRef] [PubMed]
14. Yang, Z.; Zhou, Y.; Feng, Z.; Rui, X.; Zhang, T.; Zhang, Z. A Review on Reverse Osmosis and Nanofiltration Membranes for Water Purification. *Polymers* **2019**, *11*, 1252. [CrossRef] [PubMed]
15. Nagy, E. Chapter 15—Nanofiltration. In *Basic Equations of Mass Transport Through a Membrane Layer*, 2nd ed.; Elsevier: Boston, MA, USA, 2019; pp. 417–428. [CrossRef]
16. Khan, A.A.; Boddu, S. Chapter 13—Hybrid membrane process: An emerging and promising technique toward industrial wastewater treatment. In *Membrane-Based Hybrid Processes for Wastewater Treatment*; Shah, M.P., Rodriguez-Couto, S., Eds.; Elsevier: Boston, MA, USA, 2021; pp. 257–277. [CrossRef]
17. Futselaar, H.; Schonewille, H.; van der Meer, W. Direct capillary nanofiltration—A new high-grade purification concept. *Desalination* **2002**, *145*, 75–80. [CrossRef]
18. Wang, Y.; Lee, J.; Werber, J.R.; Elimelech, M. Capillary-driven desalination in a synthetic mangrove. *Sci. Adv.* **2020**, *6*, eaax5253. [CrossRef]
19. Cai, J.; Jin, T.; Kou, J.; Zou, S.; Xiao, J.; Meng, Q. Lucas–Washburn Equation-Based Modeling of Capillary-Driven Flow in Porous Systems. *Langmuir* **2021**, *37*, 1623–1636. [CrossRef]
20. Heiranian, M.; Aluru, N.R. Modified Lucas-Washburn theory for fluid filling in nanotubes. *Phys. Rev. E* **2022**, *105*, 055105. [CrossRef]
21. Barrat, J.-L.; Bocquet, L. Influence of wetting properties on hydrodynamic boundary conditions at a fluid/solid interface. *Faraday Discuss.* **1999**, *112*, 119–128. [CrossRef]
22. Bassyouni, M.; Abdel-Aziz, M.H.; Zoromba, M.S.; Abdel-Hamid, S.M.S.; Drioli, E. A review of polymeric nanocomposite membranes for water purification. *J. Ind. Eng. Chem.* **2019**, *73*, 19–46. [CrossRef]
23. Sharma, V.; Borkute, G.; Gumfekar, S.P. Biomimetic nanofiltration membranes: Critical review of materials, structures, and applications to water purification. *Chem. Eng. J.* **2021**, *433*, 133823. [CrossRef]
24. Johnson, D.J.; Hilal, N. Nanocomposite nanofiltration membranes: State of play and recent advances. *Desalination* **2022**, *524*, 115480. [CrossRef]
25. Wang, Z.; Wang, Z.; Lin, S.; Jin, H.; Gao, S.; Zhu, Y.; Jin, J. Nanoparticle-templated nanofiltration membranes for ultrahigh performance desalination. *Nat. Commun.* **2018**, *9*, 2004. [CrossRef] [PubMed]
26. Guo, H.; Li, X.; Yang, W.; Yao, Z.; Mei, Y.; Peng, L.E.; Yang, Z.; Shao, S.; Tang, C.Y. Nanofiltration for drinking water treatment: A review. *Front. Chem. Sci. Eng.* **2021**, *16*, 681–698. [CrossRef] [PubMed]
27. Wang, C.; Park, M.J.; Seo, D.H.; Drioli, E.; Matsuyama, H.; Shon, H. Recent advances in nanomaterial-incorporated nanocomposite membranes for organic solvent nanofiltration. *Sep. Purif. Technol.* **2021**, *268*, 118657. [CrossRef]
28. Khraisheh, M.; Elhenawy, S.; AlMomani, F.; Al-Ghouti, M.; Hassan, M.K.; Hameed, B.H. Recent Progress on Nanomaterial-Based Membranes for Water Treatment. *Membranes* **2021**, *11*, 995. [CrossRef]
29. Lakhotia, S.R.; Mukhopadhyay, M.; Kumari, P. Cerium oxide nanoparticles embedded thin-film nanocomposite nanofiltration membrane for water treatment. *Sci. Rep.* **2018**, *8*, 4976. [CrossRef]
30. Irshad, M.A.; Nawaz, R.; Rehman, M.Z.U.; Adrees, M.; Rizwan, M.; Ali, S.; Ahmad, S.; Tasleem, S. Synthesis, characterization and advanced sustainable applications of titanium dioxide nanoparticles: A review. *Ecotoxicol. Environ. Saf.* **2021**, *212*, 111978. [CrossRef]
31. Thomas, M.S.; Koshy, R.R.; Mary, S.K.; Thomas, S.; Pothan, L.A. *Starch, Chitin and Chitosan Based Composites and Nanocomposites*; Springer International Publishing: Cham, Switzerland, 2019; pp. 1–65. [CrossRef]
32. Pandey, J.K.; Takagi, H.; Nakagaito, A.N.; Kim, H.-y. Processing, Performance and Application: Volume C: Polymer Nanocomposites of Cellulose Nanoparticles. In *Handbook of Polymer Nanocomposites*; Springer: Heidelberg, Germany, 2015; pp. 1–518. [CrossRef]
33. Drobny, J.G. Electron Beam Processing of Commercial Polymers, Monomers, and Oligomers. In *Plastics Design Library, Ionizing Radiation and Polymers*; William Andrew Publishing: Norwich, NY, USA, 2013; pp. 101–147. [CrossRef]
34. Jye, L.W.; Ismail, A.F. *Nanofiltration Membranes Synthesis, Characterization, and Applications*, 1st ed.; CRC Press: Boca Raton, FL, USA; Taylor & Francis Group: New York, NY, USA, 2017; pp. 1–184. ISBN 9781498751377.
35. Jährig, J.; Vredenbregt, L.; Wicke, D.; Miehe, U.; Sperlich, A. Capillary Nanofiltration under Anoxic Conditions as Post-Treatment after Bank Filtration. *Water* **2018**, *10*, 1599. [CrossRef]

36. de Gennes, P.-G.; Brochard-Wyart, F.; Quere, D. Capillarity and Gravity. In *Capillarity and Wetting Phenomena Drops, Bubbles, Pearls, Waves*, 1st ed.; Springer Science & Business Media: New York, NY, USA, 2003; p. 33. [CrossRef]
37. Collins, R.E.; Cooke, C.E. Fundamental basis for the contact angle and capillary pressure. *Trans. Faraday Soc.* **1959**, *55*, 1602–1606. [CrossRef]
38. Jurin, J., II. An account of some experiments shown before the Royal Society; with an enquiry into the cause of the ascent and suspension of water in capillary tubes. *Philos. Trans. R. Soc. Lond.* **1719**, *30*, 739–747. [CrossRef]
39. Washburn, E.W. The Dynamics of Capillary Flow. *Phys. Rev.* **1921**, *17*, 273–283. [CrossRef]
40. Benavente, D.; Lock, P.; García del Cura, M.Á.; Ordóñez, S. Predicting the Capillary Imbibition of Porous Rocks from Microstructure. *Transp. Porous Media* **2002**, *49*, 59–76. [CrossRef]
41. Matyka, M.; Khalili, A.; Koza, Z. Tortuosity-porosity relation in porous media flow. *Phys. Rev. E* **2008**, *78*, 026306. [CrossRef]
42. Young, T., III. An essay on the cohesion of fluids. *Philos. Trans. R. Soc. Lond.* **1805**, *95*, 65–87. [CrossRef]
43. Israelachvili, J.N. *Intermolecular and Surface Forces*, 3rd ed.; Elsevier: Amsterdam, The Netherlands; Academic Press: Burlington, NJ, USA, 2011; p. 196. ISBN 978-0-12-375182-9.
44. Keesom, W.H. The cohesion forces in the theory of Van Der Waals. *Phys. Z.* **1921**, *22*, 129–141.
45. Debye, P. Molecular forces and their electric explanation. *Phys. Z.* **1921**, *22*, 302–308.
46. London, F. The general theory of molecular forces. *Trans. Faraday Soc.* **1937**, *33*, 8–26. [CrossRef]
47. Karkkainen, K.K.; Sihvola, A.H.; Nikoskinen, K.I. Effective permittivity of mixtures: Numerical validation by the FDTD method. *IEEE Trans. Geosci. Remote Sens.* **2000**, *38*, 1303–1308. [CrossRef]
48. Garnett, J.C.M. VII—Colours in Metal Glasses, in Metallic Films, and in Metallic Solutions–II; Series A Containing Papers of a Mathematical or Physical Character. *Philos. Trans. Royal Soc. Lond.* **1906**, *205*, 237–288.
49. Leite, F.L.; Bueno, C.C.; Da Róz, A.L.; Ziemath, E.C.; Oliveira, O.N., Jr. Theoretical Models for Surface Forces and Adhesion and Their Measurement Using Atomic Force Microscopy. *Int. J. Mol. Sci.* **2012**, *13*, 12773–12856. [CrossRef] [PubMed]
50. Mahdhi, N.; Alsaiari, N.S.; Alzahrani, F.M.; Katubi, K.M.; Amari, A.; Hammami, S. Theoretical Investigation of the Adsorption of Cadmium Iodide from Water Using Polyaniline Polymer Filled with TiO_2 and ZnO Nanoparticles. *Water* **2021**, *13*, 2591. [CrossRef]
51. Drummond, C.J.; Chan, D.Y.C. van der Waals Interaction, Surface Free Energies, and Contact Angles: Dispersive Polymers and Liquids. *Langmuir* **1997**, *13*, 3890–3895. [CrossRef]
52. Aishwarya, S.; Shanthi, J.; Swathi, R. Surface energy calculation using Hamaker's constant for polymer/silane hydrophobic thin films. *Mater. Lett.* **2019**, *253*, 409–411. [CrossRef]
53. Saud, A.; Saleem, H.; Zaidi, S.J. Progress and Prospects of Nanocellulose-Based Membranes for Desalination and Water Treatment. *Membranes* **2022**, *12*, 462. [CrossRef] [PubMed]
54. Matei, E.; Predescu, A.M.; Râpă, M.; Țurcanu, A.A.; Mateș, I.; Constantin, N.; Predescu, C. Natural Polymers and Their Nanocomposites Used for Environmental Applications. *Nanomaterials* **2022**, *12*, 1707. [CrossRef] [PubMed]
55. Divya, S.; Oh, T.H. Polymer Nanocomposite Membrane for Wastewater Treatment: A Critical Review. *Polymers* **2022**, *14*, 1732. [CrossRef]
56. Yaqoob, A.A.; Parveen, T.; Umar, K.; Mohamad Ibrahim, M.N. Role of Nanomaterials in the Treatment of Wastewater: A Review. *Water* **2020**, *12*, 495. [CrossRef]
57. Waghmode, M.S.; Gunjal, A.B.; Mulla, J.A.; Patil, N.N.; Nawani, N.N. Studies on the titanium dioxide nanoparticles: Biosynthesis, applications and remediation. *SN Appl. Sci.* **2019**, *1*, 310. [CrossRef]
58. Ekambaram, K.; Doraisamy, M. Surface modification of PVDF nanofiltration membrane using Carboxymethylchitosan-Zinc oxide bionanocomposite for the removal of inorganic salts and humic acid. *Colloids Surfaces A Physicochem. Eng. Asp.* **2017**, *525*, 49–63. [CrossRef]
59. Shi, F.; Ma, Y.; Ma, J.; Wang, P.; Sun, W. Preparation and characterization of PVDF/TiO2 hybrid membranes with different dosage of nano-TiO2. *J. Membr. Sci.* **2012**, *389*, 522–531. [CrossRef]
60. Wang, D.; Yuan, H.; Chen, Y.; Ni, Y.; Huang, L.; Mondal, A.K.; Lin, S.; Huang, F.; Zhang, H. A cellulose-based nanofiltration membrane with a stable three-layer structure for the treatment of drinking water. *Cellulose* **2020**, *27*, 8237–8253. [CrossRef]
61. Bonardd, S.; Robles, E.; Barandiaran, I.; Saldías, C.; Leiva, A.; Kortaberria, G. Biocomposites with increased dielectric constant based on chitosan and nitrile-modified cellulose nanocrystals. *Carbohydr. Polym.* **2018**, *199*, 20–30. [CrossRef]
62. Azofeifa, D.E.; Arguedas, H.J.; Vargas, W.E. Optical properties of chitin and chitosan biopolymers with application to structural color analysis. *Opt. Mater.* **2012**, *35*, 175–183. [CrossRef]
63. Diantoro, M.; A Mustikasari, A.; Wijayanti, N.; Yogihati, C.; Taufiq, A. Microstructure and dielectric properties of cellulose acetate-ZnO/ITO composite films based on water hyacinth. *J. Physics: Conf. Ser.* **2017**, *853*, 012047. [CrossRef]
64. Sultanova, N.G.; Kasarova, S.N.; Nikolov, I.D. Characterization of optical properties of optical polymers. *Opt. Quantum Electron.* **2013**, *45*, 221–232. [CrossRef]
65. Arshad, A.N.; Wahid, M.H.M.; Rusop, M.; Majid, W.H.A.; Subban, R.H.Y.; Rozana, M.D. Dielectric and Structural Properties of Poly(vinylidene fluoride) (PVDF) and Poly(vinylidene fluoride-trifluoroethylene) (PVDF-TrFE) Filled with Magnesium Oxide Nanofillers. *J. Nanomater.* **2019**, *2019*, 5961563. [CrossRef]

66. Gibbons, J.; Patterson, S.B.; Zhakeyev, A.; Vilela, F.; Marques-Hueso, J. Spectroscopic ellipsometric study datasets of the fluorinated polymers: Bifunctional urethane methacrylate perfluoropolyether (PFPE) and polyvinylidene fluoride (PVDF). *Data Brief* **2021**, *39*, 107461. [CrossRef]
67. Manke, F.; Frost, J.M.; Vaissier, V.; Nelson, J.; Barnes, P.R.F. Influence of a nearby substrate on the reorganization energy of hole exchange between dye molecules. *Phys. Chem. Chem. Phys.* **2015**, *17*, 7345–7354. [CrossRef] [PubMed]
68. Fernández, D.P.; Mulev, Y.; Goodwin, A.R.H.; Sengers, J.M.H.L. A Database for the Static Dielectric Constant of Water and Steam. *J. Phys. Chem. Ref. Data* **1995**, *24*, 33–70. [CrossRef]
69. Cunha, A.G.; Fernandes, S.C.M.; Freire, C.S.R.; Silvestre, A.J.D.; Neto, C.P.; Gandini, A. What Is the Real Value of Chitosan's Surface Energy? *Biomacromolecules* **2008**, *9*, 610–614. [CrossRef] [PubMed]
70. Lepoittevin, B.; Elzein, T.; Dragoe, D.; Bejjani, A.; Lemée, F.; Levillain, J.; Bazin, P.; Roger, P.; Dez, I. Hydrophobization of chitosan films by surface grafting with fluorinated polymer brushes. *Carbohydr. Polym.* **2019**, *205*, 437–446. [CrossRef] [PubMed]
71. Tungprapa, S.; Puangparn, T.; Weerasombut, M.; Jangchud, I.; Fakum, P.; Semongkhol, S.; Meechaisue, C.; Supaphol, P. Electrospun cellulose acetate fibers: Effect of solvent system on morphology and fiber diameter. *Cellulose* **2005**, *14*, 563–575. [CrossRef]
72. Hamilton, W. A technique for the characterization of hydrophilic solid surfaces. *J. Colloid Interface Sci.* **1972**, *40*, 219–222. [CrossRef]
73. Wu, Q.; Tiraferri, A.; Li, T.; Xie, W.; Chang, H.; Bai, Y.; Liu, B. Superwettable PVDF/PVDF-*g*-PEGMA Ultrafiltration Membranes. *ACS Omega* **2020**, *5*, 23450–23459. [CrossRef]
74. Deka, B.J.; Guo, J.; Khanzada, N.K.; An, A.K. Omniphobic re-entrant PVDF membrane with ZnO nanoparticles composite for desalination of low surface tension oily seawater. *Water Res.* **2019**, *165*, 114982. [CrossRef]
75. Cai, J.; Chen, Y.; Liu, Y.; Li, S.; Sun, C. Capillary imbibition and flow of wetting liquid in irregular capillaries: A 100-year review. *Adv. Colloid Interface Sci.* **2022**, *304*, 102654. [CrossRef]
76. Jamil, T.S.; Dijkstra, I.; Sayed, S. Usage of permeate water for treated domestic wastewater by direct capillary nanofiltration membrane in agriculture reuse. *Desalination Water Treat.* **2013**, *51*, 2584–2591. [CrossRef]

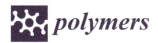

Article

Fe-Immobilised Catechol-Based Hypercrosslinked Polymer as Heterogeneous Fenton Catalyst for Degradation of Methylene Blue in Water

Thanchanok Ratvijitvech

Department of Chemistry, Faculty of Science, Mahidol University, Bangkok 10400, Thailand; thanchanok.rat@mahidol.edu

Abstract: Clean water is one of the sustainable development goals. Organic dye is one of the water pollutants affecting water quality. Hence, the conversion of dyes to safer species is crucial for water treatment. The Fenton reaction using Fe as a catalyst is a promising process. However, homogeneous catalysts are normally sensitive, difficult to separate, and burdensome to reuse. Therefore, a catechol-based hypercrosslinked polymer (catechol-HCP) was developed as an inexpensive solid support for Fe (catechol-HCP-Fe) and applied as a heterogenous Fenton catalyst. The good interaction of the catechol moiety with Fe, as well as the porous structure, simple preparation, low cost, and high stability of catechol-HCP, make it beneficial for Fe-loading in the polymer and Fenton reaction utilisation. The catechol-HCP-Fe demonstrated good catalytic activity for methylene blue (MB) degradation in a neutral pH. Complete decolouration of 100 ppm MB could be observed within 25 min. The rate of reaction was influenced by H_2O_2 concentration, polymer dose, MB concentration, pH, and temperature. The catechol-HCP-Fe could be reused for at least four cycles. The dominant reactive species of the reaction was considered to be singlet oxygen ($^1O_2$), and the plausible mechanism of the reaction was proposed.

Keywords: hypercrosslinked polymer; porous polymer; catechol; Fenton catalyst; dye degradation; methylene blue

Citation: Ratvijitvech, T. Fe-Immobilised Catechol-Based Hypercrosslinked Polymer as Heterogeneous Fenton Catalyst for Degradation of Methylene Blue in Water. *Polymers* **2022**, *14*, 2749. https://doi.org/10.3390/polym14132749

Academic Editors: Irene S. Fahim, Ahmed K. Badawi and Hossam E. Emam

Received: 17 May 2022
Accepted: 29 June 2022
Published: 5 July 2022

Publisher's Note: MDPI stays neutral with regard to jurisdictional claims in published maps and institutional affiliations.

Copyright: © 2022 by the author. Licensee MDPI, Basel, Switzerland. This article is an open access article distributed under the terms and conditions of the Creative Commons Attribution (CC BY) license (https://creativecommons.org/licenses/by/4.0/).

1. Introduction

Clean water is one of the global sustainable development goals. The contamination by dyes in wastewater from industrial and domestic uses is one of the causes of water pollution [1]. Many dyes are non-biodegradable and toxic to both humans and aquatic life. Moreover, coloured water can prevent the penetration of light into bodies of water, leading to decreases in photosynthesis and oxygen in the water, which can affect the water ecosystem and further decrease water quality [2,3]. Therefore, the development of effective approaches to remove dyes from water is crucial.

Methods such as adsorption [4–8], coagulation [9–14], and membrane filtration [14–18] have been used for dye removal. However, such methods only transfer the pollutants from one place to another. Another interesting approach is the conversion of the dyes to safer species. The Fenton reaction is a promising process for degrading dyes into other species. The reaction involves the catalytic oxidation of organic compounds, including organic dyes, by using iron (Fe) and hydrogen peroxide (H_2O_2) as reagents [19–23]. Fenton catalysis has the advantages of being inexpensive, simple, and easy to use. Fe and H_2O_2 are inexpensive, and Fe is a highly abundant element on earth. Generally, the Fenton reaction can occur under UV or visible light [24–27]. Thus, no extra energy is required. Both homogeneous and heterogeneous catalysts can be applied in the Fenton reaction. Nonetheless, heterogeneous catalysts can overcome some limitations of homogeneous catalysts, such as the pH operating range and the stability of catalysts under the oxidation conditions. The undesired Fe sludge, which is normally generated in the homogeneous

Fenton process, can be diminished by immobilising the Fe in heterogenous solid supports to prevent the leakage of the Fe into the system [28]. Moreover, solid catalysts are also easily separated from the reaction and have more potential for reuse. However, the reactivity of heterogeneous catalysts and the degradation of materials due to the harsh reaction conditions of the Fenton process are still challenging.

Hypercrosslinked polymers (HCPs) are amorphous polymeric materials with good stability [29,30]. HCPs typically have highly porous structures and large surface areas. HCPs with surface areas greater than 1000 m^2/g can be obtained using benzene as a monomer [31]. However, the surface areas of HCPs vary depending on the monomers used and the structures of the HCPs formed. HCPs can be prepared using the Friedel–Crafts reaction, which is cheap and simple in comparison to many other methods. The pores and voids of HCPs can be designed to have the desired functionalities to enhance the interactions with guest molecules [32]. A wide range of monomers with different functional groups, e.g., hydrocarbons, halides, alcohols, and amines, can be used to prepare HCPs [31,33–35]. Moreover, HCPs also demonstrate high chemical and physical stabilities. Due to these properties, HCPs are utilised in many applications, such as gas capture and storage [36–40], pollutant adsorption [41–44], antibacterial [45], catalysis [46–48], and energy storage [49–51]. HCPs are considered to be potential candidates for heterogenous Fenton catalysis. The porous structure and interconnected pores of HCPs can promote the Fe-loading in the polymers, the accessibility of the active sites, and the diffusion of the guest molecules through the materials. Moreover, the porous structure of HCPs allows for the adsorption of dyes and acts in nanoconfinement to increase the concentration of the dyes in materials, which can enhance the degradation rate. The highly stable properties of HCPs are also beneficial in preventing the degradation of materials during the strong conditions of the Fenton reaction and in increasing the potential for reusing the materials so as to reduce material preparation costs. Porous organic polymers, the similar types of materials, based on ferrocene [52] and porphyrin [24] have been reported to have high catalytic efficiency. However, expensive chemicals and complex monomer and polymer preparations were employed. Due to the large scale of industrial waste, materials with low costs are required. Therefore, the development of materials using inexpensive starting materials and simple preparation methods, which can preserve catalytic efficiency, is needed.

The expense of materials can be defined by their abundance, preparation process, efficiency, and reusability [53]. Catechol is an inexpensive and abundant compound widely found in nature and used in industries. The catechol structure has also been found to interact well with Fe. Recently, our group reported the enhancement of Fe adsorption in HCP using the catechol moiety [54]. Apart from increasing the interaction with Fe, the catechol moiety can also behave as a chelating agent to increase the regeneration of Fe^{2+} in the material, which can increase the catalytic activity of the Fenton reaction [55,56]. Thus, herein, the HCP synthesised by catechol monomer, catechol-based HCP (catechol-HCP), using inexpensive and simple chemicals and methods, was utilised as a solid support for immobilising Fe in Fenton dye-degradation catalysis. Methylene blue was used as a dye model. The Fe-immobilised catechol-HCP (catechol-HCP-Fe) was investigated for its efficiency in the degradation of MB. The catechol-HCP-Fe demonstrated good catalytic activity in MB degradation. Factors, including H_2O_2 concentration, polymer dose, initial MB concentration, pH, and reaction temperature, were found to influence the rate of reaction. The reusability of the material was also studied. The key reactive species and the plausible mechanism of the reaction were also investigated.

2. Materials and Methods
2.1. Chemicals and Reagents

Catechol was obtained from Acros Organics (Geel, Belgium). Formaldehyde dimethyl acetal (FDA), ferrous ammonium sulphate hexahydrate ($Fe(NH_4)_2(SO_4)_2 \cdot 6H_2O$), and hydrogen peroxide (H_2O_2) were purchased from Merck (Darmstadt, Germany). Iron (III) chloride ($FeCl_3$), 1,2-dichloroethane (DCE), sodium acetate anhydrous, and sodium hydroxide (NaOH) were purchased from Carlo Erba Reagents (Paris, France). Methylene blue (MB), ethylenediaminetetraacetic acid disodium salt (EDTA·2Na), 1,10-phenanthroline (*o*-phen), and hydroxylamine hydrochloride were purchased from Ajax Finechem (New South Wales, Australia). *Tert*-butanol (TBA), *p*-benzoquinone (*p*-BQ), and L-histidine (L-His) were purchased from TCI (Tokyo, Japan). Hydrochloric acid (HCl, 37%) was purchased from RCI Labscan (Bangkok, Thailand). All commercially available chemicals and solvents were used as received.

2.2. Characterisation Apparatus

Fourier transform infrared (FT-IR) spectra were recorded from a Frontier FT-IR Spectrometer (PerkinElmer, Waltham, MA, USA), using KBr disks. CHN elemental analysis was performed using a CHNS/O Analyser (Thermo Scientific™ FLASH 2000, Thermo Fisher Scientific, Waltham, MA, USA). Material surface morphology images and elemental analysis were collected by a field emission scanning electron microscope (FESEM, SU-8010, Hitachi, Tokyo, Japan) equipped with an EDAX Element Energy Dispersive Spectroscopy System. Thermogravimetric analysis (TGA) was conducted via SDT 2960 DSC-TGA (TA Instruments, New Castle, DE, USA) in an air atmosphere by heating to 800 °C at a rate of 10 °C/min. Nitrogen isotherms were collected on a 3Flex gas sorption analyser (Micromeritics, Norcross, GA, USA) at 77 K. Samples were degassed at 120 °C overnight. Surface areas were calculated using the Brunauer–Emmett–Teller (BET) theory in the relative pressure range of 0.05–0.5. UV–Visible spectra were recorded by a UV-1800 UV–Vis spectrophotometer (Shimadzu, Tokyo, Japan).

2.3. Preparation of Fe-Immobilised HCP (Catechol-HCP-Fe)

Catechol (0.1 mol, 11 g) in 1,2-dichloroethane (100 mL) was added to formaldehyde dimethyl acetal (FDA) (0.2 mol, 18 mL) and iron (III) chloride ($FeCl_3$) (0.2 mol, 32 g). After heating at 80 °C for 24 h, the reaction was filtered. The solid was washed with methanol until the filtrate was colourless and then dried in an oven at 80 °C overnight. An insoluble black powder of catechol-HCP was obtained (13 g, 99%).

Before the Fe-immobilisation process, the catechol-HCP was washed with 1 M HCl solution to remove the residual Fe from the synthetic process. The polymer was repeatedly washed until less than 1 ppm of Fe leaching into the solution was observed. The polymer was further washed with distilled water until the filtrate was neutral. After that, the washed polymer was dried in an oven at 80 °C overnight to obtain catechol-HCP-w (12 g, 94%); elemental analysis: C 62.68%, H 3.56%.

After the catechol-HCP was washed, Fe was loaded into the polymer by immersing the prepared polymer (1 g) in the Fe^{2+} solution (500 ppm, 100 mL). The solution was stirred at room temperature for 24 h and then filtered to obtain catechol-HCP-Fe as a black powder (1 g, 97%); elemental analysis: C 58.83%, H 3.44%.

The catechol-HCP and catechol-HCP-Fe were characterised by Fourier transform infrared (FT-IR) spectroscopy, elemental analysis, scanning electron microscopy (SEM), energy dispersive X-ray analysis (EDX), thermogravimetric analysis (TGA), and surface area analysis.

2.4. Fe-Loading Determination in Catechol-HCP-Fe

The amount of Fe loading in the catechol-HCP-Fe was calculated by the difference of the Fe concentration in the solution, before and after filtration. The concentration of Fe in the solution was determined by the colourimetry technique. The Fe in the solution was complexed with 1,10-phenanthroline (*o*-phen) to form the colour complex. Hydroxylamine hydrochloride and sodium acetate buffer pH 5 were used as a reducing agent and a pH-adjusting solution, respectively. An UV-Visible spectrophotometer was used to measure the absorbance of the solutions at 510 nm related to the absorption maximum wavelength of the Fe complex. The Fe loading in the catechol-HCP-Fe was calculated as being around 25 mg/g.

2.5. Dye Degradation Experiments

The Fenton catalytic activity of the materials was evaluated by their methylene blue (MB) degradation efficiency. A known concentration of dye solution (50 mL) was added to the prepared polymer. The mixture was added to a hydrogen peroxide (H_2O_2) solution and stirred (500 rpm) at room temperature. The typical experiments were done at a neutral pH (pH = 6) using deionised water. Due to the fact that the pH of deionised water is slightly lower than that of normal water (pH = 7), a pH of 6 was considered as a neutral pH in this work. The pH was further adjusted to obtain a pH range of 3–11 by using 1 M HCl and 1 M NaOH. The reaction solution was collected at a specific time so as to study the reaction kinetics. The MB concentration was determined by an UV-Visible spectrophotometer at 664 nm. The percentage of degradation was calculated using Equation (1):

$$\text{Degradation (\%)} = (C_0 - C_t) \times 100 / C_0 \tag{1}$$

where C_0 and C_t are the concentrations of MB at 0 and t min, respectively.

For the reuse experiments, the catechol-HCP-Fe was filtered from the reaction, washed with water several times, and dried in an oven at 80 °C for 24 h. The material was then reused for the next cycle.

For the mechanistic study, 1 mM of scavenger was added to the reaction mixture using a 1 g/L catechol-HCP-Fe, 0.25 M H_2O_2, 50 mL (100 ppm) MB solution, pH 6 (neutral pH), which was stirred (500 rpm) at room temperature. Tert-butanol (TBA), p-benzoquinone (p-BQ), L-histidine (L-His), and ethylenediaminetetraacetic acid disodium salt (EDTA) were used as the ·OH, ·O_2^-, $^1O_2$, and photo-generated h^+ scavengers, respectively.

3. Results and Discussion

3.1. Preparation and Characterisation of Fe-Immobilised HCP (Catechol-HCP-Fe)

3.1.1. Preparation of Fe-Immobilised HCP (Catechol-HCP-Fe)

Fe-immobilised HCP (catechol-HCP-Fe) was prepared via the synthesis of catechol-based HCP (catechol-HCP), followed by the immobilisation of Fe in the catechol-HCP (Scheme 1).

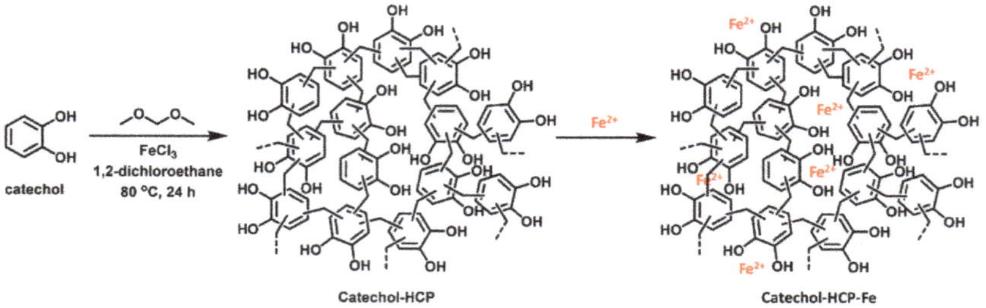

Scheme 1. Preparation of Catechol-HCP-Fe.

Catechol-HCP was synthesised by extensive crosslinking of the catechol monomer with an external crosslinker, formaldehyde dimethyl acetal (FDA), using $FeCl_3$ catalyst via a Friedel–Crafts alkylation reaction. A black powder of catechol-HCP was obtained in a good yield (99%). The result also demonstrated the possibility of scaling up the synthesis of material as a 10-g-scale of material was successfully synthesised without the reduction of the yield as compared to the smaller-scale synthesis reported in the literature [54].

Before the immobilisation process, the synthesised catechol-HCP was washed with HCl to remove the residual Fe from the synthetic process. To immobilise the Fe in the material, the washed catechol-HCP (catechol-HCP-w) was immersed in the Fe solution for 24 h and filtered to obtain the catechol-HCP-Fe. The amount of Fe adsorbed in the polymer was calculated by the difference of the Fe concentration in the solution, before and after filtration. By using 100 mL of 500 ppm Fe^{2+} solution and 1 g of catechol-HCP, around

250 ppm was found adsorbed in the polymer, calculated as being approximately 25 mg/g Fe-loading in the material.

3.1.2. Fourier Transform Infrared (FT-IR) Spectroscopy

Chemical structures of the prepared polymers were characterised by IR spectroscopy (Figure 1). IR spectra of catechol-HCP (as synthesised), catechol-HCP-w (after washed), and catechol-HCP-Fe (after Fe immobilisation) illustrated similar peaks, indicating that the chemical structure of the material did not significantly change during the modification process. Aromatic C=C stretching peaks around 1600 cm^{-1} and C-H stretching and bending peaks around 2900 and 1440 cm^{-1} demonstrated the aromatic backbone of the catechol moiety and the methylene group of the crosslinker, respectively. A broad O-H stretching peak around 3200 cm^{-1} and C-O stretching peaks around 1090 cm^{-1} indicated the presence of alcohol (-OH) groups in the materials. Other O-H and C-O stretching peaks observed at around 3400 and 1250 cm^{-1} could be attributed to the moisture, which might be adsorbed in the porous structure of the material, and the ether functional group, which might remain after the incomplete crosslinking reaction, respectively. Nevertheless, the intensity of the peak around 570 cm^{-1}, corresponding to Fe-O bonds, slightly decreased after the material was washed, and it increased again after the Fe was immobilised, suggesting the success of Fe removal and incorporation in the material [57,58].

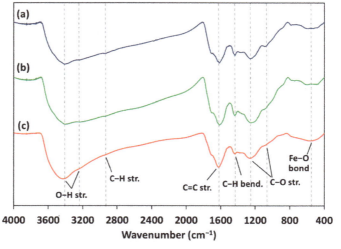

Figure 1. IR spectra of (**a**) catechol-HCP, (**b**) catechol-HCP-w, and (**c**) catechol-HCP-Fe.

3.1.3. Elemental Analysis

Elemental contents in the materials were analysed by CHN elemental analysis. The results are demonstrated in Table 1. C and H contents lower than the theoretical values were typically found in the HCPs, which could possibly be due to the adsorption ability of the materials in adsorbing molecules and moisture from the air and the higher-than-expected O content in the HCPs, resulting from the incomplete crosslinking reaction. However, comparing the catechol-HCP-w and the catechol-HCP-Fe, decreases in C and H contents in the polymer after Fe immobilisation were observed as expected due to the incorporation of Fe in the polymer.

Table 1. Elemental analysis of HCPs.

Polymer	% Carbon	% Hydrogen
Catechol-HCP-w	62.68 ± 0.01	3.56 ± 0.12
(Theoretical)	(71.64)	(4.51)
Catechol-HCP-Fe	58.83 ± 0.29	3.44 ± 0.11
(Theoretical) [1]	(69.85)	(4.40)

[1] Calculated as 2.5% Fe content (25 mg/g) in the polymer.

3.1.4. Scanning Electron Microscopy (SEM) and Energy Dispersive X-ray (EDX) Analysis

The surface morphology of the materials was perceived by a scanning electron microscope. Figure 2 illustrates the SEM images of the catechol-HCP-w (a,b) and the catechol-HCP-Fe (c,d). The polymer, after four reuse cycles, catechol-HCP-reused, was also investigated as demonstrated in Figure 2e,f. The surface morphology of the materials did not significantly change after the Fe immobilisation and reuse steps, implying the endurance of the materials under the operating conditions. While the EDX spectrum of the catechol-HCP-w (Figure S1 from the Supplementary Materials) did not show Fe peaks, the peaks appeared in the spectrum of catechol-HCP-Fe (Figure S2) after Fe immobilisation. As predicted, the EDX elemental analysis (Table S1) also elucidated a decrease in the carbon ratio and an increase in the Fe ratio in the catechol-HCP-Fe as compared to those in the catechol-HCP-w.

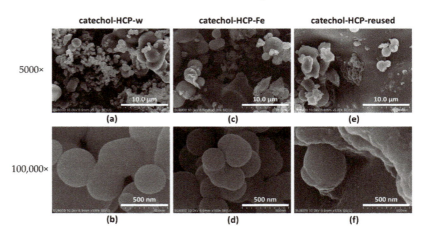

Figure 2. SEM images of catechol-HCP-w at (**a**) 5000× and (**b**) 100,000×; catechol-HCP-Fe at (**c**) 5000× and (**d**) 100,000×; and catechol-HCP-reused at (**e**) 5000× and (**f**) 100,000×.

3.1.5. Thermogravimetric Analysis (TGA)

The thermal stability of the prepared materials was analysed using the TGA technique. The TGA spectra of catechol-HCP-w and catechol-HCP-Fe are shown in Figures S3 and S4, respectively. The weight loss around 100 °C resulted from the loss of water moisture adsorbed in the polymers. The materials, both with and without Fe, were thermally stable with heating up to around 250 °C in an air atmosphere. The decomposition of materials took place at around 250–600 °C. Upon reaching 800 °C, the catechol-HCP-w produced char around 0.5%, while around 1.8% of char was obtained in the case of the catechol-HCP-Fe. The higher amount of char left, which could be from the unburned Fe, also pointed to the incorporation of Fe in the catechol-HCP-Fe.

3.1.6. Surface Area Analysis

The porosity of the materials was analysed by nitrogen sorption isotherms at 77 K (Figure 3). The isotherms of the catechol-HCP before (catechol-HCP-w, green triangle) and after (catechol-HCP-Fe, red circle) Fe immobilisation demonstrated the similar shapes and gas sorption capabilities. The surface areas, calculated by the Brunauer–Emmet–Teller (BET) theory, of the catechol-HCP-w and catechol-HCP-Fe were around 2.4 and 1.9, respectively, comparable to the reported values [54]. After the Fe was immobilised, a slight decline in surface area was observed, implying the adsorption of Fe into the porous structure. The pore volume of the material also diminished after Fe immobilisation. The total pore volume declined slightly, from 2.57 to 2.46 mm^3/g, while the micropore volume declined more significantly, from 0.78 to 0.32 mm^3/g, suggesting that the Fe was better-adsorbed into the microporous structure of the material. Despite the low measured surface area, the material showed a good Fe-adsorption capacity. This could imply the existence of the porous structure of the material. The low value of surface areas could possibly be due to many reasons. The -OH substituents on the catechol monomer could have formed H-bonding or have filled the pores, leading to the lower surface areas [59]. Moreover, the pores might collapse in the dry state. However, HCPs were found to be swellable when adsorbing the solvents or guest molecules [37,60]. Thus, although the low surface areas were observed, the material is still considered as the good adsorbent.

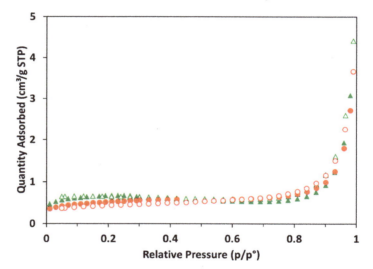

Figure 3. Nitrogen isotherms of catechol-HCP-w (green triangle) and catechol-HCP-Fe (red circle). The closed and open symbols represent the adsorption and desorption isotherms, respectively.

3.2. Catalytic Property of Catechol-HCP-Fe

The Fenton catalytic property of the prepared catechol-HCP-Fe was examined by adding catechol-HCP-Fe into the MB solution before adding H_2O_2 to the reaction mixture. For comparison, experiments using catechol-HCP-Fe mixed with MB without H_2O_2, MB solution with H_2O_2, and MB solution were conducted to confirm that the dye was not simply adsorbed into the polymer, degraded with H_2O_2, or self-degraded in the experimental environment, respectively.

The results are shown in Figure 4. The concentrations of the MB in the MB solution (MB) and the MB mixed with H_2O_2 solution (MB + H_2O_2) were only slightly changed, indicating that the self-degradation of MB, both with and without H_2O_2, occurred only minimally in the experimental environment. The adsorption of the MB by the catechol-HCP-Fe was observed, as shown by the slight decrease in the MB concentration when the catechol-HCP-Fe was added to the solution without H_2O_2 (HCP-Fe + MB). The ability of the HCP in adsorbing MB would be beneficial in the mass transportation of the guest molecules to the active sites of the material. However, compared to the experiment with the catechol-HCP-Fe in the presence of H_2O_2 (HCP-Fe + MB + H_2O_2), which demonstrated a significant decrease in the MB concentration, the adsorption process was much slower than the catalytic process. This could confirm the catalytic ability of the catechol-HCP-Fe in the degradation of the MB.

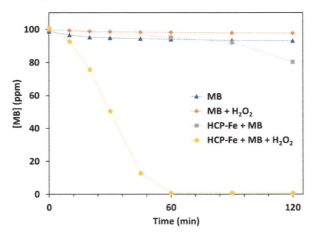

Figure 4. Concentrations of MB at different times under different conditions. Reaction conditions: [HCP-Fe] = 1 g/L, [H_2O_2] = 0.5 M, [MB] = 100 ppm, pH = 6 (neutral pH), room temperature (27 °C).

To accurately calculate the kinetics of the MB degradation reaction, the data were fitted with zero-order, first-order, and second-order kinetic models using Equations (2)–(4), respectively:

$$C_t = -k_0 t + C_0 \quad (2)$$

$$\ln C_t = -k_1 t + \ln C_0 \quad (3)$$

$$1/C_t = k_2 t + 1/C_0 \quad (4)$$

where C_0 and C_t are the concentrations of the MB at 0 and t min, and k_0, k_1, and k_2 are zero-order, first-order, and second-order rate constants, respectively. Considering the curve-fitting (Figure S5) and R^2 values (Table 2), the zero-order kinetic model demonstrated the best fit to the experimental data.

Table 2. Kinetic constants and R^2 calculated by different kinetic models.

Kinetic Model	Rate Constant (k)	Rate Constant Unit	R^2
Zero-order	2.3175	ppm/min	0.9936
First-order	0.0569	min^{-1}	0.9125
Second-order	0.0019	$ppm^{-1} min^{-1}$	0.8000

It is worth noting that most of the previously reported kinetic results followed the first-order kinetic model, where the reaction was slower with the time of reaction. This could be caused by the accumulation of intermediates in the material, which could hinder the active sites, leading to the lower reaction rate [52,58,61]. In comparison, the zero-order kinetics in this work suggested that the reaction was not slower with the reaction time and the porous structure of the HCP could promote mass transportation in the material. A similar result was also observed in the porphyrin-based porous material [24].

3.3. Effect of Parameters on MB Degradation

To optimise the catalytic efficiency of the catechol-HCP-Fe in the dye-degradation process, parameters that could affect the catalytic activity, including the H_2O_2 concentration, catechol-HCP-Fe dose, initial dye concentration, pH, and temperature, were studied.

3.3.1. Effect of H_2O_2 Concentration

To investigate the effect of the H_2O_2 concentration on the catalytic efficiency, different concentration of H_2O_2 (0.05, 0.1, 0.25, 0.5, 1, and 2 M) were added to the reactions. The concentrations were the final concentrations of H_2O_2 in the reaction mixtures. The dye-degradation efficiency, using various H_2O_2 concentrations, is demonstrated in Figure 5. The rate of reaction increased with the increase in the H_2O_2 concentrations, as shown in Figure 6 (left). The estimated reaction time of the reaction, illustrated in Figure 6 (right), was also calculated. The dye degradation occurred faster with a higher concentration of H_2O_2. The reaction time was reduced from more than 160 min to less than 30 min when the H_2O_2 concentration was increased from 0.05 to 2 M. However, the reaction time significantly decreased when the H_2O_2 concentration was increased from 0.05 to 0.5 M, but increased slightly when the H_2O_2 concentration was further increased to 1 and to 2 M. Therefore, considering the efficiency of the reagent-spending, a H_2O_2 concentration of 0.5 M was used in further experiments.

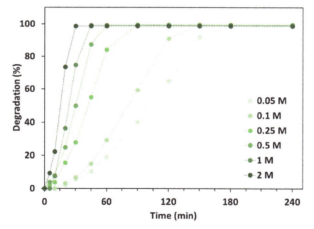

Figure 5. Degradation efficiency using different H_2O_2 concentrations. Reaction conditions: [catechol-HCP-Fe] = 1 g/L, [MB] = 100 ppm, pH = 6 (neutral pH), room temperature (27 °C).

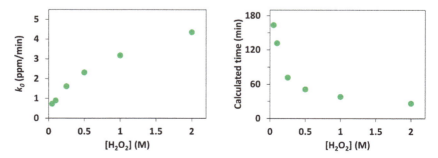

Figure 6. Rate constant (**left**) and calculated time (**right**) of MB degradation using different H_2O_2 concentrations.

3.3.2. Effect of Polymer Dose

The amount of catalyst could also influence the rate of the catalytic reaction. Therefore, the amount of the catechol-HCP-Fe was varied from 0.2 to 3 g/L. The results are illustrated in Figure 7. The dye degradation occurred faster with increases in the catechol-HCP-Fe dose. The rate of reaction could be accelerated by using a higher dose of the catalyst. The plateau was reached at 2 g/L of catechol-HCP-Fe, as demonstrated in Figure 8 (left). More than 99% of 100 ppm MB could be degraded in around 30 min using 2 g/L of catechol-HCP-Fe and 0.5 M H_2O_2. However, considering the time efficiency, (see Figure 8 (right)), the reaction time significantly decreased when the dose of catechol-HCP-Fe was increased from 0.2 to 1 g/L and levelled off when the dose was further increased to 2 and to 3 g/L. Thus, a catechol-HCP-Fe dose of 1 g/L was used in further investigations. An experiment using 2 g/L of catechol-HCP-Fe and 2 M of H_2O_2 was also carried out to optimise the reaction time; the complete degradation of the MB was observed within 25 min.

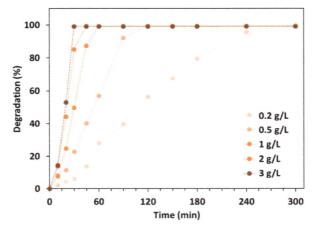

Figure 7. Degradation efficiency using different catechol-HCP-Fe doses. Reaction conditions: $[H_2O_2]$ = 0.5 M, [MB] = 100 ppm, pH = 6 (neutral pH), room temperature (27 °C).

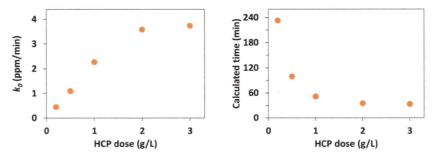

Figure 8. Rate constant (**left**) and calculated time (**right**) of MB degradation using different catechol-HCP-Fe doses.

3.3.3. Effect of Initial Dye Concentration

Another factor that could affect the dye-degradation efficiency is the initial dye concentration. Initial MB concentrations of 10, 50, 100, and 500 ppm were compared. The dye-degradation efficiency, the rate constant, and the estimated time of the reaction are shown in Figures 9 and 10. With an increase in the initial MB concentration, a longer time for the dye degradation was required, owing to a larger number of dye molecules needing to be transformed while a limited number of active sites were available at one time. However, the rate constant increased with the increase in initial dye concentration as a greater number of molecules were residing near the active sites, along with the good mass transportation in the material, leading to a faster rate of reaction. Interestingly, the time of the reaction seemed to be linearly correlated to the initial dye concentration used in the reaction. Thus, the reaction time could be estimated if the initial MB concentration were known. Moreover, water contaminated with MB in amounts of up to 500 ppm could be treated by catechol-HCP-Fe within a couple of hours.

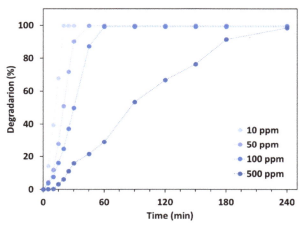

Figure 9. Degradation efficiency using different initial MB concentrations. Reaction conditions: [catechol-HCP-Fe] = 1 g/L, [H_2O_2] = 0.5 M, pH = 6 (neutral pH), room temperature (27 °C).

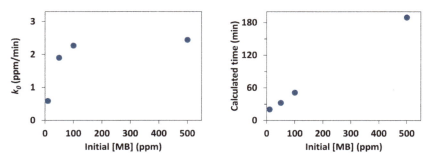

Figure 10. Rate constant (**left**) and calculated time (**right**) of MB degradation using different initial MB concentrations.

3.3.4. Effect of Initial pH of Dye Solution and Reaction Temperature

The pH and temperature of wastewater might be unpredictable. Therefore, the effects of these factors on the Fenton-reaction efficiency were studied. Nearly 100% of 100 ppm of MB could be degraded by the catechol-HCP-Fe within a pH range of 3 to 11 (Figure S6) and a reaction temperature range of around 27 °C (room temperature) to 75 °C (Figure S7). Many have reported an optimum pH of around 3 for the Fenton process [62–64]. Therefore, the extra step and reagents to acidify the system were needed. However, our catalyst demonstrated good catalytic efficiency in a wide pH range. Thus, the extra cost and step of adjusting the pH could be omitted. More surprisingly, unlike traditional Fenton reactions where the optimum pH is in acidic range, the catechol-HCP-Fe demonstrated a slightly slower reaction rate in acidic conditions and a faster reaction rate in basic conditions, as shown in Figure 11. The reaction temperature also affected the rate of MB degradation. As expected, increasing the reaction temperature accelerated the reaction rate (Figure 11) due to the increase in kinetic energy in the system.

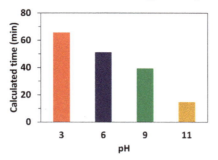
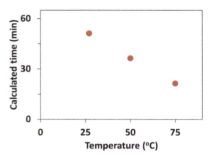

Figure 11. Calculated time of MB degradation using different pH levels (**left**) and temperatures (**right**). Reaction conditions: [catechol-HCP-Fe] = 1 g/L, [H_2O_2] = 0.5 M, [MB] = 100 ppm, pH = 6 (neutral pH) or as specified, room temperature (27 °C) or as specified.

The experiments showed that the MB-degradation efficiency of the catechol-HCP-Fe was affected by the reaction conditions. The degradation rate increased with increases in the H_2O_2 concentration, catalyst dose, and initial dye concentration. The optimum condition concerning reagent spending was 1 g/L catechol-HCP-Fe and 0.5 M H_2O_2, in which 100 ppm of MB could be degraded in around 50 min at room temperature and with a neutral pH. However, an optimum time of around 23 min was obtained using 2 g/L catechol-HCP-Fe and 2 M H_2O_2 for 100 ppm of MB at room temperature and with a neutral pH. The catechol-HCP-Fe was also proven to be efficiently used in a wide operating range, pH of 3–11 and temperature of 27–75 °C. Compared with the reported materials (Table 3), catechol-HCP-Fe is considered to be an inexpensive material with a good catalytic property, and thus having a potential use in industrial wastewater management.

Table 3. MB degradation efficiency using different catalysts under visible light.

Material	Dose (g/L)	[MB] (ppm)	[H_2O_2] (M)	Efficiency (%)	Time (min)	pH	Reference
$FeCl_2 \cdot 4H_2O$	0.08	70	0.03	99	120	7	[24]
$FeCl_3 \cdot 6H_2O$	0.08	70	0.03	99	120	7	[24]
Fe_2O_3	0.08	70	0.03	~30	120	7	[24]
Fe_3O_4	0.08	70	0.03	~25	120	7	[24]
FePPOP-1	0.08	70	0.03	99	80	7	[24]
FePPOP-1	0.08	100	0.03	99	>120	7	[24]
FcTz-POP	0.2	8	1	99	20	7	[52]
wSF-DA/Fe	0.1	20	1	98 *	10–40 *	7	[58]
Fe_2O_3@FCNT-H	0.015	3	0.05	99	60	5	[61]
M-NPs	2	100	0.56	~100	90	3.5	[63]
C-Fe_2O_3-2	0.5	50	0.0075	96	420	-	[65]
Fe_3O_4@rGO@TiO_2	1.5	10	0.176	99	120	7	[66]
MIL-53(Fe)	0.01	128	10^{-5}	20	20	7	[67]
NTU-9 (Ti(IV)-MOF)	0.5	32	~0.25	99	20–40	7	[68]
Fe, N-CDs	0.5	20	0.147	97 *	60 *	8	[69]
Catechol-HCP-Fe	1	10	0.5	99	21	6	This work
Catechol-HCP-Fe	1	100	0.5	99	51	6	This work
Catechol-HCP-Fe	2	100	2	99	23	6	This work

* Reaction temperature = 50 °C.

3.4. Reusability of Catechol-HCP-Fe for Dye Degradation

To evaluate the stability and reusability of the catalyst, the catechol-HCP-Fe was filtered from the solution after the reaction was done, washed several times with water, and dried in an oven before reuse. More than 99% of the MB could be degraded using the reused catalyst for at least 4 cycles (Figure 12). However, the rate of reaction dropped when the catalyst was reused, and a longer time was required for complete degradation (Figure S8). This could be responsible from the leaching of Fe into the solution during the reaction, causing less Fe to remain in the catalyst. However, only 0.5–2 mg/g of the Fe was found in the solution, calculated as only 2–8% Fe leaching, as compared to the amount of Fe loaded in the polymer. To improve the rate of the reaction, the Fe was reloaded into the reused catalyst prior to the new reuse cycle. By using the catalyst with reloaded Fe, it was found that the reaction time was comparable to the freshly prepared catalyst, as shown in Figure S9. These results could demonstrate that catechol-HCP-Fe can be reused and regenerated for catalytic dye-degradation reactions.

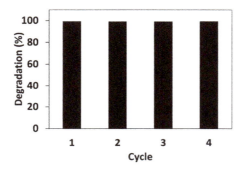

Figure 12. Degradation efficiency of the reused catalyst. Reaction conditions: [catechol-HCP-Fe] = 1 g/L, [H_2O_2] = 0.5 M, [MB] = 100 ppm, pH = 6 (neutral pH), room temperature (27 °C).

3.5. Mechanistic Study of Dye Degradation Process

The catalytic mechanism of the catechol-HCP-Fe was also studied by using trapping reagents. The major active species in the Fenton catalytic process are believed to be ·OH, ·O_2^-, and $^1O_2$. Thus, *tert*-butanol (TBA), the ·OH scavenger, *p*-benzoquinone (*p*-BQ), the ·O_2^- scavenger, L-histidine (L-His), the $^1O_2$ scavenger, and ethylenediaminetetraacetic acid disodium salt (EDTA), the photo-generated h^+ scavenger, were added to the reaction mixtures. As demonstrated in the results shown in Figure 13, the rate of the reaction decreased when the scavengers were added to the reaction, indicating the responsibility of the active species in the Fenton catalytic process using catechol-HCP-Fe. Interestingly, the trapping of ·OH, the typical prominent species in Fenton catalysis, slightly affected the reaction rate, suggesting that the ·OH might not be the main active species in this reaction mechanism. Instead, the dominant active species was considered to be $^1O_2$, owing to the significantly longer reaction time when L-His was added to the reaction. The ·O_2^- and photo-generated h^+ were also involved in the reaction mechanism.

According to the experimental results, the postulated reaction mechanism for the MB degradation of catechol-HCP-Fe could be as follows:

$$Fe^{3+} + H_2O_2 \rightarrow Fe^{2+} + \cdot O_2^- + 2H^+ \tag{5}$$

$$2 \cdot O_2^- + 2H_2O \rightarrow 2\,^1O_2 + H_2O_2 + 2H^+ \tag{6}$$

$$Fe^{2+} + 2 \cdot O_2^- + 4H^+ \rightarrow Fe^{3+} + \,^1O_2 + 2H_2O \tag{7}$$

$$MB + \,^1O_2 \rightarrow CO_2 + H_2O + \text{other products} \tag{8}$$

It is worth noting that, even though ·OH has commonly been found to be the main active species for Fenton reactions [52,58,70], some research has also reported $^1O_2$ to be the key species [24,61,71]. The nanoconfinement of the Fe in the porous structure of catechol-HCP-Fe might stimulate the generation of $^1O_2$, as Yang et al. demonstrated that $^1O_2$ was the preferably predominant species in the nanoconfined structure of Fe$_2$O$_3$@FCNT, while the ·OH was the main active species when the Fe$_2$O$_3$ was anchored on the outer surface of the FCNT [61]. The catechol moiety was also found to induce ^{1}O$_2$ production, as reported by Bokare and Choi [72]. This diverse mechanism could extend the opportunities for employing the material in broader reaction conditions and utilisations.

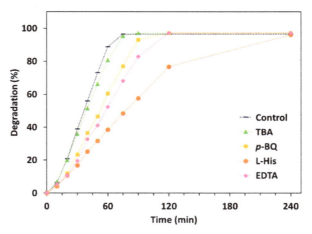

Figure 13. Effect of scavengers on the MB-degradation efficiency of catechol-HCP-Fe. Reaction conditions: [scavenger] = 1 mM, [catechol-HCP-Fe] = 1 g/L, [H$_2$O$_2$] = 0.5 M, [MB] = 100 ppm, pH = 6 (neutral pH), room temperature (27 °C).

4. Conclusions

The Fe-immobilised catechol-HCP (catechol-HCP-Fe) was successfully prepared by the synthesis of catechol-HCP, followed by the immobilisation of Fe in the material. The large-scale synthesis of catechol-HCP could provide good yields, and the catechol-HCP was fully characterised by FT-IR spectroscopy, CHN elemental analysis, SEM, EDX, TGA, and surface area analysis. The amount of Fe immobilised in the catechol-HCP-Fe was approximately 25 mg/g.

The prepared material was studied for its catalytic activity in a Fenton reaction for the degradation of methylene blue (MB), the dye model. The parameters, including H$_2$O$_2$ concentration, catechol-HCP-Fe dose, initial dye concentration, pH, and temperature, were found to affect the reaction rate. Increasing the H$_2$O$_2$ concentration and dose of the catechol-HCP-Fe were found to enhance the rate of the reaction. The complete decolouration of 100 ppm MB was observed within 25 min, using 2 M H$_2$O$_2$ and 2 g/L of catechol-HCP-Fe at room temperature and at a neutral pH. However, considering the reagents' cost efficiency, the optimised condition for MB degradation was 0.5 M H$_2$O$_2$ and 1 g/L of catechol-HCP-Fe, which could degrade 100 ppm of MB in around 50 min. The reaction time was discovered to be linearly correlated to the initial dye concentration. The catalyst could also remove MB at high concentrations, up to 500 ppm, within a couple of hours. Catechol-HCP-Fe can be efficiently utilised in wide pH (3–11) and temperature (27–75 °C) ranges. Surprisingly, unlike typical Fenton reactions, the reaction rate increased with an increase in pH. The material can be reused for at least four cycles. Fe can simply be reloaded into the polymer to regenerate the catalyst and improve catalytic efficiency. The plausible reaction mechanism was proposed. The singlet oxygen (^{1}O$_2$) was presumed to be the dominant reactive species responsible for the MB degradation in the reaction.

This research demonstrated the use of catechol-HCP as an inexpensive material for metal ion nanoconfinement and its application in Fenton catalysis. By using abundant and low-cost chemicals and simple preparation methods, together with its good catalytic property in dye degradation and reusability, catechol-HCP and catechol-HCP-Fe are promising candidates as efficient and inexpensive novel materials for wastewater treatment.

Supplementary Materials: The following supporting information can be downloaded at: https://www.mdpi.com/article/10.3390/polym14132749/s1. Figure S1: EDX spectrum and EDX elemental analysis of Catechol-HCP-w; Figure S2: EDX spectrum and EDX elemental analysis of Catechol-HCP-Fe; Figure S3: TGA spectrum of Catechol-HCP-w; Figure S4: TGA spectrum of Catechol-HCP-Fe; Figure S5: Kinetic plots of MB degradation using (a) zero-order, (b) first-order, (c) second-order kinetic model; Figure S6: Degradation efficiency using different initial pH; Figure S7: Degradation efficiency using different reaction temperature; Figure S8: Degradation efficiency of the reused catalyst without Fe reloading; Figure S9: Degradation efficiency of the reused catalyst with Fe reloading; Table S1: EDX elemental analysis (weight%) of HCPs.

Funding: This research is supported by Mahidol University (Basic Research Fund: fiscal year 2021), grant number BRF1-A25/2564; and the CIF and CNI Grant, Faculty of Science, Mahidol University, grant number CH-TR:4/64#002.

Institutional Review Board Statement: Not applicable.

Informed Consent Statement: Not applicable.

Data Availability Statement: Not applicable.

Acknowledgments: The author gratefully appreciates the financial support from Mahidol University (Basic Research Fund: fiscal year 2021), grant number BRF1-A25/2564; and the CIF and CNI Grant from the Faculty of Science, Mahidol University, grant number CH-TR:4/64#002. TR thanks the staff and students from the Department of Chemistry, Faculty of Science, and Mahidol University, especially TR's lab members, and also the staff of many instrument centres for all support and assistance.

Conflicts of Interest: The authors declare no conflict of interest.

References

1. Katheresan, V.; Kansedo, J.; Lau, S.Y. Efficiency of various recent wastewater dye removal methods: A review. *J. Environ. Chem. Eng.* **2018**, *6*, 4676–4697. [CrossRef]
2. Lellis, B.; Fávaro-Polonio, C.Z.; Pamphile, J.A.; Polonio, J.C. Effects of textile dyes on health and the environment and bioremediation potential of living organisms. *Biotechnol. Res. Innov.* **2019**, *3*, 275–290. [CrossRef]
3. Khan, S.; Malik, A. Toxicity evaluation of textile effluents and role of native soil bacterium in biodegradation of a textile dye. *Environ. Sci. Pollut. Res.* **2018**, *25*, 4446–4458. [CrossRef] [PubMed]
4. Shanker, U.; Rani, M.; Jassal, V. Degradation of hazardous organic dyes in water by nanomaterials. *Environ. Chem. Lett.* **2017**, *15*, 623–642. [CrossRef]
5. Jiang, D.; Deng, R.; Li, G.; Zheng, G.; Guo, H. Constructing an ultra-adsorbent based on the porous organic molecules of noria for the highly efficient adsorption of cationic dyes. *RSC Adv.* **2020**, *10*, 6185–6191. [CrossRef]
6. Su, P.; Zhang, X.; Xu, Z.; Zhang, G.; Shen, C.; Meng, Q. Amino-functionalized hypercrosslinked polymers for highly selective anionic dye removal and CO_2/N_2 separation. *New J. Chem.* **2019**, *43*, 17267–17274. [CrossRef]
7. Yagub, M.T.; Sen, T.K.; Afroze, S.; Ang, H.M. Dye and its removal from aqueous solution by adsorption: A review. *Adv. Colloid Interface Sci.* **2014**, *209*, 172–184. [CrossRef]
8. Motejadded Emrooz, H.B.; Maleki, M.; Rashidi, A.; Shokouhimehr, M. Adsorption mechanism of a cationic dye on a biomass-derived micro- and mesoporous carbon: Structural, kinetic, and equilibrium insight. *Biomass Convers. Biorefin.* **2020**, *11*, 943–954. [CrossRef]
9. Wu, H.; Liu, Z.; Li, A.; Yang, H. Evaluation of starch-based flocculants for the flocculation of dissolved organic matter from textile dyeing secondary wastewater. *Chemosphere* **2017**, *174*, 200–207. [CrossRef]
10. Zhao, C.; Zheng, H.; Sun, Y.; Zhang, S.; Liang, J.; Liu, Y.; An, Y. Evaluation of a novel dextran-based flocculant on treatment of dye wastewater: Effect of kaolin particles. *Sci. Total Environ.* **2018**, *640–641*, 243–254. [CrossRef]
11. Li, H.; Liu, S.; Zhao, J.; Feng, N. Removal of reactive dyes from wastewater assisted with kaolin clay by magnesium hydroxide coagulation process. *Colloids Surf. A Physicochem. Eng. Asp.* **2016**, *494*, 222–227. [CrossRef]
12. Lau, Y.-Y.; Wong, Y.-S.; Teng, T.-T.; Morad, N.; Rafatullah, M.; Ong, S.-A. Coagulation-flocculation of azo dye Acid Orange 7 with green refined laterite soil. *Chem. Eng. J.* **2014**, *246*, 383–390. [CrossRef]
13. Sadri Moghaddam, S.; Alavi Moghaddam, M.R.; Arami, M. Coagulation/flocculation process for dye removal using sludge from water treatment plant: Optimization through response surface methodology. *J. Hazard. Mater.* **2010**, *175*, 651–657. [CrossRef]

14. Kasperchik, V.P.; Yaskevich, A.L.; Bil'dyukevich, A.V. Wastewater treatment for removal of dyes by coagulation and membrane processes. *Pet. Chem.* **2012**, *52*, 545–556. [CrossRef]
15. Abid, M.F.; Zablouk, M.A.; Abid-Alameer, A.M. Experimental study of dye removal from industrial wastewater by membrane technologies of reverse osmosis and nanofiltration. *Iran. J. Environ. Health Sci. Eng.* **2012**, *9*, 17. [CrossRef]
16. Rashidi, H.R.; Sulaiman, N.M.N.; Hashim, N.A.; Hassan, C.R.C.; Ramli, M.R. Synthetic reactive dye wastewater treatment by using nano-membrane filtration. *Desalin. Water Treat.* **2015**, *55*, 86–95. [CrossRef]
17. Lau, W.-J.; Ismail, A.F. Polymeric nanofiltration membranes for textile dye wastewater treatment: Preparation, performance evaluation, transport modelling, and fouling control—A review. *Desalination* **2009**, *245*, 321–348. [CrossRef]
18. Lin, J.; Ye, W.; Huang, J.; Ricard, B.; Baltaru, M.-C.; Greydanus, B.; Balta, S.; Shen, J.; Vlad, M.; Sotto, A.; et al. Toward Resource Recovery from Textile Wastewater: Dye Extraction, Water and Base/Acid Regeneration Using a Hybrid NF-BMED Process. *ACS Sustain. Chem. Eng.* **2015**, *3*, 1993–2001. [CrossRef]
19. Hartmann, M.; Kullmann, S.; Keller, H. Wastewater treatment with heterogeneous Fenton-type catalysts based on porous materials. *J. Mater. Chem.* **2010**, *20*, 9002–9017. [CrossRef]
20. Jain, B.; Singh, A.K.; Kim, H.; Lichtfouse, E.; Sharma, V.K. Treatment of organic pollutants by homogeneous and heterogeneous Fenton reaction processes. *Environ. Chem. Lett.* **2018**, *16*, 947–967. [CrossRef]
21. Li, X.; Huang, X.; Xi, S.; Miao, S.; Ding, J.; Cai, W.; Liu, S.; Yang, X.; Yang, H.; Gao, J.; et al. Single Cobalt Atoms Anchored on Porous N-Doped Graphene with Dual Reaction Sites for Efficient Fenton-like Catalysis. *J. Am. Chem. Soc.* **2018**, *140*, 12469–12475. [CrossRef] [PubMed]
22. Macías-Sánchez, J.; Hinojosa-Reyes, L.; Guzmán-Mar, J.L.; Peralta-Hernández, J.M.; Hernández-Ramírez, A. Performance of the photo-Fenton process in the degradation of a model azo dye mixture. *Photochem. Photobiol. Sci.* **2011**, *10*, 332–337. [CrossRef] [PubMed]
23. Xu, X.-R.; Li, H.-B.; Wang, W.-H.; Gu, J.-D. Degradation of dyes in aqueous solutions by the Fenton process. *Chemosphere* **2004**, *57*, 595–600. [CrossRef] [PubMed]
24. Gao, W.; Tian, J.; Fang, Y.; Liu, T.; Zhang, X.; Xu, X.; Zhang, X. Visible-light-driven photo-Fenton degradation of organic pollutants by a novel porphyrin-based porous organic polymer at neutral pH. *Chemosphere* **2020**, *243*, 125334. [CrossRef]
25. Wang, Y.; Lin, X.; Shao, Z.; Shan, D.; Li, G.; Irini, A. Comparison of Fenton, UV-Fenton and nano-Fe$_3$O$_4$ catalyzed UV-Fenton in degradation of phloroglucinol under neutral and alkaline conditions: Role of complexation of Fe^{3+} with hydroxyl group in phloroglucinol. *Chem. Eng. J.* **2017**, *313*, 938–945. [CrossRef]
26. Ali, A.S.; Khan, I.; Zhang, B.; Nomura, K.; Homonnay, Z.; Kuzmann, E.; Scrimshire, A.; Bingham, P.A.; Krehula, S.; Musić, S.; et al. Photo-Fenton degradation of methylene blue using hematite-enriched slag under visible light. *J. Radioanal. Nucl. Chem.* **2020**, *325*, 537–549. [CrossRef]
27. Kusic, H.; Koprivanac, N.; Srsan, L. Azo dye degradation using Fenton type processes assisted by UV irradiation: A kinetic study. *J. Photochem. Photobiol. A Chem.* **2006**, *181*, 195–202. [CrossRef]
28. Faheem, M.; Jiang, X.; Wang, L.; Shen, J. Synthesis of Cu$_2$O–CuFe$_2$O$_4$ microparticles from Fenton sludge and its application in the Fenton process: The key role of Cu$_2$O in the catalytic degradation of phenol. *RSC Adv.* **2018**, *8*, 5740–5748. [CrossRef]
29. Dawson, R.; Cooper, A.I.; Adams, D.J. Nanoporous organic polymer networks. *Prog. Polym. Sci.* **2012**, *37*, 530–563. [CrossRef]
30. Tan, L.; Tan, B. Hypercrosslinked porous polymer materials: Design, synthesis, and applications. *Chem. Soc. Rev.* **2017**, *46*, 3322–3356. [CrossRef]
31. Li, B.; Gong, R.; Wang, W.; Huang, X.; Zhang, W.; Li, H.; Hu, C.; Tan, B. A New Strategy to Microporous Polymers: Knitting Rigid Aromatic Building Blocks by External Cross-Linker. *Macromolecules* **2011**, *44*, 2410–2414. [CrossRef]
32. Wu, J.; Xu, F.; Li, S.; Ma, P.; Zhang, X.; Liu, Q.; Fu, R.; Wu, D. Porous Polymers as Multifunctional Material Platforms toward Task-Specific Applications. *Adv. Mater.* **2019**, *31*, 1802922. [CrossRef] [PubMed]
33. Dawson, R.; Ratvijitvech, T.; Corker, M.; Laybourn, A.; Khimyak, Y.Z.; Cooper, A.I.; Adams, D.J. Microporous copolymers for increased gas selectivity. *Polym. Chem.* **2012**, *3*, 2034–2038. [CrossRef]
34. Alahmed, A.H.; Briggs, M.E.; Cooper, A.I.; Adams, D.J. Post-synthetic fluorination of Scholl-coupled microporous polymers for increased CO$_2$ uptake and selectivity. *J. Mater. Chem. A* **2019**, *7*, 549–557. [CrossRef]
35. Long, C.; Liu, P.; Li, Y.; Li, A.; Zhang, Q. Characterization of Hydrophobic Hypercrosslinked Polymer as an Adsorbent for Removal of Chlorinated Volatile Organic Compounds. *Environ. Sci. Technol.* **2011**, *45*, 4506–4512. [CrossRef] [PubMed]
36. Chang, Z.; Zhang, D.-S.; Chen, Q.; Bu, X.-H. Microporous organic polymers for gas storage and separation applications. *Phys. Chem. Chem. Phys.* **2013**, *15*, 5430–5442. [CrossRef] [PubMed]
37. Woodward, R.T.; Stevens, L.A.; Dawson, R.; Vijayaraghavan, M.; Hasell, T.; Silverwood, I.P.; Ewing, A.V.; Ratvijitvech, T.; Exley, J.D.; Chong, S.Y.; et al. Swellable, Water- and Acid-Tolerant Polymer Sponges for Chemoselective Carbon Dioxide Capture. *J. Am. Chem. Soc.* **2014**, *136*, 9028–9035. [CrossRef]
38. Wood, C.D.; Tan, B.; Trewin, A.; Niu, H.; Bradshaw, D.; Rosseinsky, M.J.; Khimyak, Y.Z.; Campbell, N.L.; Kirk, R.; Stöckel, E.; et al. Hydrogen Storage in Microporous Hypercrosslinked Organic Polymer Networks. *Chem. Mater.* **2007**, *19*, 2034–2048. [CrossRef]
39. Wood, C.D.; Tan, B.; Trewin, A.; Su, F.; Rosseinsky, M.J.; Bradshaw, D.; Sun, Y.; Zhou, L.; Cooper, A.I. Microporous Organic Polymers for Methane Storage. *Adv. Mater.* **2008**, *20*, 1916–1921. [CrossRef]

40. Errahali, M.; Gatti, G.; Tei, L.; Paul, G.; Rolla, G.A.; Canti, L.; Fraccarollo, A.; Cossi, M.; Comotti, A.; Sozzani, P.; et al. Microporous Hyper-Cross-Linked Aromatic Polymers Designed for Methane and Carbon Dioxide Adsorption. *J. Phys. Chem. C* **2014**, *118*, 28699–28710. [CrossRef]
41. Shen, X.; Ma, S.; Xia, H.; Shi, Z.; Mu, Y.; Liu, X. Cationic porous organic polymers as an excellent platform for highly efficient removal of pollutants from water. *J. Mater. Chem. A* **2018**, *6*, 20653–20658. [CrossRef]
42. Zhang, H.-J.; Wang, J.-H.; Zhang, Y.-H.; Hu, T.-L. Hollow porous organic polymer: High-performance adsorption for organic dye in aqueous solution. *J. Polym. Sci. Part A Polym. Chem.* **2017**, *55*, 1329–1337. [CrossRef]
43. Li, B.; Su, F.; Luo, H.-K.; Liang, L.; Tan, B. Hypercrosslinked microporous polymer networks for effective removal of toxic metal ions from water. *Microporous Mesoporous Mater.* **2011**, *138*, 207–214. [CrossRef]
44. Rao, K.V.; Mohapatra, S.; Maji, T.K.; George, S.J. Guest-Responsive Reversible Swelling and Enhanced Fluorescence in a Super-Absorbent, Dynamic Microporous Polymer. *Chem. Eur. J.* **2012**, *18*, 4505–4509. [CrossRef] [PubMed]
45. Ratvijitvech, T.; Na Pombejra, S. Antibacterial efficiency of microporous hypercrosslinked polymer conjugated with biosynthesized silver nanoparticles from Aspergillus niger. *Mater. Today Commun.* **2021**, *28*, 102617. [CrossRef]
46. Jia, Z.; Wang, K.; Tan, B.; Gu, Y. Hollow Hyper-Cross-Linked Nanospheres with Acid and Base Sites as Efficient and Water-Stable Catalysts for One-Pot Tandem Reactions. *ACS Catal.* **2017**, *7*, 3693–3702. [CrossRef]
47. Li, J.; Wang, X.; Chen, G.; Li, D.; Zhou, Y.; Yang, X.; Wang, J. Hypercrosslinked organic polymer based carbonaceous catalytic materials: Sulfonic acid functionality and nano-confinement effect. *Appl. Catal. B Environ.* **2015**, *176–177*, 718–730. [CrossRef]
48. Feng, L.-J.; Wang, M.; Sun, Z.-Y.; Hu, Y.; Deng, Z.-T. Hypercrosslinked porous polyporphyrin by metal-free protocol: Characterization, uptake performance, and heterogeneous catalysis. *Des. Monomers Polym.* **2017**, *20*, 344–350. [CrossRef]
49. Jiang, X.; Liu, Y.; Liu, J.; Fu, X.; Luo, Y.; Lyu, Y. Hypercrosslinked conjugated microporous polymers for carbon capture and energy storage. *New J. Chem.* **2017**, *41*, 3915–3919. [CrossRef]
50. Tan, L.; Li, B.; Yang, X.; Wang, W.; Tan, B. Knitting hypercrosslinked conjugated microporous polymers with external crosslinker. *Polymer* **2015**, *70*, 336–342. [CrossRef]
51. Li, B.; Guan, Z.; Yang, X.; Wang, W.D.; Wang, W.; Hussain, I.; Song, K.; Tan, B.; Li, T. Multifunctional microporous organic polymers. *J. Mater. Chem. A* **2014**, *2*, 11930–11939. [CrossRef]
52. Yang, Y.; Lai, Z. Ferrocene-based porous organic polymer for photodegradation of methylene blue and high iodine capture. *Microporous Mesoporous Mater.* **2021**, *316*, 110929. [CrossRef]
53. Hegazi, H.A. Removal of heavy metals from wastewater using agricultural and industrial wastes as adsorbents. *HBRC J.* **2013**, *9*, 276–282. [CrossRef]
54. Ratvijitvech, T. Bio-inspired Catechol-based Hypercrosslinked Polymer for Iron (Fe) Removal from Water. *J. Polym. Environ.* **2020**, *28*, 2211–2218. [CrossRef]
55. Lin, J.; Chen, S.; Xiao, H.; Zhang, J.; Lan, J.; Yan, B.; Zeng, H. Ultra-efficient and stable heterogeneous iron-based Fenton nanocatalysts for degrading organic dyes at neutral pH via a chelating effect under nanoconfinement. *Chem. Commun.* **2020**, *56*, 6571–6574. [CrossRef] [PubMed]
56. Contreras, D.; Rodríguez, J.; Freer, J.; Schwederski, B.; Kaim, W. Enhanced hydroxyl radical production by dihydroxybenzene-driven Fenton reactions: Implications for wood biodegradation. *JBIC J. Biol. Inorg. Chem.* **2007**, *12*, 1055–1061. [CrossRef] [PubMed]
57. Zárate-Guzmán, A.I.; González-Gutiérrez, L.V.; Godínez, L.A.; Medel-Reyes, A.; Carrasco-Marín, F.; Romero-Cano, L.A. Towards understanding of heterogeneous Fenton reaction using carbon-Fe catalysts coupled to in-situ H_2O_2 electro-generation as clean technology for wastewater treatment. *Chemosphere* **2019**, *224*, 698–706. [CrossRef]
58. Mia, M.S.; Yan, B.; Zhu, X.; Xing, T.; Chen, G. Dopamine Grafted Iron-Loaded Waste Silk for Fenton-like Removal of Toxic Water Pollutants. *Polymers* **2019**, *11*, 2037. [CrossRef]
59. Ratvijitvech, T.; Dawson, R.; Laybourn, A.; Khimyak, Y.Z.; Adams, D.J.; Cooper, A.I. Post-synthetic modification of conjugated microporous polymers. *Polymer* **2014**, *55*, 321–325. [CrossRef]
60. Wilson, C.; Main, M.J.; Cooper, N.J.; Briggs, M.E.; Cooper, A.I.; Adams, D.J. Swellable functional hypercrosslinked polymer networks for the uptake of chemical warfare agents. *Polym. Chem.* **2017**, *8*, 1914–1922. [CrossRef]
61. Yang, Z.; Qian, J.; Yu, A.; Pan, B. Singlet oxygen mediated iron-based Fenton-like catalysis under nanoconfinement. *Proc. Natl. Acad. Sci. USA* **2019**, *116*, 6659–6664. [CrossRef] [PubMed]
62. Li, M.; Qiang, Z.; Pulgarin, C.; Kiwi, J. Accelerated methylene blue (MB) degradation by Fenton reagent exposed to UV or VUV/UV light in an innovative micro photo-reactor. *Appl. Catal. B Environ.* **2016**, *187*, 83–89. [CrossRef]
63. Rivera, F.L.; Recio, F.J.; Palomares, F.J.; Sánchez-Marcos, J.; Menéndez, N.; Mazarío, E.; Herrasti, P. Fenton-like degradation enhancement of methylene blue dye with magnetic heating induction. *J. Electroanal. Chem.* **2020**, *879*, 114773. [CrossRef]
64. Pariente, M.I.; Martínez, F.; Melero, J.A.; Botas, J.Á.; Velegraki, T.; Xekoukoulotakis, N.P.; Mantzavinos, D. Heterogeneous photo-Fenton oxidation of benzoic acid in water: Effect of operating conditions, reaction by-products and coupling with biological treatment. *Appl. Catal. B Environ.* **2008**, *85*, 24–32. [CrossRef]
65. Ren, B.; Xu, Y.; Zhang, C.; Zhang, L.; Zhao, J.; Liu, Z. Degradation of methylene blue by a heterogeneous Fenton reaction using an octahedron-like, high-graphitization, carbon-doped Fe_2O_3 catalyst. *J. Taiwan Inst. Chem. Eng.* **2019**, *97*, 170–177. [CrossRef]
66. Yang, X.; Chen, W.; Huang, J.; Zhou, Y.; Zhu, Y.; Li, C. Rapid degradation of methylene blue in a novel heterogeneous Fe_3O_4@rGO@TiO_2-catalyzed photo-Fenton system. *Sci. Rep.* **2015**, *5*, 10632. [CrossRef]

67. Du, J.-J.; Yuan, Y.-P.; Sun, J.-X.; Peng, F.-M.; Jiang, X.; Qiu, L.-G.; Xie, A.-J.; Shen, Y.-H.; Zhu, J.-F. New photocatalysts based on MIL-53 metal–organic frameworks for the decolorization of methylene blue dye. *J. Hazard. Mater.* **2011**, *190*, 945–951. [CrossRef]
68. Gao, J.; Miao, J.; Li, P.-Z.; Teng, W.Y.; Yang, L.; Zhao, Y.; Liu, B.; Zhang, Q. A p-type Ti(iv)-based metal–organic framework with visible-light photo-response. *Chem. Commun.* **2014**, *50*, 3786–3788. [CrossRef]
69. Beker, S.A.; Khudur, L.S.; Cole, I.; Ball, A.S. Catalytic degradation of methylene blue using iron and nitrogen-containing carbon dots as Fenton-like catalysts. *New J. Chem.* **2022**, *46*, 263–275. [CrossRef]
70. Xu, N.; Ren, M. Iron Species-Supporting Hydrophobic and Nonswellable Polytetrafluoroethylene/Poly(acrylic acid-co-hydroxyethyl methacrylate) Composite Fiber and Its Stable Catalytic Activity for Methylene Blue Oxidative Decolorization. *Polymers* **2021**, *13*, 1570. [CrossRef]
71. Hong, P.; Wu, Z.; Yang, D.; Zhang, K.; He, J.; Li, Y.; Xie, C.; Yang, W.; Yang, Y.; Kong, L.; et al. Efficient generation of singlet oxygen ($^1O_2$) by hollow amorphous Co/C composites for selective degradation of oxytetracycline via Fenton-like process. *Chem. Eng. J.* **2021**, *421*, 129594. [CrossRef]
72. Bokare, A.D.; Choi, W. Singlet-Oxygen Generation in Alkaline Periodate Solution. *Environ. Sci. Technol.* **2015**, *49*, 14392–14400. [CrossRef] [PubMed]

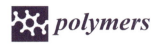

Article

Selective Removal of Iron, Lead, and Copper Metal Ions from Industrial Wastewater by a Novel Cross-Linked Carbazole-Piperazine Copolymer

Majed Al Anazi [1], Ismail Abdulazeez [2] and Othman Charles S. Al Hamouz [1,3,*]

[1] Chemistry Department, King Fahd University of Petroleum and Minerals, Dhahran 31261, Saudi Arabia; alanazimajed1@gmail.com
[2] Interdisciplinary Research Center for Membranes and Water Security, King Fahd University of Petroleum and Minerals, Dhahran 31261, Saudi Arabia; ismail.abdulazeez@kfupm.edu.sa
[3] Interdisciplinary Research Center for Hydrogen and Energy Storage, King Fahd University of Petroleum and Minerals, Dhahran 31261, Saudi Arabia
* Correspondence: othmanc@kfupm.edu.sa; Tel.: +966-552709130

Citation: Al Anazi, M.; Abdulazeez, I.; Al Hamouz, O.C.S. Selective Removal of Iron, Lead, and Copper Metal Ions from Industrial Wastewater by a Novel Cross-Linked Carbazole-Piperazine Copolymer. *Polymers* 2022, 14, 2486. https:// doi.org/10.3390/polym14122486

Academic Editors: Irene S. Fahim, Ahmed K. Badawi, Hossam E. Emam and George Z. Kyzas

Received: 19 May 2022
Accepted: 16 June 2022
Published: 18 June 2022

Publisher's Note: MDPI stays neutral with regard to jurisdictional claims in published maps and institutional affiliations.

Copyright: © 2022 by the authors. Licensee MDPI, Basel, Switzerland. This article is an open access article distributed under the terms and conditions of the Creative Commons Attribution (CC BY) license (https:// creativecommons.org/licenses/by/ 4.0/).

Abstract: A novel cross-linked Copolymer (*MXM*) was synthesized by the polycondensation reaction of 3,6-Diaminocarbazole and piperazine with *p*-formaldehyde as a cross-linker. The Copolymer was fully characterized by solid ^{13}C-NMR and FT-IR. The thermal stability of *MXM* was investigated by TGA and showed that the Copolymer was stable up to 300 °C. The synthesized polyamine was tested for the removal of iron (Fe^{2+}), lead (Pb^{2+}), and copper (Cu^{2+}) ions from aqueous and industrial wastewater solutions. The effect of pH, concentration and time on the adsorption of iron (Fe^{2+}), lead (Pb^{2+}), and copper (Cu^{2+}) ions was investigated. The adsorption of the studied ions from aqueous solutions onto the *MXM* polymer occurs following the Freundlich isotherm and pseudo-second-order kinetic models. The intraparticle diffusion model showed that the adsorption mechanism is controlled by film diffusion. The regeneration of *MXM* showed practical reusability with a loss in capacity of 2–5% in the case of Fe^{2+} and Cu^{2+} ions. The molecular simulation investigations revealed similarities between experimental and theoretical calculations. Industrial wastewater treatment revealed the excellent capabilities and design of *MXM* to be a potential adsorbent for the removal of heavy metal ions.

Keywords: polyamine; industrial wastewater treatment; heavy metal ions

1. Introduction

The growth of population and industrialization has resulted in a massive quantity of hazardous chemicals being released into the environment, posing a warning to human life and the quality of the urban environment [1]. Due to their apparent cytotoxicity, even at low quantity levels, heavy metal pollution has been a major and serious problem globally in the last several decades. As a result, it is critical to continually monitor heavy metal pollution and health risks in cities, particularly mining and industrial areas [2].

Although iron is not harmful to one's health, it is regarded as a secondary or cosmetic contaminant. An increase in iron concentration in the human body could cause health issues, although it is necessary for health at a certain level since it aids in transferring oxygen in the blood [3]. However, iron contamination in the oil industry can be devastating as iron poisoning affects fluid catalytic-cracking catalysts. After iron contamination, a thick coating can build on the catalyst's surface, which prevents reactants from diffusing into the catalyst's inner structures. This poisoning has a significant detrimental impact on the catalyst's capacity to convert heavy oils, and it might significantly reduce gasoline output while significantly boosting dry gas, coke, and slurry yields [4,5].

On the other hand, copper becomes poisonous in high quantities and can damage the brain, heart, and kidneys of human beings [6]. Moreover, it can disrupt the metabolic activities of marine species' where copper concentrations greater than 0.08 µM, for example, have a deleterious effect on the completion of several life periods in brown macroalgae [7]. In the industry, the presence of copper ions can decrease the selectivity of extracting nickel and iron in the mineral processing industry [8].

While pollution with lead is considered a serious and avoidable environmental health issue for children, infants are more vulnerable to lead exposure than adults [9]. Multiple organ systems are affected by lead exposure, resulting in significant morphological, biochemical, and physiological alterations. Chronic lead poisoning has been linked to fatigue, sleep difficulties, headaches, stupor, and anemia [10]. In the industry, catalysts that contain noble metals such as Rhodium (Rh), Palladium (Pd), and Platinum (Pt) in their structure are sensitive to lead [11].

Many techniques for water/wastewater treatment have been developed and used by scientists; adsorption, evaporation, reverse osmosis, filtration, electrolysis, flocculation, sedimentation/gravity extraction, screening, precipitating, oxidation, coagulation, solvent extraction, distillation, solidification, ion exchange, and centrifugation are some of the techniques used [12,13]. However, adsorption remains a preferable approach to other techniques due to the simplicity of obtaining a large field of adsorbents that are significantly valuable, cost-effective, ecologically acceptable, and simple to use. However, effective adsorption is dependent on the use of an effective adsorbent, necessitating ongoing studies in this area [14–16].

Polymers are an important type of adsorbent that have recently captured significant attention in the removal of heavy metal ions from wastewater solutions. Polymers have shown resilience and selectivity in removing heavy metals from aqueous solutions due to their effectiveness, durability, variability, ease of design, and low cost. Such polymers are carbazole containing polymers that have been used to remove heavy metal ions. Recently, carbazole-containing polymers have been used as sensors for detecting mercury (II) ions in aqueous solutions [17]. Hypercrosslinked microporous carbazole-based polymer has been investigated for the removal of lead ions with a % removal of 99.8% [18]; another carbazole based porous organic framework (CzBPOF) has also been used for the removal of lead (II) ions from water with a % removal of 92.56% within 80 min [19].

Most reported carbazole-based polymers have never been tested on industrial wastewater treatment under actual conditions. In our endeavor in the design and application of polymers for the treatment of heavy metal ions, a novel cross-linked Copolymer containing diaminocarbazole and piperazine has been synthesized as an effective adsorbent for the removal of iron (Fe^{2+}), lead (Pb^{2+}), and copper (Cu^{2+}) ions from wastewater solutions.

2. Experimental

2.1. Materials and Equipment

3,6-Diaminocarbazole, piperazine, *p*-formaldehyde, glacial acetic acid, methanol, and DMF were purchased from Sigma Aldrich and were used without further purification. Metal adsorption analysis was performed using ICP-OES Optima 8000 Perkin Elmer. Polymer structure was determined by ^{13}C-NMR spectra using solid-state type and was recorded by Bruker Avance III—400 WB. FTIR spectra were produced on iTR Nicolet is 10 spectrometer. pH measurements were performed using HACH HQ411D. Thermal stability was measured using (TGA) Q600 TA instruments at a heating rate of 20 °C/min under a nitrogen atmosphere.

2.2. Polymer Synthesis (**MXM**)

In a typical reaction [20], 3,6-Diaminocarbazole (0.01 mmol, 1.97 g), piperazine (0.04 mmol, 3.44 g), and *p*-formaldehyde (0.04 mmol, 1.201 g) in DMF (25 mL) were stirred for 10 min and in a 100 mL PTFE liner hydrothermal autoclave vessel. The vessel was purged with nitrogen and sealed. The reaction vessel was transferred into a stainless-steel reaction vessel and heated in an oven at 100 °C for 24 h. Once completed, the reaction mixture was filtered and washed with methanol. The dark black solid powder was then dried under vacuum at 70 °C for 24 h (Yield: 5.1 g, 77%).

2.3. Adsorption Experiments

Batch adsorption studies were performed on the capability of **MXM** to remove three metal ions Fe^{2+}, Cu^{2+}, and Pb^{2+}. In a typical experiment [21], 30 mg of the polymer was added to a 20 mL metal solution at a certain pH, initial concentration, and temperature and stirred for a specific amount of time. Once the adsorption experiment was completed, the reaction mixture was filtered, and the concentration of the metal ions after adsorption was measured. The capability of **MXM** was found by calculating the adsorption capacity as described in the following equation:

$$Q_e = \frac{(C_o - C_f) \times V}{W} \quad (1)$$

where Q_e is the adsorption capacity (mg/g), and C_o and C_f are the initial and final concentrations of metal ions (mg/L) in the solution, respectively. W is the mass of **MXM** (mg), and V is the volume of the metal ion solution (L).

2.4. Molecular Simulation

Molecular simulations were conducted using the COMPASS II forcefield [22] on Materials Studio 8.0 suite to determine the density and the fractional free volume (FFV) of the material, **MXM**. An amorphous cell comprising 20 units of **MXM** was constructed and minimized using the Forcite module, followed by dynamics simulations on the NPT and the NVT ensembles each for 1000 ps at a time step of 1.0 fs. The Nose–Hoover thermostat and the Berendsen barostat were used to control the temperature (at 298 K) and pressure (at 8 bar), respectively. The Ewald summation method was used for long-range Coulombic interactions, while Lennard-Jones interactions were estimated with a cut-off range of 12.5 Å. The interaction of **MXM** and metal ions were simulated by conducting quantum mechanical DFT simulations on Gaussian 09 [23], at the B3LYP exchange-correlation functional, in combination with the Pople's split valence basis set, 6-311G*, and the SDD pseudopotential basis set, for the non-metal and metal atoms, respectively [24,25]. The polarized continuum model–self-consistent reaction field (PCM-SCRF) model was used to simulate the aqueous media, with the solvent depicted as water [26]. After full geometry optimization, the adsorption energies of **MXM** and the metal ions were computed using the following equation:

$$E_{ad} = E_{M-MXM} - E_M - E_{MXM} \quad (2)$$

where $E_{M\text{-}MXM}$ represents the total energies of the metal ion-**MXM** complexes, while E_M and E_{MXM} represent the isolated energies of the metal ions and **MXM**, respectively. The nature of the interactions was explored using the reduced density gradient (RDG) isosurfaces plots as implemented on the multi-wavefunction analysis code [27].

3. Results and Discussion

A novel cross-linked Copolymer has been synthesized by the polycondensation reaction of 3,6-Diaminocarbazole and piperazine with p-formaldehyde to produce the **MXM** Copolymer (Scheme 1). The monomeric moieties of diaminocarbazole and piperazine were chosen to be typical for an adsorbent of heavy metal ions as they provide the rigidity of aromatic moieties in diaminocarbazole and the flexibility of piperazine units. The high amount of nitrogen atoms present in these monomers provides a high concentration of attractive adsorption sites for metal ions to be attached and removed.

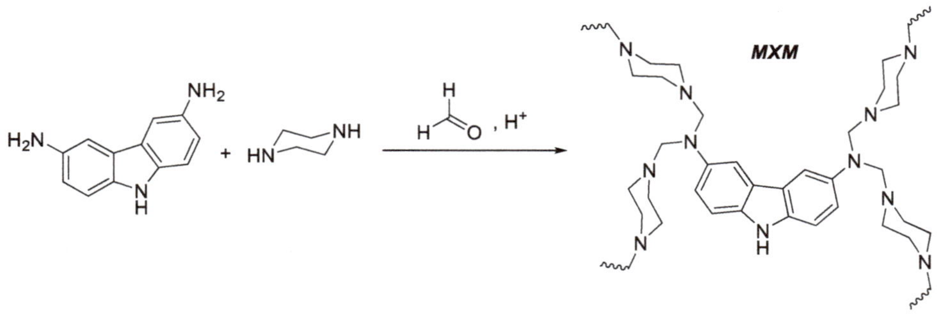

Scheme 1. Synthesis of **MXM** Copolymer.

3.1. MXM Copolymer Structure Characterization

Copolymer **MXM** was characterized by solid ^{13}C-NMR and FT-IR, as shown in Figure 1a,b. The Figure elucidates the characteristic features of the Copolymer. The FTIR spectrum in Figure 1a shows an absorption band at ~3400 cm^{-1}, which represents the –N–H stretching frequency of **MXM**, and an absorption band at ~2980 cm^{-1}, which represents the –CH$_2$– stretching vibration of the methylene units in piperazine moiety, and the methylene bridge that links the diaminocarbazole with the piperazine monomers. An absorption band at ~1600 cm^{-1} represents the –C=C– stretching vibrations of the benzene rings found in the diaminocarbazole monomer. An absorption band at ~1440 cm^{-1} represents the –C–N– stretching vibration in both monomeric units [28,29]. Figure 1b represents the solid ^{13}C-NMR spectrum for **MXM**; the spectrum shows multiple peaks ~30–50 ppm, referred to as the methylene units in the piperazine monomer and the methylene bridge between the diaminocarbazole and piperazine monomer. The peaks of ~100–150 ppm are characteristic of the aromatic carbons found in diaminocarbazole [30–32]. The peak present ~165 ppm is due to the formation of imine linkage between terminal amines with *p*-formaldehyde. Figure 1c shows the thermogravimetric analysis of the **MXM** Copolymer. The thermogram shows two degradation steps; the first thermal degradation step at ~300 °C corresponds to the thermal degradation of aliphatic chains in the piperazine units and methylene bridges between monomeric units, followed by a sharp thermal degradation at ~500 °C due to the carbonization of diaminocarbazole aromatic units [33,34].

3.2. MXM Copolymer Adsorption Properties

To fully comprehend the capability of the **MXM** Copolymer to adsorb heavy metal ions, the Copolymer was studied under different adsorption conditions experimentally. Moreover, the Copolymer was studied under molecular simulation to study the affinity of the **MXM** Copolymer toward the adsorption of lead, iron, and copper metal ions.

3.2.1. Effect of pH on the Adsorption Capacity of **MXM**

The synthesized Copolymer **MXM** was tested at different pHs (2.34, 3.57, 5.73, and 6.67) to find out the adsorption efficiency of **MXM** on a 1 ppm solution of iron (Fe^{2+}), copper (Cu^{2+}), and lead (Pb^{2+}) ions. In a typical experiment, 30 mg of **MXM** was added to a 20 mL metal ion and stirred for 1 h. Once completed, the solution was filtered, and the concentration of metal ions was measured. Figure 2 depicts the influence of pH on the adsorption capacity of **MXM**. As shown in the figure, the adsorption capacity of **MXM** increases as pH increases. In high acidic conditions, the increased protonation of the adsorbent surface results in lower metal ion adsorption, as shown in Scheme 2. The pH also has an impact on the speciation of heavy metal ions. The adsorption experiments were performed in the pH range of 2–6; after pH = 6, the metal ions start precipitating, which is expected due to the formation of metal hydroxides M(OH)$_2$. As a result, pH = 5.73 was utilized for the rest of the adsorption studies.

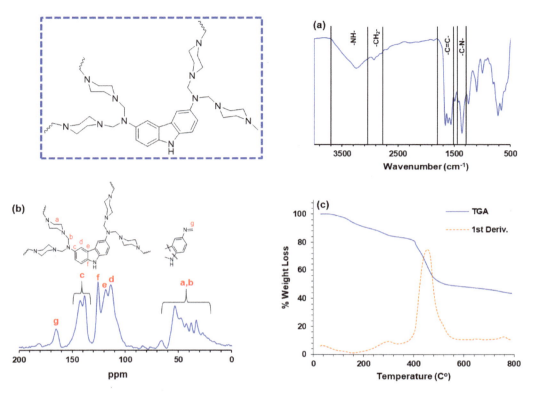

Figure 1. (a) FT−IR spectrum of *MXM* Copolymer. (b) The solid ^{13}C−NMR spectrum of *MXM* Copolymer. (c) Thermogravimetric analysis (TGA) and the first-derivative of *MXM* Copolymer.

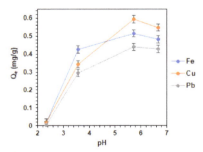

Figure 2. The effect of pH on the adsorption capacity of the *MXM* Copolymer.

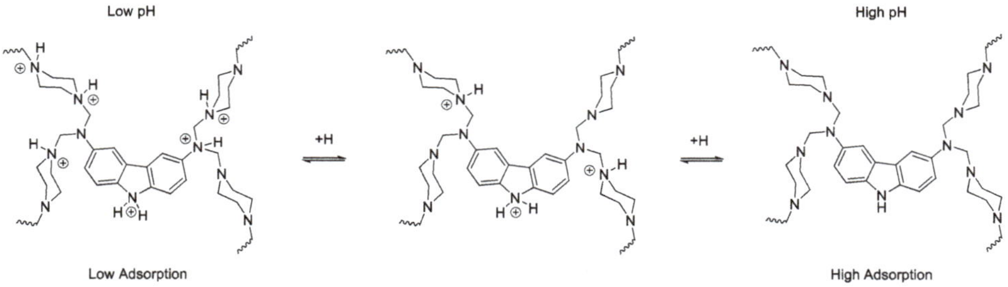

Scheme 2. Effect of pH on the protonation of **MXM** the Copolymer.

3.2.2. Effect of Initial Metal Ion Concentration on the Adsorption Capacity of MXM Copolymer

Figure 3a demonstrates the influence of the initial metal ion concentration (mg/L) on the adsorption capacity of **MXM**. The adsorption investigation was carried out on three metal ion solutions (Cu^{2+}, Fe^{2+}, and Pb^{2+}) with a concentration ranging from 0.2 to 1 (mg/L) at pH = 5.73 and a specific time of 1 h. The figure shows that the adsorption capacity increases with increases in the initial concentration of the metal ion. Three adsorption isotherm models were used to analyze the experimental data acquired by the batch adsorption studies; Langmuir, Freundlich, and the Dubinin–Kaganer–Radushkevich (DKR) isotherm model. The Langmuir isotherm model describes adsorption as a single-layer adsorption where one metal is attached to one adsorption site (Figure 3b). It depicts the uniform adsorption of metal ions on the surface of the adsorbent [35]. The negative values shown in Table 1 reveal that the adsorption process of the three metal ions does not follow the Langmuir isotherm model. In contrast to the Langmuir isotherm, the Freundlich isotherm model explains the adsorption behavior when the surface is heterogeneous, where one active site can accommodate more than one metal ion [36,37]. The heterogeneity of the surface can be determined from the slope ($1/n$). High heterogeneity is described when the value of $1/n$ is ~0. When the value of $1/n < 1$, adsorption is considered favorable. If the slope ($1/n$) has a value above one, this indicates cooperative adsorption. As shown in Table 1 (Figure 3c), the values of $1/n$ are higher than 1, indicating that the adsorption process is cooperative in nature [38]. The DKR model [39], which is applied to single solute systems, describes the adsorption mechanism as physical or chemical in nature [40]. The results shown in Table 1 (Figure 3d) show that adsorption energy E is exothermic, indicating favorable adsorption [40].

3.2.3. The Effect of Time on the Adsorption Capacity of MXM

The adsorption rate, which determines how long it takes for the adsorption process to reach equilibrium, is also influenced by adsorption kinetics and is shown in Figure 4. This information is critical for the process of innovation and adsorption system design. In this work, three kinetic models were used to analyze experimental data: pseudo first-order (PFO), pseudo second-order (PSO), and intraparticle diffusion (IPD), which were used to define the adsorption kinetics of Fe^{2+}, Cu^{2+}, and Pb^{2+} by the **MXM** Copolymer.

PFO kinetic model posits that the rate of change of adsorbed solute by time is precisely comparative to the change in saturation concentration and the quantity of solid adsorption with time, and it may be applied to any adsorption process in the first stage [41,42] from the data shown in Table 2 (Figure 4b). The $Q_{e(exp)}$ values are not close to the values of Q_e found by PFO, which implies that the experimental data does not fit the PFO kinetic model. PSO's premise is that the rate-limiting phase is chemisorption, and it predicts the behavior over the entire adsorption range [43,44]. In this situation, the adsorption rate is governed by the adsorption capacity rather than the adsorbate concentration. The model can determine the equilibrium adsorption, which is a key advantage of PSO over PFO; hence, no requirement is needed to assess the adsorption equilibrium capacity from the experiment.

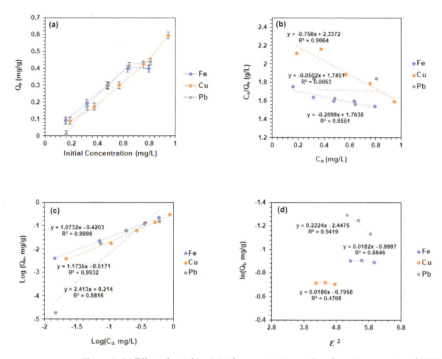

Figure 3. (**a**) Effect of metal ion initial concentration on the adsorption capacity of *MXM*; (**b**) Langmuir isotherm model for the adsorption of iron(II), Cu(II), and lead(II) ions by *MXM*; (**c**) Freundlich isotherm model for the adsorption of iron(II), Cu(II), and lead(II) ions by *MXM*; (**d**) DKR isotherm model for the adsorption of iron(II), Cu(II), and lead(II) ions by *MXM*.

Table 1. The isothermal constants of the three models Langmuir, Freundlich, and DKR.

Metal Ion	Langmuir Isotherm Model			
		Q_m	b	R^2
Fe^{2+}		−3.45	−0.164	0.8551
Cu^{2+}		−1.32	−0.323	0.9064
Pb^{2+}		−19.9	−0.029	0.0053
	Freundlich isotherm model			
		k_f	$1/n$	R^2
Fe^{2+}		0.380	1.07	0.9998
Cu^{2+}		1.17	0.304	0.9932
Pb^{2+}		2.41	1.64	0.8816
	DKR isotherm model			
	Q_m	β	E	R^2
Fe^{2+}	1.02	0.999	−0.707	0.6646
Cu^{2+}	1.02	0.796	−0.793	0.4708
Pb^{2+}	1.25	2.44	−0.452	0.9419

Table 2. PFO and PSO kinetic model constants for the adsorption of the Fe^{2+}, Cu^{2+}, and Pb^{2+} by *MXM*.

Metal Ion	PFO Kinetic Model			
	Q_e exp	q_e	k_1	R^2
Fe^{2+}	0.407	0.087	4.345	0.9159
Cu^{2+}	0.484	0.228	1.087	0.8632
Pb^{2+}	0.420	0.046	3.078	0.6873
	PSO Kinetic Model			
	Q_e exp	q_e	k_2	R^2
Fe^{2+}	0.407	0.408	466.8	1.000
Cu^{2+}	0.484	0.636	1.952	0.8343
Pb^{2+}	0.420	0.445	7.165	0.9803

As shown in Table 2 (Figure 4c), $Q_{e(exp)}$ is close to the Q_e values calculated in the PSO model with good regression values, implying that the adsorption process is chemisorption [45,46].

Weber and Morris devised an intraparticle diffusion model to determine the adsorption diffusion process. Nonlinearity, on the other hand, is occasionally found. When the data show several linear plots, it implies that adsorption was caused by more than one mechanism. As a result, these rate-determining processes may be split into multiple linear curve segments throughout time, each having control over the entire process [47,48]. The discovery of the rate-limiting step provides new insights into the mechanism of the adsorption process. According to the intraparticle diffusion model, first, metal ions move interfacially (i.e., film diffusion) between the absorbent and solution. Second, a rate-limiting intraparticle diffusion phase carries the ions into the adsorbent sites, and then adsorption reaches equilibrium [49]. From Figure 4d, the adsorption of Fe^{2+}, Cu^{2+}, and Pb^{2+} metal ions by Copolymer *MXM* proceeds through two steps, with film diffusion followed by equilibrium.

3.3. Reusability of MXM

The reusability of *MXM* is a vital factor for its use in industrial wastewater treatment; for that, *MXM* was subjected to two cycles of adsorption, as shown in Figure 5. In a typical experiment, 100 mg of *MXM* was placed in a 20 mL 1 ppm solution of Pb^{2+}, Fe^{2+}, and Cu^{2+} for 1 h under stirring. Once the experiment was completed, the adsorption capacity was measured. The polymer was soaked in a 1M solution of nitric acid for 15 min, followed by washing with water and 1M sodium hydroxide; then, it was thoroughly washed with water. The experiment was repeated, and the adsorption capacity of *MXM* was measured again. The adsorption experiments reveal the usability of *MXM* with small loss of activity in the adsorption of Fe^{2+} and Cu^{2+}, which proves the suitability of *MXM* to be used as an adsorbent.

3.4. Molecular Simulation

The results of MD simulations on 20 units of *MXM* are shown in Figure 6. The grey and blue isosurfaces represent the accessible free volume on *MXM* at a probe radius of 1.09 Å. The system was stabilized mainly by intermolecular hydrogen bonds and possible π-π stacking between the aromatic rings. The estimated FFV of *MXM* (0.15) indicates moderate hydrogen bonding within the system and suggests higher water permeance and greater mobility of the metal ions in good correlation with fully reported aromatic systems [50].

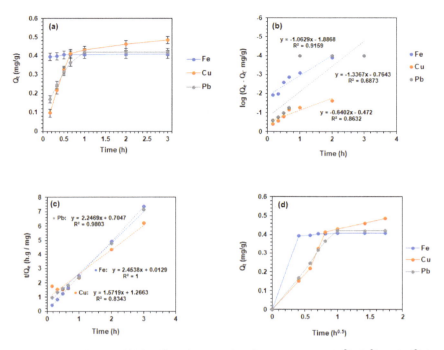

Figure 4. (**a**) The effect of time on the adsorption capacity Fe^{2+}, Cu^{2+}, and Pb^{2+} by **MXM**. (**b**) PFO kinetic model on the adsorption capacity Fe^{2+}, Cu^{2+}, and Pb^{2+} by **MXM**. (**c**) PSO kinetic model on the adsorption capacity Fe^{2+}, Cu^{2+}, and Pb^{2+} by **MXM**. (**d**) Intraparticle diffusion model on the adsorption capacity Fe^{2+}, Cu^{2+}, and Pb^{2+} by **MXM**.

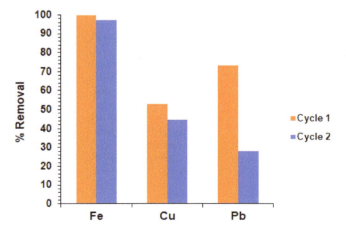

Figure 5. Regeneration capability of **MXM** polymer.

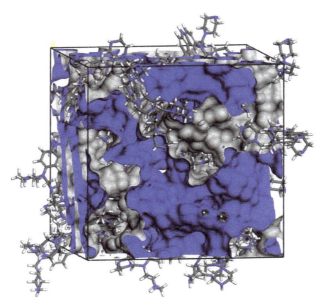

Figure 6. Energy-minimized amorphous cell comprising **MXM**.

Molecular level quantum simulations of the adsorption of Cu^{2+}, Pb^{2+}, and Fe^{2+} ions on **MXM** molecules were conducted and presented in Figure 7. Three possible adsorption modes, the in-plane, the stack, and grip configurations, were identified and geometrically optimized. The adsorption energies, E_{ad}, revealed that **MXM** exhibited higher adsorption capacity toward Cu^{2+} ions with energies of -22.3, -7.55, and -7.99 eV for the in-plane, the stack, and the grip configurations, respectively. The order of E_{ad}, $Cu^{2+} > Pb^{2+} > Fe^{2+}$, was consistent with the change in the frontier molecular orbital (FMO) energy separation of **MXM** upon the adsorption of the metal ions (Figure S1), which revealed the stabilization of the nucleophilic character of the Copolymer material by lowering the energy of the highest occupied molecular orbital. Consequently, a significant fraction of charges was transferred to the Copolymer materials when interacting with Cu^{2+} ions, resulting in higher adsorption capacity that is consistent with the experimental data.

The RDG isosurface plots of the various optimized adsorption modes of the metal ions on the active sites of **MXM** are presented in Figure S2. It mainly depicts non-covalent interactions (NCIs) characterized by electron density ρ [51,52]. RDG isosurface plots revealed predominantly weak Van der Waals interactions between **MXM** and the metal ions. In-plane geometry exhibited higher ρ along the plane of interactions, suggesting stronger adsorption as revealed by E_{ad}, and this is consistent with the change in FMO distribution. Meanwhile, partial repulsions were exhibited by Pb^{2+} and Fe^{2+} ions, resulting in slightly weaker adsorptions, in good correlation with the experimental adsorption data.

3.5. Wastewater Sample Treatment

The significant element in synthesizing an adsorbent polymer is to find out its ability to treat wastewater. The effectiveness of **MXM** in removing heavy metals was tested using two samples of wastewater (unspiked, spiked each by 1 mg/L iron, copper, and lead ions). The wastewater sample had a pH of 7.9, total dissolved solids of 1096.4 mg/L, total suspended solids of 11 mg/L, chlorides of 254 mg/L, and organic carbon of 9.8 mg/L. A 20 mL sample of unspiked and spiked samples containing the three metal ions was treated with **MXM**. The results in Table 3 reveal the efficiency of **MXM** as an adsorbent for removing Fe^{2+}, Cu^{2+}, and Pb^{2+} from wastewater solutions. A comparison Table between

MXM and other adsorbents reported in the literature shows the efficiency of MXM as an adsorbent for the removal of heavy metal ions (Table S1).

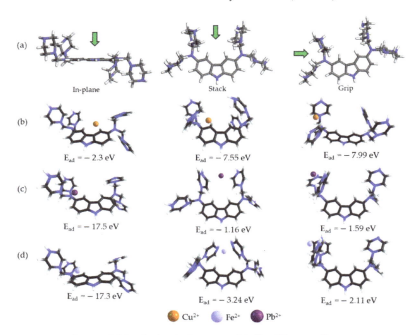

Figure 7. Molecular level adsorption of Cu^{2+}, Pb^{2+}, and Fe^{2+} ions on **MXM** at the B3LYP/6-311G* and SDD levels of theory. (**a**) The optimized adsorption configurations of the metal ions on **MXM**, and the molecular level adsorption of (**b**) Cu^{2+}, (**c**) Pb^{2+} and (**d**) Fe^{2+} ions at the B3LYP/6-311G* and SDD levels of theory.

Table 3. Comparison of the metal ions concentrations before and after treatment of wastewater sample (unspiked and spiked) and spiked DI water.

Metal Ions	Unspiked Wastewater Sample		
	Original sample	After Treatment	% Removal
Fe^{2+}	0.061	0.00	100
Cu^{2+}	0.003	0.002	33.3
Pb^{2+}	-	-	-
	Spiked Wastewater Sample		
	Original sample	After Treatment	% Removal
Fe^{2+}	0.454	0	100
Cu^{2+}	0.485	0.219	54.8
Pb^{2+}	0.376	0.015	96

4. Conclusions

A novel cross-linked carbazole-piperazine Copolymer has been synthesized, and the adsorption properties were investigated under different conditions such as pH, initial metal ion concentration, and time. The adsorption results revealed that the **MXM** polymer followed the Freundlich isotherm model, where the adsorption process is heterogeneous, whereas the adsorption kinetics followed the pseudo-second-order kinetic model indicating chemisorption. The intraparticle diffusion model indicated that adsorption experiences film diffusion followed by equilibrium. The regeneration of **MXM** polymer showed its ability to be used again with little losses of 2–5% in capacity in the case of iron and copper, which makes it effective as an adsorbent. Interestingly, the performance of **MXM** was the most efficient in the case of industrial wastewater treatment, where **MXM** showed higher adsorption capacity toward iron and lead than copper, which could be due to the different environments, pH, and salinity of the industrial wastewater sample with a % removal of 55–100% removal capacity. The results indicate the strong potential of **MXM** to be an efficient adsorbent for the removal of heavy metal ions from wastewater solutions.

Supplementary Materials: The following supporting information can be downloaded at: https://www.mdpi.com/article/10.3390/polym14122486/s1, Figure S1: The change in frontier molecular orbital distribution of (a) **MXM** when interacting with (b) Cu^{2+}, (c) Pb^{2+}, and (d) Fe^{2+} ions at the B3LYP/6-311G* and SDD levels of theory; Figure S2: The reduced density gradient (RDG) isosurface plots of the interactions of **MXM** with (a) Cu^{2+}, (b) Pb^{2+}, and (c) Fe^{2+} ions; Table S1: Comparison between **MXM** and adsorbents for the removal of heavy metal ions [53–59].

Author Contributions: M.A.A. was involved methodology of the work, investigation, formal analysis, resources, and software. I.A. was involved in molecular simulation. O.C.S.A.H. helped in visualization, supervision, investigation, writing—review and editing. All authors have read and agreed to the published version of the manuscript.

Funding: This research received no external funding.

Institutional Review Board Statement: Not applicable.

Informed Consent Statement: Not applicable.

Data Availability Statement: Data presented in this study are available upon fair request to the corresponding author.

Acknowledgments: This publication is based upon work supported by King Fahd University of Petroleum and Minerals and the authors at KFUPM acknowledge the support.

Conflicts of Interest: The authors declare no conflict of interest.

References

1. Begum, S.; Yuhana, N.Y.; Md Saleh, N.; Kamarudin, N.H.N.; Sulong, A.B. Review of chitosan composite as a heavy metal adsorbent: Material preparation and properties. *Carbohydr. Polym.* **2021**, *259*, 117613. [CrossRef] [PubMed]
2. Long, Z.; Huang, Y.; Zhang, W.; Shi, Z.; Yu, D.; Chen, Y.; Liu, C.; Wang, R. Effect of different industrial activities on soil heavy metal pollution, ecological risk, and health risk. *Environ. Monit. Assess.* **2021**, *193*, 20. [CrossRef] [PubMed]
3. Hashim, K.S.; Shaw, A.; Al Khaddar, R.; Pedrola, M.O.; Phipps, D. Iron removal, energy consumption and operating cost of electrocoagulation of drinking water using a new flow column reactor. *J. Environ. Manag.* **2017**, *189*, 98–108. [CrossRef] [PubMed]
4. Liao, Y.; Liu, T.; Du, X.; Gao, X. Distribution of Iron on FCC Catalyst and Its Effect on Catalyst Performance. *Front. Chem.* **2021**, *9*, 640413. [CrossRef]
5. Jiang, H.; Livi, K.J.; Kundu, S.; Cheng, W.-C. Characterization of iron contamination on equilibrium fluid catalytic cracking catalyst particles. *J. Catal.* **2018**, *361*, 126–134. [CrossRef]
6. Taylor, A.A.; Tsuji, J.S.; Garry, M.R.; McArdle, M.E.; Goodfellow, W.L.; Adams, W.J.; Menzie, C.A. Critical Review of Exposure and Effects: Implications for Setting Regulatory Health Criteria for Ingested Copper. *Environ. Manag.* **2020**, *65*, 131–159. [CrossRef]
7. Leal, P.P.; Hurd, C.L.; Sander, S.G.; Armstrong, E.; Fernández, P.A.; Suhrhoff, T.J.; Roleda, M.Y. Copper pollution exacerbates the effects of ocean acidification and warming on kelp microscopic early life stages. *Sci. Rep.* **2018**, *8*, 14763. [CrossRef]
8. Wang, H.; Han, J.; Manica, R.; Qi, C.; Liu, Q. Effect of Cu(II) ions on millerite (β-NiS) flotation and surface properties in alkaline solutions. *Miner. Eng.* **2022**, *180*, 107443. [CrossRef]
9. Green, R.E.; Pain, D.J. Risks to human health from ammunition-derived lead in Europe. *Ambio* **2019**, *48*, 954–968. [CrossRef]
10. Jaishankar, M.; Tseten, T.; Anbalagan, N.; Mathew, B.B.; Beeregowda, K.N. Toxicity, mechanism and health effects of some heavy metals. *Interdiscip. Toxicol.* **2014**, *7*, 60–72. [CrossRef]
11. Karim, S.; Ting, Y.-P. Recycling pathways for platinum group metals from spent automotive catalyst: A review on conventional approaches and bio-processes. *Resour. Conserv. Recycl.* **2021**, *170*, 105588. [CrossRef]
12. Fu, F.; Wang, Q. Removal of heavy metal ions from wastewaters: A review. *J. Environ. Manag.* **2011**, *92*, 407–418. [CrossRef] [PubMed]
13. Qasem, N.A.A.; Mohammed, R.H.; Lawal, D.U. Removal of heavy metal ions from wastewater: A comprehensive and critical review. *Npj Clean Water* **2021**, *4*, 36. [CrossRef]

14. Chakraborty, R.; Asthana, A.; Singh, A.K.; Jain, B.; Susan, A.B.H. Adsorption of heavy metal ions by various low-cost adsorbents: A review. *Int. J. Environ. Anal. Chem.* **2022**, *102*, 342–379. [CrossRef]
15. Ahmad, S.Z.N.; Wan Salleh, W.N.; Ismail, A.F.; Yusof, N.; Mohd Yusop, M.Z.; Aziz, F. Adsorptive removal of heavy metal ions using graphene-based nanomaterials: Toxicity, roles of functional groups and mechanisms. *Chemosphere* **2020**, *248*, 126008. [CrossRef]
16. Wan Ngah, W.S.; Hanafiah, M.A.K.M. Removal of heavy metal ions from wastewater by chemically modified plant wastes as adsorbents: A review. *Bioresour. Technol.* **2008**, *99*, 3935–3948. [CrossRef]
17. Giri, D.; Bankura, A.; Patra, S.K. Poly(benzodithieno-imidazole-alt-carbazole) based π-conjugated copolymers: Highly selective and sensitive turn-off fluorescent probes for Hg^{2+}. *Polymer* **2018**, *158*, 338–353. [CrossRef]
18. Rahnama Haratbar, P.; Ghaemi, A.; Nasiri, M. Potential of hypercrosslinked microporous polymer based on carbazole networks for Pb(II) ions removal from aqueous solutions. *Environ. Sci. Pollut. Res.* **2022**, *29*, 15040–15056. [CrossRef]
19. Melhi, S. Novel carbazole-based porous organic frameworks (CzBPOF) for efficient removal of toxic Pb(II) from water: Synthesis, characterization, and adsorption studies. *Environ. Technol. Innov.* **2022**, *25*, 102172. [CrossRef]
20. Lohse, M.S.; Bein, T. Covalent Organic Frameworks: Structures, Synthesis, and Applications. *Adv. Funct. Mater.* **2018**, *28*, 1705553. [CrossRef]
21. Al-hamouz, O.C.S.O.; Saleh, T.A.; Garrison, T.F.; Fraim, M.L.; Habib, M.A.-A. Mercury Removal from Liquid Hydrocarbons by 1,4-Benzenediamine Alkyldiamine Cross-Linked Polymersmercury Removal from Liquid Hydrocarbons by 1,4-Benzenediamine Alkyldiamine Cross-Linked Polymers. U.S. Patent No. 11,236,274, 1 February 2022.
22. Sun, H.; Jin, Z.; Yang, C.; Akkermans, R.L.; Robertson, S.H.; Spenley, N.A.; Miller, S.; Todd, S.M. COMPASS II: Extended coverage for polymer and drug-like molecule databases. *J. Mol. Model.* **2016**, *22*, 47. [CrossRef] [PubMed]
23. Frisch, M.J.; Trucks, G.W.; Schlegel, H.B.; Scuseria, G.E.; Robb, M.A.; Cheeseman, J.R.; Scalmani, G.; Barone, V.; Mennucci, B.; Petersson, G.A.; et al. *Gaussian 09, Revision B.01*. Wallingford CT. 2009, Gaussian Inc.: Wallingford, CT, USA, 2009.
24. Martin, J.M.L.; Sundermann, A. Correlation consistent valence basis sets for use with the Stuttgart–Dresden–Bonn relativistic effective core potentials: The atoms Ga–Kr and In–Xe. *J. Chem. Phys.* **2001**, *114*, 3408–3420. [CrossRef]
25. Krishnan, R.; Binkley, J.S.; Seeger, R.; Pople, J.A. Self-consistent molecular orbital methods. XX. A basis set for correlated wave functions. *J. Chem. Phys.* **1980**, *72*, 650–654. [CrossRef]
26. Tomasi, J.; Mennucci, B.; Cammi, R. Quantum Mechanical Continuum Solvation Models. *ChemInform* **2005**, *36*, 2999–3093. [CrossRef]
27. Lu, T.; Chen, F. Multiwfn: A multifunctional wavefunction analyzer. *J. Comput. Chem.* **2012**, *33*, 580–592. [CrossRef]
28. Albakri, M.A.; Abdelnaby, M.M.; Saleh, T.A.; Al Hamouz, O.C.S. New series of benzene-1,3,5-triamine based cross-linked polyamines and polyamine/CNT composites for lead ion removal from aqueous solutions. *Chem. Eng. J.* **2018**, *333*, 76–84. [CrossRef]
29. Chang, G.; Yang, L.; Yang, J.; Huang, Y.; Cao, K.; Ma, J.; Wang, D. A nitrogen-rich, azaindole-based microporous organic network: Synergistic effect of local dipole–π and dipole–quadrupole interactions on carbon dioxide uptake. *Polym. Chem.* **2016**, *7*, 5768–5772. [CrossRef]
30. Dangsopon, A.; Poomsuk, N.; Siriwong, K.; Vilaivan, T.; Suparpprom, C. Synthesis and fluorescence properties of 3,6-diaminocarbazole-modified pyrrolidinyl peptide nucleic acid. *RSC Adv.* **2016**, *6*, 74314–74322. [CrossRef]
31. Al Hamouz, O.C.S.; Estatie, M.K.; Morsy, M.A.; Saleh, T.A. Lead ion removal by novel highly cross-linked Mannich based polymers. *J. Taiwan Inst. Chem. Eng.* **2017**, *70*, 345–351. [CrossRef]
32. Patil, S.A.; Scherf, U.; Kadashchuk, A. New Conjugated Ladder Polymer Containing Carbazole Moieties. *Adv. Funct. Mater.* **2003**, *13*, 609–614. [CrossRef]
33. Muylaert, I.; Verberckmoes, A.; De Decker, J.; Van Der Voort, P. Ordered mesoporous phenolic resins: Highly versatile and ultra stable support materials. *Adv. Colloid Interface Sci.* **2012**, *175*, 39–51. [CrossRef] [PubMed]
34. Ahmad, E.E.M.; Luyt, A.S.; Djoković, V. Thermal and dynamic mechanical properties of bio-based poly(furfuryl alcohol)/sisal whiskers nanocomposites. *Polym. Bull.* **2013**, *70*, 1265–1276. [CrossRef]
35. Boparai, H.K.; Joseph, M.; O'Carroll, D.M. Kinetics and thermodynamics of cadmium ion removal by adsorption onto nano zerovalent iron particles. *J. Hazard. Mater.* **2011**, *186*, 458–465. [CrossRef]
36. Huang, Y.; Tang, J.; Gai, L.; Gong, Y.; Guan, H.; He, R.; Lyu, H. Different approaches for preparing a novel thiol-functionalized graphene oxide/Fe-Mn and its application for aqueous methylmercury removal. *Chem. Eng. J.* **2017**, *319*, 229–239. [CrossRef]
37. Mohapatra, M.; Rout, K.; Mohapatra, B.K.; Anand, S. Sorption behavior of Pb(II) and Cd(II) on iron ore slime and characterization of metal ion loaded sorbent. *J. Hazard. Mater.* **2009**, *166*, 1506–1513. [CrossRef]
38. Sahu, R.C.; Patel, R.; Ray, B.C. Adsorption of Zn(II) on activated red mud: Neutralized by CO_2. *Desalination* **2011**, *266*, 93–97. [CrossRef]
39. Lasheen, M.R.; Ammar, N.S.; Ibrahim, H.S. Adsorption/desorption of Cd(II), Cu(II) and Pb(II) using chemically modified orange peel: Equilibrium and kinetic studies. *Solid State Sci.* **2012**, *14*, 202–210. [CrossRef]
40. Shen, S.; Guishen, L.; Pan, T.; He, J.; Guo, Z. Selective adsorption of Pt ions from chloride solutions obtained by leaching chlorinated spent automotive catalysts on ion exchange resin Diaion WA21J. *J. Colloid Interface Sci.* **2011**, *364*, 482–489. [CrossRef]
41. Ozcan, A.; Ozcan, A.S.; Gok, O. *Adsorption Kinetics and Isotherms of Anionic Dye of Reactive Blue 19 from Aqueous Solutions onto DTMA-Sepiolite*; Nova Science Publishers, Inc.: Hauppauge, NY, USA, 2007; pp. 225–249.

42. Wang, X.S.; Miao, H.H.; He, W.; Shen, H.L. Competitive adsorption of Pb(II), Cu(II), and Cd(II) ions on wheat-residue derived black carbon. *J. Chem. Eng. Data* **2011**, *56*, 444–449. [CrossRef]
43. Saood Manzar, M.; Haladu, S.A.; Zubair, M.; Dalhat Mu'azu, N.; Qureshi, A.; Blaisi, N.I.; Garrison, T.F.; Charles, S. Al Hamouz, O. Synthesis and characterization of a series of cross-linked polyamines for removal of Erichrome Black T from aqueous solution. *Chin. J. Chem. Eng.* **2020**, *32*, 341–352. [CrossRef]
44. Xiong, Y.Y.; Li, J.Q.; Gong, L.L.; Feng, X.F.; Meng, L.N.; Zhang, L.; Meng, P.P.; Luo, M.B.; Luo, F. Using MOF-74 for Hg2+ removal from ultra-low concentration aqueous solution. *J. Solid State Chem.* **2017**, *246*, 16–22. [CrossRef]
45. Al-Yaari, M.; Saleh, T.A.; Saber, O. Removal of mercury from polluted water by a novel composite of polymer carbon nanofiber: Kinetic, isotherm, and thermodynamic studies. *RSC Adv.* **2021**, *11*, 380–389. [CrossRef] [PubMed]
46. Cegłowski, M.; Gierczyk, B.; Frankowski, M.; Popenda, Ł. A new low-cost polymeric adsorbents with polyamine chelating groups for efficient removal of heavy metal ions from water solutions. *React. Funct. Polym.* **2018**, *131*, 64–74. [CrossRef]
47. Hu, L.; Yang, Z.; Cui, L.; Li, Y.; Ngo, H.H.; Wang, Y.; Wei, Q.; Ma, H.; Yan, L.; Du, B. Fabrication of hyperbranched polyamine functionalized graphene for high-efficiency removal of Pb(II) and methylene blue. *Chem. Eng. J.* **2016**, *287*, 545–556. [CrossRef]
48. Samuel, M.S.; Shang, M.; Klimchuk, S.; Niu, J. Novel Regenerative Hybrid Composite Adsorbent with Improved Removal Capacity for Lead Ions in Water. *Ind. Eng. Chem. Res.* **2021**, *60*, 5124–5132. [CrossRef]
49. Moshari, M.; Mehrehjedy, A.; Heidari-Golafzania, M.; Rabbani, M.; Farhadi, S. Adsorption study of lead ions onto sulfur/reduced graphene oxide composite. *Chem. Data Collect.* **2021**, *31*, 100627. [CrossRef]
50. Ali, Z.; Ghanem, B.S.; Wang, Y.; Pacheco, F.; Ogieglo, W.; Vovusha, H.; Genduso, G.; Schwingenschlögl, U.; Han, Y.; Pinnau, I. Finely Tuned Submicroporous Thin-Film Molecular Sieve Membranes for Highly Efficient Fluid Separations. *Adv. Mater.* **2020**, *32*, 2001132. [CrossRef]
51. Bader, R.F.W. *Atoms in Molecules: A Quantum Theory*; Clarendon Press: Oxford, UK, 1994.
52. Abdelnaby, M.M.; Cordova, K.E.; Abdulazeez, I.; Alloush, A.M.; Al-Maythalony, B.A.; Mankour, Y.; Alhooshani, K.; Saleh, T.A.; Al Hamouz, O.C.S. Novel Porous Organic Polymer for the Concurrent and Selective Removal of Hydrogen Sulfide and Carbon Dioxide from Natural Gas Streams. *ACS Appl. Mater. Interfaces* **2020**, *12*, 47984–47992. [CrossRef]
53. Li, Y.; Liu, F.; Xia, B.; Du, Q.; Zhang, P.; Wang, D.; Wang, Z.; Xia, Y. Removal of copper from aqueous solution by carbon nanotube/calcium alginate composites. *J. Hazard. Mater.* **2010**, *177*, 876–880. [CrossRef] [PubMed]
54. Esmat, M.; Farghali, A.A.; Khedr, M.H.; El-Sherbiny, I.M. Alginate-based nanocomposites for efficient removal of heavy metal ions. *Int. J. Biol. Macromol.* **2017**, *102*, 272–283. [CrossRef] [PubMed]
55. Baba, Y.; Masaaki, K.; Kawano, Y. Synthesis of a chitosan derivative recognizing planar metal ion and its selective adsorption equilibria of copper(II) over iron(III)1. *React. Funct. Polym.* **1998**, *36*, 167–172. [CrossRef]
56. Limsuwan, Y.; Rattanawongwiboon, T.; Lertsarawut, P.; Hemvichian, K.; Pongprayoon, T. Adsorption of Cu(II) ions from aqueous solution using PE/PP non-woven fabric grafted with poly(bis[2-(methacryloyloxy) ethyl] phosphate). *J. Environ. Chem. Eng.* **2021**, *9*, 106440. [CrossRef]
57. Lin, X.; Shen, T.; Li, M.; Shaoyu, J.; Zhuang, W.; Li, M.; Xu, H.; Zhu, C.; Ying, H.; Ouyang, P. Synthesis, characterization, and utilization of poly-amino acid-functionalized lignin for efficient and selective removal of lead ion from aqueous solution. *J. Clean. Prod.* **2022**, *347*, 131219. [CrossRef]
58. Mzinyane, N.N.; Ofomaja, A.E.; Naidoo, E.B. Synthesis of poly (hydroxamic acid) ligand for removal of Cu (II) and Fe (II) ions in a single component aqueous solution. *S. Afr. J. Chem. Eng.* **2021**, *35*, 137–152. [CrossRef]
59. Ben Ali, M.; Wang, F.; Boukherroub, R.; Xia, M. High performance of phytic acid-functionalized spherical poly-phenylglycine particles for removal of heavy metal ions. *Appl. Surf. Sci.* **2020**, *518*, 146206. [CrossRef]

Article

Synthesis and Characterization of Thiol-Functionalized Polynorbornene Dicarboximides for Heavy Metal Adsorption from Aqueous Solution

Alejandro Onchi [1], Carlos Corona-García [1], Arlette A. Santiago [2], Mohamed Abatal [3], Tania E. Soto [1], Ismeli Alfonso [1] and Joel Vargas [1,*]

1. Instituto de Investigaciones en Materiales, Unidad Morelia, Universidad Nacional Autónoma de México, Antigua Carretera a Pátzcuaro No. 8701, Col. Ex Hacienda de San José de la Huerta, C.P. 58190 Morelia, Michoacán, Mexico; alejandro.onchi@gmail.com (A.O.); carcor93@gmail.com (C.C.-G.); tania@iim.unam.mx (T.E.S.); ialfonso@iim.unam.mx (I.A.)
2. Escuela Nacional de Estudios Superiores, Unidad Morelia, Universidad Nacional Autónoma de México, Antigua Carretera a Pátzcuaro No. 8701, Col. Ex Hacienda de San José de la Huerta, C.P. 58190 Morelia, Michoacán, Mexico; arlette_santiago@enesmorelia.unam.mx
3. Facultad de Ingeniería, Universidad Autónoma del Carmen, Avenida Central S/N Esq. con Fracc. Mundo Maya, C.P. 24115 Ciudad del Carmen, Campeche, Mexico; mabatal@pampano.unacar.mx
* Correspondence: jvargas@iim.unam.mx; Tel.: +52-4431-477-887

Abstract: The contamination of water resources with heavy metals is a very serious concern that demands prompt and effective attention due to the serious health risks caused by these contaminants. The synthesis and ring-opening metathesis polymerization (ROMP) of norbornene dicarboximides bearing thiol pendant groups, specifically, N-4-thiophenyl-*exo*-norbornene-5,6-dicarboximide (**1a**), N-4-(methylthio)phenyl-*exo*-norbornene-5,6-dicarboximide (**1b**) and N-4-(trifluoromethylthio)phenyl-*exo*-norbornene-5,6-dicarboximide (**1c**), as well as their assessment for the removal of heavy metals from aqueous systems, is addressed in this work. The polymers were characterized by NMR, SEM and TGA, among others. Single and multicomponent aqueous solutions of Pb^{2+}, Cd^{2+} and Ni^{2+} were employed to perform both kinetic and isothermal adsorption studies taking into account several experimental parameters, for instance, the initial metal concentration, the contact time and the mass of the polymer. In general, the adsorption kinetic data fit the pseudo-second-order model more efficiently, while the adsorption isotherms fit the Freundlich and Langmuir models. The maximum metal uptakes were 53.7 mg/g for Pb^{2+}, 43.8 mg/g for Cd^{2+} and 29.1 mg/g for Ni^{2+} in the SH-bearing polymer **2a**, 46.4 mg/g for Pb^{2+}, 32.9 mg/g for Cd^{2+} and 27.1 mg/g for Ni^{2+} in the SCH_3-bearing polymer **2b** and 40.3 mg/g for Pb^{2+}, 35.9 mg/g for Cd^{2+} and 27.8 mg/g for Ni^{2+} in the SCF_3-bearing polymer **2c**, correspondingly. The better performance of polymer **2a** for the metal uptake was ascribed to the lower steric hindrance and higher hydrophilicity imparted by –SH groups to the polymer. The results show that these thiol-functionalized polymers are effective adsorbents of heavy metal ions from aqueous media.

Keywords: ROMP; thiol-functionalized polymer; polynorbornene dicarboximide; heavy metal adsorption; lead; cadmium; nickel

Citation: Onchi, A.; Corona-García, C.; Santiago, A.A.; Abatal, M.; Soto, T.E.; Alfonso, I.; Vargas, J. Synthesis and Characterization of Thiol-Functionalized Polynorbornene Dicarboximides for Heavy Metal Adsorption from Aqueous Solution. *Polymers* **2022**, *14*, 2344. https://doi.org/10.3390/polym14122344

Academic Editors: Irene S. Fahim and Dimitrios Bikiaris

Received: 5 May 2022
Accepted: 7 June 2022
Published: 9 June 2022

Publisher's Note: MDPI stays neutral with regard to jurisdictional claims in published maps and institutional affiliations.

Copyright: © 2022 by the authors. Licensee MDPI, Basel, Switzerland. This article is an open access article distributed under the terms and conditions of the Creative Commons Attribution (CC BY) license (https://creativecommons.org/licenses/by/4.0/).

1. Introduction

In the last decades, there has been a concerning rise in the number of heavy metals found in affluents due to the steady rise of industrial activities [1,2]. The pollution attributed to the industrial development is present in different ways, and among all the activities, metal industry for construction, battery manufacturing, mining and refining processes are the prime sources of releasing heavy metals to the environment, such as Pb^{2+}, Cd^{2+} and Ni^{2+} [3,4]. The presence of these contaminants is known to be a serious harm to human health, besides being a considerable hazard to animals in the affected area. There are several techniques that are known to be effective in the elimination of heavy metals in aqueous media, for instance, ion exchange, osmosis, chemical precipitation and adsorption, among others [5–7]. In fact, the development and improvement of these and other new techniques are of the utmost importance for the urgent remediation of the environment.

Adsorption is a very interesting method due to its adaptability, good performance and the availability of a large number of materials with adsorption capacities [8,9]. In this sense, polymers have shown high effectiveness in the adsorption of heavy metals from aqueous systems due to their easy chemical functionalization, which allows adjusting the affinity towards heavy metal ions [10–13]. Norbornene-based monomers are readily functionalized, making the preparation of systematic series of norbornene derivatives an attainable goal [14]. Additionally, ring-opening metathesis polymerization (ROMP) is a useful tool for preparing and functionalizing polymers with outstanding physical and chemical integrity that can withstand harsh environments [15–17]. Furthermore, the information at hand points out that using functionalized ROMP-prepared polymers in adsorption processes is a fairly viable option for capturing Pb^{2+}, Cd^{2+}, Ni^{2+}, Cr^{3+} and Cr^{6+} from aqueous systems [18,19].

Among the many functional group options used to endow materials with adsorption properties, thiols have drawn much attention and are well known for their ability to coordinate with divalent ions and have already been used successfully in the treatment of poisoning by Pb, Hg and As [20–22]. Since the ROMP affords high molecular weight polymers in high yields [23,24], ROMP-prepared thiol-containing polymers are expected to show high degrees of thiol moieties which in turn will likely result in high adsorption capacities. Based on the latter, ROMP-prepared polynorbornene derivatives could be used in the development of adsorbents bearing thiol groups as coordination points in order to tune the heavy metal adsorption capacity.

Hence, in this work, we carried out the synthesis and ROMP of three thiol-containing norbornene dicarboximide monomers along with the assessment of these polymers as adsorbents for capturing Pb^{2+}, Cd^{2+} and Ni^{2+} from aqueous media. The physical properties, the thermal properties, and the surface morphology of the polymers are studied as well. As far as we know, the use of thiol-functionalized polynorbornene dicarboximides for heavy metal capture from water has not been reported, evidencing the necessity of exploring the assessment of this kind of macromolecular materials for such application.

2. Experimental Part
2.1. Characterization Techniques

The monomers and polymers obtained were characterized using the following techniques: Fourier-transform infrared spectroscopy (FT-IR), nuclear magnetic resonance (NMR), differential scanning calorimetry (DSC), thermomechanical analysis (TMA), thermogravimetric analysis (TGA) and X-ray diffraction (XRD). The FT-IR spectra were recorded in a Thermo Scientific Nicolet iS10 FT-IR spectrometer using an attenuated total reflectance (ATR) accessory with a diamond crystal. The final spectrum is an average of thirty-two spectra collected for each sample in a range of 4000–650 cm^{-1} at a spectral resolution of 4 cm^{-1}. The ^{1}H-NMR, ^{13}C-NMR and ^{19}F-NMR spectra were recorded on a Bruker Avance III HD at 400, 100 and 376 MHz, respectively. Tetramethylsilane (TMS) and hexafluorobenzene (HFB) were used as internal standards for the NMR analysis. The samples were measured in CDCl$_3$ for **1a**, **1b**, **1c**, **2b** and **2c**, while the polymer **2a** was measured in DMF-d$_7$. DSC was carried out in a SENSYS evo DSC, with samples encapsulated in standard aluminum DSC pans, at a scanning rate of 10 °C min^{-1} under a nitrogen atmosphere in a temperature range between 30 °C and 500 °C, and it was used for determining the glass transition temperature (T_g) of the polymers. The TMA was carried out in a TA Instruments Thermomechanical Analyzer TMA Q400, at a rate of 10 °C min^{-1} under a nitrogen atmosphere to corroborate the T_g values. TGA was conducted to determine the onset of decomposition temperature (T_d) of the polymers. The samples, around 10 mg, were heated at a rate of 10 °C min^{-1} from 30 to 600 °C under a nitrogen atmosphere in a TA Instruments Thermogravimetric

Analyzer TGA 5500. XRD measurements were performed in polymer films on a Bruker D2-Phaser 2nd Generation diffractometer between 7 and 70° 2θ, at 30 KV 10 mA, using CuK$_\alpha$ radiation (1.54 Å). The polymer density (ρ) was measured in film form in ethanol at room temperature using the analytical balance model Sartorius Quintix 124-1s by the flotation method.

The heavy metal adsorption capacity was calculated through a series of experiments using Pb^{2+}, Cd^{2+} and Ni^{2+} aqueous solutions where $PbCl_2$, $CdCl_2 \cdot 2.5H_2O$ y $NiCl_2 \cdot 6H_2O$ were used to prepare the respective solutions. For the study of the adsorption kinetics, 0.01 g of each polymer were put in contact with a 100-ppm solution for 5, 10, 15, 30, 60, 120, 180, 360, 720 and 1440 min, respectively. For the study of the adsorption isotherms, 0.01 g of each polymer were put in contact with a solution of 10, 20, 40, 60, 80, 100, 200, 300, 400 and 500 ppm for 24 h, respectively. For the study of the mass effect, 0.01, 0.02, 0.05, 0.08 and 0.10 g of each polymer were put in contact with a 100-ppm solution of each metal for 24 h. For the study of the multicomponent adsorption, a 100 ppm of ternary heavy metal solution using a 100-ppm solution of each metal ion was prepared, and 0.01 g of each polymer was put in contact with it for 24 h. The changes in heavy metal concentration in the solution after the contact with the polymers were quantified using atomic absorption spectroscopy (AAS) in a Thermo Scientific iCE 3000 Series.

2.2. Reagents

The *exo*-norbornene-5,6-dicarboxylic anhydride (NDA) was obtained via the Diels-Alder cycloaddition reaction of maleic anhydride and cyclopentadiene in accordance with the literature [14]. 4-Aminothiophenol, 4-(methylthio)aniline, 4-(trifluoromethylthio)aniline and tricyclohexylphosphine [1,3-bis(2,4,6-trimethylphenyl)-4,5-dihydroimidazol-2-ylide ne][benzylidene] ruthenium dichloride (2nd Generation Grubbs catalyst) were employed as they were received. Lead(II) chloride ($PbCl_2$), cadmium(II) chloride hemi(pentahydrate) ($CdCl_2 \cdot 2.5H_2O$), nickel(II) chloride hexahydrate ($NiCl_2 \cdot 6H_2O$), chloroform and methanol were employed as they were received. Dichloromethane and 1,2-dichloroethane were dried over anhydrous $CaCl_2$, then distilled over CaH_2 and used as solvents. All reagents were purchased from Merck Sigma-Aldrich.

2.3. Synthesis and Characterization of Monomers

2.3.1. Monomer *N*-4-thiophenyl-*exo*-norbornene-5,6-dicarboximide (**1a**)

In total, 30 mL of dichloromethane were used to dissolve 0.30 g (0.0023 mol) of 4-aminothiophenol and 0.39 g (0.0023 mol) of NDA. The solution was stirred for 12 h at 60 °C to give an amic acid. Next, the amic acid was filtered and dissolved in 30 mL of $(CH_3CO)_2O$, then 0.50 g of CH_3COONa were added to the mixture and kept with constant stirring for 12 h at 70 °C. The product precipitated on pouring the reaction mixture into iced water and was then filtered. Finally, the monomer was recrystallized twice from a toluene:hexane (1:1) solution and dried at 50 °C for 12 h under vacuum. The pure monomer was obtained in 72% yield. Melting point (MP): 166–168 °C (Scheme 1a).

FT-IR (powder, cm^{-1}): υ 3070 (C=C–H aromatic (ar.) stretching (str.)), 2990 (C–H asymmetric (asym.) str.), 2885 (C–H symmetric (sym.) str.), 1774 (C=O), 1697 (C=O), 1492 (C=C str.), 1373 (C–N) and 684 (C–S).

^{1}H-NMR (400 MHz, CDCl$_3$, ppm): δ 7.55–7.37 (4H, m, ar.), 6.39 (2H, s), 3.42 (2H, s), 2.88 (2H, s), 2.45 (1H, s), 1.64 (1H, m) and 1.46 (1H, m).

^{13}C-NMR (100 MHz, CDCl$_3$, ppm): δ 176.6 (C=O), 138.0 (C=C), 126.7 (C–S) 47.8, 45.9 and 43.0.

2.3.2. Monomer *N*-4-(methylthio)phenyl-*exo*-norbornene-5,6-dicarboximide (**1b**)

In total, 30 mL of dichloromethane were used to dissolve 0.30 g (0.0021 mol) of 4-(methylthio)aniline and 0.35 g (0.0021 mol) of NDA. The solution was stirred for 12 h at 60 °C to give an amic acid. Next, the amic acid was filtered and dissolved in 30 mL of $(CH_3CO)_2O$, then 0.50 g of CH_3COONa were added to the mixture and kept with constant stirring for 12 h at 70 °C. The product precipitated on pouring the reaction mixture into iced water and was then filtered. Finally, the monomer was recrystallized twice from ethanol and dried at 50 °C for 12 h under vacuum. The pure monomer was obtained in 61% yield. Melting point (MP): 164–166 °C (Scheme 1a).

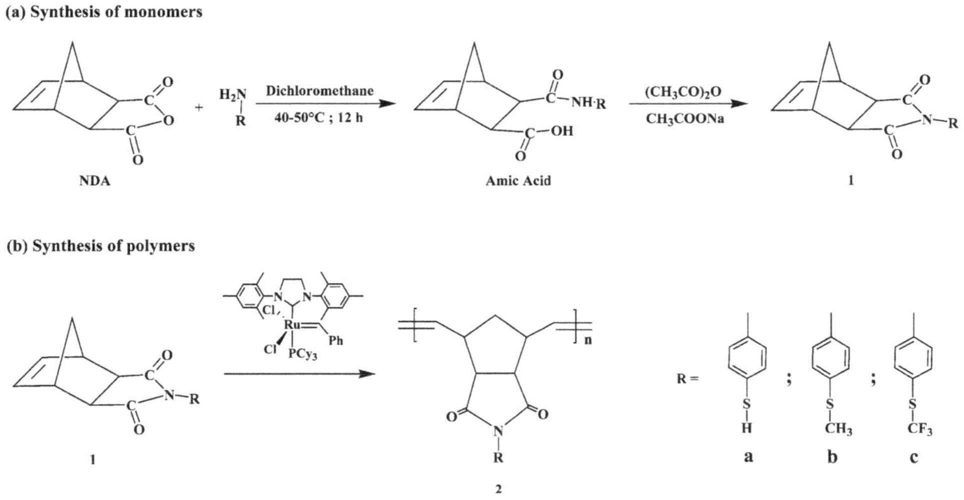

Scheme 1. (a) Synthesis of the thiol-functionalized monomers and (b) synthesis of the thiol-containing polymers via ROMP.

FT-IR (powder, cm^{-1}): υ 3073 (C=C–H ar. str.), 2980 (C–H asym. str.), 2945 (C–H sym. str.), 1773 (C=O), 1698 (C=O), 1493 (C=C str.), 1438 (C–H), 1385 (C–N) and 681 (C–S).

^{1}H-NMR (400 MHz, CDCl$_3$, ppm): δ 7.36–7.18 (4H, m, ar.), 6.36 (2H, s), 3.42 (2H, s), 2.87 (2H, s), 2.51 (3H, s), 1.65 (1H, m) and 1.47 (1H, m).

^{13}C-NMR (100 MHz, CDCl$_3$, ppm): δ 177.0 (C=O), 138.0 (C=C), 125.3 (C–S), 47.8, 45.8 and 43.0.

2.3.3. Monomer N-4-(trifluoromethylthio)phenyl-*exo*-norbornene-5,6-dicarboximide (**1c**)

In total, 30 mL of dichloromethane were used to dissolve 0.20 g (0.0010 mol) of 4-(trifluoromethylthio)aniline and 0.16 g (0.0010 mol) of NDA. The solution was stirred for 12 h at 60 °C to give an amic acid. Next, the amic acid was filtered and dissolved in 30 mL of (CH$_3$CO)$_2$O, then 0.50 g of CH$_3$COONa were added to the mixture and kept with constant stirring for 12 h at 70 °C. The product precipitated on pouring the reaction mixture into iced water and was then filtered. Finally, the monomer was recrystallized twice from hexane and dried at 50 °C for 12 h under vacuum. The pure monomer was obtained in 49% yield. Melting point (MP): 157–159 °C (Scheme 1a).

FT-IR (powder, cm^{-1}): υ 3068 (C=C–H ar. str.), 2988 (C–H asym. str.), 2970 (C–H sym. str.), 1775 (C=O), 1706 (C=O), 1493 (C=C str.), 1382 (C–N), 1109 (C–F) and 683 (C–S).

^{1}H-NMR (400 MHz, CDCl$_3$, ppm): δ 7.77–7.38 (4H, m, ar.), 6.38 (2H, s), 3.42 (2H, s), 2.88 (2H, s), 1.65 (1H, m) and 1.43 (1H, m).

^{13}C-NMR (100 MHz, CDCl$_3$, ppm): δ 172.5 (C=O), 138.0 (C=C), 47.8, 45.8 and 43.0.

^{19}F-NMR: (376 MHz, CDCl$_3$, ppm): δ −44.75 (CF$_3$).

2.4. *ROMP of the Thiol-Functionalized Monomers*

Monomer polymerizations were carried out in 25 mL round bottom flasks under N$_2$ atmosphere. C$_2$H$_5$OCH=CH$_2$ was employed as an inhibitor in the final stage. The polymers were obtained by pouring the reaction solution into 60 mL of acidified methanol under constant stirring at 50 °C, precipitating in the form of white fibers. The resulting polymers were collected by filtration, then dissolved in chloroform and poured again into hot stirring methanol for being purified. Next, the products were dried for 12 h at 50 °C under vacuum.

2.4.1. Synthesis and Characterization of Poly(N-4-thiophenyl-*exo*-norbornene-5,6-dicarboximide) (**2a**)

In total, 10 mL of 1,2-dichloroethane were employed to dissolve 0.2 g (0.73 mmol) of monomer **1a** and 6.1×10^{-4} g (7.1×10^{-4} mmol) of Grubbs catalyst. The polymerization reaction was stirred at room temperature for 2 h (Scheme 1b). The thiol-functionalized polymer **2a** was soluble in N,N-dimethylformamide (DMF) and 1,2-dichloroethane.

FT-IR (thin film, cm^{-1}): υ 3472 (C=C–H ar. str.), 2922 (C–H asym. str.), 2852 (C–H sym. str.), 1776 (C=O), 1703 (C=O), 1494 (C=C str.), 1367 (C–N) and 659 (C–S).

^{1}H-NMR (400 MHz, DMF-d$_7$, ppm): δ 7.61−7.28 (4H, m, ar.), 5.87 (1H, s, *trans*), 5.63 (1H, s, *cis*), 3.45 (1H, s), 3.18 (2H, s), 2.76 (1H, s), 2.46 (1H, s), 2.18 (1H, s) and 1.69 (1H, s).

^{13}C-NMR (100 MHz, DMF-d$_7$, ppm): δ 176.6 (C=O), 138.0 (C=C), 134.8, 126.7, 77.3, 47.8, 45.9 and 43.0.

2.4.2. Synthesis and Characterization of Poly(N-4-(methylthio)phenyl-*exo*-norbornene-5,6-dicarboximide) (**2b**)

In total, 2 mL of 1,2-dichloroethane were employed to dissolve 0.2 g (0.51 mmol) of monomer **1b** and 7.6×10^{-4} g (8.9×10^{-4} mmol) of Grubbs catalyst. The polymerization reaction was stirred at 45 °C for 2 h (Scheme 1b). The thiol-functionalized polymer **2b** was soluble in N,N-dimethylformamide (DMF), chloroform and 1,2-dichloroethane.

FT-IR (thin film, cm^{-1}): υ 3627 (C=C–H ar. str.), 2919 (C–H asym. str.), 2855 (C–H sym. str.), 1773 (C=O), 1701 (C=O), 1495 (C=C str.), 1436 (C–H), 1373 (C–N) and 660 (C–S).

^{1}H-NMR (400 MHz, CDCl$_3$, ppm): δ 7.31−7.05 (4H, m, ar.), 5.78 (1H, s, *trans*), 5.52 (1H, s, *cis*), 3.48 (1H, s), 3.26 (2H, s), 2.83 (1H, s), 2.43 (3H, s), 2.09 (1H, s) and 1.60 (1H, s).

^{13}C-NMR (100 MHz, CDCl$_3$, ppm): δ 177.5 (C=O), 139.7 (C=C), 134.4, 127.0, 125.0, 77.9, 53.8, 51.4 and 46.9.

2.4.3. Synthesis and Characterization of Poly(N-4-(trifluoromethylthio)phenyl-*exo*-norbornene-5,6-dicarboximide) (**2c**)

In total, 2 mL of 1,2-dichloroethane were employed to dissolve 0.2 g (0.58 mmol) of monomer **1c** and 9.5×10^{-4} g (1.1×10^{-3} mmol) of Grubbs catalyst. The polymerization reaction was stirred at 45 °C for 2 h (Scheme 1b). The thiol-functionalized polymer **2c** was soluble in N,N-dimethylformamide (DMF), chloroform and 1,2-dichloroethane.

FT-IR (thin film, cm^{-1}): υ 3480 (C=C–H ar. str.), 2954 (C–H asym. str.), 2859 (C–H sym. str.), 1779 (C=O), 1707 (C=O), 1495 (C=C str.), 1366 (C–N), 1109 (C–F) and 659 (C–S).

^{1}H-NMR (400 MHz, CDCl$_3$, ppm): δ 7.77−7.19 (4H, m, ar.), 5.84 (1H, s, *trans*), 5.59 (1H, s, *cis*), 3.51 (1H, s), 3.22 (2H, s) 2.91 (1H, s), 2.12 (1H, s) and 1.61 (1H, s).

^{13}C-NMR (100 MHz, CDCl$_3$, ppm): δ 176.9 (C=O), 147.2 (C=C), 137.1, 134.5, 127.4, 124.8, 77.7, 54.0, 51.2 and 47.1.

^{19}F-NMR: (376 MHz, CDCl$_3$, ppm): δ −46.8 (CF$_3$).

2.5. Membrane Preparation

Membranes were cast from polymeric 1,2-dichloroethane solutions (~8 wt%) at room temperature. The solution was filtered and poured onto a glass plate, and the solvent was allowed to evaporate slowly under a controlled 1,2-dichloroethane atmosphere. Then, the membranes were dried under a vacuum at 100 °C for 24 h. The average thickness of the films was around 50 µm. The prepared films were only used for FT-IR, XRD, thermal and density measurement purposes.

2.6. Heavy Metal Adsorption

Pb^{2+}, Cd^{2+} and Ni^{2+} stock solutions of 1000 mg/L were prepared by dissolving PbCl$_2$, CdCl$_2$·2.5H$_2$O and NiCl$_2$·6H$_2$O, respectively, in deionized water. Before each series of

AAS measurements, the deionized water was first used as a blank to discard any heavy metal trace that could be in the remaining liquid. Subsequently, each heavy metal solution concentration was corroborated using AAS before the respective experiments.

2.7. Adsorption Kinetics

For the adsorption kinetics experiments, solutions of Pb^{2+}, Cd^{2+} and Ni^{2+} with concentrations of 100 ppm were prepared, respectively. An amount of 0.01 g of each polymer was added, separately, to a 10 mL aliquot of individual heavy metal solution. The mixtures of each polymer with each solution were kept in constant stirring using a rotatory shaker for 5, 10, 15, 30, 60, 120, 180, 360, 720 and 1440 min. After a specific contact time, the samples were centrifuged at 3500 rpm for 10 min to separate the polymer from the solution. Then, AAS was used to measure the final solution concentration. The quantity of heavy metal adsorbed on the surface of each polymer at a specific time, denoted as q_t (mg/g), was calculated employing Equation (1), knowing that m is the amount of polymer used for each sample (g), V is the solution volume (L), C_0 is the initial heavy metal concentration (mg/L) and C_f is the final heavy metal concentration (mg/L).

$$q_t = V(C_0 - C_f)/m \tag{1}$$

To describe the adsorption kinetics of the respective metals on the polymer surface, pseudo-first and pseudo-second-order models were applied to the analysis of the experimental data. Both mathematical expressions were linearized and are shown in Equations (2) and (3), respectively.

$$\log(q_e - q_t) = \log(q_e) - (k_1/2.303)t \tag{2}$$

$$t/q_t = \{1/(k_2 q_e^2)\} + (1/q_e)t \tag{3}$$

q_t (mg/g) is the concentration of the heavy metal at a time t (min) and q_e (mg/g) is the concentration of the heavy metal in equilibrium. Likewise, k_1 (1/min) and k_2 (g/mg·min) are the pseudo-first and pseudo-second-order model equilibrium rate constants, respectively.

2.8. Isotherm Study

Regarding the isotherm studies, solutions of Pb^{2+}, Cd^{2+} and Ni^{2+} were prepared with the following concentrations: 10, 20, 40, 60, 80, 100, 200, 300, 400 and 500 ppm, then, 0.01 g of each polymer was added to every solution separately and was kept with constant stirring in a rotatory shaker for 24 h. Once the contact time was reached, the samples were subjected to centrifugation for 10 min at 3500 rpm to separate the polymer from the solution. Then, AAS was used to measure the final solution concentration.

For the analysis of the experimental data, the Freundlich and Langmuir models were used to explain the equilibrium adsorption of the heavy metals on the polymeric surface. The equations of these models were linearized and are shown in Equations (4) and (5), respectively.

$$\ln(q_e) = \ln(K_F) + \{(1/n) \ln(C_e)\} \tag{4}$$

$$1/q_e = 1/q_m + \{1/(q_m K_L)\}\{1/C_e\} \tag{5}$$

q_m (mg/g) is the maximum concentration of heavy metal, q_e (mg/g) is the equilibrium concentration of heavy metal and C_e (mg/L) is the heavy metal concentration in the equilibrium of the solution. Likewise, K_F (1/min) and K_L (g/mg·min) are the Freundlich and Langmuir model equilibrium rate constants, respectively.

2.9. Mass Effect Study

The effect of the mass on the adsorption capacity was assessed by adding 0.01, 0.02, 0.05, 0.08 and 0.10 g of each polymer to 100 ppm solutions of Pb^{2+}, Cd^{2+} and Ni^{2+}, respectively. The polymer was left in contact with the solution and stirred for 24 h in a rotatory shaker; next, the samples were centrifuged at 3500× *g* rpm for 10 min to separate the polymer from the solution. Then, AAS was used to measure the final solution concentration.

2.10. Multicomponent Adsorption Study

For studying the adsorption capacity of the polymers in a multicomponent solution, a unique 100 ppm solution of Pb^{2+}, Cd^{2+} and Ni^{2+} was prepared and 0.01 g of each polymer was put in contact, with constant stirring in a rotatory shaker, with the solution for 24 h. Once the contact time was reached, the samples were subjected to centrifugation for 10 min at 3500× *g* rpm to separate the polymer from the solution. Then, AAS was used to measure the final solution concentration.

3. Results and Discussions

Monomers **1a**, **1b** and **1c** were prepared successfully in a two-step reaction in 72%, 61% and 49% yield, respectively. NDA reacted with 4-aminothiophenol, 4-(methylthio)aniline and 4-(trifluoromethylthio)aniline, respectively, to yield amic acids, which were dehydrated employing acetic anhydride and anhydrous sodium acetate to afford the corresponding thiol-functionalized imide monomers (Scheme 1). The melting points of the monomers were in the range of 166–168 °C, 164–166 °C and 157–159 °C for **1a**, **1b** and **1c**, respectively. Photographic images of the thiol-functionalized norbornene dicarboximide monomers synthesized in this study as well as their corresponding raw polymers prepared by ROMP are shown in Figure 1. It can be seen that the monomers exhibit different colorations depending on the thiol substituent attached to the aromatic ring being pale yellow for monomer **1a**, slightly violet for monomer **1b** and white for monomer **1c**. Despite the monomer coloration, all of them afforded white fibrous polymers capable of forming tough transparent films.

Figure 1. Photographic images of the thiol-functionalized norbornene dicarboximide monomers (**top**) and their corresponding ROMP-prepared raw polymers (**bottom**).

The chemical structures of the thiol-functionalized norbornene dicarboximide monomers were confirmed by FT-IR spectroscopy and the spectra are shown in Figure 2. Spectra are very similar to each other with slight differences due to the thiol group substituents. In general, the following characteristic absorption bands can be observed: the absorption band of the H–C bond in the aromatic groups is seen at about 3070 cm^{-1}; the bands of the C–H bond in the methylene groups corresponding to the antisymmetric and symmetric vibration modes are observed at 2990 and 2945 cm^{-1}, respectively; likewise, the bands of the C=O bond associated to the antisymmetric and symmetric vibration modes are shown at 1774 and 1697 cm^{-1}; correspondingly, the absorption band of the C=C bond related to the stretching mode is displayed at 1492 cm^{-1}; the absorption band of the C–H bond in the methyl groups corresponding to the stretching mode is observed around 1438 cm^{-1}; the band of the C–N bond attributed to the stretching mode is seen about 1382 cm^{-1}; the band of the C–F bond assigned to the stretching mode was found around 1109 cm^{-1}; the signal observed about 683 cm^{-1} is ascribed to the stretching vibration mode of the C–S bond.

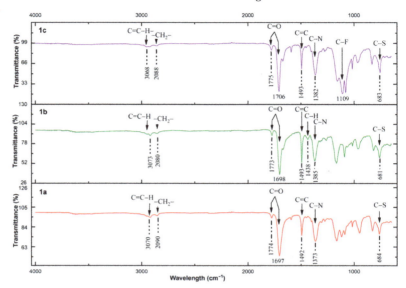

Figure 2. FT-IR spectra of the thiol-functionalized norbornene dicarboximide monomers.

The monomer's chemical structures were also corroborated by ^{1}H-NMR, ^{13}C-NMR and ^{19}F-NMR. The ^{1}H-NMR spectra of all monomers exhibited aromatic proton signals (H$_f$ and H$_g$) in the range of 7.61–7.18 ppm (Figure 3). The signals corresponding to the olefinic protons (H$_a$) ranged from 6.39 to 6.36 ppm, while the signals associated with the protons of the –CH– groups (H$_b$ and H$_c$) were found in the range of 3.42–2.87 ppm. For all monomers, the signals ascribed to the methylene protons (H$_d$ and H$_e$) were seen ranging from 1.65 to 1.43 ppm. Finally, the signals associated with the protons of the thiol groups (H$_h$) were observed at 2.45 ppm for monomer **1a** and at 2.51 ppm for monomer **1b**. The ^{13}C-NMR spectra of the monomers showed characteristic signals for the carbon atoms in the C=O groups ranging between 172.5 and 177.0 ppm while the signals associated with the carbon atoms in the C=C groups were observed at 138.0 ppm for all monomers. Likewise, the ^{19}F-NMR analysis indicated that fluorine atoms in the –CF$_3$ groups of monomer **1c** are magnetically equivalent, leading to the appearance of one signal peak around −46.8 ppm.

The polymers ^{1}H-NMR spectra are shown in Figure 4. As it is seen, the ^{1}H-NMR spectra of the polymers are very similar to those of the monomers. For all polymers, the signals attributed to the aromatic protons (H$_f$ and H$_g$) appeared between 7.77 and 7.05 ppm. In general, the olefinic proton signals of the monomers around δ = 6.38 ppm are substituted by new olefinic proton (H$_a$) signals that arise in the range of δ = 5.87–5.78 ppm and 5.63–5.52 ppm, corresponding, respectively, to the *trans* and *cis* double bonds of the polymer backbone. The proton signals of the –CH– groups (H$_b$ and H$_c$) were observed in the range of 3.51–2.76 ppm. The signals associated with the methylene protons (H$_d$ and H$_e$) were seen ranging from 2.18 to 1.60 ppm, with slight signal overlapping arising in this area owing to the signals from the protons (H$_h$) attached to the thiol substituents that appeared at 2.46 and 2.43 ppm for the polymers **2a** and **2b**, respectively.

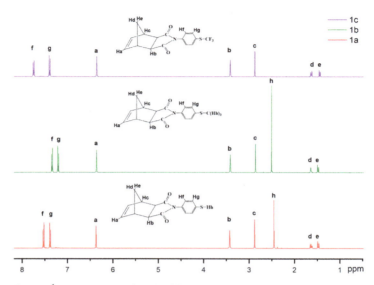

Figure 3. ^{1}H-NMR spectra of the thiol-functionalized norbornene dicarboximide monomers.

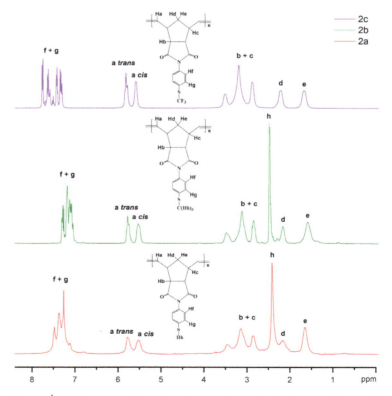

Figure 4. ^{1}H-NMR spectra of the polynorbornene dicarboximides bearing thiol pendant groups.

Table 1 summarizes the physical properties of the polymers bearing thiol pendant groups. The glass transition temperature (T_g) was measured by DSC and the values were corroborated by TMA. The T_g values obtained were 225 °C, 216 °C and 215 °C for the polymers **2a**, **2b** and **2c**, respectively. The decrease in the T_g value for polymers **2b** and **2c** compared to that of **2a** is attributed to the substituent on the thiol group; since the methylthiol and trifluoromethylthiol groups introduce more free volume into the polymer chains, higher conformational mobility is achieved, which in turn lowers the T_g. The polymer's thermal stability was studied under a nitrogen atmosphere by TGA. The thermal decomposition curves, shown in Figure 5, were shifted by 2% from each other starting from that corresponding to the polymer bearing –SH groups, **2a**, to clearly track the decomposition profiles of all the thiol-functionalized polynorbornene dicarboximides. From Figure 5, it can be seen that all the thiol-functionalized polymers show the onset temperature for decomposition (T_d) in the range of 409–459 °C, indicating that the norbornene dicarboximide monomers reported here yield polymers of relatively high thermal stability. The polymer density, ρ, was measured in film form by the flotation method at ambient conditions in ethanol [25]. In Table 1, it is noticed that polymer **2a** exhibits the higher density of all the polymers studied here, which could be associated with a higher polymer chain packing efficiency promoted by the thiol groups. In polymer **2b**, the presence of methylthiol groups causes greater steric hindrance than the thiol groups in polymer **2a**, which decreases the chain packing efficiency and, in turn, the polymer density. Likewise, the presence of bulky trifluoromethylthiol groups in polymer **2c** also diminishes the polymer packing efficiency, which leads to a lower density than that of polymer **2a**.

Table 1. Physical parameters of the thiol-functionalized polymers.

Polymer	T_g [a] (°C)	T_d [b] (°C)	ρ [c] (g/cm^3)	$d_{spacing}$ [d] (Å)
2a	225	459	1.38	4.83
2b	216	409	1.23	4.88
2c	215	415	1.33	5.01

[a] glass transition temperature measured by DSC. [b] onset of decomposition temperature estimated by TGA. [c] density calculated by the flotation method in ethanol. [d] mean intersegmental distance determined by XRD.

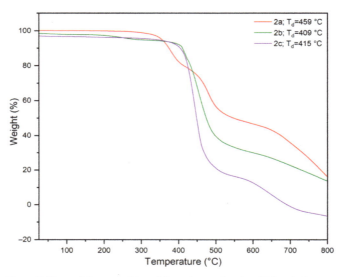

Figure 5. Thermal decomposition of the polymers bearing thiol moieties.

XRD measurements carried out in films of the thiol-functionalized polynorbornene dicarboximide revealed amorphous polymers with no regions of crystallinity. The diffraction patterns showed a characteristic broad peak for polynorbornene dicarboximides with the highest intensity of reflection in the range of 17–20°, in the 2θ scale (Figure 6) [19]. Using Bragg's equation, $n\lambda = 2d\sin\theta$, the d-spacing is determined at the maximum reflection angle of the amorphous curve [26]. This parameter is considered a measure of the mean intersegmental distance of the polymer chains and can be used to understand the effect of the different thiol substituents on the chain packing efficiency. In Table 1, it can be noticed that d-spacing decreases in the order **2c > 2b > 2a**, which suggests that the trifluoromethylthiol and methylthiol groups exhibit higher steric hindrance than the thiol group. This trend in the d-spacing values correlates quite well with the values found for the density since both parameters are inversely affected by the chain packing efficiency.

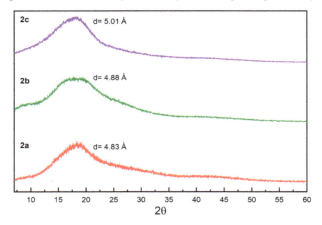

Figure 6. X-ray diffraction patterns of the polymers bearing thiol moieties.

3.1. Kinetics

The adsorption kinetics of the different polymers can be observed in Figure 7, where it is shown the adsorption capacity (q_t) as a function of the time of contact (t). Due to a large number of available sites for coordination at the beginning of the experiment, it is possible to observe a rapid increase in the adsorption of the metallic ions up to 180 min; once the divalent ions start to occupy several adsorption sites, the adsorption rate slowly stabilized until equilibrium was reached. The high heavy metal concentration at the beginning of the experiment also can produce a concentration gradient that contributes to the fast adsorption in the early stages. It is also possible to observe a tendency in the quantity of heavy metal adsorbed for each polymer, where Pb^{2+} is the most adsorbed ion, followed by Cd^{2+} and finally Ni^{2+}. It is seen that the higher the atomic number, the more probability of coordinating with the thiol group; thus, lead, having a higher atomic number than cadmium, is adsorbed in greater quantity. In the same way, cadmium is adsorbed in greater quantity than nickel.

The pseudo-first-order and pseudo-second-order models were applied to estimate the kinetic parameters and the theoretical adsorption capacity at equilibrium ($q_{e,\,the}$); the data are presented in Table 2 with the maximum experimental adsorption capacity ($q_{t,\,exp}$). The $q_{t,\,exp}$ for the polymer **2a** was 19.07 mg/g for Pb^{2+}, 15.49 mg/g for Cd^{2+} and 12.12 mg/g for Ni^{2+}. The maximum experimental capacity for the polymer **2b** was 15.98 mg/g for Pb^{2+}, 13.71 mg/g for Cd^{2+} and 10.95 mg/g for Ni^{2+}. The maximum experimental capacity for the polymer **2c** was 13.84 mg/g for Pb^{2+}, 11.12 mg/g for Cd^{2+} and 10.47 mg/g for Ni^{2+}. The correlation coefficients (R^2) allow us to determine which model is more appropriate for each type of adsorption. The R^2 values point out that the pseudo-second-order model better describes the heavy metal adsorption on these polymers and that the theoretical maximum capacity of adsorption ($q_{e,\,the}$) of this model fits better with the experimental maximum adsorption capacity ($q_{t,\,exp}$) obtained in this study. This theoretical model suggests that the determining step for the three different polymers is not the mass transfer but rather the adsorption reaction. The velocity constants observed in the theoretical predictions allow us to know the rate at which the adsorption takes place, being Ni^{2+} the fastest ion to be adsorbed by polymer **2a**, Cd^{2+} is the fastest ion to be adsorbed by polymer **2b** and Pb^{2+} is the fastest ion to be adsorbed by polymer **2c**.

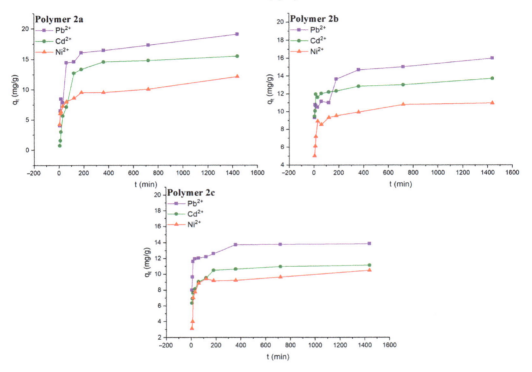

Figure 7. Amount of heavy metal uptaken by the polymers **2a**, **2b** and **2c** as a function of time (t).

Table 2. Kinetic parameters of the heavy metal uptake by the thiol-functionalized polymers.

Polymer	Metal	$q_{t,\,exp}$ (mg/g)	Pseudo-First Order			Pseudo-Second Order		
			k_1 (min^{-1})	$q_{e,\,the}$ (mg/g)	R^2	k_2 (g/mg·min)	$q_{e,\,the}$ (mg/g)	R^2
2a	Pb^{2+}	19.07	0.0027	9.04	0.69	0.0026	17.70	0.99
	Cd^{2+}	15.49	0.0044	9.58	0.79	0.0008	16.66	0.98
	Ni^{2+}	12.12	0.0016	5.22	0.68	0.0078	10.14	0.99
2b	Pb^{2+}	15.98	0.0027	5.49	0.87	0.0646	15.20	0.99
	Cd^{2+}	13.71	0.0020	2.38	0.67	0.2190	13.03	0.99
	Ni^{2+}	10.95	0.0043	3.74	0.91	0.0756	10.81	0.99
2c	Pb^{2+}	13.84	0.0057	2.97	0.83	0.1324	13.86	0.99
	Cd^{2+}	11.12	0.0043	3.15	0.87	0.1017	11.03	0.99
	Ni^{2+}	10.47	0.0023	3.16	0.47	0.1113	9.69	0.99

3.2. Isotherms

The maximum equilibrium adsorption capacity (q_e) of the three polymers as a function of the heavy metal solution concentration (C_e) can be observed in Figure 8. The results show that the q_e increases while C_e also increases, where Pb^{2+} has the largest variation of the three divalent ions. The q_e for polymer **2a** was found to be 53.78 mg/g for Pb^{2+}, 43.80 mg/g for Cd^{2+} and 29.10 mg/g for Ni^{2+}. The q_e for polymer **2b** was found to be 46.45 mg/g for Pb^{2+}, 32.95 mg/g for Cd^{2+} and 27.10 mg/g for Ni^{2+}. The q_e for polymer **2c** was found to be 40.31 mg/g for Pb^{2+}, 35.93 mg/g for Cd^{2+} and 27.84 mg/g for Ni^{2+}. The higher heavy metal adsorption capacity of polymer **2a** could be ascribed to the lower steric hindrance imparted by –SH groups to the polymer backbone compared to those of –SCH$_3$ and –SCF$_3$ groups in polymers **2b** and **2c**, respectively. The aforementioned, in conjunction with the higher hydrophilic character of the –SH groups could favor the water absorption by the polymer matrix, thus promoting the coordination between the polymer and the heavy metals during the ion adsorption process.

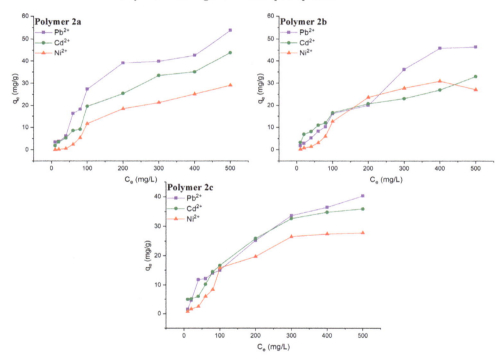

Figure 8. Amount of Pb^{+2}, Cd^{2+} and Ni^{2+} adsorbed by the polymers **2a**, **2b** and **2c** as a function of the heavy metal solution concentration (C_e).

The correlation coefficient (R^2) is useful to determine which model can best describe the equilibrium adsorption on the surface of polymers. The data with coefficient $R^2 < 0.92$ were discarded due to poor fit of the data with the respective model, while other experimental data were adjusted acceptably ($R^2 > 0.95$) to both models, so it is necessary to take more information, obtained from the theoretical fits, into account to suggest the adsorption behavior. The maximum theoretical adsorption capacity (q_m) was used to decide which model best fit the experimental values. In this regard, the data acquired on the adsorption of Cd^{2+} in the polymer **2a**, Ni^{2+} and Cd^{2+} in the polymer **2b** and Pb^{2+} and Ni^{2+} in the polymer **2c**, suggest that the Langmuir model better describes the equilibrium adsorption, while the adsorption data for Pb^{2+} and Ni^{2+} in the polymer **2a**, Ni^{2+} in the polymer **2b** and

Cd^{2+} in the polymer **2c**, suggest that the Freundlich model better describes the equilibrium adsorption. The parameters of the two models are shown in Table 3. The K_F parameter shows the affinity of the ions for the polymer, while the constant n lets us know that the adsorption process is favorable due to the high adsorbent-adsorbate affinity, in addition to suggesting that the adsorption is carried out by the phenomenon of chemisorption [27–29]. We can see in Table 3 that K_F for Ni^{2+} in polymers **2a** and **2b** is high, which suggests that the adsorption of the divalent ion should be higher than that of the other metals, but $n < 1$, which means that the adsorption of Ni^{2+} on the surface of these polymers is considerably weak, which may explain why the adsorption of Ni^{2+} is less than that of Pb^{2+} and Cd^{2+}. The K_L parameter also shows the affinity of the ions to the binding sites of the polymers.

Table 3. Parameters of the Langmuir and Freundlich models for the adsorption of individual Pb^{2+}, Cd^{2+} and Ni^{2+} ions by the thiol-functionalized polymers.

Polymer	Metal	$q_{e,\,exp}$ (mg/g)	Langmuir Isotherm			Freundlich Isotherm		
			K_L (L/mg)	q_m (mg/g)	R^2	K_F (mg/g) (L/g)$^{1/n}$	n	R^2
2a	Pb^{2+}	53.78	0.0082	39.06	0.84	1.91	1.29	0.92
	Cd^{2+}	43.80	0.0043	44.84	0.98	1.24	1.21	0.97
	Ni^{2+}	29.10	0.0034	3.85	0.90	267.73	0.65	0.92
2b	Pb^{2+}	46.45	0.0038	45.87	0.98	4.43	1.14	0.97
	Cd^{2+}	32.95	0.0134	27.70	0.97	1.12	1.83	0.97
	Ni^{2+}	27.10	0.0080	1.91	0.90	89.38	0.73	0.94
2c	Pb^{2+}	40.31	0.0015	108.69	0.95	2.24	1.31	0.93
	Cd^{2+}	35.93	0.0240	21.32	0.74	1.03	1.67	0.95
	Ni^{2+}	27.84	0.0017	41.66	0.98	1.27	1.00	0.94

3.3. Mass Effect

The equilibrium adsorption capacity behavior of the three polymers was studied by varying the amount of each polymer to adsorb heavy metals in an aqueous media. The heavy metal removal percentage against the mass of polymer is shown in Figure 9. A clear increase in adsorption capacity can be observed when more polymer is used, but this increase was not proportional to the increase in mass. The latter could be attributed to the ratio of mass to the surface area since the increase in surface area is not proportional to the increase in mass due to the physical properties of polymers. For instance, polymer **2a** had an increase of about 65% in the adsorption of Pb^{2+} when the amount of polymer was increased ten times; likewise, polymer **2b** had a 71% increase and polymer **2c** had a 56% increase in adsorption of the same ion. Polymer **2a** shows an increase of 54% and 47% for Cd^{2+} and Ni^{2+}, respectively, while polymers **2b** and **2c** show a smaller increase in the adsorption of these divalent ions.

3.4. Multicomponent Adsorption

The effect of ion competition was studied by adding each polymer to a mixture of Pb^{2+}, Cd^{2+} and Ni^{2+}. The results are shown in Figure 10, where it is possible to observe a decreasing trend in the removal of all ions. The decrease in the uptake of these heavy metal ions in the multicomponent mixture compared to that achieved in non-competitive circumstances could be due to the content for the same available adsorption sites on the thiol-functionalized polymer surface, the gradual active sites saturation and the shielding effect generated by the other competing metals. The removal efficiency order found is as follows: $Pb^{2+} > Cd^{2+} > Ni^{2+}$; which is the same as that attained in the individual ions assessment. This trend in the removal efficiency presented by the polymers for Pb^{2+} over Cd^{2+} and Ni^{2+} can be ascribed to Pb^{2+} having a smaller hydration radius and hydration energy.

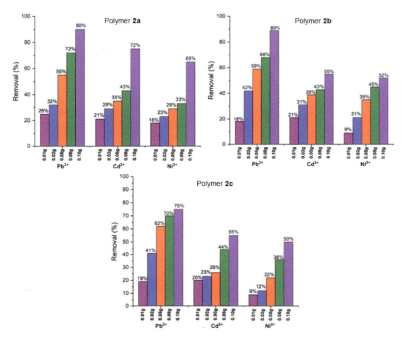

Figure 9. Percent removal of heavy metal ions versus the amount of polymer.

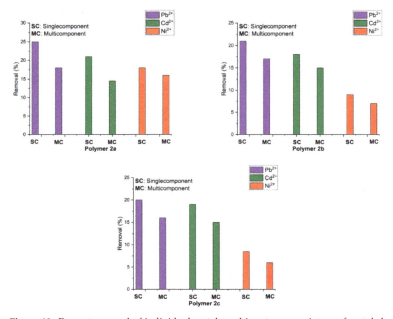

Figure 10. Percent removal of individual metals and in a ternary mixture of metals by the thiol-functionalized polymers.

3.5. Polymer Surface

Topographic images of raw polymer samples were obtained using a scanning electron microscope and are presented in Figure 11. It is possible to see at low magnification (top) that polymers **2a** and **2b** have much smoother surfaces than that of polymer **2c**, where a porous structure prevails. In this sense, the bulky –CF$_3$ groups are known to increase the polymer-free volume, thus promoting a material with microvoids. Using a higher resolution (bottom), it is possible to see that the surface of polymer **2a** is composed of a large number of microvoids that augment the surface area compared to polymer **2b**, which appears to maintain a regular morphology. The higher ion adsorption capacity of **2a** compared to **2b** could be attributed to its also higher surface area. Despite having an apparently increased surface area, it is likely that the hydrophobic character of the –CF$_3$ groups in polymer **2c** restricts the water absorption by the polymer matrix, thus decreasing the coordination between the polymer and the heavy metals during the ion adsorption processes, which in turn is reflected in the lower adsorption capacity of all the polymeric materials examined here. More investigation is required to elucidate this subject.

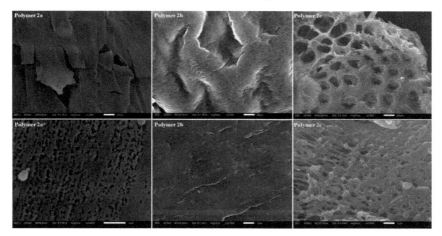

Figure 11. SEM images of the raw polymers bearing thiol moieties. Low image magnification (**top**) and high image magnification (**bottom**). The scales are 10 µm and 1 µm, respectively.

Table 4 shows the maximum adsorption amounts of Pb^{2+}, Cd^{2+} and Ni^{2+} for the thiol-functionalized polynorbornene dicarboximides reported in the present study and those of other polymeric adsorbents previously reported in the literature. It is seen that the outcomes found for the polymers bearing thiol pendant groups are comparable with those results of polymers functionalized with amide and sulfonic groups. It is worth mentioning that these results are even higher than those of some chelating polymers, which are adsorbents frequently employed for the heavy metal uptake from aqueous media. As the outcomes of this study indicate, the thiol group and its derivates show competitive adsorption capacities to be used effectively in the removal of Pb^{2+}, Cd^{2+} and Ni^{2+} from aqueous solutions. This kind of polymer has several advantages over other materials; for instance, norbornene dicarboximide monomers are easily functionalized and obtained from economically accessible raw materials, thus leading to low-cost polymers. The polynorbornene dicarboximides bear hydrophobic cyclopentane rings that allow tailoring the mechanical performance, as well as hydrophilic imide moieties, which endow the polymer with ionic properties. This chemical structure favors the segregation of phases which in turn increases the adsorption area and the diffusion of the adsorbate within the material; thus, the balance in the structure and composition of the polymers could be used to tune the heavy metal removal capacity systematically. In addition, these polymers are easily processable and handled; therefore, they can be adapted to different environments to carry out adsorption in the most efficient way; that is, they can be used in the form of membranes, fibers or particles with a specific and homogeneous size. Likewise, these polymers are not soluble in water, so their separation, recovery and regeneration are very simple, thus increasing their economic viability.

Table 4. Maximum amounts of adsorption of Pb^{2+}, Cd^{2+} and Ni^{2+} by the thiol-functionalized polymers studied in this work and other previously published adsorbent polymers.

Polymer	q_e (Pb^{2+}) (mg/g)	q_e (Cd^{2+}) (mg/g)	q_e (Ni^{2+}) (mg/g)
2a	53.7	43.8	29.1
2b	46.4	32.9	27.1
2c	40.3	35.9	27.8
5b [a] [19]	31.5	26.6	7.0
HDI-IC-PEHA Resin [b] [30]		69.4	26.4
Nafion 117 [c] [31]	58.0		64.0
XAD-4-NH-CH2-Cat [d] [32]	6.5	2.9	2.8

[a] sulfonated ROMP-prepared polymer. [b] polyamide bearing amine chelating moieties. [c] sulfonic acid-containing fluoropolymer. [d] chloromethylated polystyrene-divinylbenzene with dithiooxamide.

4. Conclusions

Thiol-functionalized polynorbornene dicarboximides were successfully synthesized by ROMP and fully characterized by FT-IR, NMR, TGA and TMA, among others. The adsorptions kinetics of Pb^{2+}, Cd^{2+} and Ni^{2+} in the polymers **2a**, **2b** and **2c** were investigated by batch adsorption experiments showing that the maximum uptake and the affinity in the thiol-functionalized polymers decreased in the order $Pb^{2+} > Cd^{2+} > Ni^{2+}$ that could be ascribed to the higher atomic number and increasing heavy metals ionic radii. The adsorption kinetic data fit the pseudo-second-order model more efficiently, while the adsorption isotherms fit the Freundlich and Langmuir models. The adsorption capacity is attributed to the coordination that occurs between the divalent ions and the sulfur atom of the thiol groups. Moreover, the higher heavy metal uptake of polymer **2a** was attributed to the lower steric hindrance and higher hydrophilicity imparted by –SH groups to the polymer backbone compared to those of –SCH$_3$ and –SCF$_3$ groups in polymers **2b** and **2c**, respectively. According to the outcomes, polynorbornene dicarboximides containing thiol groups could potentially be used in ecosystem protection because of their effectiveness as heavy metal adsorbents, specifically to remove Pb^{2+}, Cd^{2+} and Ni^{2+} from aqueous systems. The thiol-functionalized polymeric adsorbents reported in this work have higher uptakes than other adsorbents previously reported in the literature.

Author Contributions: Conceptualization, J.V. and M.A.; Data curation, A.O., C.C.-G. and A.A.S.; Investigation, A.O. and C.C.-G.; Methodology, A.O., A.A.S., T.E.S. and I.A.; Writing—original draft, A.O., A.A.S. and J.V.; Writing—review and editing, A.O. and J.V. All authors have read and agreed to the published version of the manuscript.

Funding: Financial support from DGAPA-UNAM PAPIIT through the project IN108022 is gratefully acknowledged.

Institutional Review Board Statement: Not applicable.

Informed Consent Statement: Not applicable.

Data Availability Statement: The datasets used and/or analyzed during the current study are available from the corresponding author on reasonable request.

Acknowledgments: We are grateful to Gerardo Cedillo Valverde and Karla Eriseth Reyes Morales for their assistance in NMR and thermal properties, respectively.

Conflicts of Interest: The authors declare no conflict of interest.

References

1. Jabeen, A.; Huang, X.; Aamir, M. The challenges of water pollution, threat to public health, flaws of water laws and policies in Pakistan. *J. Water Resour. Prot.* **2015**, *7*, 1516–1526. [CrossRef]
2. World Health Organization. *Guidelines for Drinking-Water Quality*, 4th ed.; WHO: Geneva, Switzerland, 2017.
3. Liu, G.; Yu, Y.; Hou, J.; Xue, W.; Liu, X.; Liu, Y.; Wang, W.; Alsaedi, A.; Hayat, T.; Liu, Z. An ecological risk assessment of heavy metal pollution of the agricultural ecosystem near a lead-acid battery factory. *Ecol. Indic.* **2014**, *47*, 210–218. [CrossRef]
4. Schreck, E.; Foucault, Y.; Sarret, G.; Sobanska, S.; Cécillon, L.; Castrec-Rouelle, M.; Uzu, G.; Dumat, C. Metal and metalloid foliar uptake by various plant species exposed to atmospheric industrial fallout: Mechanisms involved for lead. *Sci. Total Environ.* **2012**, *427–428*, 253–262. [CrossRef] [PubMed]
5. Ghosh, P.; Samanta, A.N.; Ray, S. Reduction of COD and removal of Zn^{2+} from rayon industry wastewater by combined electro-fenton treatment and chemical precipitation. *Desalination* **2011**, *266*, 213–217. [CrossRef]
6. Huang, J.-H.; Zeng, G.-M.; Zhou, C.-F.; Li, X.; Shi, L.-J.; He, S.-B. Adsorption of surfactant micelles and Cd^{2+}/Zn^{2+} in micellar-enhanced ultrafiltration. *J. Hazard. Mater.* **2010**, *183*, 287–293. [CrossRef] [PubMed]
7. Pruvot, C.; Douay, F.; Hervé, F.; Waterlot, C. Heavy metals in soil, crops and grass as a source of human exposure in the former mining areas (6 Pp). *J. Soils Sediments* **2006**, *6*, 215–220. [CrossRef]
8. Ipek, U. Removal of Ni(II) and Zn(II) from an aqueous solutionby reverse osmosis. *Desalination* **2005**, *174*, 161–169. [CrossRef]
9. Razzak, S.A.; Faruque, M.O.; Alsheikh, Z.; Alsheikhmohamad, L.; Alkuroud, D.; Alfayez, A.; Hossain, S.M.Z.; Hossain, M.M. A comprehensive review on conventional and biological-driven heavy metals removal from industrial wastewater. *Environ. Adv.* **2022**, *7*, 100168. [CrossRef]
10. Dutta, D.; Borah, J.P.; Puzari, A. Adsorption of Mn^{2+} from aqueous solution using manganese oxide-coated hollow polymethyl-methacrylate microspheres (MHPM). *Adsorpt. Sci. Technol.* **2021**, *2021*, 5597299. [CrossRef]
11. Wang, W.; Yu, F.; Ba, Z.; Qian, H.; Zhao, S.; Liu, J.; Jiang, W.; Li, J.; Liang, D. In-depth sulfhydryl-modified cellulose fibers for efficient and rapid adsorption of Cr(VI). *Polymers* **2022**, *14*, 1482. [CrossRef]
12. Ibrahim, M.; Tashkandi, N.; Hadjichristidis, N.; Alkayal, N.S. Synthesis of naphthalene-based polyaminal-linked porous polymers for highly effective uptake of CO_2 and heavy metals. *Polymers* **2022**, *14*, 1136. [CrossRef] [PubMed]
13. Santiago, A.A.; Ibarra-Palos, A.; Cruz-Morales, J.A.; Sierra, J.M.; Abatal, M.; Alfonso, I.; Vargas, J. Synthesis, characterization, and heavy metal adsorption properties of sulfonated aromatic polyamides. *High Perform. Polym.* **2018**, *30*, 591–601. [CrossRef]
14. Cruz-Morales, J.A.; Vargas, J.; Santiago, A.A.; Vásquez-García, S.R.; Tlenkopatchev, M.A.; de Lys, T.; López-González, M. Synthesis and gas transport properties of new polynorbornene dicarboximides bearing trifluoromethyl isomer moieties. *High Perform. Polym.* **2016**, *28*, 1246–1262. [CrossRef]
15. Aranda-Suárez, I.; Corona-García, C.; Santiago, A.A.; López Morales, S.; Abatal, M.; López-González, M.; Vargas, J. Synthesis and gas permeability of chemically cross-linked polynorbornene dicarboximides bearing fluorinated moieties. *Macromol. Chem. Phys.* **2019**, *220*, 1800481. [CrossRef]
16. Chen, Y.; Abdellatif, M.M.; Nomura, K. Olefin metathesis polymerization: Some recent developments in the precise polymerizations for synthesis of advanced materials (by ROMP, ADMET). *Tetrahedron* **2018**, *74*, 619–643. [CrossRef]
17. McQuade, J.; Serrano, M.I.; Jäkle, F. Main group functionalized polymers through ring-opening metathesis polymerization (ROMP). *Polymer* **2022**, *246*, 124739. [CrossRef]
18. Maya, V.G.; Contreras, A.P.; Canseco, M.-A.; Tlenkopatchev, M.A. Synthesis and chromium complexation properties of a ionic polynorbornene. *React. Funct. Polym.* **2001**, *49*, 145–150. [CrossRef]
19. Ruiz, I.; Corona-García, C.; Santiago, A.A.; Abatal, M.; Téllez Arias, M.G.; Alfonso, I.; Vargas, J. Synthesis, characterization, and assessment of novel sulfonated polynorbornene dicarboximides as adsorbents for the removal of heavy metals from water. *Environ. Sci. Pollut. Res.* **2021**, *28*, 52014–52031. [CrossRef]
20. Choi, H.Y.; Bae, J.H.; Hasegawa, Y.; An, S.; Kim, I.S.; Lee, H.; Kim, M. Thiol-functionalized cellulose nanofiber membranes for the effective adsorption of heavy metal ions in water. *Carbohydr. Polym.* **2020**, *234*, 115881. [CrossRef]
21. Xia, Z.; Baird, L.; Zimmerman, N.; Yeager, M. Heavy metal ion removal by thiol functionalized aluminum oxide hydroxide nanowhiskers. *Appl. Surf. Sci.* **2017**, *416*, 565–573. [CrossRef]
22. Zhang, W.; Cai, Y.; Downum, K.R.; Ma, L.Q. Thiol synthesis and arsenic hyperaccumulation in pteris vittata (Chinese brake fern). *Environ. Pollut.* **2004**, *131*, 337–345. [CrossRef] [PubMed]
23. Alentiev, D.A.; Dzhaparidze, D.M.; Chapala, P.P.; Bermeshev, M.V.; Belov, N.A.; Nikiforov, R.Y.; Starannikova, L.E.; Yampolskii, Y.P.; Finkelshtein, E.S. Synthesis and properties of metathesis polymer based on 3-silatranyltricyclo[4.2.1.02.5]non-7-ene. *Polym. Sci. Ser. B* **2018**, *60*, 612–620. [CrossRef]
24. Morozov, O.S.; Babkin, A.V.; Ivanchenko, A.V.; Shachneva, S.S.; Nechausov, S.S.; Alentiev, D.A.; Bermeshev, M.V.; Bulgakov, B.A.; Kepman, A.V. Ionomers Based on addition and ring opening metathesis polymerized 5-phenyl-2-norbornene as a membrane material for ionic actuators. *Membranes* **2022**, *12*, 316. [CrossRef] [PubMed]
25. Corona-García, C.; Onchi, A.; Santiago, A.A.; Martínez, A.; Pacheco-Catalán, D.E.; Alfonso, I.; Vargas, J. Synthesis and characterization of partially renewable oleic acid-based ionomers for proton exchange membranes. *Polymers* **2020**, *13*, 130. [CrossRef]
26. Charati, S.G.; Houde, A.Y.; Kulkarni, S.S.; Kulkarni, M.G. Transport of gases in aromatic polyesters: Correlation with WAXD studies. *J. Polym. Sci. Part B Polym. Phys.* **1991**, *29*, 921–931. [CrossRef]

27. An, S.; Jeon, B.; Bae, J.H.; Kim, I.S.; Paeng, K.; Kim, M.; Lee, H. Thiol-based chemistry as versatile routes for the effective functionalization of cellulose nanofibers. *Carbohydr. Polym.* **2019**, *226*, 115259. [CrossRef]
28. Mu, T.-H.; Sun, H.-N. Sweet potato leaf polyphenols: Preparation, individual phenolic compound composition and antioxidant activity. In *Polyphenols in Plants*; Elsevier: Amsterdam, The Netherlands, 2019; pp. 365–380.
29. Proctor, A.; Toro-Vazquez, J.F. The Freundlich isotherm in studying adsorption in oil processing. *J. Am. Oil Chem. Soc.* **1996**, *73*, 1627–1633. [CrossRef]
30. Cegłowski, M.; Gierczyk, B.; Frankowski, M.; Popenda, Ł. A new low-cost polymeric adsorbents with polyamine chelating groups for efficient removal of heavy metal ions from water solutions. *React. Funct. Polym.* **2018**, *131*, 64–74. [CrossRef]
31. Nasef, M.M.; Yahaya, A.H. Adsorption of some heavy metal ions from aqueous solutions on nafion 117 membrane. *Desalination* **2009**, *249*, 677–681. [CrossRef]
32. Bernard, J.; Branger, C.; Nguyen, T.L.A.; Denoyel, R.; Margaillan, A. Synthesis and characterization of a polystyrenic resin functionalized by catechol: Application to retention of metal ions. *React. Funct. Polym.* **2008**, *68*, 1362–1370. [CrossRef]

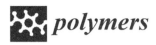

Article

Dried Brown Seaweed's Phytoremediation Potential for Methylene Blue Dye Removal from Aquatic Environments

Abdallah Tageldein Mansour [1,2,*], Ahmed E. Alprol [3], Khamael M. Abualnaja [4], Hossam S. El-Beltagi [5,6], Khaled M. A. Ramadan [7,8] and Mohamed Ashour [3,*]

1. Animal and Fish Production Department, College of Agricultural and Food Sciences, King Faisal University, P.O. Box 420, Al-Ahsa 31982, Saudi Arabia
2. Fish and Animal Production Department, Faculty of Agriculture (Saba Basha), Alexandria University, Alexandria 21531, Egypt
3. National Institute of Oceanography and Fisheries (NIOF), Cairo 11516, Egypt; ah831992@gmail.com
4. Department of Chemistry, College of Science, Taif University, P.O. Box 11099, Taif 21944, Saudi Arabia; k.ala@tu.edu.sa
5. Agricultural Biotechnology Department, College of Agriculture and Food Sciences, King Faisal University, P.O. Box 420, Al-Ahsa 31982, Saudi Arabia; helbeltagi@kfu.edu.sa
6. Biochemistry Department, Faculty of Agriculture, Cairo University, Giza 12613, Egypt
7. Central Laboratories, Department of Chemistry, King Faisal University, P.O. Box 420, Al-Ahsa 31982, Saudi Arabia; kramadan@kfu.edu.sa
8. Department of Biochemistry, Faculty of Agriculture, Ain Shams University, Cairo 11566, Egypt
* Correspondence: amansour@kfu.edu.sa (A.T.M.); microalgae_egypt@yahoo.com (M.A.)

Abstract: The dried form of the brown seaweed *Sargassum latifolium* was tested for its ability to remove toxic Methylene Blue Dye (MBD) ions from aqueous synthetic solutions and industrial wastewater effluents. In a batch adsorption experiment, different initial concentrations of MBD (5, 10, 20, 30, and 40 mg L^{-1}), sorbent dosages (0.025, 0.05, 0.1, 0.2, 0.3, 0.4, and 0.5 g L^{-1}), contact time (5, 10, 15, 30, 60, 120 min), pH (3, 5, 8, 10, and 12), and temperature (30, 40, 50, 60 °C) were observed. Dried powder of *S. latifolium* was characterized before and after adsorption of MBD using different techniques, such as FTIR, SEM, UV visible spectral examination, and BET techniques. The BET surface area suggests the formation of *S. latifolium* was 111.65 m^2 g^{-1}, and the average pore size was 2.19 nm. The obtained results showed that at an MBD concentration of 40 mg L^{-1}, the adsorption was rapid in the first 5, 10, and 15 min of contact time, and an equilibrium was reached in about 60 and 120 min for the adsorption. At the optimum temperature of 30 °C and the adsorbent dose of 0.1 g L^{-1}, approximately 94.88% of MBD were removed. To find the best-fit isotherm model, the error function equations are applied to the isotherm model findings. Both Tempkin and Freundlich isotherm models could appropriate the equilibrium data, as well as the pseudo 2nd order kinetics model due to high correlation coefficients (R^2). Thermodynamic and Freundlich model parameters were assessed and showed that the mechanism of the sorption process occurs by an endothermic and physical process. According to the results of the experiments, *S. latifolium* is a promising environmentally friendly approach for eliminating MBD from the aqueous solution that is also cost-effective. This technology could be useful in addressing the rising demand for adsorbents employed in environmental protection processes.

Keywords: *Sargassum latifolium*; methylene blue dye; adsorption; water pollution; equilibrium isotherm

Citation: Mansour, A.T.; Alprol, A.E.; Abualnaja, K.M.; El-Beltagi, H.S.; Ramadan, K.M.A.; Ashour, M. Dried Brown Seaweed's Phytoremediation Potential for Methylene Blue Dye Removal from Aquatic Environments. *Polymers* **2022**, *14*, 1375. https://doi.org/10.3390/polym14071375

Academic Editors: Irene S. Fahim, Ahmed K. Badawi and Hossam E. Emam

Received: 20 February 2022
Accepted: 23 March 2022
Published: 28 March 2022

Publisher's Note: MDPI stays neutral with regard to jurisdictional claims in published maps and institutional affiliations.

Copyright: © 2022 by the authors. Licensee MDPI, Basel, Switzerland. This article is an open access article distributed under the terms and conditions of the Creative Commons Attribution (CC BY) license (https:// creativecommons.org/licenses/by/ 4.0/).

1. Introduction

A pollutant is a substance that changes the environment's nature through chemical, biological, or physical mechanisms, resulting in contamination of water, soil, and/or air [1]. Several synthetic dyes are primary sources of high effluent pollutants produced as a result of industrial pollution [2]. Textile effluents are considered one of the most environmentally harmful pollutants due to their massive discharge volume and verity in composition, as well as the high volumes of poisonous dye that are lost through the dyeing operation [3]. Several industries, including paper, dyeing, pulp, textiles, paint, and industrial effluent, have produced over 8000 dyes, both insoluble and soluble [1]. Because of their carcinogenic and mutagenic properties, these poisonous dyes pose a threat to human health [4–6]. Furthermore, these toxic dyes have had devastating effects on aquatic ecosystems, as well as a significant impact on the distributions and compositions of aquatic animals, zooplanktonic, and phytoplanktonic organisms [7–11]. The dyes alter the characteristics of water and prevent sunlight from penetrating, decreasing photosynthetic activity [2,12,13]. With the harmful effects of several toxic dyes on ecosystem components, it is necessary to find appropriate and effective substances to eliminate these toxic dyes from industrial waste. Despite there being various physical, chemical, and biological processes available for the elimination of these toxic dyes, their efficiency still needs improvement [14–17]. Globally, the annual production of a large assortment of these toxic dyes is estimated at 1.6 million tons [18].

As a result, the developed techniques that are capable of eliminating dyes that have not adhered to tissue fibers over the process procedures and are emitted along with the effluents of such industries are a constant concern [19–22]. The chemical stability of synthetic textile dyes makes existing wastewater treatment technologies incapable of the treatment of effluents that contain these contaminants [23]. Adsorption methods have high adequacy in industrial wastewater treatment, owing to their significant capacity for eliminating organic matter and low cost [24–27]. Methylene blue dye (MBD) is one of the most extensively used industrial dyes, especially in the textile industry. $C_{16}H_{18}N_3SCl$ is the formula for MBD, a heterocyclic chemical molecule [28]. MBD appears as a dark green powder at room temperature and as a blue hue in the water [29]. MBD was first produced in 1876 as an aniline-based synthetic dye to stain cotton for the textile industry [30]. The maximum absorption of MBD was at around 665 nm [18]. Flocculation [31], coagulation [32], photocatalysis chemical precipitation [33,34], chemical oxidation [35], and ion exchange [36] are some of the treatment procedures used to remove organic pollutants from wastewater effluents.

Adsorbent refers to either adsorption or absorption and is currently one of the most widely used adsorbents. The amalgamation of a material into another material of an unlike condition is known as absorption (liquids are absorbed by gases or/and a solid is absorbed via a liquid). However, the adsorption process is a physical attachment of molecules and ions to another molecule's surface. Sorption has several advantages, including efficiency, simplicity of operation, regeneration of the adsorbents, and processing [37]. Other advantages of the adsorption process, such as the potential for cost-effective adsorbent regeneration and less sludge generation, making it one of the most important treatment methods for water-intensive sectors like textiles and paints. The efficacy of the adsorption techniques for textile and dye wastewater treatment has been demonstrated through extensive previous work [38–40].

Numerous microorganisms' biomass, for example, fungi, yeast, microalgae, seaweed, and bacteria, have been used as biological, sustainable, and low-cost efficient adsorption materials for the absorption of hazardous dyes, according to the literature [41–46]. Among all microorganisms, algal cells are the most promising, sustainable, and low-cost biomaterials for adsorptions [47]. In general, aquatic organisms are the richest sources of biologically active compounds on the planet [48–50]. However, amongst all aquatic organisms, algal cells are the richest sources of biomolecules, making algae a vital player in a variety of bioindustries [51–56]. Depending on the algal strain, algal cells contain proteins (EAAs), lipids (PUFA, AA, EPA, and DHA), and carbohydrates (polysaccharides, etc.) [57]. However, algal cells' adsorption ability is linked to the cell wall's heteropolysaccharide and lipid molecules, which contain several functional groups, for example, phosphate, carboxyl, amino, carbonyl, and hydroxyl groups [46].

Sargassum species are brown macroalgae (seaweed) found in shallow marine meadows in the tropical and subtropical regions. These are a good source of vitamins, carotenoids, dietary fibers, proteins, and minerals, as well as other bioactive components. Biologically active substances such as terpenoids, flavonoids, sterols, sulfated polysaccharides, polyphenols, sargaquinoic acids, sargachromenol, and pheophytine have also been identified in various Sargassum species. [58].

This work aimed to investigate the efficacy of the brown seaweed, Sargassum latifolium, as a suggested MBD removal substance. This dye is globally utilized in textile effluents. Characterizations of materials and understanding of the mechanisms comprised in the sorption process of dyes through the identification of functional groups were determined using different characterization techniques (SEM, FTIR, BET, and UV visible). The parameters (pH, contact time, sorbent dose, temperature, and initial MBD concentration) controlling the sorption process were examined. Moreover, equilibrium and kinetics studies were conducted using adsorption experiments. The experimental data have been used to adjust the Langmuir, Freundlich, Smith Halsey, Henderson, Tempkin, and Harkins–Jura models.

2. Materials and Methods

2.1. Brown Seaweeds (Sargassum latifolium)

Brown seaweed, *S. latifolium*, samples were collected in May 2021 from the intertidal zone of Suez Bay, Red Sea, Egypt. To remove salts and particle materials adhering to the surface, samples were first washed with seawater, then transferred to the laboratory and cleaned with tap water, followed by deionized water [59]. After sample cleaning, *S. latifolium* was sun-dried, and the obtained biomass were powdered using a grinder, sieved to gain a uniform particle size, and stored in plastic bags until further use. Approximately one gram of the brown algae powder was soaked in 100 mL of deionized water and boiled at 45 °C for 1 h to obtain the extract.

2.2. Methylene Blue Dye (MBD)

The solid form of the studied dye (MBD) was obtained from Sigma-Aldrich (Milano, Italy). MBD (Figure 1) is a cationic dye with the chemical formula $C_{16}H_{18}N_3SCl$, molar mass of 319.85 g mol^{-1}, and a maximum wavelength of 665 nm [60]. MBD has a positively charged S atom and is highly water soluble at 293 °K. The stock solutions of MBD were made via dissolving one gram of MBD in one liter of distilled water.

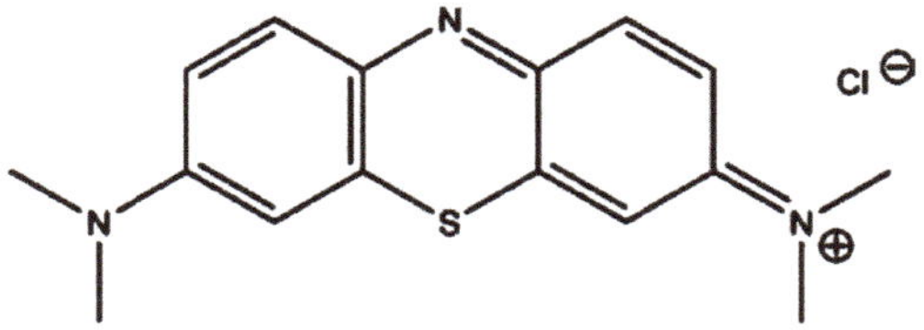

Figure 1. Chemical structure of the Methylene Blue Dye (MBD).

2.3. Adsorption Experimentation

The seaweed, *S. latifolium*, was utilized as a low-cost adsorbent in this investigation to remove MBD from a synthetic aqueous solution. In distilled water, dried powdered *S. latifolium* was suspended and then exposed to a vortex (Dremel, 1100-01, São Paulo, Brazil) at 10,000 rpm for 20 min [47]. Using batch ion exchange, the experiments were carried out in flasks (100 mL of distilled water, three replicates of each treatment) in a constant shaker (120 rpm) under different initial MBD concentrations (5, 10, 20, 30, and 40 mg L^{-1}), adsorbent doses of *S. latifolium* (0.025, 0.05, 0.1, 0.2, 0.3, 0.4, and 0.5 g L^{-1}), different contact times (5, 10, 15, 30, 60, and 120 min), temperature (30, 40, 50, and 60 °C), and pH (3, 5, 8, 10, and 12). The adsorption capacity (q_e) and MBD elimination percentage were calculated using Equations (1) and (2), as follows [61]:

$$q_e = \frac{(C_i - C_f) \times V}{W} \qquad (1)$$

$$\text{Percentage removal (\%)} = \frac{(C_i - C_f)}{C_i} \times 100 \qquad (2)$$

where C_i and C_f (mg L^{-1}): the early concentration at the primary contact time in addition to, the final concentration of MBD at a definite time, respectively, the volume of the MBD mixture (L) is represented by V, while the weight of the dry adsorbent is represented by W (g).

2.4. Adsorption Isotherm Studies

The amount of MBD removed from aqueous solutions is highly dependent on the initial MBD concentrations. To measure the isotherm study, various MBD concentrations (5 to 40 mg L^{-1}) were examined at constant factors of 0.01 g of *S. latifolium* adsorbents, pH 6, 30 °C, and at 150 rpm, 120 min, and mixed with 50 mL of MBD solution [62]. The data were fitted and calculated in terms of the isotherm models of Freundlich, Langmuir, Henderson, Halsey, Harkins–Jura, Smith, and Tempkin.

2.4.1. The Freundlich Model

By plotting a curve of log q_e with relation to log C_e, the efficiency of the Freundlich model to describe the experimental data was used to produce a slope of n and an intercept value of K_f. By plotting the Freundlich model in logarithmic form, it is easy to linearize [63]:

$$\log q_e = \log K_f + 1/n \log C_e \qquad (3)$$

Freundlich parameters K_f and n were calculated from the isotherm equation by Equation (3).

2.4.2. The Langmuir Model

The model of Langmuir predicts uniform adsorption energies of a solute from a liquid solution onto a surface with a finite number of equal sites as monolayer adsorption with no adsorbate movement in the surface plane [64]. To estimate the highest adsorption capacity similar to complete monolayer adsorption on the sorbent surface, the Langmuir isotherm model was used. The mathematical expression represents the Langmuir model as the following equation [65]:

$$q_e = q_{max} bC_e / (1 + bC_e) \qquad (4)$$

where q_{max} (mg g^{-1}) considers the highest sorption capacity in proportion to the saturation capacity, and b (L mg^{-1}) shows a coefficient relating to the affinity between *S. latifolium* and dye ions. The values of q_{max} and b calculated from the intercept and slope, respectively, by plotting curve (1/q_e) vs. (1/C_e) yields the following linear relationship:

$$1/q_e = 1/(bq_{max} C_e) + 1/q_{max} \qquad (5)$$

2.4.3. The Henderson and Halsey Isotherm Models

These models are suitable for heterosporous solids and the multilayer sorption technique [66,67]. These models were calculated using the following equations:

$$\ln q_e = \frac{1}{n} \ln K + \frac{1}{n} \ln C_e \qquad (6)$$

where: n and K are Halsey constants. Meanwhile, the Henderson model was obtained from the following equation:

$$\text{Ln}[n\text{Ln}(1nC_e)] = \text{Ln}K + \left(\frac{1}{n}\right)\text{Ln}q_e \tag{7}$$

while: the Henderson constants are nh and Kh.

2.4.4. The Harkins–Jura Model

This model explains multilayer sorption and the presence of heterogeneous pore scattering in a sorbent [68]. This model was obtained from the following equation:

$$\frac{1}{q_e^2} = \left(\frac{B_2}{A}\right) - \left(\frac{1}{A}\right)\log C_e \tag{8}$$

where: the isotherm constants are A and B.

2.4.5. The Smith Model

For heteroporous solids and multilayer adsorption, the Smith model is acceptable. This model is usually obtained from the following equation:

$$q_e = W_{bS} - W_S \text{Ln}(1 - C_e) \tag{9}$$

where: the Smith model parameters are W_{bs} and W_s.

2.4.6. The Tempkin Model

The Tempkin isotherm model assumes that owing to adsorbate–adsorbent interactions, a uniform distribution of binding energies up to greater binding energies characterizes the sorption process, and the heat of adsorption of each particle in the layer decreases with saturation [69,70]. The following equation was used to calculate the Tempkin model:

$$q_e = B \text{Ln} A + B \text{Ln} C_e \tag{10}$$

where: A is the equilibrium binding constant (L g^{-1}), b (J mol^{-1}) is a constant that corresponds to the heat of sorption, and B = (RT/b) (J mol^{-1}) is the Tempkin constant, which is connected to the heat of adsorption and the gas constant (R = 8.314 J mol^{-1} K^{-1}).

2.5. Error Functions Tests

Various error functions were examined to find the best and most fitting model for analyzing the equilibrium data. The following models were used for the error function test [40,71].

2.5.1. Fractional Error Hybrid (HYBRID)

Because it adjusts for small concentrations by balancing absolute deviation versus fractional error and is more dependable than other error functions, the hybrid fractional error function is used. The following equation represents the hybrid error (11).

$$\text{HYBRID} = \frac{100}{N-P} \sum \left|\frac{q_{e,\exp} - q_{e,\text{calc}}}{q_{e,\exp}}\right|_i \tag{11}$$

where: the $q_{e,\exp}$ and $q_{e,\text{calc}}$ denote the experimental and the calculated data (mg g^{-1}), in addition to N is the number of parameters of the isotherm equation, and P the number of data points.

2.5.2. Average Percentage Error (APE)

The ABE model shows a tendency or appropriateness among the predicted and experimental values of the sorption capacity used for plotting model curves (APE) and can be designed consistent with the following equation:

$$\text{APE}(\%) = \frac{100}{N} \sum_{i=1}^{N} \left| \frac{q_{e,\text{isotherm}} - q_{e,\text{calc}}}{q_{e,\text{isotherm}}} \right|_i \tag{12}$$

The number of data points under investigation is denoted by the letter N.

2.5.3. Nonlinear Chi-Square Analysis (χ^2)

The nonlinear chi-square experiment is a statistical factor for evaluating which treatment system is best. The approach of the chi-square error, χ^2, and model is assumed as the following equation:

$$\chi^2 = \frac{\left(q_{e,\text{isotherm}} - q_{e,\text{calc}}\right)^2}{q_{e,\text{isotherm}}} \tag{13}$$

2.5.4. Sum Squares of the Errors (ERRSQ)

The following Equation (14) gives the sum of the squares of the errors (ERRSQ) [72].

$$\text{ERRSQ} = \sum_{i=1}^{P} \left(q_{e,\text{calc}} - q_{e,\text{isotherm}}\right)^2_i \tag{14}$$

2.5.5. Sum of Absolute Errors (EABS)

An increase in errors increases the fit, resulting in a bias toward high concentration data. EABS examinations can be evaluated [73] using the following equation:

$$\text{EABS} = \sum_{i=1}^{P} \left| q_{e,\text{calc}} - q_{e,\text{isotherm}} \right| \tag{15}$$

where P is the number of data points.

2.6. Adsorption Kinetics Studies

2.6.1. Pseudo-First-Order (PFO)

The model equation of the generalized pseudo-first-order equation is given by the following equation [74].

$$Dq/d_t = K_1 (q_e - q_t) \tag{16}$$

where q_e represents the quantity of dye adsorbed at equilibrium (mg g^{-1}), q_t represents the number of dyes adsorbed at time t (mg g^{-1}), and K_1 represents a pseudo-first-order rate constant (min^{-1}). The integrating equation was assessed [75] as follows:

$$\text{Log}(q_e/q_e - q_t) = k_1 t / 2.303 \tag{17}$$

In a linear equation, the PFO equation is given by the formula:

$$\text{Log}(q_e - q_t) = \log q_e - k_1 t / 2.303 \tag{18}$$

Log($q_e - q_t$) against (t) plots should show a linear relationship between k_1 and q_e, as measured by the slope and intercept, respectively.

2.6.2. Pseudo-Second-Order (PSO)

The PSO equation is expressed as follows [76].

$$dq_t/d_t = K_2(q_e - q_t)^2 \qquad (19)$$

where: K_2 indicates the constant of the second-order rate (g mg^{-1} min). The integrated equation is presented as follows:

$$1/(q_e - q_t) = 1/q_e + K_2 \qquad (20)$$

Following Ho et al. [76] and the following linear form of the pseudo-second-order equation:

$$t/q_t = 1/K_2\, q_e^2 + t/q_e \qquad (21)$$

Plotting (t/q_t) versus (t) would yield a linear relationship, and the values of (t/qt) would be the values of the q_e from the slope and intercept can be used to calculate the K2 parameters, respectively.

2.6.3. The Intraparticle Diffusion Model

The following is the intraparticle diffusion equation [14]:

$$q_t = K_{dif}\, t^{1/2} + C \qquad (22)$$

where C denotes the intercept, and K_{dif} (mg g^{-1} min$^{0.5}$) is the intraparticle diffusion rate constant, which is calculated using the regression lines of the slope.

2.7. Characterization of Adsorbents

The morphological characterization (before and after the experiment) of the adsorbent brown seaweed, *S. latifolium*, was performed via Scanning Electron Microscope, SEM (JEOL JSM 6360, Peabody, MA, USA), followed by Fourier transform infrared (Shimadzu FTIR-8400 S, Kyoto, Japan). A UV–vis spectrophotometer (UV-2550, Shimadzu, OSLO, Kyoto, Japan) was used to evaluate the absorption spectrum of *S. latifolium* marine algae extract in the range of 200–800 nm, with deionized water as a blank.

3. Results and Discussion

3.1. Characterizations

3.1.1. Functional Groups

The infrared spectrum of biomass in nature is shown in Figure 2. The stretching of the extension vibrations of the –NH and O–H groups was attributed to the inter- and intermolecular hydrogen bonding of polymeric compounds such as phenols or alcohols, which resulted in large and intense peaks at 3112 and 3729 cm^{-1} [77]. Similarly, the band at 2925 cm^{-1} corresponds to methylene's symmetric and asymmetric stretch (H-C-H), whereas the peaks at 1649 cm^{-1} are C=C stretches, which could be owing to the presence of aromatic, olefinic, or N-H bending bands.

Only dye saturated biomass showed a peak, as the 1547 cm^{-1} band, indicating the presence of C=N and C=C stretch and the typical absorptions. Stretching of the sp3 C-H bond is responsible for the band at 1419 cm^{-1}. The C-O stretching of acyl or phenol is related to the sharp peak at 1251 cm^{-1}. The absorption band of the –OH and –NH groups at 3733 cm^{-1} changed to 3436 cm^{-1} after interaction with MBD adsorption, as shown in Figure 2. The C–H stretching vibrations of sp^3 hybridized C in CH$_3$ and CH$_2$ functional groups is assigned to another peak with a shoulder at 2937 cm^{-1} and 2939 cm^{-1} [78]. CH$_3$ bending and C–H stretching vibrations are responsible for the other absorption peaks, which occurred at 1419, 1467 cm^{-1}, and 1361 cm^{-1}, respectively. Algal cells have many functional groups on their surfaces, such as phosphate, hydroxyl, amino, and carboxylate, which are responsible for removing pollutants from various wastewaters. The accumulation of dye ions on the biopolymers of the algal surface, followed by dye migration from the aqueous phase to the biopolymer solid phase, could be the source of dye removal using algae [79].

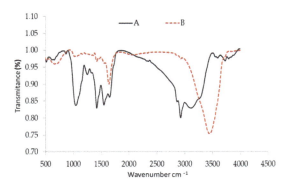

Figure 2. FTIR analysis of *Sargassum latifolium* before (A) and safter adsorption (B) of methylene blue dye.

3.1.2. Surface Morphology

SEM analysis of algal cells showed the surface shape of the adsorbents before and after MBD adsorption. Figure 3 presents SEM pictures of *S. latifolium* before and after dye uptake at 2000 magnifications as an example. Figure 3a shows that the adsorbent has a uniform porosity structure and that salt crystals are present on the biosorbent's outside because of natural mineral deposition. There are also spots related to crystalline salts and the irregularity of the algal cell surface. The cross-linked micropores on the adsorbent surface may enhance the efficient uptake of the liquid electrolyte and improve the adsorbent's ionic conductivity. In Figure 3b there are significant changes to the surface morphology of the adsorbents, as well as the formation of separate aggregates on their surfaces following MBD adsorption. In addition, after contact with MBD, the porous texture was filled and vanished. Interaction with MBD changed the surface texture of the algal cell surface from smooth to rough and irregular. Meanwhile, tiny swellings were observed after contact with MBD.

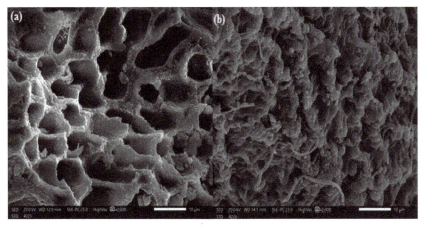

Figure 3. Scanning electron microscope of *Sargassum latifolium* at (**a**) before and (**b**) after adsorption of methylene blue dye.

3.1.3. UV—Visible Spectral Examination

Spectral analysis in the ultraviolet (UV) ranges according to the sharpness of the peaks, the UV-VIS spectrum summary of *S. latifolium* extract was chosen between 200 and 800 nm, as shown in Figure 4. Flavonoids and their derivatives have unique absorption spectra. The spectral bands of flavonoids are typically absorption spectra with maximal values in the range of 300–360 nm [80]. The absorption maximum bands for anthocyanins are at 460–560 nm, while flavones and flavonols are around 310–370 nm, as seen in peaks (350, 375, 399, 461,485, 547, and 559 nm) [81,82]. Furthermore, the profile revealed that the compounds were separated at 608, 657, and 694 nm. The exact location and virtual intensities of these peaks provide a wealth of information about the flavonoids' nature [83]. These peaks at 234–676 nm indicate the presence of phenolic and alkalidic chemicals in *S. latifolium*. The extract has some comparable alkaloid, flavonoids, and glycoside components when compared to the spectra of seeds and flowers [84,85]. The absorption spectra of *S. latifolium* seaweed extract revealed a broad absorbance range between 300 and 700 nm [86] with absorption maxima in the visible area of the solar spectrum at 412.5, 608, and 657 nm. The characteristic absorption of chlorophylls can be related to these absorption peaks [86,87].

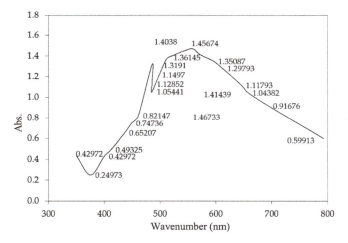

Figure 4. UV-Visible spectrum of *S. linifolium* extracts obtained with aqueous extract.

3.1.4. BET Characterization

To investigate any changes in the porosity of the sample under investigation, the specific surface area and pore volumes of the adsorbent were investigated as presented in Table 1. The *S. latifolium* isotherm displays quick N_2 uptake at low pressures (P/P_0 0.01) and constant high adsorption with hysteresis at higher pressures. The surface area of the BET is estimated to be 111.65 m^2/g. Because the pores operate as binding or receptor sites during the adsorption phase, this is critical for pollutant trapping. Additionally, the pore volume is 0.122 cc/g.

Table 1. Textural properties of *S. latifolium* as determined by N_2 adsorption.

Surface Area Results (m^2/g)					Total Pore Volume	Average Pore Size	Average Particle Radius
Single Point BET	Multipoint BET	Langmuir Method	BJH Adsorption	BJH Desorption			
100.78	111.65	178.40	67.87	58.486	0.12 cc/g	2.19 nm	1.22 + 001 nm

3.2. Study of Batch Adsorption Process

3.2.1. pH

The pH value of the MBD solution has a significant effect on the adsorption mechanism because it affects both the adsorbent's surface binding sites and the dye molecules' ionization process. The net charge on the adsorbent is also pH-dependent since the adsorbent surface contains polymers with numerous unique functional groups such as phosphates, carboxyl, amino, and hydroxyl. In this work, the influence of pH was determined for pH values ranging from 3 to 12, as shown in Figure 5.

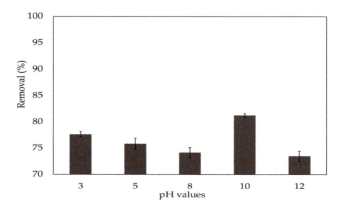

Figure 5. Impact of the initial pH on removal efficiency of methylene blue dye.

The *S. latifolium* can be employed in a wide pH range, and the rate of MBD removal increases as the pH value increases. When the pH was 10, the removal rate was 81.3%, demonstrating that the pH of the solution had an important influence on MBD adsorption. The presence of a positive charge on the nitrogen in the MB dye's structure indicates that it is a basic and cationic dye. The surface charge on the adsorbent can be used to explain the influence of pH. However, the poor MBD adsorption has been seen at low pH values (pH 3), increasing the density of positive charge (protons) at surface biomass sites. The electrostatic repulsion that occurs between the MBD cations and the positive charges on the adsorbent's surface explains this. The rate of negative surface charge of the biomass increases at a higher pH value (pH > 3), which electrostatically attracts cationic dyes [88], resulting in good MB adsorption on the surface of the adsorbent. Furthermore, the strong physical interactions (such as H-bonds) between both the adsorbate molecules (MBD) and various functional groups, namely hydroxyl, esters, alkynes, carbonyl, and carboxyl present in the phenols and flavonoids present in the noncellulosic cells of adsorbents of brown algal Sargassum, may describe the adsorption of MBD on *S. latifolium*.

The surface charge on the adsorbent, which is affected by the solution pH, is the most important factor in the adsorption of these positively charged dye groups on the surface of the adsorbent [89]. Methylene blue adsorption by dried *Ulothrix* sp. Biomass was examined by Doğar, et al. [90], who showed that the adsorption increased as the pH was increased.

3.2.2. MDB Concentration

The initial dye concentration works as a forceful driving force, increasing the dye's mass transfer resistance between the aqueous and solid phases to its limit. The effect of concentrations of MBD on the adsorption was studied at 5, 10, 20, 30, and 40 mg L^{-1} with the initial pH value of 10. Figure 6 shows the relationship between the percentage removals of MBD and adsorbent capacity (q_e) and the equilibrium dye concentration in the liquid phase (C_e). When the MBD initial concentration is augmented from 5 to 40 mg L^{-1}, the quantity of MBD adsorbed increases from 68.42% to 84.96%. In addition, the adsorption capacity is improved from 7.8 to 25.9 by increasing the initial dye concentration from 5 to 40 mg L^{-1} as shown in Figure 6. The total number of dye molecule collisions and macrophytes increases when the initial dye concentration is increased. As a result, increasing the initial dye concentration could accelerate the adsorption process [91]. Furthermore, it might be attributed to an increase in the driving force of the concentration gradient with the increase in the initial concentration [92]. This is in agreement with results reported by other studies [2].

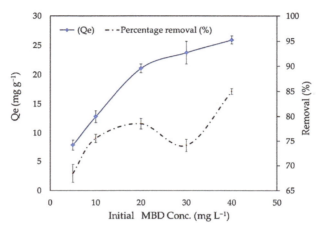

Figure 6. Impact of different methylene blue dye concentrations on adsorption capacity and removal percentages.

3.2.3. Sorbent Loading

At an MBD concentration of 10 mg L^{-1}, the impact of the adsorbent amount on methylene blue removal was examined. In the MBD solution, *S. latifolium* concentrations varying from 0.025 to 0.5 g L^{-1} were incorporated. Figure 7 shows that when the adsorbent loading was raised from 0.025 to 0.1 g, the elimination effectiveness increased from 86.53 to 95.97 percent, then dropped after this value. The availability of additional active adsorption sites, as well as an increase in the adsorptive surface area, is the key reason for this. Furthermore, adsorption onto the adsorbent surface happens very quickly at higher adsorbent dosages. The amounts of MBD adsorbed per unit weight of seaweed were lowered when the adsorbent loading was augmented from 0.2 to 0.5 g. Increased biomass loading may lead to a decrease in the removal percentage value due to the complicated interaction of numerous factors, for example, solute availability, binding site interference, and electrostatic interactions [93]. At higher adsorbent dosages, the available MBD molecules are inadequate to cover all of the exchangeable sites on the adsorbent, resulting in the limited dye uptake. Additionally, as adsorbent mass increases, the amount of dye adsorbed q_e (mg g^{-1}) decreases due to a split in the flow or concentration gradient between the solute concentration in the solution and solute concentration on the adsorbent surface [2].

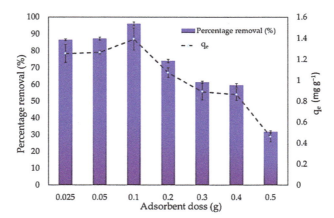

Figure 7. Impact of adsorbent dose on adsorption capacity and removal percentages of methylene blue dye.

3.2.4. Contact Time

The equilibrium adsorption of *S. latifolium* was measured at various contact time intervals to investigate the influence of different contact times on adsorption. The percentage of elimination efficiency increases as contact time increases (Figure 8). The result showed that the percentage of removal efficiency increased from 12.7% to 80.43% with increasing the heating time from 5 to 15 min, then decreased to 76% after the first and the second hour of contact time (Figure 8).

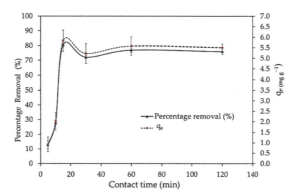

Figure 8. Impact of different contact times on adsorption percentages of the methylene blue dye.

The rapid adsorption at the initial contact time is due to the availability of the positively charged surface of the *S. latifolium* for the adsorption of MBD. Moreover, it is clear that MBD ions adsorption mainly consists of two stages, an initial rapid stage related to the instantaneous external surface adsorption of MBD ions with the more available reactive groups, which allows quick binding of ions on the biomass [94], whereas fast adsorption equilibrium was reached in the first 15 min. This could be due to the high specific surface area of the adsorbent particles and the absence of internal diffusion resistance at this stage [95]. As these sites became progressively covered, the rate of adsorption decreased. The second stage is much slower, with gradual adsorption levels taking place and lasting until adsorbate ions adsorption attains equilibrium, which is attributed to the diffusion of dye ions into the adsorbent layers for binding on the inner reactive groups, with the aid of hydrophilic and swelling characteristics of the adsorbent material. Therefore, the initial adsorption uptake rate is increased while the second step is slowed down [96]. Additionally, Khan et al. [97] reported that the transfer of the dye from the solution phase into the pores of the adsorbent is the rate-controlling stage in batch experiments under rapid stirring circumstances. Likewise, Sheen [98] found that adsorption was rapid in the first 5 min of contact with an uptake of more than 90%, and equilibrium was reached in 60 min of agitation time as the binding sites became exhausted. As the binding sites became exhausted, the uptake rate slowed due to metal ions competing for the decreasing availability of active sites. According to the test results, the rest of the batch experiment's shaking period was set at 120 min to ensure that equilibrium was achieved. This quick (or rapid) adsorption phenomenon is advantageous in applications because the short contact time effectively allows for a smaller contact equipment size, which has a direct effect on both the capacity and operation cost of the process.

3.2.5. Temperature

The equilibrium uptake of dye into the biomass of *S. latifolium* was also influenced by the temperature parameter (Figure 9). The uptake of MBD onto the dried biomass was found to be most effective at 50 °C, with a removal percentage of 89.71%. With a decrease in temperature, adsorption decreased. The viscosity of the dye-containing solution is reduced as the temperature rises, allowing more dye molecules to diffuse through the external boundary layer and into the internal pores of the adsorbent particles [99]. Temperature also has a major influence on dye degradation, first by directly changing the chemistry of the pollutant, and then by influencing its physiology.

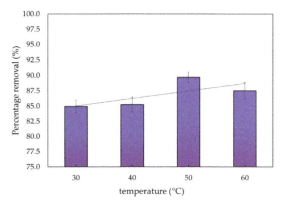

Figure 9. Impact temperature on the percentages of adsorption of methylene blue dye.

3.3. Equilibrium Adsorption

Sorption equilibrium is reached when the concentration of solutes adsorbent onto matches the amount of associated development. When this happens, the equilibrium solution concentration remains constant. By graphing the concentration of the solid phase versus the concentration of the liquid phase, the equilibrium adsorption isotherm will be seen. The Freundlich, Langmuir, Harkins–Jura, Halsey, Henderson, Smith, and Tempkin models were used to analyze equilibrium data.

3.3.1. Freundlich Isotherm

The Freundlich adsorption isotherms with the correlation coefficient are shown in Figure 10 and Table 2.

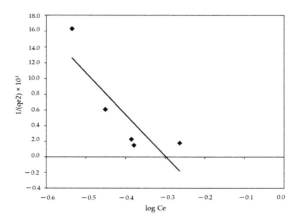

Figure 10. Linear Freundlich isotherm for MBD sorption.

Table 2. Factors of isotherm models from non-linear and linear solvation.

Isotherm Model	Isotherm Parameter	Value
Freundlich	$1/n$	1.115
	K_F (mg$^{1-1/n}$L$^{1/n}$g^{-1})	31.6
	R^2	0.906
Langmuir	Q_{max} (mg g^{-1})	0.819
	b (L mg^{-1})	0.068
	R_L	0.969
	R^2	0.751
Tempkin	A_T	2.47
	B_T	14.67
	b_T	168.88
	R^2	0.916
Harkins–Jura	A	0.02
	B	0.3
	R^2	0.719
Halsey	$1/n_H$	1.912
	K_H	11
	R^2	0.768
Henderson	$1/n_h$	0.519
	K_h	0.119
	R^2	0.735
Smith	W_{bs}	0.768
	W_s	36.01
	R^2	0.605

The R^2 (0.906) and $1/n$ (1.115) results indicate that adsorption isotherms and monolayer coverage on the adsorbent surface are applicable, based on the discovered linear relationships. The value of $1/n$, also known as the heterogeneity factor, describes the deviation from sorption linearity as follows: if $1/n = 1$, the adsorption is linear, and the dye particle concentration does not affect the division between the two stages. Chemical adsorption occurs when $1/n$ is less than 1, whereas cooperative adsorption occurs when $1/n$ is more than 1, which is more physically favorable and involves strong contacts between the adsorbate particles [100]. In this study, the values of factor "$1/n$" are more than 1, indicating that using this isotherm equation to properly perform the physical sorption mechanism on an exterior surface is preferable. The adsorbent could be used as a low-cost adsorbent for MBD removal due to its outstanding monolayer adsorption properties. Additionally, the Freundlich isotherm equation can be used to describe heterogeneous systems and can be applied to adsorption on heterogeneous surfaces with the interaction between adsorbed molecules.

3.3.2. Langmuir Isotherm

The regression data and adsorption parameters obtained are shown in Figure 11 and Table 2. The Langmuir equation's linear form has the advantage of pointing to the high success rate of using algal biomass for dye removal. $R^2 = 0.751$ for MBD shows the monolayer coverage of MBD on the outer surface of the sorbent, in which the mechanism of adsorption occurs uniformly on the reactive section of the surface. The sorbent shows that the Langmuir model does not apply to the tested system. The maximum adsorption capacity (q_{max}) is 0.819 mg g^{-1}, which corresponds to complete monolayer coverage on the surface. The adsorption capacities (q_{max}) of various adsorbents for MBD are compared in Table 3. q_{max} of *S. latifolium* is significantly higher than those of other adsorbents. Due to its low cost, reusability, and high adsorption capacity, R_L can also be used to express the fundamental features of a Langmuir isotherm, which is determined from the relation of Hall et al. [101]:

$$R_L = \frac{1}{1 + bC_i} \quad (23)$$

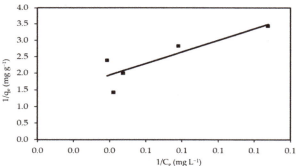

Figure 11. Linear Langmuir isotherm for MBD sorption.

Table 3. Comparison of higher monolayer sorption of MBD onto different adsorbents materials.

Adsorbent	Capacities (mg g^{-1})	Reference
Banana peel	0.124	[102]
Cotton stalk	11.6	[103]
Wheat shells	16.6	[104]
Coir pith carbon	5.9	[105]
Wheat shells	16.6	[104]
Activated date pits	12.9	[106]
Cereal chaff	20.3	[107]
Fly Ash	0.0727	[108]
S. latifolium	0.819	This study

The types of equilibrium isotherms are related to the R_L values; for $R_L > 1$, an unfavorable process dominates, but for $R_L = 1$, a favorable mechanism is available [109]. Table 2 shows that the values of R_L in this study ranged from zero to one (0.969), indicating that MBD adsorption on algal biomass was favorable.

3.3.3. Tempkin Isotherm

The Tempkin isotherm model posits that the heat of adsorption of all molecules reduces linearly rather than logarithmically, with coverage due to the interaction between adsorbate and adsorbent. From the Tempkin model (Figure 12 and Table 2), the following values were obtained: A = 2.47 L g^{-1}, B = 14.67 J mol^{-1}, and b$_T$ = 168.8. The correlation coefficient value (R^2 = 0.916) is high and demonstrates a satisfactory well to the experimental data where the sorption process of MBD of the algal biomass does obey the Tempkin isotherm. Additionally, it assumes that the heat of adsorption of all the molecules in the layer decreases linearly with coverage due to adsorbate–adsorbate repulsions and that the adsorption is a uniform distribution of maximum binding energy [110].

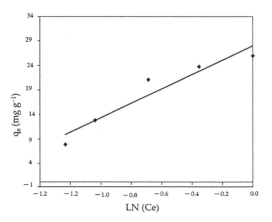

Figure 12. Linear Tempkin model for MBD sorption.

3.3.4. Harkins–Jura Isotherm

As shown in Figure 13, the Harkins–Jura adsorption isotherm can be written as Equation (8), which can be solved by plotting $1/q_e$ vs. log C_e. The presence of a heterogeneous pore distribution and multilayer adsorption can be represented by the Harkins–Jura model. The range of isotherm constants and correlation coefficients is R^2 = 0.719 (Table 2). This may indicate that the Harkins–Jura model is less applicable.

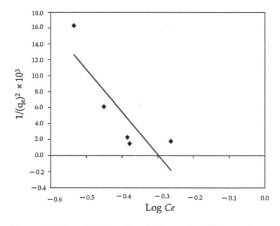

Figure 13. Linear Harkins–Jura isotherm for MBD sorption.

3.3.5. Halsey and Henderson Isotherm

Figures 14 and 15 show the Halsey and Henderson adsorption isotherms, respectively. The Halsey isotherm shows that the correlation coefficient was $R^2 = 0.768$, while Henderson shows the regression coefficient value of $R^2 = 0.735$. The results obtained from Halsey and Henderson show that both models are small and did not demonstrate a satisfactory response to the experimental data where the sorption process of MBD onto algal biomass was concerned.

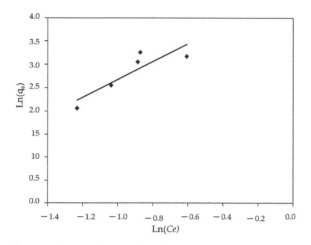

Figure 14. Linear Halsey isotherm for MBD sorption.

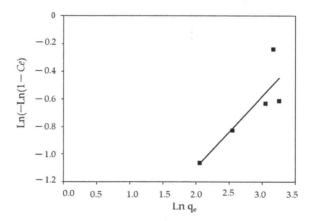

Figure 15. Linear Henderson isotherm for MBD sorption.

3.3.6. Smith Isotherm

Smith's model is useful in explaining the adsorption isotherm of biological substances, such as cellulose and starch, in addition to being appropriate for heterosporous solids and multilayer adsorption [111]. The Smith model can be solved by the plot of q_e vs. Ln $(1 - Ce)$ as presented in Figure 16 and Table 2. The sorption isotherms were accurately reproduced by Smith models across the entire range of water activity. However, Smith's equation was not effective in describing the isotherms of MBD on algal biomass because the model gave lower R^2 values (0.605).

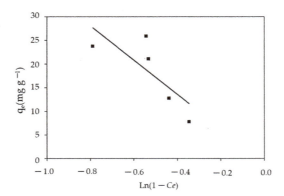

Figure 16. Smith isotherm for MBD sorption.

3.3.7. Error Functions Examination for Best-Appropriate Isotherm Model

Traditional linear regression methods for calculating isotherm parameters appear to provide a satisfactory fit to the experimental data. The R^2 is determined using the linear versions of the isotherm equations, but it eliminates the isotherm curve error [71]. To assess the fit of the isotherm equations to the experimental data, several error functions of non-linear regression were employed to establish the constant model parameters, and they were compared to those determined from less precise linearized data fitting [112]. Several error functions, such as the hybrid fractional error (Hybrid), the average percentage errors (APE) equation, the Chi-square error (X^2) equation, the sum of the squares of the errors (ERRSQ), and the sum of absolute errors (SAE), were applied to determine the most-fit model for the investigational data (EABS). Table 4 summarizes the data gathered from various error functions. From the observed data, the best appropriate isotherm models are Halsey, Henderson, Freundlich, Langmuir Harkins–Jura, Tempkin, and Smith. However, the error functions studied gave variable results for each isotherm model and the comparison between the isotherm models should be focused on each error function separately. The (EABS) equation demonstrates that all the isotherm models can be applied to the investigational data except the Harkins–Jura, Tempkin, and Smith isotherms. Analysis of Table 4 shows that the Halsey, Henderson, Freundlich, and Langmuir models have smaller errors in almost all of the cases and that the Harkins–Jura, Tempkin, and Smith models did not show high accuracy. In this case, the Freundlich isotherm model can be more useful for describing the adsorption process of MB by *S. latifolium* adsorbent.

Table 4. The values of the five various error assessments of isotherm models for adsorption.

Isotherm Model	Hybrid	APE%	X^2	ERRSQ	EABS
Freundlich	5.453	2.035	1.854	4.707	13.017
Langmuir	363.42	13.66	83.58	305.55	87.40
Harkins–Jura	202.700	10.203	46.621	170.425	65.273
Halsey	0.151	0.278	0.035	0.127	1.781
Henderson	0.605	0.557	0.139	0.508	3.565
Smith	391.312	14.177	90.002	329.004	90.692
Tempkin	359.232	13.583	82.623	302.033	86.895

3.4. Kinetic Models

Kinetic analysis was performed using pseudo-first and second-order models, as well as the intraparticle diffusion model, to evaluate the adsorption processes of MBD on *S. latifolium*. These models are commonly used to describe the sorption of dyes and other pollutants on solid sorbents, as well as to evaluate their applicability. For the adsorption of a solute from a liquid solution, the Lagergren rate equation is one of the most extensively used adsorption rate equations. The linear regressions are shown in Figures 17–19. In addition to the intraparticle diffusion model, Table 5 shows the coefficients of the pseudo-first and second-order adsorption kinetic models.

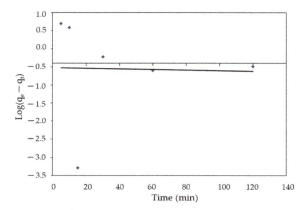

Figure 17. The pseudo-first-order equation for the elimination of MBD onto *S. latifolium*.

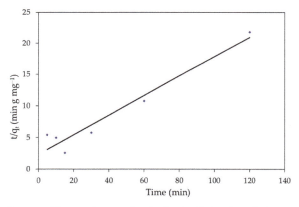

Figure 18. The pseudo-second-order for the elimination of MBD onto *S. latifolium*.

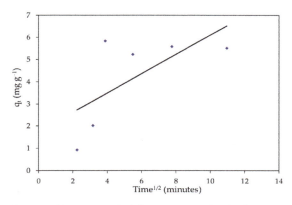

Figure 19. The intraparticle diffusion equation for the elimination of MBD onto *S. latifolium*.

Table 5. The comparison of the first order, second-order, and intraparticle diffusion for adsorption rate constants.

Model	Parameter	Value
First-order Kinetic	q_e (calc.)	3.30
	K_1	2.07
	R^2	0.001
Second-order Kinetic	q_e (calc)	6.44
	K_2	0.01
	R^2	0.936
	q_e (exp.)	4.18
Interparticle Diffusion	K_{dif}	1.747
	C	0.44
	R^2	0.45

The linearized form of the plot log ($q_e - q_t$) versus time (t) as shown in Figure 17 showed a poor pseudo-first-order linear regression coefficient (R^2) value of 0.001. The slope of the plot is used to calculate the rate constant of pseudo-first-order adsorption. Furthermore, the pseudo first-order kinetic model predicted a significantly lower equilibrium adsorption capacity (q_e = 3.30) than the experimental value (q_e = 4.181), indicating the model's impermissibility.

On the other hand, the intercept and slope of the plots of t/q_t vs. t, were used to compute the second-order rate constant, k_2, and the equilibrium adsorption capacity, qe. Figure 18 depicts pseudo-second-order kinetic graphs. In addition, the linear regression coefficient (R^2) value for the pseudo-second-order kinetic model was 0.943, indicating that the model is applicable, as shown in Table 5. The (q_e) computed value was close to the experimental data, hence a pseudo-second-order model was fitted. Generally, the kinetic studies displayed that the adsorption of MBD was followed by pseudo-second-order kinetics models and the sorption process was controlled through the chemisorption process. Similar results have been reported for the adsorption of organic pollutants [113,114]. Furthermore, understanding the adsorption system's dynamic behavior is critical for process design. The removal of MBD by adsorption on biomass was shown to be quick for the first 15 min, but thereafter became slow as contact duration increased. The dye molecules migrate from the bulk solution to the surface of the sorbent and penetrate through the boundary layer to the surface of the sorbent, where adsorption and intra-particle diffusion actually occur, according to the intra-particle diffusion mechanism for the sorption system of MBD removal from its aqueous phase [115]. The intercept is C, and the slope directly predicts the rate constant K_{dif}, as illustrated in Figure 19. The form of Figure 16 reveals that straight lines do not pass through the origin and that the removal of MBD on *S. latifolium* has a poor correlation (0.446), reflecting the fact that the intraparticle diffusion model does not suit the experimental results. Because the resistance to exterior mass transfer increases as the intercept increases, the value of the C factor provides information about the thickness of the boundary layer.

3.5. Thermodynamic Studies

Table 6 displays the values for $\Delta G°$, $\Delta H°$, and $\Delta S°$. The reaction is spontaneous and endothermic, as evidenced by the positive values of $\Delta H°$ and $\Delta S°$. The negative values of $\Delta G°$ also represent the process's spontaneity. The values of $\Delta G°$ get increasingly negative as the temperature rises, indicating that the process is becoming more spontaneous. Physical binding forces may be involved in the process, as evidenced by the low value of $\Delta H°$. The range of $\Delta G°$ values for physisorption is between −20 and 0 kJ mol^{-1}, but the range for chemisorption is between −80 and −400 kJ mol^{-1}. According to the activation energy estimations in this work, MBD sorption onto the marine algae *S. latifolium* occurred via a physisorption mechanism.

Table 6. Thermodynamic factors of the sorption of IV MB 2R onto *S. latifolium*.

Temperature (°C)	$\Delta G°$ (kJ mol^{-1})	$\Delta H°$ (kJ mol^{-1})	$\Delta S°$ (J mol^{-1})
30	−6.35	35.205	−0.131
40	−6.62		
50	−7.86		
60	−7.54		

3.6. Treatment of Real Dye Effluent

To check the efficiency of *S. latifolium* as an adsorbent for the removal of MBD from industrial effluents, simulated water was prepared as a control to study the influence of biomass on the adsorption process, and optimization conditions (at an acidic pH 10, 0.1 g of adsorbent after 120 min at a higher temperature of 50 °C) were used. Due to the low quantities of MBD, a large amount of MBD was applied to achieve a final concentration of 40 mg L^{-1}. Industrial effluent samples were obtained from a factory in Al-Mahla Al-kobra, Egypt, which specializes in dyeing and printing. To track the percentage of the dye combination eliminated from the real dye effluents, the UV–vis spectra of the simulated water and effluents treated with *S. latifolium* were recorded at 665 nm. The results confirmed that changing the type of water affected the largest removal, with simulated water having the least impact on the sorption process, with a percentage removal of 89.15%. Real wastewater, on the other hand, contains very high concentrations of interfering ions from a variety of contaminants, which had a major impact on the removal effectiveness of MBD, with a percentage removal of 65.9%.

The novelty in this research lies in different values of the optimization process, mathematical models, and the highest suitable model being achieved by using error function tests such as a fractional error hybrid, average percentage error, sum squares of errors, nonlinear chi-square analysis, and the sum of absolute errors. To describe the elimination of the MBD's effects on *S. latifolium*, error functions were proposed. This study is unique in that it highlights the possibility of integrating the use of marine algae biomass into the removal of MBD from aqueous environments in accordance with circular economic principles. Thus, using a simple and abandoned, readily available, low-cost, cheap, and environmentally friendly bio-material experimental procedure, the *S. latifolium* material, which can be more efficiently used to retain MBD, is obtained from a low-performance adsorbent material, such as marine brown algae biomass. Therefore, *S. latifolium* was chosen to prepare the adsorbent based on the above parameters. A step was taken to prepare the adsorbent, which was then utilized to remove the MBD from the water.

4. Conclusions

In the current study, brown seaweed, *S. latifolium*, was applied as an adsorbent for the elimination of methylene blue dye (MBD) from an aqueous solution. Initial pH, initial MBD concentration, adsorbent dose, and adsorption temperature all had an impact on adsorption. The highest removal of MB was 95.97% under optimal operating conditions, which included an initial MBD concentration of 40 mg L^{-1}, a temperature of 50 °C, and a pH of 10. Increases in the adsorbent dose and starting MBD concentrations increased MBD removal effectiveness. Furthermore, MBD removal using *S. latifolium* is also influenced by the presence of functional groups such as carbonyl, hydroxyl, amino, carboxylate groups, and other natural compounds that support ions binding like lignins, cellulose, and lipids. Additionally, SEM analysis demonstrates the ability of *S. latifolium* biomass to adsorb and remove MBD from an aqueous solution. Moreover, the absorption bands for photosynthetic pigments, such as flavonoids and their derivatives, were found in the UV-visible spectra of seaweed. In addition, the quality of *S. latifolium* was also assessed using BET characterization, which shows that *S. latifolium* has a greater surface area. MBD adsorption on *S. latifolium* followed Halsey, Henderson, Harkins–Jura, Freundlich, Tempkin, Smith, and Langmuir isotherms revealed in the research. *S. latifolium* was found to be a good adsorbent for removing MBD from wastewater through the adsorption process.

Author Contributions: Conceptualization, A.E.A. and M.A.; methodology, A.E.A. and M.A.; software, A.T.M., H.S.E.-B., A.E.A. and M.A.; validation, A.E.A. and M.A.; formal analysis, A.E.A. and M.A.; investigation, A.E.A. and M.A.; resources, A.T.M., H.S.E.-B., A.E.A., K.M.A., K.M.A.R. and M.A.; data curation, A.E.A. and M.A.; writing—original draft preparation, A.E.A. and M.A.; writing—review and editing, A.T.M., H.S.E.-B., A.E.A., K.M.A., K.M.A.R. and M.A.; visualization, A.E.A. and M.A.; supervision, A.T.M., H.S.E.-B., A.E.A. and M.A.; project administration, A.T.M., H.S.E.-B., K.M.A. and K.M.A.R.; funding acquisition, A.T.M., H.S.E.-B., A.E.A., K.M.A., K.M.A.R. and M.A. All authors have read and agreed to the published version of the manuscript.

Funding: This work was supported by the Deanship of Scientific Research, Vice Presidency for Graduate Studies and Scientific Research, King Faisal University, Saudi Arabia [Project No. 10]. Taif University Researchers Supporting Project number TURSP-2020/267, Taif University, Taif, Saudi Arabia.

Institutional Review Board Statement: Not applicable.

Informed Consent Statement: Not applicable.

Data Availability Statement: The data that support the findings of this study are available from the authors upon reasonable request.

Acknowledgments: The authors gratefully appreciate Deanship of Scientific Research, Vice Presidency for Graduate Studies and Scientific Research, King Faisal University, Saudi Arabia [Project No. 10], for supporting the present study. Also, the authors appreciated Taif University Researchers Supporting Project number TURSP-2020/267, Taif University, Taif, Saudi Arabia.

Conflicts of Interest: The authors declare no conflict of interest.

References

1. Badawi, A.K.; Ismail, B.; Baaloudj, O.; Abdalla, K.Z. Advanced wastewater treatment process using algal photo-bioreactor associated with dissolved-air flotation system: A pilot-scale demonstration. *J. Water Process Eng* **2022**, *46*, 102565. [CrossRef]
2. Abualnaja, K.M.; Alprol, A.E.; Abu-Saied, M.A.; Mansour, A.T.; Ashour, M. Studying the Adsorptive Behavior of Poly(Acrylonitrile-co-Styrene) and Carbon Nanotubes (Nanocomposites) Impregnated with Adsorbent Materials towards Methyl Orange Dye. *Nanomaterials* **2021**, *11*, 1144. [CrossRef] [PubMed]
3. Al-Ghouti, M.A.; Al-Degs, Y.S.; Khraisheh, M.A.; Ahmad, M.N.; Allen, S.J. Mechanisms and chemistry of dye adsorption on manganese oxides-modified diatomite. *J. Environ. Manag.* **2009**, *90*, 3520–3527. [CrossRef] [PubMed]
4. Sakkayawong, N.; Thiravetyan, P.; Nakbanpote, W. Adsorption mechanism of synthetic reactive dye wastewater by chitosan. *J. Colloid Interface Sci* **2005**, *286*, 36–42. [CrossRef]
5. Manzoor, J.; Sharma, M. Impact of Textile Dyes on Human Health and Environment. In *Impact of Textile Dyes on Public Health and the Environment*; IGI Global: Hershey, PA, USA, 2020; pp. 162–169. [CrossRef]
6. Chung, K.-T. Azo dyes and human health: A review. *J. Environ. Sci. Health* **2016**, *34*, 233–261. [CrossRef]
7. Abbas, E.M.; Ali, F.S.; Desouky, M.G.; Ashour, M.; El-Shafei, A.; Maaty, M.M.; Sharawy, Z.Z. Novel Comprehensive Molecular and Ecological Study Introducing Coastal Mud Shrimp (*Solenocera Crassicornis*) Recorded at the Gulf of Suez, Egypt. *J. Mar. Sci. Eng.* **2020**, *9*, 9. [CrossRef]
8. Alprol, A.E.; Heneash, A.M.M.; Soliman, A.M.; Ashour, M.; Alsanie, W.F.; Gaber, A.; Mansour, A.T. Assessment of Water Quality, Eutrophication, and Zooplankton Community in Lake Burullus, Egypt. *Diversity* **2021**, *13*, 268. [CrossRef]
9. Ashour, M.; Alprol, A.E.; Heneash, A.M.M.; Saleh, H.; Abualnaja, K.M.; Alhashmialameer, D.; Mansour, A.T. Ammonia Bioremediation from Aquaculture Wastewater Effluents Using *Arthrospira platensis* NIOF17/003: Impact of Biodiesel Residue and Potential of Ammonia-Loaded Biomass as Rotifer Feed. *Materials* **2021**, *14*, 5460. [CrossRef]
10. Ashour, M.; Mabrouk, M.M.; Abo-Taleb, H.A.; Sharawy, Z.Z.; Ayoub, H.F.; Van Doan, H.; Davies, S.J.; El-Haroun, E.; Goda, A.M.S.A. A liquid seaweed extract (TAM®) improves aqueous rearing environment, diversity of zooplankton community, whilst enhancing growth and immune response of Nile tilapia, *Oreochromis niloticus*, challenged by *Aeromonas hydrophila*. *Aquaculture* **2021**, *543*, 736915. [CrossRef]
11. Alprol, A.E.; Ashour, M.; Mansour, A.T.; Alzahrani, O.M.; Mahmoud, S.F.; Gharib, S.M. Assessment of Water Quality and Phytoplankton Structure of Eight Alexandria Beaches, Southeastern Mediterranean Sea, Egypt. *J. Mar. Sci Eng.* **2021**, *9*, 1328. [CrossRef]
12. da Silva, G.L.; Silva, V.L.; Vieira, M.G.; da Silva, M.G. Solophenyl navy blue dye removal by smectite clay in a porous bed column. *Adsorpt. Sci. Technol.* **2009**, *27*, 861–875. [CrossRef]
13. Sardar, M.; Manna, M.; Maharana, M.; Sen, S. Remediation of dyes from industrial wastewater using low-cost adsorbents. In *Green Adsorbents to Remove Metals, Dyes and Boron from Polluted Water*; Springer: New York, NY, USA, 2021; pp. 377–403. [CrossRef]
14. Raza, W.; Lee, J.; Raza, N.; Luo, Y.; Kim, K.-H.; Yang, J. Removal of phenolic compounds from industrial waste water based on membrane-based technologies. *J. Ind. Eng. Chem.* **2019**, *71*, 1–18. [CrossRef]

15. Dawood, S.; Sen, T. Review on dye removal from its aqueous solution into alternative cost effective and non-conventional adsorbents. *J. Chem. Process Eng.* **2014**, *1*, 1–11.
16. Kumar, P.S.; Joshiba, G.J.; Femina, C.C.; Varshini, P.; Priyadharshini, S.; Karthick, M.A.; Jothirani, R. A critical review on recent developments in the low-cost adsorption of dyes from wastewater. *Desalin. Water Treat* **2019**, *172*, 395–416. [CrossRef]
17. Mehrotra, T.; Dev, S.; Banerjee, A.; Chatterjee, A.; Singh, R.; Aggarwal, S. Use of immobilized bacteria for environmental bioremediation: A review. *J. Environ. Chem. Eng.* **2021**, *9*, 105920. [CrossRef]
18. Sarma, G.; Gupta, S.S.; Bhattacharyya, K. Removal of hazardous basic dyes from aqueous solution by adsorption onto kaolinite and acid-treated kaolinite: Kinetics, isotherm and mechanistic study. *SN Appl. Sci.* **2019**, *1*, 211. [CrossRef]
19. Souza, M.; Lenzi, G.; Colpini, L.; Jorge, L.; Santos, O. Photocatalytic discoloration of reactive blue 5G dye in the presence of mixed oxides and with the addition of iron and silver. *Braz. J. Chem. Eng.* **2011**, *28*, 393–402. [CrossRef]
20. Katheresan, V.; Kansedo, J.; Lau, S.Y. Efficiency of various recent wastewater dye removal methods: A review. *J. Environ. Chem. Eng.* **2018**, *6*, 4676–4697. [CrossRef]
21. Rahimian, R.; Zarinabadi, S. A review of studies on the removal of methylene blue dye from industrial wastewater using activated carbon adsorbents made from almond bark. *Prog Chem. Biochem. Res.* **2020**, *3*, 251–268.
22. El Haggar, S. *Sustainable Industrial Design and Waste Management: Cradle-to-Cradle for Sustainable Development*; Academic Press: Cambridge, MA, USA, 2010.
23. Forgacs, E.; Cserhati, T.; Oros, G. Removal of synthetic dyes from wastewaters: A review. *Environ. Int.* **2004**, *30*, 953–971. [CrossRef]
24. Ncibi, M.; Hamissa, A.B.; Fathallah, A.; Kortas, M.; Baklouti, T.; Mahjoub, B.; Seffen, M. Biosorptive uptake of methylene blue using Mediterranean green alga *Enteromorpha* spp. *J. Hazard. Mater.* **2009**, *170*, 1050–1055. [CrossRef] [PubMed]
25. Ali, I.; Gupta, V. Advances in water treatment by adsorption technology. *Nat. Protoc.* **2006**, *1*, 2661–2667. [CrossRef] [PubMed]
26. Bonilla-Petriciolet, A.; Mendoza-Castillo, D.I.; Reynel-Ávila, H.E. *Adsorption Processes for Water Treatment and Purification*; Springer: NewYork, NY, USA, 2017. [CrossRef]
27. Lakherwal, D. Adsorption of heavy metals: A review. *Int. J. Environ. Res. Dev.* **2014**, *4*, 41–48.
28. Mashkoor, F.; Nasar, A. Magsorbents: Potential candidates in wastewater treatment technology–A review on the removal of methylene blue dye. *J. Magn. Magn. Mater.* **2020**, *500*, 166408. [CrossRef]
29. Oz, M.; Lorke, D.E.; Hasan, M.; Petroianu, G.A. Cellular and molecular actions of methylene blue in the nervous system. *Med. Res. Rev.* **2011**, *31*, 93–117. [CrossRef]
30. Brooks, M.M. Methylene blue as antidote for cyanide and carbon monoxide poisoning. *J. Am. Med. Assoc.* **1933**, *100*, 59. [CrossRef]
31. Buthelezi, S.P.; Olaniran, A.O.; Pillay, B. Textile dye removal from wastewater effluents using bioflocculants produced by indigenous bacterial isolates. *Molecules* **2012**, *17*, 14260–14274. [CrossRef]
32. Zahrim, A.; Hilal, N. Treatment of highly concentrated dye solution by coagulation/flocculation–sand filtration and nanofiltration. *Water Res. Ind.* **2013**, *3*, 23–34. [CrossRef]
33. Baaloudj, O.; Nasrallah, N.; Bouallouche, R.; Kenfoud, H.; Khezami, L.; Assadi, A.A. High efficient Cefixime removal from water by the sillenite Bi12TiO20: Photocatalytic mechanism and degradation pathway. *J. Clean. Prod.* **2022**, *330*, 129934. [CrossRef]
34. Abul, A.; Samad, D.; Huq, D.; Moniruzzaman, M.; Masum, M. Textile dye removal from wastewater effluents using chitosan-ZnO nanocomposite. *J. Text. Sci. Eng.* **2015**, *5*, e1000200. [CrossRef]
35. El-Hamid, A.; Al-Prol, A.; El-Alfy, M.A. Remediation of extracted water from El-Burullus drains sediments using chemical oxidation. *J. Environ. Sci. Mansoura Univ.* **2021**, *50*, 20–26.
36. Mansour, A.T.; Ashour, M.; Alprol, A.E.; Alsaqufi, A.S. Aquatic Plants and Aquatic Animals in the Context of Sustainability: Cultivation Techniques, Integration, and Blue Revolution. *Sustainability* **2022**, *14*, 3257. [CrossRef]
37. Babel, S.; Opiso, E.M. Removal of Cr from synthetic wastewater by sorption into volcanic ash soil. *Int. J. Environ. Sci. Technol.* **2007**, *4*, 99–107. [CrossRef]
38. Badawi, A.K.; Abd Elkodous, M.; Ali, G.A. Recent advances in dye and metal ion removal using efficient adsorbents and novel nano-based materials: An overview. *RSC Adv.* **2021**, *11*, 36528–36553. [CrossRef]
39. Abdelwahab, O.; Amin, N. Adsorption of phenol from aqueous solutions by Luffa cylindrica fibers: Kinetics, isotherm and thermodynamic studies. *Egypt. J. Aquat. Res.* **2013**, *39*, 215–223. [CrossRef]
40. Amrhar, O.; Nassali, H.; Elyoubi, M. Modeling of adsorption isotherms of methylene blue onto natural illitic clay: Nonlinear regression analysis. *Moroc. J. Chem.* **2015**, *3*, 582–593.
41. Yagub, M.T.; Sen, T.K.; Afroze, S.; Ang, H.M. Dye and its removal from aqueous solution by adsorption: A review. *Adv. Colloid Interface Sci.* **2014**, *209*, 172–184. [CrossRef]
42. Khan, R.; Bhawana, P.; Fulekar, M. Microbial decolorization and degradation of synthetic dyes: A review. *Rev. Environ. Sci. Bio/Technol.* **2013**, *12*, 75–97. [CrossRef]
43. Mishra, S.; Cheng, L.; Maiti, A. The utilization of agro-biomass/byproducts for effective bio-removal of dyes from dyeing wastewater: A comprehensive review. *J. Environ. Chem. Eng.* **2021**, *9*, 104701. [CrossRef]
44. Kapoor, R.T.; Danish, M.; Singh, R.S.; Rafatullah, M.; HPS, A.K. Exploiting microbial biomass in treating azo dyes contaminated wastewater: Mechanism of degradation and factors affecting microbial efficiency. *J. Water Process Eng.* **2021**, *43*, 102255. [CrossRef]
45. Blaga, A.C.; Zaharia, C.; Suteu, D. Polysaccharides as Support for Microbial Biomass-Based Adsorbents with Applications in Removal of Heavy Metals and Dyes. *Polymers* **2021**, *13*, 2893. [CrossRef] [PubMed]

46. El-Sheekh, M.M.; Gharieb, M.; Abou-El-Souod, G. Biodegradation of dyes by some green algae and cyanobacteria. *Int. Biodeterior. Biodegrad.* **2009**, *63*, 699–704. [CrossRef]
47. Alprol, A.E.; Heneash, A.M.M.; Ashour, M.; Abualnaja, K.M.; Alhashmialameer, D.; Mansour, A.T.; Sharawy, Z.Z.; Abu-Saied, M.A.; Abomohra, A.E. Potential Applications of *Arthrospira platensis* Lipid-Free Biomass in Bioremediation of Organic Dye from Industrial Textile Effluents and Its Influence on Marine Rotifer (*Brachionus plicatilis*). *Materials* **2021**, *14*, 4446. [CrossRef] [PubMed]
48. Abo-Taleb, H.A.; FZeina, A.; Ashour, M.; MMabrouk, M.; ESallam, A.; MMEl-feky, M. Isolation and cultivation of the freshwater amphipod *Gammarus pulex* (Linnaeus, 1758), with an evaluation of its chemical and nutritional content. *Egypt. J. Aquat. Biol Fish.* **2020**, *24*, 69–82. [CrossRef]
49. Metwally, A.S.; El-Naggar, H.A.; El-Damhougy, K.A.; Bashar, M.A.E.; Ashour, M.; Abo-Taleb, H.A.H. GC-MS analysis of bioactive components in six different crude extracts from the Soft Coral (*Sinularia maxim*) collected from Ras Mohamed, Aqaba Gulf, Red Sea, Egypt. *Egypt. J. Aquat. Biol. Fish.* **2020**, *24*, 425–434. [CrossRef]
50. Magouz, F.I.; Essa, M.A.; Matter, M.; Tageldein Mansour, A.; Alkafafy, M.; Ashour, M. Population Dynamics, Fecundity and Fatty Acid Composition of *Oithona nana* (Cyclopoida, Copepoda), Fed on Different Diets. *Animals* **2021**, *11*, 1188. [CrossRef] [PubMed]
51. Abomohra, A.E.-F.; Almutairi, A.W. A close-loop integrated approach for microalgae cultivation and efficient utilization of agar-free seaweed residues for enhanced biofuel recovery. *Bioresour. Technol.* **2020**, *317*, 124027. [CrossRef]
52. Ashour, M.; El-Shafei, A.A.; Khairy, H.M.; Abd-Elkader, D.Y.; Mattar, M.A.; Alataway, A.; Hassan, S.M. Effect of Pterocladia capillacea Seaweed Extracts on Growth Parameters and Biochemical Constituents of Jew's Mallow. *Agronomy* **2020**, *10*, 420. [CrossRef]
53. Ashour, M.; Hassan, S.M.; Elshobary, M.E.; Ammar, G.A.G.; Gaber, A.; Alsanie, W.F.; Mansour, A.T.; El-Shenody, R. Impact of Commercial Seaweed Liquid Extract (TAM®) Biostimulant and Its Bioactive Molecules on Growth and Antioxidant Activities of Hot Pepper (*Capsicum annuum*). *Plants* **2021**, *10*, 1045. [CrossRef]
54. Hassan, S.M.; Ashour, M.; Sakai, N.; Zhang, L.; Hassanien, H.A.; Gaber, A.; Ammar, G. Impact of Seaweed Liquid Extract Biostimulant on Growth, Yield, and Chemical Composition of Cucumber (*Cucumis sativus*). *Agriculture* **2021**, *11*, 320. [CrossRef]
55. Hassan, S.M.; Ashour, M.; Soliman, A.A.F.; Hassanien, H.A.; Alsanie, W.F.; Gaber, A.; Elshobary, M.E. The Potential of a New Commercial Seaweed Extract in Stimulating Morpho-Agronomic and Bioactive Properties of *Eruca vesicaria* (L.) Cav. *Sustainability* **2021**, *13*, 4485. [CrossRef]
56. Shao, W.; Ebaid, R.; El-Sheekh, M.; Abomohra, A.; Eladel, H. Pharmaceutical applications and consequent environmental impacts of *Spirulina* (*Arthrospira*): An overview. *Grasasy Aceites* **2019**, *70*, 292. [CrossRef]
57. Raja, R.; Coelho, A.; Hemaiswarya, S.; Kumar, P.; Carvalho, I.S.; Alagarsamy, A. Applications of microalgal paste and powder as food and feed: An update using text mining tool. *Beni-Suef Univ. J. Basic Appl. Sci.* **2018**, *7*, 740–747. [CrossRef]
58. Al Prol, A.E.; EAEl-Metwally, M.; Amer, A. Sargassum latifolium as eco-friendly materials for treatment of toxic nickel (II) and lead (II) ions from aqueous solution. *Egypt. J. Aquat. Biol. Fish.* **2019**, *23*, 285–299. [CrossRef]
59. Ashour, M.; Mabrouk, M.M.; Ayoub, H.F.; El-Feky, M.M.M.M.; Zaki, S.Z.; Hoseinifar, S.H.; Rossi, W.; Van Doan, H.; El-Haroun, E.; Goda, A.M.A.S. Effect of dietary seaweed extract supplementation on growth, feed utilization, hematological indices, and non-specific immunity of Nile Tilapia, *Oreochromis niloticus* challenged with *Aeromonas hydrophila*. *J. Appl. Phycol.* **2020**, *32*, 3467–3479. [CrossRef]
60. Maurya, R.; Ghosh, T.; Paliwal, C.; Shrivastav, A.; Chokshi, K.; Pancha, I.; Ghosh, A.; Mishra, S. Biosorption of methylene blue by de-oiled algal biomass: Equilibrium, kinetics and artificial neural network modelling. *PLoS ONE* **2014**, *9*, e109545. [CrossRef]
61. Ghoneim, M.M.; El-Desoky, H.S.; El-Moselhy, K.M.; Amer, A.; Abou El-Naga, E.H.; Mohamedein, L.I.; Al-Prol, A.E. Removal of cadmium from aqueous solution using marine green algae, *Ulva lactuca*. *Egypt. J. Aquat. Res.* **2014**, *40*, 235–242. [CrossRef]
62. Dada, A.; Olalekan, A.; Olatunya, A.; Dada, O. Langmuir, Freundlich, Temkin and Dubinin–Radushkevich isotherms studies of equilibrium sorption of Zn2+ unto phosphoric acid modified rice husk. *IOSR J. Appl. Chem.* **2012**, *3*, 38–45.
63. Freundlich, H. Over the adsorption in solution. *J. Phys. Chem.* **1906**, *57*, 1100–1107.
64. Langmuir, I. The constitution and fundamental properties of solids and liquids. Part, I. Solids. *J. Am. Chem. Soc.* **1916**, *38*, 2221–2295. [CrossRef]
65. Langmuir, I. The constitution and fundamental properties of solids and liquids. II. Liquids. *J. Am. Chem. Soc.* **1917**, *39*, 1848–1906. [CrossRef]
66. Halsey, G. Physical adsorption on non-uniform surfaces. *J. Chem. Phys.* **1948**, *16*, 931–937. [CrossRef]
67. Harkins, W.D.; Jura, G. An adsorption method for the determination of the area of a solid without the assumption of a molecular area, and the area occupied by nitrogen molecules on the surfaces of solids. *J.Chem.Phys.* **1943**, *11*, 431–432. [CrossRef]
68. Harkins, W.D.; Jura, G. Surfaces of solids. XIII. A vapor adsorption method for the determination of the area of a solid without the assumption of a molecular area, and the areas occupied by nitrogen and other molecules on the surface of a solid. *J. Am. Chem. Soc.* **1944**, *66*, 1366–1373. [CrossRef]
69. Mall, I.D.; Srivastava, V.C.; Agarwal, N.K. Removal of Orange-G and Methyl Violet dyes by adsorption onto bagasse fly ash—kinetic study and equilibrium isotherm analyses. *Dyes Pigment.* **2006**, *69*, 210–223. [CrossRef]
70. Temkin, M.; Pyzhev, V. Recent modifications to Langmuir isotherms. *Chemosphere* **1940**, *12*, 217–222.
71. Foo, K.Y.; Hameed, B.H. Insights into the modeling of adsorption isotherm systems. *Chem. Eng. J.* **2010**, *156*, 2–10. [CrossRef]
72. Ng, J.; Cheung, W.; McKay, G. Equilibrium studies of the sorption of Cu (II) ions onto chitosan. *J. Colloid Interface Sci.* **2002**, *255*, 64–74. [CrossRef]

73. Kumar, Y.P.; King, P.; Prasad, V. Removal of copper from aqueous solution using *Ulva fasciata* sp.—A marine green algae. *J. Hazard. Mater.* **2006**, *137*, 367–373. [CrossRef]
74. Levankumar, L.; Muthukumaran, V.; Gobinath, M. Batch adsorption and kinetics of chromium (VI) removal from aqueous solutions by Ocimum americanum L. seed pods. *J. Hazard. Mater.* **2009**, *161*, 709–713. [CrossRef]
75. Lagergren, S. Zur theorie der sogenannten adsorption geloster stoffe. *Kungliga Svenska Vetenskapsakademiens Handlingar* **1898**, *24*, 1–39.
76. Ho, Y.; McKay, G.; Wase, D.; Forster, C. Study of the sorption of divalent metal ions on to peat. *Adsorpt. Sci. Technol.* **2000**, *18*, 639–650. [CrossRef]
77. Abualnaja, K.M.; Alprol, A.E.; Abu-Saied, M.A.; Ashour, M.; Mansour, A.T. Removing of Anionic Dye from Aqueous Solutions by Adsorption Using of Multiwalled Carbon Nanotubes and Poly (Acrylonitrile-styrene) Impregnated with Activated Carbon. *Sustainability* **2021**, *13*, 7077. [CrossRef]
78. Inyinbor, A.; Adekola, F.; Olatunji, G.A. Kinetics, isotherms and thermodynamic modeling of liquid phase adsorption of Rhodamine B dye onto Raphia hookerie fruit epicarp. *Water Res. Ind.* **2016**, *15*, 14–27. [CrossRef]
79. Dönmez, G.; Aksu, Z. Removal of chromium (VI) from saline wastewaters by Dunaliella species. *Process Biochem.* **2002**, *38*, 751–762. [CrossRef]
80. Rajeswari, R.; Jeyaprakash, K. Bioactive potential analysis of brown seaweed *Sargassum wightii* using UV-VIS and FT-IR. *J. Drug Deliv. Ther* **2019**, *9*, 150–153. [CrossRef]
81. Santos-Buelga, C.; González-Paramás, A.M.; González-Manzano, S.; Dueñas, M. Analysis and occurrence of flavonoids in foods and biological samples. In *Recent Advances in Medicinal Chemistry*; Atta-ur-Rahman, M., Iqbal, C., Perry, G., Eds.; Bentham Science Publisher: Shajah, United Arab Emirates, 2015; pp. 10–58.
82. Vihakas, M. Flavonoids and Other Phenolic Compounds: Characterization and Interactions with Lepidopteran and Sawfly Larvae. Ph.D. Thesis, University of Turku, Turku, Finland, 2014.
83. Sahu, N.; Saxena, J. Phytochemical analysis of Bougainvillea glabra Choisy by FTIR and UV-VIS spectroscopic analysis. *Int. J. Pharm. Sci. Rev. Res* **2013**, *21*, 196–198.
84. Sofowora, A. *Medicinal Plants and Traditional Medicine in Africa*; Karthala: Paris, France, 1996.
85. Jasper, C.; Maruzzella, J.; Henry, P. The antimicrobial activity of perfume oils. *J. Am. Pharm. Assoc.* **1958**, *47*, 471–476.
86. Anand, M.; Suresh, S. Marine seaweed Sargassum wightii extract as a low-cost sensitizer for ZnO photoanode based dye-sensitized solar cell. *Adv. Nat. Sci. Nanosci. Nanotechnol.* **2015**, *6*, 035008. [CrossRef]
87. Lai, W.H.; Su, Y.H.; Teoh, L.G.; Hon, M.H. Commercial and natural dyes as photosensitizers for a water-based dye-sensitized solar cell loaded with gold nanoparticles. *J. Photochem. Photobiol. A Chemistry* **2008**, *195*, 307–313. [CrossRef]
88. Ali, H. Biodegradation of synthetic dyes—A review. *Water Air Soil Pollut.* **2010**, *213*, 251–273. [CrossRef]
89. Sivaprakasha, S.; Kumarb, P.S.; Krishnac, S. Adsorption study of various dyes on Activated Carbon Fe_3O_4 Magnetic Nano Composite. *Int. J. Appl. Chem.* **2017**, *13*, 255–266.
90. Doğar, Ç.; Gürses, A.; Açıkyıldız, M.; Özkan, E. Thermodynamics and kinetic studies of biosorption of a basic dye from aqueous solution using green algae *Ulothrix* sp. *Colloids. Surf. B Biointerfaces* **2010**, *76*, 279–285. [CrossRef] [PubMed]
91. Maleki, A.; Mahvi, A.H.; Ebrahimi, R.; Zandsalimi, Y. Study of photochemical and sonochemical processes efficiency for degradation of dyes in aqueous solution. *Korean J. Chem. Eng.* **2010**, *27*, 1805–1810. [CrossRef]
92. Yao, Y.; Xu, F.; Chen, M.; Xu, Z.; Zhu, Z. Adsorption behavior of methylene blue on carbon nanotubes. *Bioresour. Technol.* **2010**, *101*, 3040–3046. [CrossRef]
93. Konicki, W.; Pełech, I.; Mijowska, E.; Jasińska, I. Adsorption of anionic dye Direct Red 23 onto magnetic multi-walled carbon nanotubes-Fe3C nanocomposite: Kinetics, equilibrium and thermodynamics. *Chem. Eng. J.* **2012**, *210*, 87–95. [CrossRef]
94. Aravindhan, R.; Rao, J.R.; Nair, B.U. Removal of basic yellow dye from aqueous solution by sorption on green alga *Caulerpa scalpelliformis*. *J. Hazard Mater.* **2007**, *142*, 68–76. [CrossRef]
95. Gupta, S.; Kumar, D.; Gaur, J. Kinetic and isotherm modeling of lead (II) sorption onto some waste plant materials. *Chem. Eng. J.* **2009**, *148*, 226–233. [CrossRef]
96. Xue, X.; Wang, J.; Mei, L.; Wang, Z.; Qi, K.; Yang, B. Recognition and enrichment specificity of Fe_3O_4 magnetic nanoparticles surface modified by chitosan and Staphylococcus aureus enterotoxins A antiserum. *Colloids Surf. B Biointerfaces* **2013**, *103*, 107–113. [CrossRef]
97. Tolba, A.A. Evaluation of uranium adsorption using magnetic-polyamine chitosan from sulfate leach liquor of sela ore material, South Eastern Desert, Egypt. *Egypt. J. Chem.* **2020**, *63*, 5219–5238. [CrossRef]
98. Khan, M.M.R.; Ray, M.; Guha, A.K. Mechanistic studies on the binding of Acid Yellow 99 on coir pith. *Bioresour. Technol.* **2011**, *102*, 2394–2399. [CrossRef]
99. Ong, P.S.; Ong, S.T.; Hung, Y.T. Utilization of mango leaf as low-cost adsorbent for the removal of Cu (II) ion from aqueous solution. *Asian J. Chem.* **2011**, *25*, 6141–6145. [CrossRef]
100. Khalaf, M.A. Biosorption of reactive dye from textile wastewater by non-viable biomass of Aspergillus niger and *Spirogyra* sp. *Biores. Technol.* **2008**, *99*, 6631–6634. [CrossRef]
101. Al Prol, A.E. Study of environmental concerns of dyes and recent textile effluents treatment technology: A Review. *Asian J. Fish. Aquat. Res.* **2019**, *3*, 1–18. [CrossRef]

102. Sampranpiboon, P.; Charnkeitkong, P.; Feng, X. Equilibrium isotherm models for adsorption of zinc (II) ion from aqueous solution on pulp waste. *WSEAS Transac. Environ. Dev.* **2014**, *10*, 35–47.
103. Gautam, S.; Khan, S.H. Removal of methylene blue from waste water using banana peel as adsorbent. *Int. J. Sci. Environ. Technol.* **2016**, *5*, 3230–3236.
104. Tahir, K.; Nazir, S.; Li, B.; Khan, A.U.; Khan, Z.U.H.; Ahmad, A.; Khan, F.U. An efficient photo catalytic activity of green synthesized silver nanoparticles using Salvadora persica stem extract. *Sep. Purif. Technol.* **2015**, *150*, 316–324. [CrossRef]
105. Bulut, Y.; Aydın, H. A kinetics and thermodynamics study of methylene blue adsorption on wheat shells. *Desalination* **2006**, *194*, 259–267. [CrossRef]
106. Kavitha, D.; Namasivayam, C. Experimental and kinetic studies on methylene blue adsorption by coir pith carbon. *Bioresour. Technol.* **2007**, *98*, 14–21. [CrossRef]
107. Banat, F.; Al-Asheh, S.; Al-Makhadmeh, L. Evaluation of the use of raw and activated date pits as potential adsorbents for dye containing waters. *Process. Biochem.* **2003**, *39*, 193–202. [CrossRef]
108. Han, R.; Wang, Y.; Han, P.; Shi, J.; Yang, J.; Lu, Y. Removal of methylene blue from aqueous solution by chaff in batch mode. *J. Hazard. Mater.* **2006**, *137*, 550–557. [CrossRef] [PubMed]
109. Hall, K.R.; Eagleton, L.C.; Acrivos, A.; Vermeulen, T. Pore-and solid-diffusion kinetics in fixed-bed adsorption under constant-pattern conditions. *Ind. Eng Chem. Fundam.* **1966**, *5*, 212–223. [CrossRef]
110. Abdelwahab, O.; El Nemr, A.; El Sikaily, A.; Khaled, A. Use of rice husk for adsorption of direct dyes from aqueous solution: A case study of Direct, F. Scarlet. *Egypt. J. Aquat. Res.* **2005**, *31*, 1–11.
111. El Nemr, A.; Khaled, A.; Abdelwahab, O.; El-Sikaily, A. Treatment of wastewater containing toxic chromium using new activated carbon developed from date palm seed. *J. Hazard. Mat.* **2008**, *152*, 263–275. [CrossRef]
112. Abualnaja, K.M.; Alprol, A.E.; Ashour, M.; Mansour, A.T. Influencing Multi-Walled Carbon Nanotubes for the Removal of Ismate Violet 2R Dye from Wastewater: Isotherm, Kinetics, and Thermodynamic Studies. *Appl. Sci.* **2021**, *11*, 4786. [CrossRef]
113. Potgieter, J.; Pearson, S.; Pardesi, C. Kinetic and thermodynamic parameters for the adsorption of methylene blue using fly ash under batch, column, and heap leaching configurations. *Coal Combust. Gasif. Prod.* **2018**, *10*, 23–33.
114. Baaloudj, O.; Assadi, I.; Nasrallah, N.; El Jery, A.; Khezami, L.; Assadi, A.A. Simultaneous removal of antibiotics and inactivation of antibiotic-resistant bacteria by photocatalysis: A review. *J. Water Process Eng.* **2021**, *42*, 102089. [CrossRef]
115. Vimonses, V.; Lei, S.; Jin, B.; Chow, C.W.; Saint, C. Kinetic study and equilibrium isotherm analysis of Congo Red adsorption by clay materials. *Chem. Eng. J.* **2009**, *148*, 354–364. [CrossRef]

Article

Lignocellulosic Based Biochar Adsorbents for the Removal of Fluoride and Arsenic from Aqueous Solution: Isotherm and Kinetic Modeling

Iram Ayaz [1], Muhammad Rizwan [1,*], Jeffery Layton Ullman [1,2,3,*], Hajira Haroon [4], Abdul Qayyum [5,*], Naveed Ahmed [1], Basem H. Elesawy [6], Ahmad El Askary [7], Amal F. Gharib [7] and Khadiga Ahmed Ismail [7]

[1] US Pakistan Center for Advanced Studies in Water, Mehran University of Engineering and Technology, Jamshoro 76062, Pakistan; iree.az308@gmail.com (I.A.); naveed.uspcasw@faculty.muet.edu.pk (N.A.)
[2] Department of Civil and Environmental Engineering, University of Utah, 201 Presidents Circle, Room 201, Salt Lake City, UT 84112, USA
[3] Rwanda Institute of Conservation Agriculture, Kagasa-Batima Rd, Gashora, Bugesera, Rwanda
[4] Department of Environmental Sciences, The University of Haripur, Haripur 22620, Pakistan; hajira@uoh.edu.pk
[5] Department of Agronomy, The University of Haripur, Haripur 22620, Pakistan
[6] Department of Pathology, College of Medicine, Taif University, P.O. Box 11099, Taif 21944, Saudi Arabia; basemelesawy2@gmail.com
[7] Department of Clinical Laboratory Sciences, College of Applied Medical Sciences, Taif University, P.O. Box 11099, Taif 21944, Saudi Arabia; ahmedelaskary3@gmail.com (A.E.A.); dr.amal.f.gharib@gmail.com (A.F.G.); khadigaah.aa@tu.edu.sa (K.A.I.)
* Correspondence: drmrizwan.uspcasw@faculty.muet.edu.pk (M.R.); jeffery.layton.ullman@utah.edu (J.L.U.); aqayyum@uoh.edu.pk (A.Q.)

Abstract: *Eucalyptus* wood is made up of lignocellulosic material; this lignocellulosic material contains two types of biopolymers, i.e., carbohydrate and aromatic polymers. In this study, this lignocellulosic material was used to prepare biochar. Three biochar, i.e., laboratory-based (B1), barrel-based (B2), and brick kiln-biochar (B3), were used for fluoride and arsenic removal from aqueous solution. Barrel-based biochar was prepared by using the two-barrel method's alteration. The highest fluoride removal (99%) was attained at pH 2 in the presence of B1, while in the presence of B2 and B3, maximum fluoride removal was 90% and 45.7%, respectively. At pH 10, the maximum arsenic removal in the presence of B1, B2, and B3 was 96%, 94%, and 93%, respectively. The surface characteristics obtained by Fourier-transform infrared spectroscopy (FTIR) showed the presence of carbonyl group (C-O), and alkene (C=C) functional groups on all the three studied biochars. Isotherm studies showed that the adsorption was monolayered (all the adsorbed molecules were in contact with the surface layer of the adsorbent) as the Langmuir isotherm model best fits the obtained data. Adsorption kinetics was also performed. The R^2 value supports the pseudo-second-order kinetics, which means that chemisorption was involved in adsorbing fluoride and arsenic. It is concluded that B1 gives maximum removal for both fluoride (99%) and arsenic (96%). The study shows that lignocellulose-based biochar can be used for arsenic and fluoride removal from water.

Keywords: kinetic study; *Eucalyptus*; biochar; arsenic; fluoride

1. Introduction

The presence of toxic substances in drinking water can cause a risk to the human health [1,2]. Pakistan Council of Research in Water Resources (PCRWR) monitored the water quality in some major cities of Pakistan. The study found the presence of arsenic, fluoride, and bacteria in drinking water [3].

Clean water is a fundamental human right, but pollution with metals, non-metals, natural processes, and some inorganic components such as fluoride and arsenic poses serious health problems. Arsenic is an element found in a natural setting, organisms, soil, and aquatic environment. Arsenic enters into the soil and water by biological processes, volcanos, rocks weathering, mining, pesticides with arsenic, and burning fossil fuels [4]. The permissible limit of arsenic in air, freshwater, soil, and seawater is 3 ng m^{-3}, 10 mg L^{-1}, 100 mg kg^{-1}, and 1.5 mg L^{-1}, respectively [5]. Arsenic is among the prominent four non-essential elements: toxic arsenic, mercury, cadmium, and lead. It is also considered the potential carcinogenic element [6]. According to the WHO, it is among group 1 human carcinogens [7].

The well-known active pollutant found in water in numerous areas in Pakistan is fluoride; this is a threat to water quality consumed by the masses. The permissible limit of fluoride in drinking water is 0.5–1.5 mg L^{-1}. Unfortunately, 260 million people drink fluoride contaminated (>1.5 mg L^{-1}) water. Fluoride is essential for teeth and bones to prevent tooth decay and protect bones. However, a higher fluoride concentration can result in dental and skeletal fluorosis [8]. The general issues of drinking water containing fluoride results in fluorosis, teeth mottles, bones weakness, and it also affects the human's nervous system [1]. These diseases primarily affect children, who are more susceptible to fluoride than grownups.

Thus, removing these harmful components from water is essential to make it potable. Many available techniques are already used for this purpose: ion exchange, reverse osmosis (RO), and coagulation. The drawbacks of using these methods are their high operation and maintenance costs [9–14].

Another emergent technique is known globally; biochar is an active adsorbent to decontaminate various pollutants from water. Biochar is a carbonaceous compound synthesized by pyrolyzing various biomasses, i.e., wood, leaves, vegetable wastes, and seeds [15–17].

Adsorption is a process well known for its cost-effectiveness and ability to remove metals and non-metals from water [18]. Adsorption is defined as the adhesion of adsorbate on the surface of the adsorbent, either by physical or chemical adsorption. The main features of biochar are its inorganic constituents, functional groups, pores on its surface, and greater surface area; thus, it can be used as an effective adsorbent material [19,20]. This active material can be produced from various renewable resources [21]. Cocos-nucifera core, pinewood and pine bark, *eucalyptus* bark, and tea waste are the raw materials on which several studies are already conducted to prepare biochar [22–25].

In this study, biopolymer lignocellulosic material, i.e., eucalyptus wood was used to prepare biochar under three different conditions. These biochars were then evaluated as an an effective adsorbent for fluoride and arsenic removal from synthetic water. This is the first time that three different methods were employed for the preparation of biochar from eucalyptus wood and were compared to remove two pollutants, i.e., arsenic and fluoride from water. Previously, there was no study in which this type of comparison was done. Isotherm and kinetics were also carried out in the current study.

2. Materials and Methods

2.1. Preparation of Biochars

Laboratory, barrel, and brick kiln biochar were prepared by using branches of *Eucalyptus* plant. Laboratory-based biochar (B1) was prepared in a muffle furnace (Nabertherm B 180, Lilienthal, Germany) in presence of nitrogen gas, barrel-based biochar (B2) was prepared by using the two-barrel method, and the brick kiln biochar (B3) was synthesized in a brick clamp kiln [26,27]. For the preparation of B2 and B3, the branches were cut into 20 cm lengths and chopped using a hand hatchet to obtain a cross-sectional dimension of 1 cm or less. The temperatures used for the preparation of B1, B2, and B3 were 350 °C, 550 °C, and 450 °C, respectively. For B1, the split branches were cut into the size of 1 cm length. A 2.5 L airtight, stainless-steel reactor with incorporated vents for air inflow and outflow was used to prepare B1. Stainless-steel reactor was put inside a muffle furnace, and the temperature was recorded as 350 °C for 2 h. Nitrogen was continuously purged during the experiment to guarantee an oxygen-free environment. After the preparation of biochar, it was crushed by using a mortar and pestle. The biochar was sieved using 0.595 mm^{-1} mm sizes and saved in zip lock bags for further batch experimentation.

2.2. Characterization of Biopolymer Containing Biochar

SEM (scanning electron microscope) was used to generate an image of the material under study with varying magnification. Brunauer–Emmett–Teller (BET) was used to determine biochars surface area. Zeta potential technique was used to calculate the charge on the biochar surface. At the same time, Fourier-transform infrared (FTIR) spectroscopy was used to find the functional groups present on the biochar surface.

2.3. Chemicals and Reagents

Stock solution (1000 mg/L) of fluoride was prepared by using 2.2101 g of sodium fluoride (NaF) in a deionized (DI) water. The arsenic solution was prepared by using a 1000 ppm standard solution of arsenic.

The solution's pH was adjusted using 0.1 M NaOH and 0.1 M NaCl solutions.

2.4. Analysis

Fluoride was analyzed by using SPANDS method and UV-spectrophotometer (Perkinelmer model: lamda 365, Waltham, MA, USA). Arsenic was analyzed by using inductively coupled plasma mass spectrometry (ICPMS) (Perkinelmer model: Nexion 350, Waltham, MA, USA). All samples were analyzed through standard methods [28,29].

2.4.1. Batch Experiments for Fluoride and Arsenic Adsorption

In these experiments, the adsorption capacity of B1, B2 and B3 biochars were examined. Different concentrations of fluoride and arsenic (10, 30, 45, and 60 mg L^{-1}) having 10 mL volume were used with 0.1 g of biochars during the batch experiment. The effect of various parameters like pH, adsorbate concentration, and time on the adsorption of arsenic and fluoride on the three prepared biochars were examined and optimized.

The sorption efficiency or percent removal and sorption capacity were determined by using the following equations:

$$\text{Sorption efficiency}(\%) = \frac{(C_o - C_f)}{C_o} \times 100 \quad (1)$$

$$\text{Sorption capacity}(q) = \frac{(C_o - C_e)}{M} \times V \quad (2)$$

where q indicates the metal uptake (mg g^{-1}), Co and Cf indicate the initial and equilibrium concentrations (mg L^{-1}) before and after adsorption, respectively, V indicates the volume of synthetic solution (mL), and M represents the adsorbent dose (g).

Isotherm and Kinetic models:

Freundlich and Langmuir isotherm models were used in the current study. The adsorption isotherms help understand the affiliation among adsorbate concentration and its amount accumulated on the adsorbent surface [30–32].

Linear form of Langmuir isotherm was used

$$\frac{C_e}{q_e} = \frac{1}{Q_{max} b} + \frac{C_e}{Q_{max}} \quad (3)$$

where Ce is the equilibrium concentration, qe is the amount of adsorbent adsorbed, and Qmax and b are known as Langmuir constants linked to adsorption capability and sorption affinity, respectively. The slopes can measure the plot's Q max and b-intercept in Ce/qe against Ce.

The Freundlich isotherm is

$$\text{Log } q_e = \text{Log } K_f + \left(\frac{1}{n}\right) \log C_e \quad (4)$$

where qe is the adsorbed quantity of adsorbate at equilibrium, Ce is the equilibrium concentration of adsorbate, Kf is the sorption capacity, and 1/n is the heterogeneity factor.

Pseudo first order and pseudo-second-order kinetic models were applied to find out the adsorption mechanism. The equations used for adsorption kinetics were:

Pseudo First order (PFO):

$$\text{Ln}(q_e - q_t) = \ln q_e - k_1 t \tag{5}$$

Pseudo Second-order (PSO):

$$\frac{t}{q_t} = \frac{1}{k_2 q_e^2} + \frac{1}{q_e} \tag{6}$$

where k1 and k2 are the rate constants of pseudo 1st and pseudo 2nd order, qe is the amount of adsorbate adsorbed at equilibrium, and qt is the amount of adsorbate adsorbed at time t [32,33].

2.4.2. Statistical Analysis

The statistical analysis was carried out using Microsoft excel by incorporating the above equations into the software. The graphs were developed in Sigma plot.

3. Results

3.1. Characterization

3.1.1. Brunauer Emmet–Teller (BET)

Surface area (SA) is one of the main factors in determining the biochar ability to adsorb various contaminants. The surface area of all three prepared biochars was measured with the help of BET analysis [34]. The B1, B2, and B3 biochars' surface area was 0.885 ± 0.505, 99.449 ± 9.091, and 6.341 ± 0.427 m^2/g, respectively (Table 1). B2 biochar had the highest surface area than B1 and B3 and is also greater than many other biochars found in the literature. The surface areas of coffee ground carbon, perilla leaf, dry pinewood, and pine bark biochars were reported to be 5.0, 3.2, 2.73, and 1.88 mm, respectively.

Table 1. Surface area of B1, B2, and B3 biochars.

Adsorbent	Surface Area (m^2/g)
Laboratory Biochar (B1)	0.885
Barrel Biochar (B2)	99.449
Brick Biochar (B3)	6.341

3.1.2. Scanning Electron Microscope Analysis (SEM)

The monographs of B1, B2, and B3 biochars were captured using SEM. Figure 1 showed that the B1 and B2 biochars has a heterogeneous surface with a honeycomb structure. In contrast, the B3 biochar showed a rough and uneven surface. The average pore sizes of B1, B2, and B3 biochars were 0.061, 0.276, and 0.100 µm, respectively.

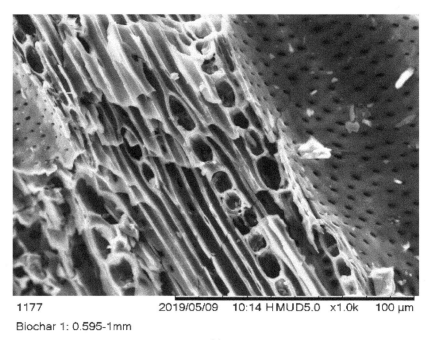

(a)

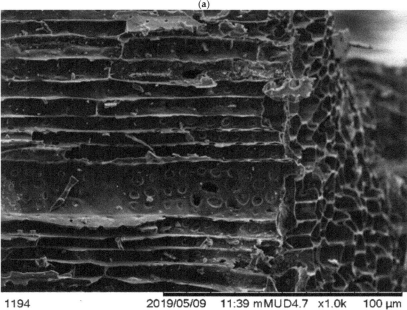

(b)

Figure 1. *Cont.*

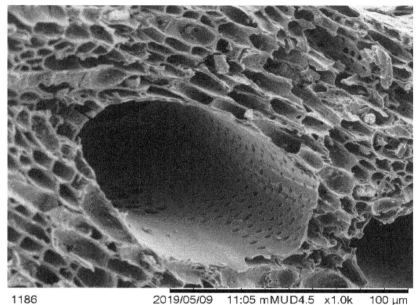

1186 2019/05/09 11:05 mMUD4.5 x1.0k 100 μm
Biochar 3: 0.595-1mm

(c)

Figure 1. SEM images of (**a**) (B1) Laboratory-biochar, (**b**) (B2) Barrel biochar, and (**c**) (B3) Brick Kiln biochar.

3.1.3. FTIR

The functional groups on the biochars surface were determined using the FTIR method. The FTIR spectroscopy measures the surface chemistry of a solid material. A band of spectra was formed by a range of functional groups present on the surface of prepared biochars. The functional groups present on the surface of B1, B2, and B3 biochars are shown in the Figure 2.

In the case of B1, B2, and B3, the absorbance peaks at 1401 cm^{-1}, 1408 cm^{-1}, and 1370 cm^{-1} indicated the presence of a C–O functional group.

B1 and B3 contain a C=C functional group at the peaks of 1670 cm^{-1} while B2 contains a C=C functional group at 1872 cm^{-1}. Thus, these C–O and C=C functional groups present on the surfaces of B1, B2, and B3 are involved in the adsorption of fluoride and arsenic.

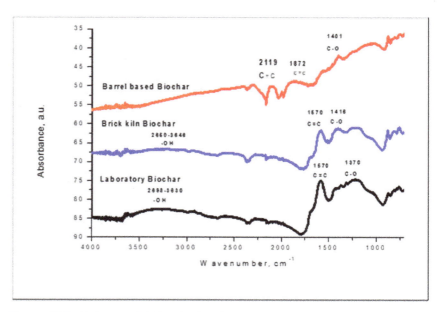

Figure 2. FTIR of Laboratory (B1), Barrel (B2), and Brick Kiln biochar (B3).

3.1.4. Zeta Potential

Zeta potential gives us information about the surface charges of a material. The B1, B2, and B3 have surface charges of −41.37 mV, −40.30 mV, and −42.67 mV, respectively. These obtained values showed that B2 has more negative zeta potential than biochars prepared by the remaining two methods. This specifies that the surfaces of biochars are considered negative.

3.2. Batch Adsorption of F^- Fluoride and As-Arsenic

3.2.1. Effect of Contact Time

The influence of time was observed in this study by varying the time from 15–60 min, whereas the initial concentration of fluoride and arsenic was kept at 10 mg/L and 0.5 mg/L, respectively. The highest removal efficiency of 95% was observed after 1 h contact time for fluoride with B1 as shown in the Figure 3a. In context with arsenic (Figure 3b), the B1 biochar has shown an adsorption efficiency of 96% in 1 h equilibrium time. At first, generally, rapid uptake of adsorbate occurs because plenty of active binding sites were present on biochars surface. In the case of fluoride removal using B2 and B3, the equilibrium time was attained at 45 min and 1 h, respectively. For arsenic, the maximum adsorption was attained at the contact time of 1h.

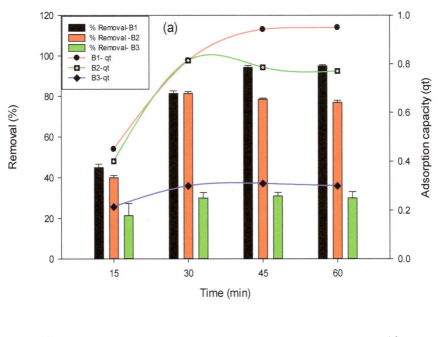

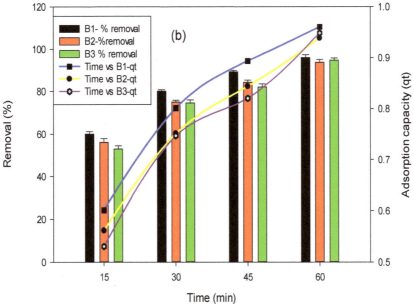

Figure 3. Effect of time on fluoride (**a**) and arsenic (**b**) removal percent and adsorption capacities using B1 (laboratory-based biochar), B2 (Barrel- biochar), and B3 (Brick kiln biochar).

3.2.2. Effect of pH

One major factor that influences the interaction of biochar with pollutants (fluoride and arsenic) is the solution's pH. In the current study, the pH of a solution played a vital role in fluoride and arsenic adsorption. The removal of fluoride and arsenic was examined by varying the pH (2–10) of the solution. Results in Figure 4a show that, in the case of B1, a maximum adsorption of 99% was achieved at pH 2 for fluoride, while in the case of B2 and B3, the maximum removal efficiency was calculated to be 90 percent and 45 percent at pH 2 for fluoride. Maximum arsenic removal of 96%, 93% and 94% was achieved at pH 10 for all three studied biochars, i.e., BI, B2 and B3, respectively, as shown in Figure 4b.

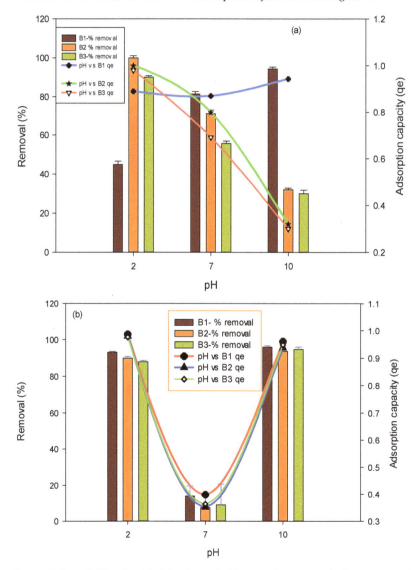

Figure 4. Effect of pH on fluoride (**a**) and arsenic (**b**) removal percent and adsorption capacity by using B1 (laboratory-based biochar), B2 (Barrel- biochar), and B3 (Brick kiln biochar).

3.2.3. Effect of Initial Concentration

To assess the effect of concentration on the removal efficiency of biochars, concentrations of fluoride (10, 30, 45, and 60 mg/L) and arsenic (0.05, 0.51, and 5 mg/L) were varied. Results in Figure 5a showed that the maximum fluoride removal of 99% was achieved in the case of B1, whereas 96% of arsenic removal was achieved (Figure 5b) in the case of B1, where the initial concentration of arsenic was 0.5 mg/L.

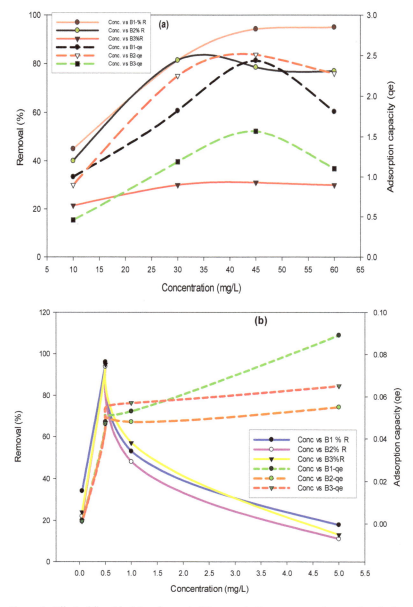

Figure 5. Effect of fluoride (**a**) and arsenic (**b**) concentration on removal percent and adsorption capacities using B1 (laboratory-based biochar), B2 (Barrel- biochar), and B3 (Brick kiln biochar).

3.3. Adsorption Isotherms

3.3.1. Isotherms and Kinetic Studies for Fluoride Removal

Langmuir and Freundlich's models were utilized to check whether the biochar surface is homogeneous or heterogeneous. Table 2 shows that the biochars B1, B2, and B3 prepared with different methods individually were identified as homogenous, because the Langmuir isotherm model fits well with the obtained data. Table 2 shows that for fluoride, the regression coefficient value was highest (0.9937) and maximum sorption capacity (q_{max}) was 2.376 in the case of B2. The absence of the heterogeneous nature of biochar was because of deficiency of active areas exponential distribution on its surface. The results showed that fluoride adsorption onto all the three biochars is monolayer because it explains the homogeneous nature of adsorbent and justifies the Langmuir isotherm with the highest R^2 values.

Table 2. Langmuir and Freundlich models' constants and correlation coefficients, fluoride and arsenic.

Biochar	Adsorbate	Langmuir			Freundlich		
		R^2	Qmax	b	R^2	Kf	n
B1	Fluoride	0.977	1.832	0.872	0.756	1.059	5
	Arsenic	0.839	0.1086	1.152	0.339	19.741	2.355
B2	Fluoride	0.993	2.376	1.559	0.835	1.0128	3.593
	Arsenic	0.806	0.066	1.149	0.281	27.646	2.322
B3	Fluoride	0.87	1.333	0.216	0.701	4.149	2.105
	Arsenic	0.776	0.079	1.096	0.284	23.660	2.3164

Moreover, B2 biochar is most appropriate for fluoride adsorption among the three studied biochars. Pseudo first and pseudo second order kinetic models in Table 3 show that the highest R^2 value (0.9862) is obtained in the case of B3. All three biochars for fluoride favored the pseudo second order rate equation, which revealed the chemisorption nature of all studied biochars for fluoride removal.

Table 3. Pseudo-first and second-order kinetics for Fluoride and Arsenic.

Biochar	Adsorbate	Pseudo 1st Order			Pseudo 2nd Order		
		R^2	qe	K1	R^2	Qe	K2
B1	Fluoride	0.1006	2.027898	0.0276	0.9088	0.123913	69.42557
	Arsenic	0.0469	2.228659	0.0107	0.9733	0.20094	24.70232
B2	Fluoride	0.2866	1.60914	0.0086	0.9249	0.147406	39.30873
	Arsenic	0.0375	2.146632	0.0083	0.9641	0.169739	34.01136
B3	Fluoride	0.1477	12.9656	0.0313	0.9862	0.125	20.20393
	Arsenic	0.0345	2.016776	0.0072	0.9516	0.149301	43.93484

3.3.2. Isotherms and Kinetic Studies for Arsenic Removal

Results (R^2 value) in Table 3 show the homogeneous (monolayer) adsorption on the surfaces of all prepared biochars for arsenic removal. The highest R^2 value (0.839) for B1 was obtained with arsenic, and its extreme adsorption capacity is 0.108. The B1 biochar is more suitable for arsenic removal from water than B2 and B3.

The kinetic studies data for arsenic removal (Table 3) showed that all prepared biochar in this study best fitted pseudo second order kinetics model. However, B1 has the highest the R^2 value as compared to B2 and B3 for arsenic removal. This again justifies the chemical adsorption nature of biochars instead of the physical sorption.

4. Discussion

In this study, the barrel biochar had a higher surface area (99.449 m^2/g) than the laboratory and brick kiln biochar. Mishra et al., 2017, reported a *Eucalyptus* biochar with a surface area of 20 m^2/g which was lower than the surface area of barrel biochar used in the current study [35]. Han et al., 2013, obtained the maximum surface area was 383.66 m^2/g, achieved in activated softwood biochar [36]. Several factors affect the surface area of biochar; these factors include the type of biomass, pyrolysis temperature, and preparation method. Greater surface area facilitates the adsorption process [37]. Various chars were studied with distinct preparatory methods and concluded that chars with greater surface area, i.e., 55.20 m^2/g tend to stand out with greater reactivity [27]. However, chars having minimum surface area, i.e., 10.53 m^2/g, adhere to low reactivity with an adsorbate. Oh et al., 2012, reported deep and variable pore sizes for granular and grounded powder biochar. Moreover, the biochar showed asymmetrical forms and sizes; these sites may provide higher internal surface areas in orange peel and sludge biochar [38].

The morphology of biochar was studied by using SEM technique. The B1 and B2 biochar structures were like a honeycomb, while the surface B3 biochar was rough and uneven. The preparation method of biochar might be the reason for the different morphologies of biochar. In a previous study performed by Hajira et al., 2016, the SEM image of *Eucalyptus camaldulensis* showed a rough and uneven surface with heterogeneous pores of different sizes [39]. Mishra et al., 2017, showed that the biochar prepared from *Eucalyptus* biomass was highly heterogeneous and had macropores [35]. Zhang et al., 2018, used *Eucalyptus* sawdust biochar for chromium removal. The surface of biochar was rough, and it contained heterogeneous particles [40].

Generally, FTIR spectra was observed in the range from 4000 to 500 cm^{-1} [36,41]. The B1 and B3 biochars of the current study showed prominent peaks at absorbance of 1570 cm^{-1}, which is characteristic of C=C, and absorbance between 2850–3000 cm^{-1} showed –OH, functional group in both B1 and B3 biochars [42]. All three studied biochars in the current study had a C-O functional group at absorbance peaks of 14416 and 1370 cm^{-1}. Biochar B2 has a prominent functional group of alkynes (C≡C) triple bond at an absorbance of 2119 cm^{-1}. In a previous study on biochar derived from *Eucalyptus* wood, the FTIR also showed the presence of –COOH, -OH, and C=O functional groups [35]. In another study, perennial grass-based activated biochar was used to remove fluoride and arsenic from an aqueous solution. The FTIR of the activated perennial grass biochar showed the presence of O–H, C–H, C=C, C–O, and -CH functional groups [43]. Papari et al., 2017, used *Conocarpus erectus* biochar for fluoride removal, and the FTIR revealed the presence of following functional groups: ≡C–H and –OH, –CH=CHR, –S=O, C–F, S–OR esters, –C–H and S–S [37]. These functional groups are different due to the biomass used in this study.

The present study achieved the maximum adsorption of fluoride and arsenic within 1 h. Activated rice straw biochar was used to remove fluoride from an aqueous solution. The maximum adsorption was achieved after 3 h [44]. In another study, *Conocarpus erectus* biochar was used by Papari et al., 2017, for maximum fluoride removal from aqueous solution, groundwater, and seawater within 1.5 h [37]. Mishra et al., 2017, used *Eucalyptus* wood for uranium removal in 20 min [35].

The initial concentration of fluoride and arsenic was inversely proportional to the maximum adsorption. A similar trend was observed in a previous study performed by Saikia et al., 2017 in which the increase of fluoride concentration resulted in a decrease in adsorption efficiency [43]. This is due to the saturation of adsorption sites of biochar. Daifullah et al., 2007, also observed that fluoride adsorption increased with the increase in fluoride concentration until the fluoride concentration reached 18 mg/L [44].

As time progresses, removal efficiency declines because of the repulsion between solute constituents on the surface of the adsorbent and in solution [45]. At lower pH, the fluoride removal was recorded to be the highest because, in this condition, the attractive forces were increased amongst biochars and fluoride; this is then due to the H$^+$ presence on biochars surface [30]. The phenomenon is known as deprotonation of the functional groups on the adsorbent's surface if the solution pH was increased, which justifies the arsenic adsorption onto biochars at high pH; this facilitates the adsorption of positively charged ions on the negatively charged surface (biochar) [46]. Moreover, the ion interchange between arsenic and –OH results in maximum arsenic removal at higher pH, i.e., 10 [47]. It is being noticed that if the initial concentration increases, the removal efficiency decreases due to the saturation of the active regions which are present on the biochar [32,44,45].

Likewise, Mohan et al., 2012, used pine wood, and Papari et al., 2017, performed monolayer adsorption experiments for fluoride removal. Langmuir isotherm model was also best fitted to the data in their results [37,48]. In comparison, Goswami et al., 2018, utilized nano-rice husk biochar to remove fluoride and showed that both Langmuir and Freundlich isotherms were best fitted with the its obtained data [33]. These results showed that the interaction of the fluoride particles and biochars is chemical, with the functional groups residing on the biochars surface [31,33]. Biochar made of oak wood, perennial grass, and pine cone [43,49,50] favored monolayer (homogenous) sorption for arsenic, because the adsorption data was well fitted into Langmuir model. There is a possibility that biochars synthesized under higher temperatures have a material with a crystalline structure. This is because of the turbostratic crystallites, which mean biochar has graphene layers ordered not properly [46]. This might be a possible reason for the best fitting of adsorption data with Langmuir model as compared to the Freundlich. A significant R^2 value was attained [43,45], which also showed a pseudo-second-order rate equation for arsenic removal. Arsenite (As III) and arsenate (As V) were removed from aqueous solution and ground water using perilla leaf biochar. Maximum arsenic removal (88–90%) was achieved at pH 7–9. Langmuir isotherm model was best fitted for both As III and V [49]. Mohan et al., 2014 used magnetic and nonmagnetic stover biochars to remove fluoride from groundwater. Nonmagnetic stover biochar showed better flouride removal capacity while magnetic stover biochar showed better biochar recovery, redispersion and washing [23]. Coffee grounds (CG) were used for the preparation of Carbonaceous material. The carbonaceous coffee grounds were used for the removal of fluoride from water. The CG calcinated at 600 °C showed the maximum fluoride removal compared to the CG calcinated at 400, 800 and 1000 °C [51]. In a study conducted by Papari et al., 2017, *Conocarpus erectus* based granular and powdered biochar was used for fluoride removal from aqueous solution. The maximum fluoride removal in the presence of granular and powdered biochar was 80% and 98.5%, respectively. Langmuir isotherm model was best fitted with the adsorption data [37].

5. Conclusions

In the present study, three different types of biochars, i.e., B1, B2, and B3 were prepared from *Eucalyptus* wood. These biochars were then analyzed for their capability/potential for adsorbing both fluoride and arsenic from water. The highest fluoride removal was attained at pH 2, biochar-adsorbent dose of 0.1 g, and a contact time of 60 min in the case of both B1 and B2. For arsenic, maximum removal was obtained at pH 10, 0.1 g (biochar dose), and at a contact time of 60 min using the B1 followed by the B2 biochar. For the B2 biochar, the assessed surface area was 99.449 m^2/g, and the average pore length was 0.275 μm. The FTIR spectra revealed the involvement of the C-O, and C=C functional group in fluoride adsorption onto B1 and B3. Highest removal of arsenic is attained using the B1 biochar, which revealed that the arsenic might bind to the C-O and C=C functional groups present on the biochar. Langmuir and pseudo second order kinetic models were most suitable for fluoride and arsenic removal for all three studied biochars. The qmax for fluoride was 2.376 mg g^{-1} in the case of B2, whereas the qmax for arsenic was 0.108 mg g^{-1} for B1. Results showed that B1 showed a maximum adsorption of 99% and 96% for both fluoride and arsenic, respectively. It is concluded that cost-effective biopolymer-based biochar could be recommended for the treatment of fluoride and arsenic polluted water.

Author Contributions: M.R. and J.L.U. conceived of the idea. I.A. conducted the experiment. A.E.A., K.A.I., A.F.G., and B.H.E. collected the literature review. H.H., A.Q., and N.A. provided technical expertise. H.H. helped in statistical analysis. M.R. and J.L.U. proofread and provided intellectual guidance. All authors read the first draft, helped in revision, and approved the article. All authors have read and agreed to the published version of the manuscript.

Funding: This work was supported by the US government and the American people through the United States Agency for International Development (USAID) and was partially supported by the Mehran University of Engineering and Technology, Jamshoro, Pakistan. This study was also supported by Taif University Researchers Supporting Project number (TURSP-2020/117), Taif University, Taif, Saudi Arabia.

Institutional Review Board Statement: Not applicable.

Informed Consent Statement: Not applicable.

Data Availability Statement: Data presented in this study are available on fair request to the corresponding author.

Conflicts of Interest: The authors declare no conflict of interest.

References

1. Tahir, M.; Rasheed, H. Fluoride in the drinking water of pakistan and the possible risk of crippling fluorosis. *Drink. Water Eng. Sci.* **2013**, *6*, 17–23. [CrossRef]
2. Haroon, H.; Shah, J.A.; Khan, M.S.; Alam, T.; Khan, R.; Asad, S.A.; Ali, M.A.; Farooq, G.; Iqbal, M.; Bilal, M. Activated carbon from a specific plant precursor biomass for hazardous cr (vi) adsorption and recovery studies in batch and column reactors: Isotherm and kinetic modeling. *J. Water Process Eng.* **2020**, *38*, 101577.
3. Arshad, N.; Imran, S. Assessment of arsenic, fluoride, bacteria, and other contaminants in drinking water sources for rural communities of kasur and other districts in punjab, pakistan. *Environ. Sci. Pollut. Res.* **2017**, *24*, 2449–2463. [CrossRef] [PubMed]
4. Smedley, P.L.; Kinniburgh, D.G. A review of the source, behaviour and distribution of arsenic in natural waters. *Appl. Geochem.* **2002**, *17*, 517–568.
5. Rieuwerts, J. *The Elements of Environmental Pollution*; Routledge: London, UK, 2017.
6. Roy, P.; Saha, A. Metabolism and toxicity of arsenic: A human carcinogen. *Curr. Sci.* **2002**, *82*, 38–45.
7. Van Halem, D.; Bakker, S.; Amy, G.; Van Dijk, J. Arsenic in drinking water: A worldwide water quality concern for water supply companies. *Drink. Water Eng. Sci.* **2009**, *2*, 29–34. [CrossRef]
8. Mondal, N.K.; Bhaumik, R.; Datta, J.K. Removal of fluoride by aluminum impregnated coconut fiber from synthetic fluoride solution and natural water. *Alex. Eng. J.* **2015**, *54*, 1273–1284. [CrossRef]
9. Chubar, N. New inorganic (an) ion exchangers based on mg–al hydrous oxides:(alkoxide-free) sol–gel synthesis and characterisation. *J. Colloid Interface Sci.* **2011**, *357*, 198–209. [CrossRef]
10. Behbahani, M.; Moghaddam, M.A.; Arami, M. Techno-economical evaluation of fluoride removal by electrocoagulation process: Optimization through response surface methodology. *Desalination* **2011**, *271*, 209–218. [CrossRef]
11. Richards, L.A.; Vuachère, M.; Schäfer, A.I. Impact of ph on the removal of fluoride, nitrate and boron by nanofiltration/reverse osmosis. *Desalination* **2010**, *261*, 331–337. [CrossRef]
12. Shrivastava, B.K.; Vani, A. Comparative study of defluoridation technologies in india. *Asian J. Exp. Sci* **2009**, *23*, 269–274.
13. Zhang, A.; Li, X.; Xing, J.; Xu, G. Adsorption of potentially toxic elements in water by modified biochar: A review. *J. Environ. Chem. Eng.* **2020**, *8*, 104196. [CrossRef]
14. Elazhar, F.; Tahaikt, M.; Achatei, A.; Elmidaoui, F.; Taky, M.; El Hannouni, F.; Laaziz, I.; Jariri, S.; El Amrani, M.; Elmidaoui, A. Economical evaluation of the fluoride removal by nanofiltration. *Desalination* **2009**, *249*, 154–157. [CrossRef]
15. Abdel-Fattah, T.M.; Mahmoud, M.E.; Ahmed, S.B.; Huff, M.D.; Lee, J.W.; Kumar, S. Biochar from woody biomass for removing metal contaminants and carbon sequestration. *J. Ind. Eng. Chem.* **2015**, *22*, 103–109. [CrossRef]
16. Usman, A.R.; Sallam, A.S.; Al-Omran, A.; El-Naggar, A.H.; Alenazi, K.K.; Nadeem, M.; Al-Wabel, M.I. Chemically modified biochar produced from conocarpus wastes: An efficient sorbent for fe (ii) removal from acidic aqueous solutions. *Adsorpt. Sci. Technol.* **2013**, *31*, 625–640. [CrossRef]
17. Bautista-Toledo, M.I.; Rivera-Utrilla, J.; Ocampo-Pérez, R.; Carrasco-Marin, F.; Sanchez-Polo, M. Cooperative adsorption of bisphenol-a and chromium (iii) ions from water on activated carbons prepared from olive-mill waste. *Carbon* **2014**, *73*, 338–350. [CrossRef]
18. Yu, Y.; Wang, C.; Guo, X.; Chen, J.P. Modification of carbon derived from sargassum sp. By lanthanum for enhanced adsorption of fluoride. *J. Colloid Interface Sci.* **2015**, *441*, 113–120. [CrossRef] [PubMed]
19. Lu, H.; Zhang, W.; Yang, Y.; Huang, X.; Wang, S.; Qiu, R. Relative distribution of pb2+ sorption mechanisms by sludge-derived biochar. *Water Res.* **2012**, *46*, 854–862. [CrossRef]

20. Li, D.; Zhao, R.; Peng, X.; Ma, Z.; Zhao, Y.; Gong, T.; Sun, M.; Jiao, Y.; Yang, T.; Xi, B. Biochar-related studies from 1999 to 2018: A bibliometrics-based review. *Environ. Sci. Pollut. Res.* **2020**, *27*, 2898–2908. [CrossRef]
21. Yao, C.; Pan, Y.; Lu, H.; Wu, P.; Meng, Y.; Cao, X.; Xue, S. Utilization of recovered nitrogen from hydrothermal carbonization process by arthrospira platensis. *Bioresour. Technol.* **2016**, *212*, 26–34. [CrossRef]
22. Halder, G.; Khan, A.A.; Dhawane, S. Fluoride sorption onto a steam-activated biochar derived from cocos nucifera shell. *CLEAN–Soil Air Water* **2016**, *44*, 124–133. [CrossRef]
23. Mohan, D.; Kumar, S.; Srivastava, A. Fluoride removal from ground water using magnetic and nonmagnetic corn stover biochars. *Ecol. Eng.* **2014**, *73*, 798–808. [CrossRef]
24. Sawood, G.M.; Mishra, A.; Gupta, S. Optimization of arsenate adsorption over aluminum-impregnated tea waste biochar using rsm–central composite design and adsorption mechanism. *J. Hazard. Toxic Radioact. Waste* **2021**, *25*, 04020075. [CrossRef]
25. Patnukao, P.; Pavasant, P. Activated carbon from eucalyptus camaldulensis dehn bark using phosphoric acid activation. *Bioresour. Technol.* **2008**, *99*, 8540–8543. [CrossRef]
26. Deal, C.; Brewer, C.E.; Brown, R.C.; Okure, M.A.; Amoding, A. Comparison of kiln-derived and gasifier-derived biochars as soil amendments in the humid tropics. *Biomass Bioenergy* **2012**, *37*, 161–168. [CrossRef]
27. Zhang, H.; Pu, W.-X.; Ha, S.; Li, Y.; Sun, M. The influence of included minerals on the intrinsic reactivity of chars prepared at 900 c in a drop tube furnace and a muffle furnace. *Fuel* **2009**, *88*, 2303–2310. [CrossRef]
28. Hosseini, S.S.; Mahvi, A.H.; Tsunodac, M. Fluoride content of coconut water and its risk assessment. *Fluoride* **2019**, *52*, 553–561.
29. Gómez-Ariza, J.L.; Sánchez-Rodas, D.; Giráldez, I.; Morales, E. A comparison between icp-ms and afs detection for arsenic speciation in environmental samples. *Talanta* **2000**, *51*, 257–268. [CrossRef]
30. Alagumuthu, G.; Rajan, M. Equilibrium and kinetics of adsorption of fluoride onto zirconium impregnated cashew nut shell carbon. *Chem. Eng. J.* **2010**, *158*, 451–457. [CrossRef]
31. Chen, G.-J.; Peng, C.-Y.; Fang, J.-Y.; Dong, Y.-Y.; Zhu, X.-H.; Cai, H.-M. Biosorption of fluoride from drinking water using spent mushroom compost biochar coated with aluminum hydroxide. *Desalination Water Treat.* **2016**, *57*, 12385–12395. [CrossRef]
32. Ibupoto, A.S.; Qureshi, U.A.; Ahmed, F.; Khatri, Z.; Khatri, M.; Maqsood, M.; Brohi, R.Z.; Kim, I.S. Reusable carbon nanofibers for efficient removal of methylene blue from aqueous solution. *Chem. Eng. Res. Des.* **2018**, *136*, 744–752. [CrossRef]
33. Goswami, R.; Kumar, M. Removal of fluoride from aqueous solution using nanoscale rice husk biochar. *Groundw. Sustain. Dev.* **2018**, *7*, 446–451. [CrossRef]
34. Brunauer, S.; Emmett, P.H. The use of low temperature van der waals adsorption isotherms in determining the surface areas of various adsorbents. *J. Am. Chem. Soc.* **1937**, *59*, 2682–2689. [CrossRef]
35. Mishra, V.; Sureshkumar, M.; Gupta, N.; Kaushik, C. Study on sorption characteristics of uranium onto biochar derived from eucalyptus wood. *Water Air Soil Pollut.* **2017**, *228*, 1–14. [CrossRef]
36. Han, Y.; Boateng, A.A.; Qi, P.X.; Lima, I.M.; Chang, J. Heavy metal and phenol adsorptive properties of biochars from pyrolyzed switchgrass and woody biomass in correlation with surface properties. *J. Environ. Manag.* **2013**, *118*, 196–204. [CrossRef]
37. Papari, F.; Najafabadi, P.R.; Ramavandi, B. Fluoride ion removal from aqueous solution, groundwater, and seawater by granular and powdered conocarpus erectus biochar. *Desal. Water Treat.* **2017**, *65*, 375–386. [CrossRef]
38. Oh, T.-K.; Choi, B.; Shinogi, Y.; Chikushi, J. Effect of ph conditions on actual and apparent fluoride adsorption by biochar in aqueous phase. *Water Air Soil Pollut.* **2012**, *223*, 3729–3738. [CrossRef]
39. Haroon, H.; Ashfaq, T.; Gardazi, S.M.H.; Sherazi, T.A.; Ali, M.; Rashid, N.; Bilal, M. Equilibrium kinetic and thermodynamic studies of cr (vi) adsorption onto a novel adsorbent of eucalyptus camaldulensis waste: Batch and column reactors. *Korean J. Chem. Eng.* **2016**, *33*, 2898–2907. [CrossRef]
40. Zhang, X.; Zhang, L.; Li, A. Eucalyptus sawdust derived biochar generated by combining the hydrothermal carbonization and low concentration koh modification for hexavalent chromium removal. *J. Environ. Manag.* **2018**, *206*, 989–998. [CrossRef]
41. Kinney, T.; Masiello, C.; Dugan, B.; Hockaday, W.; Dean, M.; Zygourakis, K.; Barnes, R. Hydrologic properties of biochars produced at different temperatures. *Biomass Bioenergy* **2012**, *41*, 34–43. [CrossRef]
42. Vaughn, S.F.; Kenar, J.A.; Thompson, A.R.; Peterson, S.C. Comparison of biochars derived from wood pellets and pelletized wheat straw as replacements for peat in potting substrates. *Ind. Crops Prod.* **2013**, *51*, 437–443. [CrossRef]
43. Saikia, R.; Goswami, R.; Bordoloi, N.; Senapati, K.K.; Pant, K.K.; Kumar, M.; Kataki, R. Removal of arsenic and fluoride from aqueous solution by biomass based activated biochar: Optimization through response surface methodology. *J. Environ. Chem. Eng.* **2017**, *5*, 5528–5539. [CrossRef]
44. Daifullah, A.; Yakout, S.; Elreefy, S. Adsorption of fluoride in aqueous solutions using kmno4-modified activated carbon derived from steam pyrolysis of rice straw. *J. Hazard. Mater.* **2007**, *147*, 633–643. [CrossRef] [PubMed]
45. Alam, M.A.; Shaikh, W.A.; Alam, M.O.; Bhattacharya, T.; Chakraborty, S.; Show, B.; Saha, I. Adsorption of as (iii) and as (v) from aqueous solution by modified cassia fistula (golden shower) biochar. *Appl. Water Sci.* **2018**, *8*, 1–14. [CrossRef]
46. Jiang, S.; Huang, L.; Nguyen, T.A.; Ok, Y.S.; Rudolph, V.; Yang, H.; Zhang, D. Copper and zinc adsorption by softwood and hardwood biochars under elevated sulphate-induced salinity and acidic ph conditions. *Chemosphere* **2016**, *142*, 64–71. [CrossRef]
47. Alkurdi, S.S.; Herath, I.; Bundschuh, J.; Al-Juboori, R.A.; Vithanage, M.; Mohan, D. Biochar versus bone char for a sustainable inorganic arsenic mitigation in water: What needs to be done in future research? *Environ. Int.* **2019**, *127*, 52–69. [CrossRef]
48. Mohan, D.; Sharma, R.; Singh, V.K.; Steele, P.; Pittman Jr, C.U. Fluoride removal from water using bio-char, a green waste, low-cost adsorbent: Equilibrium uptake and sorption dynamics modeling. *Ind. Eng. Chem. Res.* **2012**, *51*, 900–914. [CrossRef]

49. Niazi, N.K.; Bibi, I.; Shahid, M.; Ok, Y.S.; Burton, E.D.; Wang, H.; Shaheen, S.M.; Rinklebe, J.; Lüttge, A. Arsenic removal by perilla leaf biochar in aqueous solutions and groundwater: An integrated spectroscopic and microscopic examination. *Environ. Pollut.* **2018**, *232*, 31–41. [CrossRef] [PubMed]
50. Van Vinh, N.; Zafar, M.; Behera, S.; Park, H.-S. Arsenic (iii) removal from aqueous solution by raw and zinc-loaded pine cone biochar: Equilibrium, kinetics, and thermodynamics studies. *Int. J. Environ. Sci. Technol.* **2015**, *12*, 1283–1294. [CrossRef]
51. Ogata, F.; Tominaga, H.; Yabutani, H.; Kawasaki, N. Removal of fluoride ions from water by adsorption onto carbonaceous materials produced from coffee grounds. *J. Oleo Sci.* **2011**, *60*, 619–625. [CrossRef]

Communication

Synthesis of a Novel Water-Soluble Polymer Complexant Phosphorylated Chitosan for Rare Earth Complexation

Yuxin Chen, Yujuan Chen, Dandan Lu and Yunren Qiu *

School of Chemistry and Chemical Engineering, Central South University, Changsha 410083, China; yxchen1230@163.com (Y.C.); xk1257713404@163.com (Y.C.); ludandan0707@163.com (D.L.)
* Correspondence: csu_tian@csu.edu.cn

Abstract: Combining the characteristics of rare earth extractants and water-soluble polymer complexants, a novel complexant phosphorylated chitosan (PCS) was synthesized by Kabachnik–Fields reaction with alkalized chitosan, dimethyl phosphonate, and formaldehyde as raw materials and toluene-4-sulfonic acid monohydrate (TsOH) as catalyst. The complexation properties of PCS and poly (acrylic acid) sodium (PAAS) for lanthanum ions in the solution were compared at the same pH and room temperature. In addition, the frontier molecular orbital energies of polymer–La complexes were calculated by the density functional theory method, which confirmed the complexation properties of the polymers to rare earths. The results indicate that the PCS has better water solubility compared with chitosan and good complex ability to rare earths, which can be used for rare earth separation by the complexation–ultrafiltration process.

Keywords: polymer complexant; rare earth separation; phosphorylated chitosan; complexation–ultrafiltration

Citation: Chen, Y.; Chen, Y.; Lu, D.; Qiu, Y. Synthesis of a Novel Water-Soluble Polymer Complexant Phosphorylated Chitosan for Rare Earth Complexation. *Polymers* **2022**, *14*, 419. https://doi.org/10.3390/polym14030419

Academic Editors: Irene S. Fahim, Ahmed K. Badawi and Hossam E. Emam

Received: 28 November 2021
Accepted: 17 January 2022
Published: 21 January 2022

Publisher's Note: MDPI stays neutral with regard to jurisdictional claims in published maps and institutional affiliations.

Copyright: © 2022 by the authors. Licensee MDPI, Basel, Switzerland. This article is an open access article distributed under the terms and conditions of the Creative Commons Attribution (CC BY) license (https://creativecommons.org/licenses/by/4.0/).

1. Introduction

Rare earth elements, known as industrial vitamins, are wildly used in military industry, electronics, chemical industry, metallurgy, and other fields [1]. The efficient separation of rare earths is the key to the development of the rare earth industry. At present, solvent extraction is the main method for the rare earth separation in the industry [2,3]. Although a series of novel extractants containing multiple different coordination functional groups have been developed, such as α-aminophosphonate extractant [4,5], to improve the efficiency of concentration and separation, inevitable disadvantages still exist, such as low separation efficiency of single stage, large amount of consumption of acid and alkali, and the loss of extractants [6–8].

Complexation–ultrafiltration (C–UF) uses water soluble polymer complexants, of which the molecular weight is greater than the molecular weight cut-off (MWCO) of the ultrafiltration membrane. In addition, the complexants have an amount of nitrogen, sulfur, phosphorus or carboxyl groups in order that they can bond with the metal ions. Moreover, the complexed ions can be rejected by the ultrafiltration membrane, while the free ions can permeate the UF membrane [9–12]. C–UF has been widely used for the concentration and separation of heavy metal ions, and the used complexants are poly (acrylic acid) sodium (PAAS), copolymer of maleic acid and acrylic acid (PMA), chitosan (CS), etc. [13,14]. However, there are few reports regarding the separation of rare earths by C–UF. In addition, after primary exploration, it was found that the rare earth separation using PAAS and PMA as complexants was not satisfactory due to the poor complexation performance of carboxyl groups to rare earths. Therefore, the design and synthesis of novel complexants for efficient separation of rare earths is of great significance, according to the characteristics of rare earth extractants and the polymer complexants.

Chitosan is a type of non-toxic and harmless biopolymer obtained from the natural polymer chitin. It is used in the field of water treatment due to its richness in amino and hydroxyl groups and good adsorption capacity to metals [15]. However, chitosan has some defects, such as low adsorption selectivity for metals and poor solubility, which limit its application. Therefore, a modified chitosan with excellent performance is often obtained by functional group modification to expand the application range of chitosan [16]. Although there are a large number of hydroxyl groups in the structure of chitosan, they can provide fewer modified active sites due to the steric hindrance effect and the charge carried by the groups, while most of the amino groups in chitosan are in a free state and their reaction activity is significantly greater than hydroxyl sites [17].

Herein, combining the structural characteristics of rare earth extractants and chitosan, we proposed the grafting of phosphoryl groups onto amino groups of chitosan to obtain α-substituted phosphorylated chitosan (PCS), which can be used as a novel complexant for rare earth concentration and separation by C–UF. The PCS was characterized by Fourier transform infrared (FTIR), X-ray diffraction (XRD), and scanning electron microscopy (SEM), etc. The results show that the PCS has better water solubility compared with chitosan and good complex ability to rare earths. Moreover, it has a good application prospect in the field of rare earth separation.

2. Materials and Methods

2.1. Materials and Membrane

Chitosan with the average molecular weight of 300 kDa and deacetylation degree of 80% was purchased from Shanghai Macklin Biochemical Co., Ltd. (Shanghai, China). Poly (acrylic acid) sodium (PAAS, MW = 250 kDa) was provided by Tianjin Guangfu Fine Chemical Research Institute (Tianjin, China). Dimethyl phosphonate was obtained from Shanghai Yi En Chemical Technology Co., Ltd. (Shanghai, China). Toluene-4-sulfonic acid monohydrate (TsOH) and sodium hydroxide were supplied by Sinopharm Chemical Reagent Co., Ltd., Shanghai, China. Lanthanum oxide (La_2O_3), dehydrated alcohol, acetone, acetic acid, and 37~40% of formaldehyde solution (all provided by Chengdu Chron Chemicals Co., Ltd., Chengdu, China) were used. MD44 dialysis bags (average diameter of 28 mm) with the retention range from 8 to 14 kDa were bought from Shanghai Leibusi Company (Shanghai, China). All of the chemical reagents in this experiment were analytical pure.

2.2. Methods

2.2.1. Synthesis of the PCS

To weaken the intermolecular force of chitosan, 10 g of chitosan was added into a 250 mL beaker containing 50 g of sodium hydroxide and 150 mL of ultrapure water under stirring at room temperature. After mixing evenly, it was placed in the refrigerator for 8 days. The alkalized chitosan was obtained after thawing and removing the alkaline solution by suction filtration.

Thereafter, 1 g of alkalized chitosan dissolved in 150 mL of 1% acetic acid solution and 4 mL of dimethyl phosphonate were mixed in a three-port flask under stirring. Then, the temperature was raised to 65 °C. In addition, 10 mL of formaldehyde solution (37~40%) and 0.1 g of TsOH as reaction catalyst were added to the system. After refluxing for 5 h, the reaction solution was transferred to a dialysis bag at room temperature for 48 h for the removal of low molecular weight soluble impurities to obtain the high purity product. After vacuum distillation, the mixture of absolute ethanol (200 mL) and acetone (200 mL) was added to the concentrated solution to precipitate. Then, after filtering, the precipitates were washed with absolute ethanol and acetone for several times, and vacuum dried at 50 °C to obtain the pure polymer phosphorylated chitosan (PCS).

2.2.2. Procedure of C–UF

The ultrafiltration membrane device used in this study was the polyvinyl butyral (PVB) hollow fiber membrane with molecular weight cut-off (MWCO) of 20 kDa [18]. Initially, the complexants were pretreated using diafiltration to remove the small polymers. Then, the complex solution of polymer and rare earths at different P/RE values (mass ratio of complexant to rare earth ions) was prepared. After contacting for about 2 h, the obtained materials were injected into the hollow fiber membrane module by a peristaltic pump at the feed rate of 30 L h^{-1}. The permeates were collected for 30 s after 5 min and further analyzed to determine the rare earth ions and polymer concentration. All of the experiments were performed at 25 °C.

2.3. Characterization

The functional groups between the raw materials and products were characterized by Fourier transform infrared (NicoletiS50, Thermo Fisher Scientific, Waltham, MA, USA). The synthesized PCS and CS were dissolved in deuterium oxide (D$_2$O) to prepare NMR samples, and the ^{1}H NMR of samples was detected with a 500 MHz BRUKER spectrometer. The crystallinity of chitosan and PCS was tested by X-ray diffraction (Advance D8). The morphology of the products was observed by SEM on a JSM-7610F field emission scanning electron microscope.

3. Results and Discussion

Scheme 1 shows the synthesis mechanism for the PCS. The amino group on chitosan was condensed with formaldehyde under the catalysis to obtain methylene-amine positive ions, and then continued to undergo an electrophilic substitution reaction with the active hydrogen of dimethyl phosphonate to obtain the product. The solubility of PCS is higher than 2 mg mL^{-1} in water, which is significantly higher than chitosan.

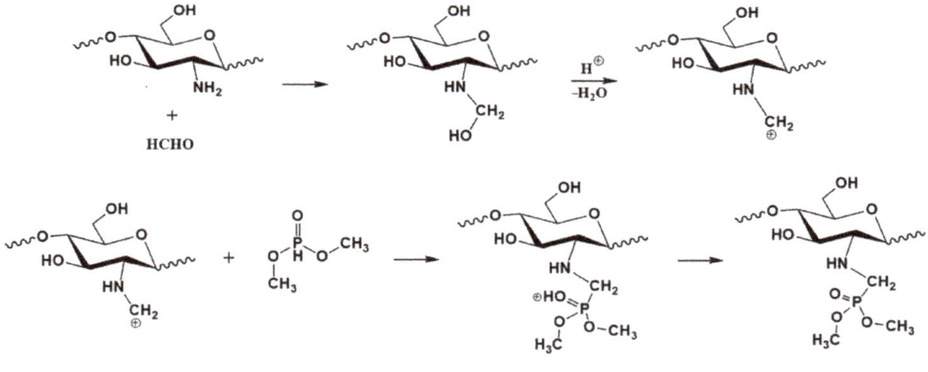

Scheme 1. Synthesis mechanism for the PCS.

Figure 1 shows the Fourier transform infrared (FTIR) absorption spectra of the PCS and chitosan. The chitosan powder exhibits characteristic absorption bands of methylene group antisymmetric stretching vibration at 2930 cm^{-1}, C–N stretching vibration (coupled with N–H bending) at 1375 cm^{-1}, N–H bending (coupled with C–N stretching) vibration of amide groups at 1639 and 1592 cm^{-1}, and C–O stretching vibration at 1061 cm^{-1}. The broad band observed at 3200–3500 cm^{-1} is attributed to the intermolecular and intramolecular hydrogen bonding of –NH$_2$ and –OH stretching vibration of chitosan, which is O–H stretching at 3440 cm^{-1}, as well as N–H asymmetric stretching at 3387 and 3284 cm^{-1} [19]. In the case of PCS, characteristic bands of P=O stretching vibration at 1202 cm^{-1} and P–O–C antisymmetric stretching vibration at 1018 cm^{-1} [20] were shown, which illustrate that the phosphoryl groups are successfully grafted onto the chitosan structure. In addition, the bond at 1375 cm^{-1} (C–N–H bending) of PCS is clearly smaller than the chitosan, suggesting that the grafting reaction mainly occurs on the amino groups.

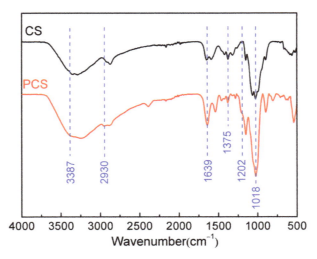

Figure 1. FTIR spectra of the PCS and chitosan.

The ^{1}H NMR spectra of the unmodified chitosan and PCS are shown in Figure 2. In the spectrum of chitosan, the signals in the range of 3.36–3.81 ppm are the resonance of H2–H6 protons in the heterocycle. In the case of PCS, the chemical shift at 3.11 ppm represents the proton signal of methylene grafted to the amino position. In addition, the emerging signal at 3.76 ppm is ascribed to the methyl group on the phosphoryl group, which makes it clear that the phosphoryl group is only grafted on the methylene.

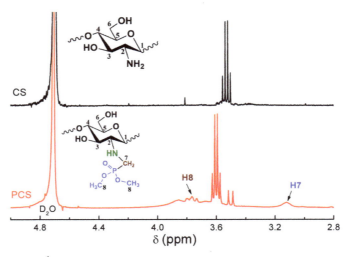

Figure 2. ^{1}H NMR spectra of the PCS and CS.

The crystallinity of chitosan and PCS is compared by X-ray diffraction (XRD) analysis, as shown in Figure 3. Intramolecular and intermolecular hydrogen bonds were found in the structure of chitosan. Therefore, chitosan shows strong crystallinity and poor solubility [21]. In addition, chitosan has a main crystalline peak at 2θ = 20° and an amorphous peak around 2θ = 10°. After modification, both peaks of chitosan are significantly reduced and tend to be flat. As a result, the grafted substituents destroyed the original crystalline structure of chitosan, mainly the hydrogen bond at the amino position, and the molecule almost lost its crystallization ability. This is the main reason for the good water solubility of the PCS.

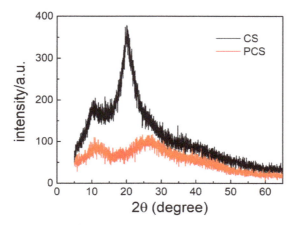

Figure 3. X-ray diffraction (XRD) patterns of the PCS and CS.

Figure 4a,b shows SEM images of raw materials of chitosan and the PCS at the same magnification, respectively. Compared with chitosan, the surface of PCS is rough and porous, and its cross section shows a porous morphology, which is related to the good water solubility of the PCS. The element distribution of PCS is analyzed by SEM–EDS mapping (Figure 4c,d). The results show that phosphorus entered the chitosan structure, and the C, O, N, and P elements of PCS are evenly distributed.

Figure 4. SEM images of chitosan (**a**) and the PCS (**b**). SEM–EDS mapping images of the PCS (**c**,**d**).

In addition, the complexation ability to rare earth ions of PCS was studied. The ultrafiltration membrane was used to reject the complex formed by the complexants and rare earths, and the rejection effect was expressed by the rejection rate (R). The complexation ability of PCS and PAAS to lanthanum ions (La^{3+}) at pH = 7.0 is shown in Figure 5. The data are the average of three experimental results. The PCS can basically complexate 98% of La^{3+} when P/Re is 10, while PAAS can only complexate 82% of lanthanum ions in the solution when P/Re is 20, which indicates that the PCS has better complexation ability to rare earths than PAAS.

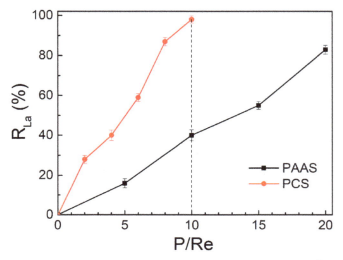

Figure 5. Comparison of complexation ability of the PCS and PAAS to La^{3+} at pH = 7.0.

By analyzing the molecular structure of rare earth extractants, it is concluded that organophosphorus extractants are generally coordinated with metal ions through the P=O group, while amine extractants are coordinated with metal ions through nitrogen atoms, and the extraction order of rare earths by these two extractants is opposite [22–24]. The α-aminophosphonate compounds containing both phosphoryl and amino groups have two coordination atoms, which can form a more stable structure with rare earths in a wide range of acidity, and have better complexation and separation properties to rare earths [25]. Therefore, it is not difficult to understand that the P=O and nitrogen atom of the PCS can coordinate with rare earths at the same time to form a stable five-membered ring structure, while PAAS only complexes with rare earths through carboxyl groups. This is the reason why the PCS shows good rare earth complexation ability. Moreover, the adsorption properties of PCS and PAAS with La(III) have been calculated with the DMol3 electronic structure code (Materials Studio 2019) founded on the density functional theory GGA/BLYP/DNP (Figure 6a,b). The energy gap ($\Delta E = E_{LUMO} - E_{HOMO}$) represents the chemical activities of the molecule. The ΔE of PCS–La and PAA–La complexes is 0.047 and 0.032 au, respectively. This confirms that the complexation ability of the PCS to La is stronger than PAAS at the molecular level.

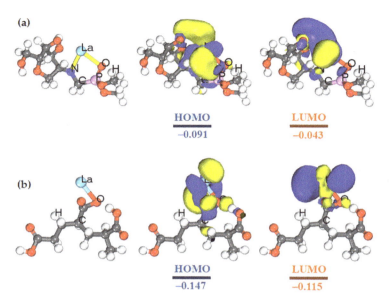

Figure 6. Isosurfaces and energy of the stable configuration (**left**), HOMOs (**middle**), and LUMOs (**right**) of (**a**) PCS–La, (**b**) PAA–La complexes.

4. Conclusions

In conclusion, phosphorylated chitosan (PCS) with good water solubility is successfully designed and synthesized. Benefiting from the amino and phosphoryl groups contained in the structure of PCS, the PCS shows excellent complexing performance to rare earth ions, which provides a good application prospect as an excellent rare earth complexant. Moreover, the green and convenient synthesis process represents a potential possibility for large-scale industrial production of the PCS in the future.

Author Contributions: Y.C. (Yuxin Chen) carried out the experiments, analyzed data, wrote the manuscript, and organized the documents; Y.C. (Yujuan Chen) and D.L. checked and modified the manuscript; Y.Q. directed and supervised the experiments, checked and improved the manuscript. All authors have read and agreed to the published version of the manuscript.

Funding: This research was funded by the National Natural Science Foundation of China, grant number 22178392, Fundamental Research Funds for the Central Universities of Central South University, grant number 2021zzts0533, and College Student Innovation Program of Central South University, grant number S2021105330343.

Institutional Review Board Statement: Not applicable.

Informed Consent Statement: Not applicable.

Data Availability Statement: Not applicable.

Conflicts of Interest: The authors declare no conflict of interest.

References

1. Balaram, V. Rare earth elements: A review of applications, occurrence, exploration, analysis, recycling, and environmental impact. *Geosci. Front.* **2019**, *10*, 1285–1303. [CrossRef]
2. Costa, T.; Silva, M.; Vieira, M. Recovery of rare-earth metals from aqueous solutions by bio/adsorption using non-conventional materials: A review with recent studies and promising approaches in column applications. *J. Rare Earths* **2020**, *38*, 339–355. [CrossRef]
3. Xie, F.; Zhang, T.; Dreisinger, D.; Doyle, F. A critical review on solvent extraction of rare earths from aqueous solutions. *Miner. Eng.* **2014**, *56*, 10–28. [CrossRef]

4. Lu, Y.; Zhang, Z.; Li, Y.; Liao, W. Extraction and recovery of cerium(IV) and thorium(IV) from sulphate medium by an α-aminophosphonate extractant. *J. Rare Earths* **2017**, *35*, 34–40. [CrossRef]
5. Zhao, Q.; Zhang, Z.; Li, Y.; Bian, X.; Liao, W. Solvent extraction and separation of rare earths from chloride media using α-aminophosphonic acid extractant HEHAMP. *Solvent Extr. Ion Exch.* **2018**, *36*, 136–149. [CrossRef]
6. Jyothi, R.K.; Thenepalli, T.; Ji, W.; Parhi, P.K.; Lee, J. Review of rare earth elements recovery from secondary resources for clean energy technologies: Grand opportunities to create wealth from waste. *J. Clean. Prod.* **2020**, *267*, 122048. [CrossRef]
7. Hidayah, N.N.; Abidin, S.Z. The evolution of mineral processing in extraction of rare earth elements using liquid-liquid extraction: A review. *Miner. Eng.* **2018**, *121*, 146–157. [CrossRef]
8. Xiao, Y.; Long, Z.; Huang, X.; Feng, Z.; Cui, D.; Wang, L. Study on non-saponification extraction process for rare earth separation. *J. Rare Earths* **2013**, *31*, 512–516. [CrossRef]
9. Desai, K.R.; Murthy, Z.V.P. Removal of silver from aqueous solutions by complexation–ultrafiltration using anionic polyacrylamide. *Chem. Eng. J.* **2012**, *185*, 187–192. [CrossRef]
10. Barakat, M.A.; Schmidt, E. Polymer-enhanced ultrafiltration process for heavy metals removal from industrial wastewater. *Desalination* **2010**, *256*, 90–93. [CrossRef]
11. Tang, S.; Qiu, Y. Removal of copper(II) ions from aqueous solutions by complexation–ultrafiltration using rotating disk membrane and the shear stability of PAA–Cu complex. *Chem. Eng. Res. Des.* **2018**, *136*, 712–720. [CrossRef]
12. Gao, J.; Qiu, Y.; Ben, H.; Zhang, Q.; Zhang, X. Treatment of wastewater containing nickel by complexation- ultrafiltration using sodium polyacrylate and the stability of PAA-Ni complex in the shear field. *Chem. Eng. J.* **2018**, *334*, 1878–1885. [CrossRef]
13. Tang, S.; Qiu, Y. Selective separation and recovery of heavy metals from electroplating effluent using shear-induced dissociation coupling with ultrafiltration. *Chemosphere* **2019**, *236*, 124330. [CrossRef] [PubMed]
14. Zhou, H.; Qiu, Y.; Le, H. Recovery of metals and complexant in wastewater by shear induced dissociation coupling with ultrafiltration. *J. Appl. Polym. Sci.* **2019**, *137*, 48–54. [CrossRef]
15. Zemmouri, H.; Drouiche, M.; Sayeh, A.; Lounici, H.; Mameri, N. Coagulation Flocculation Test of Keddara's Water Dam Using Chitosan and Sulfate Aluminium. *Procedia Eng.* **2012**, *33*, 254–260. [CrossRef]
16. Wang, J.; Wang, L.; Yu, H.; Abdin, Z.; Chen, Y.; Chen, Q.; Zhou, W.; Zhang, H.; Chen, X. Recent progress on synthesis, property and application of modified chitosan: An overview. *Int. J. Biol. Macromol.* **2016**, *88*, 333–344. [CrossRef] [PubMed]
17. Izumrudov, V.; Volkova, I.; Grigoryan, E.; Gorshkova, M. Water-soluble nonstoichiometric polyelectrolyte complexes of modified chitosan. *Polym. Sci.* **2011**, *53*, 281–288. [CrossRef]
18. Xu, J.; Tang, S.; Qiu, Y. Pretreatment of poly (acrylic acid) sodium by continuous diafiltration and time revolution of filtration potential. *J. Cent. South Univ.* **2019**, *26*, 577–586. [CrossRef]
19. Liu, X.; Zhang, L. Removal of phosphate anions using the modified chitosan beads: Adsorption kinetic, isotherm and mechanism studies. *Powder Technol.* **2015**, *277*, 112–119. [CrossRef]
20. Zhao, D.; Xu, J.; Wang, L.; Du, J.; Dong, K.; Wang, C.; Liu, X. Study of two chitosan derivatives phosphorylated at hydroxyl or amino groups for application as flocculants. *J. Appl. Polym. Sci.* **2019**, *125*, 299–305. [CrossRef]
21. Yang, T.; Huang, N.; Meng, L. Chitosan modified by nitrogen-containing heterocycle and its excellent performance for anhydrous proton conduction. *RSC Adv.* **2013**, *3*, 4341–4349. [CrossRef]
22. Lu, Y.; Wei, H.; Zhang, Z.; Li, Y.; Liao, W. Selective extraction and separation of thorium from rare earths by a phosphorodiamidate extractant. *Hydrometallurgy* **2016**, *163*, 192–197. [CrossRef]
23. Wei, H.; Li, Y.; Zhang, Z.; Xue, T.; Kuang, S.; Liao, W. Selective Extraction and Separation of Ce (IV) and Th (IV) from RE(III) in Sulfate Medium using Di(2-ethylhexyl)-N-heptylaminomethylphosphonate. *Solvent Extr. Ion Exch.* **2017**, *35*, 117–129. [CrossRef]
24. Liao, W.; Zhang, Z.; Kuang, S.; Wu, G.; Li, Y. Selective extraction and separation of Ce(IV) from thorium and trivalent rare earths in sulfate medium by an α-aminophosphonate extractant. *Hydrometallurgy* **2017**, *167*, 107–114.
25. Zhao, Q.; Li, Y.; Kuang, S.; Zhang, Z.; Bian, X.; Liao, W. Synergistic extraction of heavy rare earths by mixture of α-aminophosphonic acid HEHAMP and HEHEHP. *J. Rare Earths* **2019**, *37*, 422–428. [CrossRef]

Article

Continuous Fixed-Bed Column Studies on Congo Red Dye Adsorption-Desorption Using Free and Immobilized *Nelumbo nucifera* Leaf *Adsorbent*

Vairavel Parimelazhagan [1,*], Gautham Jeppu [1] and Nakul Rampal [2]

[1] Department of Chemical Engineering, Manipal Institute of Technology, Manipal Academy of Higher Education, Manipal 576104, Udupi District, India; gautham.jeppu@gmail.com
[2] Department of Chemical Engineering and Biotechnology, University of Cambridge, Philippa Fawcett Drive, Cambridge CB3 0AS, UK; nakulrampal@gmail.com
* Correspondence: pvairavel@gmail.com; Tel.: +91-903-627-0978

Abstract: The adsorption of Congo red (CR), an azo dye, from aqueous solution using free and immobilized agricultural waste biomass of *Nelumbo nucifera* (lotus) has been studied separately in a continuous fixed-bed column operation. The *N. nucifera* leaf powder adsorbent was immobilized in various polymeric matrices and the maximum decolorization efficiency (83.64%) of CR occurred using the polymeric matrix sodium silicate. The maximum efficacy (72.87%) of CR dye desorption was obtained using the solvent methanol. Reusability studies of free and immobilized adsorbents for the decolorization of CR dye were carried out separately in three runs in continuous mode. The % color removal and equilibrium dye uptake of the regenerated free and immobilized adsorbents decreased significantly after the first cycle. The decolorization efficiencies of CR dye adsorption were 53.66% and 43.33%; equilibrium dye uptakes were 1.179 mg g^{-1} and 0.783 mg g^{-1} in the third run of operation with free and immobilized adsorbent, respectively. The column experimental data fit very well to the Thomas and Yoon–Nelson models for the free and immobilized adsorbent with coefficients of correlation R$^2 \geq$ 0.976 in various runs. The study concludes that free and immobilized *N. nucifera* can be efficiently used for the removal of CR from synthetic and industrial wastewater in a continuous flow mode. It makes a substantial contribution to the development of new biomass materials for monitoring and remediation of toxic dye-contaminated water resources.

Keywords: Congo red dye; *Nelumbo nucifera* leaf adsorbent; fixed-bed column; immobilization; breakthrough curve; desorption

Citation: Parimelazhagan, V.; Jeppu, G.; Rampal, N. Continuous Fixed-Bed Column Studies on Congo Red Dye Adsorption-Desorption Using Free and Immobilized *Nelumbo nucifera* Leaf *Adsorbent*. *Polymers* **2022**, *14*, 54. https://doi.org/10.3390/polym14010054

Academic Editors: Ahmed K. Badawi, Hossam E. Emam and Irene S. Fahim

Received: 13 November 2021
Accepted: 6 December 2021
Published: 24 December 2021

Publisher's Note: MDPI stays neutral with regard to jurisdictional claims in published maps and institutional affiliations.

Copyright: © 2021 by the authors. Licensee MDPI, Basel, Switzerland. This article is an open access article distributed under the terms and conditions of the Creative Commons Attribution (CC BY) license (https://creativecommons.org/licenses/by/4.0/).

1. Introduction

Nowadays, environmental pollution is having an adverse effect on humans and ecosystems. The presence of toxic dye contaminants in aqueous streams, resulting from the discharge of untreated dye containing effluents into water bodies, is one of the essential global environmental problems [1]. Synthetic dyes are widely used in textile, leather, paper, food, pigments, plastics, and cosmetic industries to color their final products, and they consume substantial volumes of water in the process [2]. Rapid industrialization has resulted in the increased disposal of colored effluent into the aquatic environment. Color is the most pervasive contaminant amongst the various pollutants of wastewaters [3]. The presence of dye in effluent even at very low concentration is highly observable and undesirable [4]. The amount of dye lost is dependent upon the types of dyes, the method of application and the depth of shade required [5]. The synthetic dye Congo red (CR) is a polar diazo anionic dye that shows a high affinity for cellulose fibers and is widely used in textile processing industries. It is considered as highly toxic due to its metabolism to benzidine, a human carcinogen [6]. The discharge of CR dye effluents into natural resources has led to many problems in human beings, such as cancer, skin allergy, eye and gastrointestinal irritation [7]. Even a low concentration of CR dye causes various harmful effects, such as difficulties in breathing, diarrhea, nausea, vomiting, abdominal and chest pain, severe headache, etc. [8]. It is used as a laboratory aid for testing free hydrochloric acid in gastric contents, in the diagnosis of amyloidosis, as an indicator of pH, and also as a histological stain for amyloid [9]. The industrial effluents containing CR dye must be treated to bring down its concentration to permissible and bearable levels before discharging into water bodies [10]. Therefore, the removal of CR from industrial effluents is of great importance to present-day researchers. Various conventional treatment methods, such as chemical coagulation, electrochemical oxidation, photo-catalytic degradation, ozonation, sonication, Fenton reagent method, membrane separation, and biological degradation, have been used to treat textile effluents with varying degrees of success in dye removal [11]. However, these technologies have several disadvantages, such as high capital and operating costs, inefficiency in dye removal, complexity of the treatment processes, and the need for the appropriate treatment of residual dye sludge [12]. Amongst the above treatment methods, adsorption is one of the most widely adopted techniques because of its low initial investment, rapid process, efficacy, reliability, versatility, greater flexibility in operation and simplicity of design of equipment, applicability on a large scale, and the easy and safe recovery of the adsorbent as well as adsorbate materials [13]. Moreover, the adsorption-based processes permit the removal of toxic substances from effluent without producing any by-products as compared to conventional treatment methods [14]. Liquid-phase adsorption on powdered activated carbon is a widely used method for the removal of color from wastewater due to its better efficacy and excellent adsorption capacity [15]. However, its widespread use is still limited in large-scale applications, mainly because of the high cost of the adsorbent, besides the difficulty in regeneration and final disposal of the spent activated carbon [16]. Numerous alternative materials from plant and agricultural by-products have been studied by various researchers for adsorbing dyes from aqueous solutions. These adsorbents include wheat bran [17], papaya seeds [18], peanut husks [19], coffee husks [20], guava leaves [21], neem leaves [22], and *malesianus* leaves [23]. These waste materials are cost-effective adsorbents, available in abundant quantities, are highly effective, and the regeneration of these adsorbents may not be necessary, unlike activated carbon [24]. However, the use of agricultural waste remains limited due to the insufficient documentation of the treatment of real-time waste water [13].

Most of these batch adsorption studies focus on the use of adsorbent material in powdered form. The use of powdered adsorbent (plant and agricultural waste material) for the successful removal of toxic dye in large-scale process applications is not practical because of its smaller particle size, low mechanical strength, low density and poor rigidity. The regeneration of the adsorbent after adsorption becomes difficult, and hence results in loss of adsorbent after regeneration. These problems may be rectified by immobilizing the adsorbent in a polymeric matrix, which is used as a supporting material [25]. An immobilized adsorbent has the advantages of a better mechanical strength, offering a higher resistance to various chemical compounds, minimal clogging in continuous-flow systems, easier liquid–solid separation, and ease of regeneration and reuse as compared to a free adsorbent [26]. Batch adsorption studies are not sufficient when designing a treatment system for continuous operation. Column studies are important to obtaining the model parameters required for the design of continuous fixed-bed adsorbers in the large-scale treatment of wastewater [27]. A large volume of effluent can be treated continuously using a fixed quantity of adsorbent in an appropriately designed column, resulting in a better quality of the effluent. The column becomes saturated as the available binding sites are occupied by the target adsorbate molecules [28]. A fixed-bed column has several advantages, such as simplicity of operation, high yield, and easy scale-up [29]. However, column studies on the adsorption of CR dye from wastewater using immobilized *N. nucifera* leaf adsorbent is an area that has not been explored much. To the best of our knowledge, there has been practically no work reported describing the potential of using immobilized *Nelumbo nucifera* leaf powder (NNLP) as an adsorbent in continuous fixed-bed operations for the removal of CR dye from aqueous solutions. Therefore, the present research work focusses on the removal of CR from a synthetic and industrial effluent using free and immobilized *N. nucifera* leaf adsorbents separately in a continuous fixed-bed adsorption operation. *N. nucifera* is commonly grown in subtropical and temperate regions [24]. Its leaf, as an agricultural waste material, contains abundant floristic fiber, protein, lignin, cellulose, hemicellulose, flavonoids, alkaloids and major functional groups (hydroxyl, methyl, alkyne, carbonyl and carboxylate), which are responsible for the increase in adsorption of the various toxic pollutants [30]. The *N. nucifera* leaves are cheap, widely available in India, and can easily be cultivated from seeds or vegetative propagation. The objectives of this present study are to investigate the decolorization and desorption efficiency of free and immobilized NNLP adsorbent separately in various runs. The column experimental data are fitted to various mathematical models to predict the breakthrough curve (BTC) and to evaluate the column capacity and kinetic constants of the models.

2. Materials and Methods

2.1. N. nucifera Leaf Powder Preparation, Chemical Reagents and Analytical Methods

The mature *N. nucifera* leaves were collected from Bastar district in Chhattisgarh State, India. The leaves were dried under sunlight to remove the moisture and ground to fine powder using a pulverizer. The powdered material was thoroughly washed with distilled water several times to remove all the dirt particles and other impurities. Then, the washed material was dried in a hot-air oven at a temperature of 338 K for 8 h, ground, and screened to obtain particles <100 µm in size [31]. The fine powdered material was stored in an airtight plastic container for further use in adsorption experiments. Analytical grade anionic diazo acid dye Congo red with 99.8% purity was supplied by Sigma-Aldrich, Bengaluru, India. All other chemicals and reagents used throughout this study were of analytical grade and were taken from Merck, Mumbai, India. A stock solution of 1000 mg L^{-1} CR dye was prepared by dissolving 1 g of dye powder in 1000 mL of deionized water. Experimental solutions of the required initial dye concentrations ranging from 15 to 50 mg L^{-1} were made by further diluting the stock solution with pH-adjusted deionized water by adding 0.1 N HCl or 0.1 N NaOH. After dilution, the final pH of the dye solution was measured and was found to be 6. The adsorption experiments are carried out at pH 6 due to the stability of CR dye color [17,27]. The powdered material was weighed using a digital weighing balance (Citizen, Mumbai, India). The pH of the dye solution was measured by a digital pH-meter (Systronics 335, Bengaluru, India). After adsorption, the unknown residual CR dye concentration was determined by measuring the absorbance at 498 nm (λ_{max}) using a pre-calibrated double-beam UV/visible spectrophotometer (Shimadzu UV-1800, Tokyo, Japan). The prepared free NNLP adsorbent was characterized by particle size, surface area, and pore volume analyses.

2.2. Immobilization of NNLP Adsorbent for CR Dye Adsorption

The various matrices used for the immobilization of NNLP adsorbent are calcium alginate gel, polyvinyl alcohol, polysulfone and sodium silicate [26,32].

2.2.1. Immobilization of the NNLP Adsorbent in Calcium Alginate

A slurry of 2% (w/v) sodium alginate was prepared in hot distilled water at 333 K for 1 h, resulting in a transparent and viscous solution. After cooling, varying quantities (1–10% w/v) of NNLP adsorbent were added and stirred for 30 min [26]. For the polymerization and preparation of beads, the alginate–NNLP adsorbent slurry was extruded drop by drop into a cold, sterile 0.05 M calcium chloride solution with the help of a sterile 12 mL syringe with a 2 mm inner diameter [27]. The water-soluble sodium alginate was converted into water-insoluble calcium alginate entrapped on NNLP adsorbent beads via treatment with calcium chloride solution. The beads were hardened by re-suspending them into a fresh cold 0.05 M calcium chloride solution for 24 h with gentle agitation [33].

2.2.2. Immobilization of the NNLP Adsorbent in Polyvinyl Alcohol

A polyvinyl alcohol–sodium alginate slurry in a weight ratio of 2:1 was prepared in hot distilled water at 333 K for 1 h in a beaker. Various quantities (1–10% (w/v)) of NNLP adsorbent were added to the slurry and the mixtures were stirred for 30 min. Beads were prepared as mentioned above in Section 2.2.1 for polyvinyl alcohol–alginate–NNLP adsorbent slurry in 4% (w/v) cold calcium chloride solution. The prepared beads were re-suspended in a fresh cold 4% (w/v) calcium chloride solution for 6 h with gentle agitation to increase the hardness. After 6 h of stabilization, the beads were subjected to three cycles of freezing at <275 K and thawing at 301 K to get spherical beads [26,27].

2.2.3. Immobilization of the NNLP Adsorbent in Polysulfone

A 10% (w/v) polysulfone solution was prepared in dimethylformamide. Various amounts of the NNLP adsorbent (1–10% (w/v)) were mixed with the polysulfone slurry. Beads were prepared by following the above procedure as mentioned in Section 2.2.1. The slurry was polymerized in distilled water and the immobilized beads were cured for 16 h in distilled water [26,27].

Various immobilized beads were washed with distilled water and kept in an oven at 323 K for 24 h to remove moisture. Finally, the resultant bead diameter was determined experimentally.

2.2.4. Immobilization of the NNLP Adsorbent in Sodium Silicate

A 6% (w/v) sodium silicate solution was prepared with distilled water in an Erlenmeyer flask. The sodium silicate solution was added dropwise into 15 mL of 5% (v/v) sulfuric acid until the pH reached 2. Various quantities of NNLP adsorbent powder (1–10% (w/v)) were dissolved in 2% (v/v) acetic acid. Then, 50 mL of this adsorbent dissolved solution was added drop by drop to the silicate solution. Next, the solution was mixed for 15 min. The polymeric gel was made by the addition of sodium silicate solution to reach pH 7. The resultant gel was purified using distilled water to eliminate sulfate ions. Then, the NNLP adsorbent-immobilized sodium silicate was air dried in an oven at 333 K and powdered using a mortar and pestle [32].

2.3. Batch Experiments with Synthetic CR Dye Wastewater Using Various Immobilized NNLP Adsorbents

The required amounts of different compositions of various immobilized NNLP adsorbents were added to the CR dye solutions of 250 mg L^{-1} concentration. The dye solutions were stirred at 150 rpm for 24 h at 301 K. The effluent samples were collected and analyzed for residual dye concentration in aqueous solution. A suitable polymeric matrix for the immobilization of NNLP adsorbent was selected based on the maximum adsorption efficiency, effectiveness factor and mechanical stability. The % color removal in dye solution and effectiveness factor of various immobilized NNLP are determined using Equations (1) and (2), respectively [26,34].

$$\% \text{ CR dye color removal} = \frac{(C_o - C_e) \times 100}{C_o} \quad (1)$$

$$\text{Effectiveness factor} = \frac{\% \text{ adsorption of CR dye by immobilized NNLP adsorbent}}{\% \text{ adsorption of CR dye by free NNLP adsorbent}} \quad (2)$$

where C_o and C_e are the initial and equilibrium concentrations of CR in the aqueous solution (mg L^{-1}).

2.4. Column Experiments for Removal of Color from Synthetic Dye Wastewater Using Free and Immobilized NNLP Adsorbent in Various Runs

Continuous-flow fixed-bed adsorption experiments were performed in a Perspex glass column of 2.1 cm inner diameter and 39 cm height for the removal of CR dye from aqueous solution at 301 K using free and immobilized NNLP adsorbent separately [35]. A schematic diagram of the fixed-bed column is given in Figure 1. A rubber cork of 1.5 cm was provided at the top and bottom of the column to support the inlet and outlet pipes. The column was packed with 1.8 cm of glass wool followed by glass beads, 1.5 mm in diameter, both at the top and bottom [36]. A known amount of free (3.72 g) and sodium silicate gel-immobilized NNLP (3.12 g) adsorbent was packed separately in the column to obtain a bed height of 2.5 cm in the first run. Before the experiment started, the adsorbent packed in the column was wetted with distilled water in the upward flow direction using a peristaltic pump (Enertech, India) to withdraw the trapped air between the particles [37]. The peristaltic pump was used to provide a uniform inlet at the bottom of the column. Steady state flow was maintained by measuring the flow rate at the bottom and top of the column. After setting up the column, the water was replaced with an aqueous CR dye solution of known concentration, 15 mg L^{-1}, at pH 6, at the required flow rate of 1 mL min^{-1}. The treated dye solution was collected at uniform time intervals from the top of the column with the same flow rate as the feed stream and its concentration was measured using a UV/visible spectrophotometer. The experiments were continued until the concentration of treated dye solutions reached the feed concentration of the adsorbate [38].

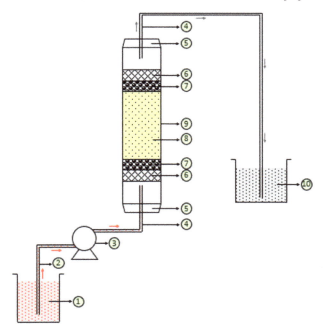

Figure 1. Schematic diagram of the fixed-bed adsorption column. (**1**) Influent CR dye solution; (**2**) silicon tube; (**3**) peristaltic pump; (**4**) glass tube; (**5**) rubber cork; (**6**) glass wool; (**7**) glass beads; (**8**) NNLP adsorbent; (**9**) glass column; (**10**) treated CR dye effluent.

2.5. Reusability of Free and Immobilized NNLP Adsorbents for CR Dye Adsorption in Column Studies

Desorption studies were performed using various desorbing agents, such as methanol, ethanol, butanol, acetone, and 1 M NaOH in separate batches to explore the possibility of the recovery of adsorbent. After desorption using a suitable desorbing reagent, the free and immobilized adsorbents were collected separately by centrifugation, washed with distilled water, and left to dry at 338 K for 8 h [24]. In order to establish the reusability of the adsorbent, consecutive adsorption–desorption cycles were repeated (three times) using the same adsorbent. The regenerated free and immobilized adsorbents were reused separately for three runs in column experiments. The regenerated adsorbent was added to a dye solution of concentration 15 mg L^{-1} at pH 6. The % color removal of each regenerated

NNLP adsorbent was tested and compared to the first run. The column experimental procedure was the same as mentioned before in Section 2.4.

2.6. Mathematical Description of Adsorption in a Continuous Fixed-Bed Column Study

The continuous adsorption of dye molecules in a fixed-bed column is an effective process for cyclic adsorption/desorption, as it makes use of the concentration difference, known to be the driving force for adsorption, and allows the more efficient utilization of the adsorbent capacity, resulting in a better-quality effluent. The performance of a fixed-bed column is described through the concept of the breakthrough curve (BTC), which is the plot of time vs. the ratio of effluent adsorbate concentration to inlet adsorbate concentration (C_t/C_o). The breakthrough time (t_b) and bed exhaustion time (t_E) were used to evaluate the BTCs. Effluent volume (V_{eff}) can be calculated as follows [39]:

$$V_{eff} = Q\, t_E \tag{3}$$

where Q is the volumetric flow rate (mL min^{-1}). The total quantity of dye adsorbed in the column (m_{ad}) is calculated from the area above the BTC multiplied by the flow rate and inlet adsorbate concentration. It is represented as

$$m_{ad} = \left[\int_0^{t_E} \left(1 - \frac{C_t}{C_o}\right) dt\right] C_o\, Q \tag{4}$$

The total amount of dye sent through the column (m_{total}) is calculated using the following equation [38]:

$$m_{total} = \frac{C_o\, Q\, t_E}{1000} \tag{5}$$

Total dye removal efficiency with respect to flow volume can be also found from the ratio of the quantity of total adsorbed dye to the total amount of dye sent to the column:

$$\text{Total dye removal}\,(\%) = \frac{m_{ad}}{m_{total}} \times 100 \tag{6}$$

Equilibrium adsorption capacity, q_e (mg g^{-1}), is defined as the ratio between the total quantity of solute adsorbed (m_{ad}) and the mass of free/immobilized dry adsorbent (W) packed in the column, expressed as [27]

$$q_e = \frac{m_{ad}}{W} \tag{7}$$

2.7. Mathematical Models Used for the Breakthrough Curve in Fixed-Bed Column Studies

Various mathematical models, such as the Adams–Bohart, bed depth service time (BDST), Thomas, Yoon–Nelson and Wolborska models, are used in this study to analyze the behavior of the free and immobilized NNLP adsorbent–adsorbate system separately, and to evaluate the kinetic model parameters at various runs [40]. These were then used for the design of the column adsorption process and to scale it up for industrial applications. Linear regression analysis was used to determine the kinetic constants [41].

2.7.1. Adams–Bohart Model

The Adams–Bohart model assumes that the rate of adsorption is proportional to the residual capacity of adsorbent and concentration of the adsorbed species, and it is given by the following linear Equation (8) [42]:

$$\ln\left(\frac{C_t}{C_o}\right) = K_{AB} C_o t - \frac{K_{AB} N_o Z}{U_o} \tag{8}$$

where t is the flow time (min), K_{AB} is the Adams–Bohart kinetic constant (L mg^{-1} min^{-1}), N_o is the maximum saturation concentration (mg L^{-1}), Z is the bed height (cm), and U_o is the superficial velocity (cm min^{-1}). The characteristic operational parameters K_{AB} and N_o can be determined from the slope and intercept of the plot of $\ln\left(\frac{C_t}{C_o}\right)$ vs. t.

2.7.2. Bed Depth Service Time (BDST) Model

The BDST model assumes that the rate of adsorption is controlled by the surface reaction between adsorbate and the unused capacity of the adsorbent. This model ignores the intraparticle diffusion and external film resistance, such that the adsorbate is accumulated onto the adsorbent particle surface directly. The service time, t, of a column in the BDST model is given by the following linear equation [43]:

$$t = \left(\frac{N_o Z}{C_o U_o}\right) - \left(\frac{1}{C_o K}\right) \ln\left(\frac{C_o}{C_t} - 1\right) \tag{9}$$

where K is the adsorption rate constant (L mg^{-1} min^{-1}). The linear plot of t vs. $\ln\left(\frac{C_o}{C_t} - 1\right)$ permits the determination of N_o and K from the intercept and slope of the plot.

2.7.3. Thomas Model

The Thomas model assumes that the external and pore diffusions are not the rate-controlling steps. The Langmuir kinetics of adsorption are valid, and this follows the pseudo-second-order reversible reaction kinetics with no axial dispersion. The linear form of the Thomas model can be expressed as follows [44]:

$$\ln\left(\frac{C_o}{C_t} - 1\right) = \frac{K_{Th} q_{oTh} W}{Q} - \frac{K_{Th} C_o V_{eff}}{Q} \tag{10}$$

where K_{Th} is the Thomas rate constant (L min^{-1} mg^{-1}) and q_{oTh} is the maximum solid phase concentration of the solute (mg g^{-1}). The plot of $\ln\left(\frac{C_o}{C_t} - 1\right)$ vs. V_{eff} yields a straight line for which the slope K_{Th} and intercept q_{oTh} can be estimated.

2.7.4. Yoon–Nelson Model

A linear form of the Yoon–Nelson model is expressed as follows [45]:

$$\ln\left(\frac{C_t}{C_o - C_t}\right) = k_{YN} t - \tau k_{YN} \tag{11}$$

where k_{YN} is the Yoon–Nelson rate constant (min^{-1}) and τ is the time required for 50% solute breakthrough (min). The values of k_{YN} and τ can be determined from the slope and intercept of the linear plot of $\ln\left(\frac{C_t}{C_o - C_t}\right)$ vs. sampling time, t. The adsorption capacity of the column in this model (q_{oYN}) can be determined as [25]:

$$q_{oYN} = \frac{C_o Q \tau}{1000 W} \tag{12}$$

2.7.5. Wolborska Model

The Wolborska model describes the adsorption dynamics at low adsorbate concentration BTCs. It assumes that the axial dispersion is negligible at a high flow rate of solution. The linearized form of the Wolborska model is expressed as [46]:

$$\ln\left(\frac{C_t}{C_o}\right) = \frac{\beta_a C_o t}{N_o} - \frac{\beta_a Z}{U_o} \tag{13}$$

where β_a is the kinetic coefficient of the external mass transfer (min^{-1}). A plot of $\ln\left(\frac{C_t}{C_o}\right)$ vs. t is expected to yield a linear curve in which the model constants β_a and N_o can be evaluated from the intercept and slope, respectively.

2.8. Physicochemical Characteristics of Textile Industrial Dye Effluent

The industrial CR dye effluent was collected from Bright Traders, Erode District, Tamil Nadu State, India. The physicochemical characteristics of real industrial CR dye effluent, such as pH, turbidity, total suspended solids, total dissolved solids, biological oxygen demand (BOD), chemical oxygen demand (COD), total alkalinity, total hardness, dissolved oxygen concentration, electrical conductivity, sulfates and chlorides, were analyzed according to standard operating procedures suitable for wastewater samples [47,48]. The optimized value of various process parameters, obtained from the central composite design (CCD), are used to analyze the physicochemical characteristics of the real textile industrial CR dye effluents in batch studies using free and sodium silicate gel-immobilized NNLP adsorbent separately [24]. The methodology used to analyze the physicochemical parameters of real industrial CR dye effluent are reported in Table 1.

Table 1. Methodology used to analyze the textile industrial Congo red (CR) dye effluent using standard operating procedures [47–51].

Sl. No	Physicochemical Parameters	Method/Instrument
1	pH	Digital pH meter, Systronics
2	Turbidity, NTU	Nephelometric turbidimeter, Systronics
3	Total suspended solids, mg L^{-1}	Gravimetric method, oven-drying at 378 K
4	Total dissolved solids, mg L^{-1}	Gravimetric method, oven-drying at 378 K
5	Biological oxygen demand, mg L^{-1}	Incubating the sample at 303 K for 5 days followed by titration
6	Chemical oxygen demand, mg L^{-1}	Closed reflux method
7	Total alkalinity, mg L^{-1}	Acid-base titration
8	Total hardness, mg L^{-1}	Complexometric titration
9	Dissolved oxygen concentration, mg L^{-1}	Dissolved oxygen meter, Systronics
10	Electrical conductivity, mS cm^{-1}	Conductivity meter, Digisun
11	Sulfates, mg L^{-1}	Titrimetric method
12	Chlorides, mg L^{-1}	Argentometric titration

3. Results and Discussions

3.1. Evaluation of a Suitable Matrix for Immobilization of the NNLP Adsorbent in Batch Studies

The Brunauer–Emmett–Teller (BET) surface area and pore volume of the NNLP adsorbent were found to be 4.72 m^2 g^{-1} and 7.1 mm^3 g^{-1}, respectively, with the average particle size of 93.80 μm. The average diameter of various immobilized NNLP adsorbent beads was found to be 2.25 mm. To optimize NNLP adsorbent loading in each polymeric matrix, immobilized adsorbents were prepared with varying compositions (1–10% (w/v)) of NNLP adsorbent in each matrix. The results of batch adsorption experiments obtained using different immobilized matrices with varying adsorbent loadings are reported in Figure 2 and Table S1. Figure 2 shows that the % dye removal increased with increase in adsorbent loading in various polymeric matrices up to an optimal limit, and beyond this it decreased, which may be attributed to the difference in porosity of the beads/gels when a higher dosage of adsorbent was loaded. The beads/gels may be less porous at higher adsorbent concentration. As the adsorbent dose was increased in the different polymeric matrices, the increase in loading may affect the free transport of CR dye anions to the interior binding sites through the formation of a physical boundary layer. This phenomenon can be explained by the agglomeration of adsorbent particles in various polymeric matrices [26]. It may also be due to the screening effect of the denser adsorbent particles at the outer layer of the immobilized matrix [27]. The optimum NNLP adsorbent loading was found to be 5% (w/v) for calcium alginate, 4% (w/v) for polyvinyl alcohol, 6% (w/v) for polysulfone and 3% (w/v) for sodium silicate. As the adsorbent dose was increased above the optimal limit in various polymeric matrices, the pore volume and surface area of the bead/gel decreased (shown in Table 2). The decrease in pore volume and surface area of the immobilized adsorbent may affect the % color removal [27,52]. The CR dye adsorption efficiencies on various immobilized NNLP adsorbent beads were compared with the adsorption efficiencies on the free adsorbent. The effectiveness factor was determined to be in the order of free NNLP adsorbent (1) > sodium silicate (0.9343) > polysulfone (0.8775) > calcium alginate (0.8583) > polyvinyl alcohol (0.7762), with an initial dye concentration of 200 mg L^{-1}. Among the various immobilized polymeric matrices, sodium silicate was chosen as the superior matrix, because it exhibits the highest adsorption efficiency (83.64%) and has the lowest cost among the different polymeric materials studied. The other three matrices were relatively less effective in preferentially adsorbing the CR dye. The calcium alginate, polyvinyl alcohol and polysulfone immobilized adsorbent matrices adhered together to form clumps during both adsorption and desorption cycles in aqueous solutions [26]. This property was encountered with the effective flow of feed solution when the immobilized adsorbent was packed in fixed-bed column reactors. Therefore, further column experiments were conducted using the optimized value of NNLP adsorbent loading in the sodium silicate matrix.

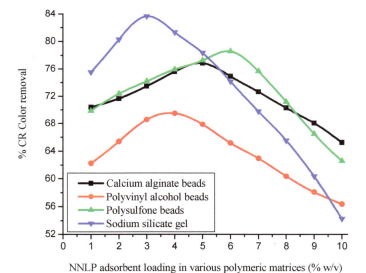

Figure 2. Effect of *Nelumbo nucifera* leaf powder (NNLP) adsorbent loading in various polymeric matrices on CR dye decolorization. (Initial pH: 6; initial dye concentration: 250 mg L^{-1}; free NNLP adsorbent particle size: 94 μm; immobilized adsorbent dosage: 6 g L^{-1}; agitation speed: 150 rpm; temperature: 301 K; contact time: 24 h)

Table 2. Physical properties of various polymeric matrices immobilized with NNLP adsorbent.

Various Polymeric Matrices	NNLP Adsorbent Optimum Loading % (w/v)	Immobilized Adsorbent Characteristics		NNLP Adsorbent Maximum Loading % (w/v)	Immobilized Adsorbent Characteristics	
		BET Surface Area (m^2 g^{-1})	Pore Volume (mm^3 g^{-1})		BET Surface Area (m^2 g^{-1})	Pore Volume (mm^3 g^{-1})
Calcium alginate	5	0.65	0.82	10	0.27	0.35
Polyvinyl alcohol	4	0.38	0.53	10	0.13	0.16
Polysulfone	6	0.72	0.94	10	0.34	0.48
Sodium silicate	3	1.84	2.18	10	1.16	1.65

3.2. Inference from Desorption Studies and Reusability of Free and Sodium Silicate Gel Immobilized NNLP Adsorbent in Batch Studies

Desorption experiments were performed for the removal of CR from sodium silicate-immobilized NNLP adsorbent, and the results are shown in Figure 3 and Table S2. They show that the amount of CR dye desorbed decreased with an increasing number of runs. The % desorption in all the runs was determined to be in the order of methanol > ethanol > 1 M NaOH > acetone > butanol, with various desorbing reagents in separate batches. It was found that a maximum of 52.67% of the dye could be desorbed using the solvent methanol in the third run, compared with other desorbing reagents. It was also found that the immobilized NNLP adsorbent was best regenerated using methanol as a solvent. The decrease in desorption efficiency with an increase in the number of runs may be due to the low volume of the desorbing reagent or the lack of agitation speed, which may prevent the further release of bound dye anions into the solvent [21,34]. Desorption may be explained based on electrostatic repulsion between the negatively charged sites of the adsorbent and anionic dye molecules [53]. Due to the strong hydrophilic attraction between desorbing reagent and dye molecules, the adsorbed dye gets solubilized in the solvent methanol [54]. This is the opposite of the electrostatic interaction effect, indicating that ion exchange is probably the major mode of the adsorption process. The very low desorption of dye suggests that chemisorption might be the major mode of dye removal from aqueous solutions by the adsorbent [22]. The dye under consideration is acidic in nature and exhibits significant attraction towards the solvent methanol [55]. Therefore, for the desorption of CR loaded in free and sodium silicate-immobilized NNLP adsorbents, the organic solvent methanol was determined to be the most efficient compared to other desorbing reagents, and hence further batch desorption experiments were carried out using the solvent methanol.

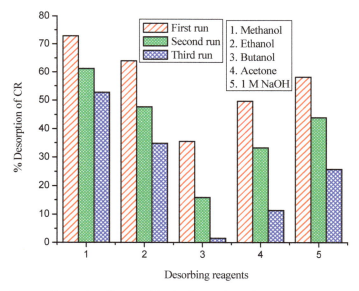

Figure 3. Desorption efficiency of CR dye from sodium silicate gel-immobilized NNLP adsorbent in various runs. (Volume of desorbing reagent: 100 mL; shaking speed: 150 rpm; temperature: 301 K; contact time: 24 h)

The regenerated sodium silicate gel-immobilized NNLP adsorbent was added to a dye solution of concentration 200 mg L^{-1} in an Erlenmeyer flask. The regenerated adsorbent was tested in the second and third runs. The results obtained from the reusability studies of the decolorization of CR in various runs are shown in Figure S1 and Table S3. They show that, in comparison to the first run, 66.75% adsorption was maintained after 24 h in the second run and 57.84% in the third run of operation.

3.3. Reusability of Free and Sodium Silicate Gel-Immobilized NNLP Adsorbents for CR Dye Adsorption in Column Studies

Desorption experiments were conducted to remove CR from free and immobilized NNLP adsorbent individually. After desorption, fixed-bed column studies of CR dye adsorption were carried out in three runs with free and sodium silicate gel-immobilized NNLP adsorbent separately. The column experimental results obtained by using free and immobilized NNLP adsorbents are reported in Table 3. The BTCs (i.e., $\frac{C_t}{C_o}$ vs. t) for the adsorption of CR onto free and immobilized NNLP adsorbents in various runs are shown in Figures 4 and 5. The results show that the saturation of binding sites for the removal of CR takes place rapidly in the initial phases of the adsorption process. Initially, the effluent was almost free of solute. As the dye solution continued to flow, the amount of solute adsorbed started decreasing because of the progressive saturation of the adsorbate on the binding sites of the NNLP adsorbent. The effluent concentration started to rise until the bed was exhausted [35]. The shape of the BTC followed a sigmoidal trend. The column reached t_B and t_E faster, and steep BTCs were observed in the second and third runs of operation. The breakthrough time and bed exhaustion time decreased significantly after the first run, showing that the treated effluent volume, total quantity of dye adsorbed in the column, % adsorption and equilibrium dye uptake of the regenerated free and immobilized NNLP adsorbent decreased during the second and third runs as compared to the first run, due to the insufficient desorption of the bound dye anions from the adsorbent surface. The free and immobilized NNLP adsorbent active sites may be blocked with CR dye molecules, leading to a reduction in the number of available binding sites on the adsorbent surface. The % color removal and equilibrium dye uptake using immobilized NNLP adsorbent in all three runs were lower than when free NNLP adsorbent was used. This may be due to the formation of a physical boundary layer around the immobilized matrix. The matrix thereby impedes the accessibility of dye anions to the binding sites of the NNLP adsorbent [26,33]. The decolorization efficiencies of CR dye adsorption were 53.66% and 43.33%; the equilibrium dye uptakes were 1.180 mg g^{-1} and 0.783 mg g^{-1} in the third run of operation with free and immobilized NNLP adsorbent, respectively. Therefore, a decrease in the % color removal was observed when increasing the number of runs. The prepared NNLP adsorbent can be used for up to three runs to adsorb CR dye in aqueous solution after regeneration using the solvent methanol.

Table 3. Column experimental parameters obtained at various runs for the decolorization of CR dye onto free and immobilized NNLP adsorbents.

Adsorbent	Column Parameters	First Run	Second Run	Third Run
Free NNLP adsorbent	Z (cm)	2.50	2.25	2.10
	W (g)	3.72	3.35	3.07
	t_b (h)	8.75	5.25	3.00
	t_E (h)	17.50	12.00	7.50
	m_{ad} (mg)	12.04	7.053	3.622
	m_{total} (mg)	15.75	10.80	6.75
	V_{eff} (L)	1.05	0.72	0.45
	q_e (mg g^{-1})	3.236	2.106	1.180
	% color removal	76.43	65.31	53.66
Sodium silicate gel-immobilized NNLP adsorbent	Z (cm)	2.50	2.10	1.75
	W (g)	3.12	2.69	2.24
	t_b (h)	6.50	3.50	1.25
	t_E (h)	13.5	9.00	4.50
	m_{ad} (mg)	8.178	4.596	1.755
	m_{total} (mg)	12.15	8.10	4.05
	V_{eff} (L)	0.81	0.54	0.27
	q_e (mg g^{-1})	2.621	1.708	0.783
	% color removal	67.30	56.74	43.33

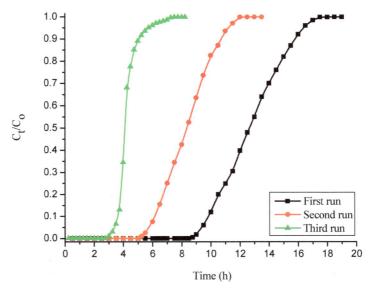

Figure 4. Breakthrough curves (BTCs) for decolorization of CR dye onto free NNLP adsorbent in various runs. (Initial pH: 6; flow rate: 1 mL min^{-1}; inlet dye concentration: 15 mg L^{-1}; temperature: 301 K)

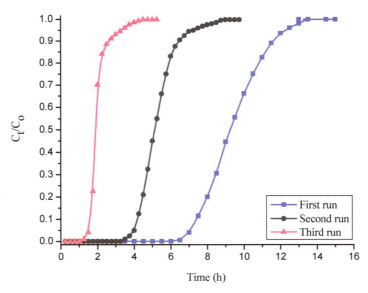

Figure 5. BTCs for decolorization of CR dye by sodium silicate gel-immobilized NNLP adsorbent in various runs. (Initial pH: 6; flow rate: 1 mL min^{-1}; inlet dye concentration: 15 mg L^{-1}; temperature: 301 K)

3.4. Analysis of BTCs for CR Dye Adsorption in Various Runs and Estimation of Kinetic Model Parameters in Different Models

The column experiments were performed at a constant flow rate of 1 mL min^{-1} and a fixed dye concentration of 15 mg L^{-1} at pH 6 in various runs. The column experimental data were fitted to various mathematical models to predict the BTCs and to evaluate the kinetic constants of the models using free and immobilized NNLP adsorbents separately.

3.4.1. Adams–Bohart Model

The Adams–Bohart model kinetic parameters in various runs are reported in Tables 4 and 5 for free and immobilized adsorbent, respectively. From Tables 4 and 5, it can be observed that as the number of runs increased, the values of kinetic constant, K_{AB}, and maximum saturation concentration, N_o, decreased. This may be due to a decrease in the number of binding sites available for CR dye adsorption. In addition, the loss of adsorbent in the second and third runs of operation may result in a shorter mass transfer zone, leading to a decrease in the t_B [56]. The predicted and experimental BTCs obtained for free and immobilized NNLP adsorbent in various runs are shown in Figures S2 and S3. Large differences were found between the predicted and experimental BTCs. The linear regression coefficient, R^2, values were in the range of 0.72–0.88 in different runs, suggesting that this model does not fit the data points very well.

Table 4. Column characteristic parameters obtained at various runs for the decolorization of CR onto free NNLP adsorbent.

Free NNLP Adsorbent		Model Parameters		
		First Run	Second Run	Third Run
Adams–Bohart model	K_{AB} (L mg^{-1} min^{-1})	4.98×10^{-4}	3.09×10^{-4}	1.34×10^{-4}
	N_o (mg L^{-1})	1624.1	1226.7	779.92
	R^2	0.85	0.88	0.78
BDST model	K (L mg^{-1} min^{-1})	1.142×10^{-3}	8.943×10^{-4}	6.152×10^{-4}
	N_o (mg L^{-1})	1329.8	966.03	560.18
	R^2	0.96	0.95	0.96
Thomas model	K_{TH} (mL mg^{-1} min^{-1})	1.069×10^{-3}	8.264×10^{-4}	5.839×10^{-4}
	q_{oTH} (mg g^{-1})	3.236	2.244	1.309
	R^2	0.98	0.98	0.99
Yoon–Nelson model	K_{YN} (min^{-1})	0.0164	0.0085	0.0057
	τ (min)	766.9	501.3	268.0
	q_{oYN} (mg g^{-1})	3.092	2.244	1.309
	R^2	0.98	0.98	0.98
Wolborska model	β_a (min^{-1})	0.809	0.773	0.658
	N_o (mg L^{-1})	1624.1	1226.7	779.92
	R^2	0.85	0.88	0.78

3.4.2. Bed Depth Service Time Model

The BDST model constants in various runs were evaluated and are reported in Tables 4 and 5 for free and immobilized adsorbent, respectively. Tables 4 and 5 show that the values of K and N_o decreased with increasing numbers of runs. This may be due to the available binding sites being saturated with CR dye molecules [56]. The rate constant K characterizes the rate of solute transfer from a liquid to a solid phase. The predicted and empirical BTCs obtained for free and immobilized NNLP adsorbent in various runs are shown in Figures S4 and S5. They show that there is a small deviation between empirical and predicted BTCs. Additionally, the R^2 values were in the range of 0.94–0.96, indicating that this model does not quite fit the column experimental data in various runs.

3.4.3. Thomas Model

The Thomas model adsorption data in various runs were evaluated, and their values are reported in Tables 4 and 5 for free and immobilized adsorbent, respectively. Tables 4 and 5 show that as the number of runs increased, the values of K_{Th} and q_{oTH} decreased significantly. This may be due to the inadequate desorption of the bound dye anions from the solid adsorbent surface and the loss of adsorbent after the first run [27,57]. The predicted and empirical BTCs obtained for free and immobilized NNLP adsorbent in various runs are shown in Figures 6 and 7. The theoretical and empirical data show a similar correlation and breakthrough trend. Additionally, the calculated values of bed capacity, q_{oTH}, are close to the experimental values of q_{oTH}, with a high R^2 value (>0.98) in various runs, suggesting that this model is valid for CR adsorption. The Thomas model assumes that the external and pore diffusion steps are not the rate limiting steps, that there is no axial dispersion, and that the Langmuir kinetics of adsorption are valid. However, the adsorption process is typically not controlled by chemical reaction kinetics and is often limited by solid–liquid interphase mass transfer, and the effect of axial dispersion may be important at lower flow rates [29,58].

Table 5. Column characteristic parameters obtained at various runs for the decolorization of CR by sodium silicate gel-immobilized NNLP adsorbent.

Immobilized NNLP Adsorbent		Model Parameters		
		First Run	Second Run	Third Run
Adams–Bohart model	K_{AB} (L mg^{-1}min^{-1})	4.17×10^{-4}	2.65×10^{-4}	1.09×10^{-4}
	N_0 (mg L^{-1})	1239.2	938.88	538.04
	R^2	0.83	0.80	0.72
BDST model	K (L mg^{-1}min^{-1})	1.086×10^{-3}	6.754×10^{-4}	4.563×10^{-4}
	N_0 (mg L^{-1})	994.68	678.16	339.04
	R^2	0.95	0.95	0.94
Thomas model	K_{TH} (mL mg^{-1} min^{-1})	9.542×10^{-4}	5.483×10^{-4}	2.256×10^{-4}
	q_{oTH} (mg g^{-1})	2.754	1.822	0.944
	R^2	0.98	0.99	0.98
Yoon–Nelson model	K_{YN} (min^{-1})	0.0118	0.0062	0.0024
	τ (min)	572.9	326.8	132.3
	q_{oYN} (mg g^{-1})	2.754	1.822	0.886
	R^2	0.98	0.98	0.99
Wolborska model	β_a (min^{-1})	0.765	0.642	0.573
	N_0 (mg L^{-1})	1239.2	938.88	538.04
	R^2	0.83	0.80	0.72

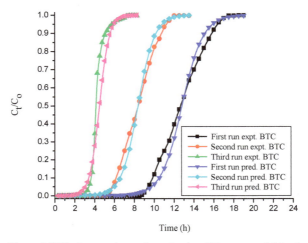

Figure 6. BTCs for experimental vs. simulated Thomas model for the decolorization of CR by free NNLP adsorbent in various runs. (Initial pH: 6; flow rate: 1 mL min^{-1}; inlet dye concentration: 15 mg L^{-1}; temperature: 301 K)

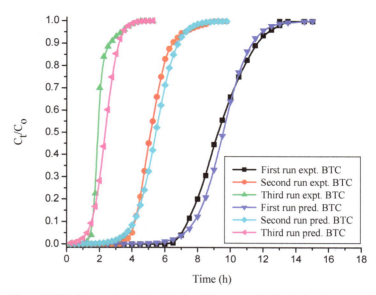

Figure 7. BTCs for experimental vs. simulated Thomas model for the decolorization of CR by sodium silicate gel-immobilized NNLP adsorbent in various runs. (Initial pH: 6; flow rate: 1 mL min^{-1}; inlet dye concentration: 15 mg L^{-1}; temperature: 301 K)

3.4.4. Yoon–Nelson Model

The Yoon–Nelson model constants are evaluated in various runs for free and immobilized adsorbent and the values are reported in Tables 4 and 5. This shows that the values of adsorption rate constant, K_{YN} and τ, decreased with increasing numbers of runs. This may be because of the inadequate desorption of the adsorbed dye molecules from the solid particle surface and the lack of binding sites on the adsorbent after the first run [27]. Due to the loss in adsorbent active sites during the second and third runs, the adsorbate molecules have less time to diffuse through the adsorbent, which may result in a reduced adsorption rate constant [59]. The predicted and empirical BTCs obtained at various runs for free and immobilized NNLP adsorbent are shown in Figures 8 and 9. The predicted BTCs, calculated bed capacity (q_{oYN}) and τ are close to experimental data in various runs, with the standard deviation of τ and q_{oYN} being less than 1.536% under the given set of operating conditions. This is proven by the high values of R^2, ranging from 0.98 to 0.99, which suggests that this model is valid for CR adsorption. The Yoon–Nelson model assumes that the rate of decrease in the probability of adsorption for each solute molecule is proportional to the probability of adsorbate adsorption and the probability of adsorbate breakthrough on the adsorbent. In addition, it assumes that there is negligible or no axial dispersion [27,60]. The advantage of applying the Yoon–Nelson model is that the mathematical application is very direct, and it provides the information of 50% column breakthrough, which enables us to predict the exhaustion period without the need for a long experimental time [35,61].

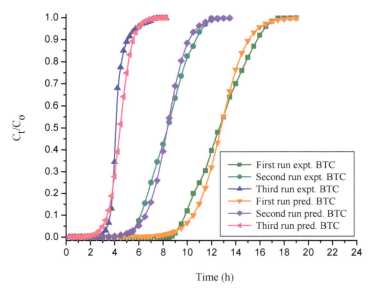

Figure 8. BTCs for experimental vs. simulated Yoon–Nelson model for the decolorization of CR by free NNLP adsorbent in various runs. (Initial pH: 6; flow rate: 1 mL min^{-1}; inlet dye concentration: 15 mg L^{-1}; temperature: 301 K)

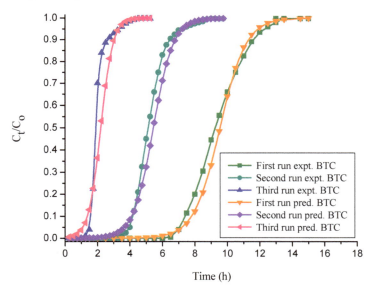

Figure 9. BTCs for experimental vs. simulated Yoon–Nelson model for the decolorization of CR dye by sodium silicate gel-immobilized NNLP adsorbent in various runs. (Initial pH: 6; flow rate: 1 mL min^{-1}; inlet dye concentration: 15 mg L^{-1}; temperature: 301 K)

3.4.5. Wolborska Model

The Wolborska model is applied to column experimental data for the description of the BTCs. The model parameters are determined in various runs for free and immobilized adsorbent, and the values are given in Tables 4 and 5. From Tables 4 and 5, we can see that the values of parameters β_a and N_o are influenced in various runs. It is inferred that as the number of runs increases, the values of model parameters β_a and N_o decrease. This may be due to the loss of active sites in the adsorbent particle surface for CR dye adsorption [46]. The predicted and experimental BTCs with respect to various runs for free and immobilized adsorbent are shown in Figures S6 and S7. As can be seen in Figures S6 and S7, there is a large discrepancy between experimental and predicted BTCs. The R^2 values range from 0.72 to 0.88, suggesting that this model does not appropriately fit the column experimental data in various runs. Therefore, the Adams–Bohart and Wolborska models' validity may be limited to the range of the process conditions used.

Out of the five kinetic models tested for CR dye adsorption, the Thomas and Yoon–Nelson models show a good fit to the experimental data, with good regression coefficients and very low standard deviations in various runs for both free and immobilized NNLP adsorbent.

3.5. Physicochemical Analysis of Textile Industrial CR Dye Effluent in Batch Studies

The physicochemical characteristics of the real and treated textile industrial CR dye effluent have been analyzed and compared with the Central Pollution Control Board (CPCB) standard limits. All of the data were recorded twice, and the average values are reported in Table 6. The optimal conditions for the maximum adsorption efficiency were obtained at an initial pH of 6, a free NNLP adsorbent dosage of 6 g L^{-1}, an adsorbent particle size of 42 µm, an agitation speed of 150 rpm, an immobilized NNLP adsorbent dosage of 6.5 g L^{-1}, and an immobilized adsorbent particle size of 94 µm. Among all the pollutants in the wastewater, the most important components were COD, BOD, pH, oil, nitrogen, phosphorus, sulfate, and suspended and dissolved solids [62]. From Table 6, we can see that the characteristics of the raw effluent are higher than the standards prescribed by the CPCB. The effluent had high turbidity and was colored with average organic and inorganic loading. The turbidity range (150 NTU) was not within the allowable limit of wastewater, and the sample was identified as having a greater volume of suspended particles [63]. The pH of the raw industrial dye effluent was 8.56, which indicates that the effluents from dyeing industries under study are alkaline in nature, indicating that different types of chemicals, such as sodium hyphochlorite, sodium carbonate, sodium bicarbonate, sodium hydroxide, surfactants and sodium phosphate, were used during the processing steps [48]. These chemicals have tremendous effects on the receiving water because they contain high concentrations of organic matter, non-biodegradable matter and toxic substances. A higher value of BOD (622 mg L^{-1}) depletes the dissolved oxygen concentration and kills aquatic fish, while the total suspended solid increases turbidity and allows less light penetration for photosynthetic activities. Most wastewater contains high pH, which is not favorable to the growth of organisms; the dye covers the surface of the water, which reflects the light and further denies aquatic organisms oxygen for their metabolism [64]. The dark color in the wastewater can increase the turbidity of the water body and affect the process of photosynthesis [64]. The higher value of electrical conductivity (5.34 mS cm^{-1}) indicates that there are more chemicals and dissolved substances present in the water. Higher amounts of these impurities will lead to a higher conductivity. The value of COD for the given dye effluent was 1845 mg L^{-1}, which is higher than the CPCB standard. The higher value of COD is primarily due to the nature of the chemicals employed in the dyeing unit of the textile processing industry. The ratio of BOD:COD obtained from the result was less than 0.5 (0.34), indicating that the effluent contains a large portion of non-biodegradable matter [65]. However, the characteristics of treated industrial CR dye effluent using free and immobilized NNLP adsorbent separately are closer to standard values. This shows that free and immobilized NNLP adsorbent can be efficiently used to remove CR from real industrial dye effluents.

Table 6. Characteristics of real and treated textile industrial CR dye effluent with CPCB acceptable limits.

Sl. No	Physicochemical Parameters	Raw Effluent Value	Treated Effluent Value		CPCB Standard
			Free NNLP Adsorbent	Sodium Silicate-Immobilized NNLP Adsorbent	
1	pH	8.56	8.50	8.52	6–9
2	Turbidity, NTU	150	18	24	<10
3	Total suspended solids, mg L^{-1}	134	26	45	<100
4	Total dissolved solids, mg L^{-1}	3824	965	1012	<2000
5	Biological oxygen demand, mg L^{-1}	622	42	58	<30
6	Chemical oxygen demand, mg L^{-1}	1845	238	246	<250
7	Total alkalinity, mg L^{-1}	418	87	102	<200
8	Total hardness, mg L^{-1}	756	415	420	<300
9	Dissolved oxygen concentration, mg L^{-1}	1.12	1.06	1.08	<4.00
10	Electrical conductivity, mS cm^{-1}	5.34	4.92	5.13	<2.25
11	Sulfates, mg L^{-1}	560	138	165	<250
12	Chlorides, mg L^{-1}	1524	236	274	<500

3.6. Analysis of Column Experiments with Textile Industrial CR Dye Effluent Using Free and Sodium Silicate Gel-Immobilized NNLP Adsorbents

The fixed-bed column experiments were conducted separately with industrial CR dye effluents using free and sodium silicate gel-immobilized NNLP adsorbents. The inlet adsorbate concentration was measured and it was found to be 215 mg L^{-1}. The bed height (2.5 cm) and flow rate (1 mL min^{-1}) were kept constant. The column experimental procedure is the same as that given before in Section 2.4. The column experimental results are reported in Table 7. While using free and immobilized adsorbents, the decolorization efficiency of the industrial dye effluent was more than 60%. The efficiencies of solute adsorption were 76.25% and 62.18%; equilibrium solute uptakes were 8 mg g^{-1} and 5.84 mg g^{-1} with free and immobilized NNLP adsorbent, respectively. While using the immobilized NNLP adsorbent, the breakthrough point time, bed exhaustion time, treated effluent volume, equilibrium dye uptake, mass of solute adsorbed and decolorization efficiency were lower than when free NNLP adsorbent was used. The BTCs for the adsorption of solute from industrial dye effluent using free and sodium silicate gel-immobilized NNLP adsorbents are shown in Figure 10. The column reached t_B and t_E faster, and a steep BTC was observed with the immobilized adsorbent at high inlet adsorbate concentrations. The intensity of the peaks of dye effluent was measured before and after adsorption. The intensity of peaks declined considerably after treatment with both the free and immobilized NNLP adsorbents (shown in Figure S8).

Table 7. Column experimental parameters for the adsorption of solute onto free and sodium silicate gel-immobilized NNLP adsorbents with industrial CR dye effluent.

Adsorbent	W (g)	t_b (h)	t_E (h)	m_{ad} (mg)	m_{total} (mg)	V_{eff} (L)	q_e (mg g^{-1})	% Color Removal
Free NNLP adsorbent	3.72	0.375	3.00	29.783	39.06	0.180	8.00	76.25
Immobilized NNLP adsorbent	3.12	0.20	2.25	18.215	29.295	0.135	5.84	62.18

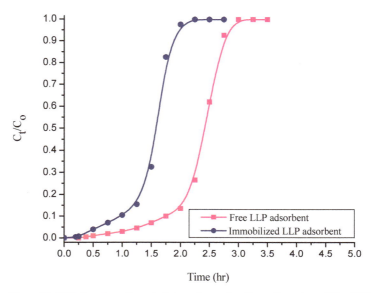

Figure 10. BTCs for adsorption of solute on free and sodium silicate-immobilized NNLP adsorbents with industrial CR dye effluent. (Initial pH: 6; bed height: 2.5 cm; flow rate: 1 mL min^{-1}; inlet adsorbate concentration: 215 mg L^{-1}; temperature: 301 K)

4. Conclusions

The adsorbent prepared from *N. nucifera* leaf is an effective adsorbent for the removal of CR from an aqueous solution in continuous-mode experiments. The NNLP adsorbent loading in various polymeric matrices was optimized. The maximum adsorption efficiency (83.64%) of CR occurred using the polymeric matrix sodium silicate. Desorption studies on loaded free and immobilized adsorbent showed that the maximum % of CR dye can be desorbed using the solvent methanol. They show that the % desorption decreased with the increase in the number of runs for all desorbing reagents. The regenerated free and immobilized NNLP adsorbent used separately in column studies showed that the % color removal and equilibrium dye uptake decreased with increases in the number of runs. The regenerated adsorbent can be used effectively for up to three cycles to adsorb CR dye in aqueous solutions, with considerable reductions in adsorption efficiency. The efficiencies of CR dye adsorption were 53.66% and 43.33%; the equilibrium dye uptakes were 1.180 mg g^{-1} and 0.783 mg g^{-1} in the third run of operation with free and immobilized NNLP adsorbent, respectively. The column reached t_B and t_E faster, and steep BTCs were observed in the second and third runs of operation. Various models were applied to the column empirical data to estimate the breakthrough curves and evaluate the model parameters in various runs using free and immobilized adsorbent separately. While considering the regression coefficient, R^2, predicted breakthrough curves and adsorption capacity, it could be said that the Thomas and Yoon–Nelson models best fit the real-time column experimental data in various runs. The physicochemical characteristics of the real textile industrial CR dye effluent obtained for free and immobilized NNLP adsorbents were within the limits specified by the CPCB standards. The adsorption of solute from textile industrial CR dye effluent studies shows that the decolorization efficiencies of solute adsorption were 76.25% and 62.18%; the equilibrium solute uptakes were 8 mg g^{-1} and 5.84 mg g^{-1} with the free and immobilized NNLP adsorbent, respectively. While using the immobilized NNLP adsorbent, the equilibrium dye uptake and % color removal were lower than when free NNLP adsorbent was used. The bed was exhausted in a short period (2.25 h) and a steep BTC was observed for an immobilized adsorbent at a high inlet adsorbate concentration.

The experimental results conclude that NNLP is an effective adsorbent for the removal of color from synthetic and industrial CR dye effluent in fixed-bed column studies. The higher adsorption efficiency of industrial CR dye effluent suggests that free and immobilized *N. nucifera* leaf fine powders can be used effectively to decolorize other anionic dyes from industrial effluents.

Supplementary Materials: The following are available online at https://www.mdpi.com/article/10.3390/polym14010054/s1, Figure S1: Reusability of immobilized NNLP adsorbent for the adsorption of CR dye in various runs. (Initial pH: 6; initial dye concentration: 200 mg L^{-1}; volume of dye solution: 100 mL; agitation speed: 150 rpm; temperature: 301 K; contact time: 24 h), Figure S2: BTCs for experimental vs. simulated Adams-Bohart model for the decolorization of CR by free NNLP adsorbent in various runs (Initial pH: 6; flow rate: 1 mL min^{-1}; inlet dye concentration: 15 mg L^{-1}; temperature: 301 K), Figure S3: BTCs for experimental vs. simulated Adams-Bohart model for the decolorization of CR dye by sodium silicate gel immobilized NNLLP adsorbent in various runs (Initial pH: 6; flow rate: 1 mL min^{-1}; inlet dye concentration: 15 mg L^{-1}; temperature: 301 K), Figure S4: BTCs for experimental vs. simulated BDST model for the decolorization of CR by free NNLP adsorbent in various runs (Initial pH: 6; flow rate: 1 mL min^{-1}; inlet dye concentration: 15 mg L^{-1}; temperature: 301 K), Figure S5: BTCs for experimental vs. simulated BDST model for the decolorization of CR dye by sodium silicate gel immobilized NNLP adsorbent in various runs (Initial pH: 6; flow rate: 1 mL min^{-1}; inlet dye concentration: 15 mg L^{-1}; temperature: 301 K), Figure S6: BTCs for experimental vs. simulated Wolborska model for the decolorization of CR by free NNLP adsorbent in various runs (Initial pH: 6; flow rate: 1 mL min^{-1}; inlet dye concentration: 15 mg L^{-1}; temperature: 301 K), Figure S7: BTCs for experimental vs. simulated Wolborska model for the decolorization of CR dye by sodium silicate gel immobilized NNLP adsorbent in various runs (Initial pH: 6; flow rate: 1 mL min^{-1}; inlet dye concentration: 15 mg L^{-1}; temperature: 301 K), Figure S8: Industrial CR dye effluent decolorization profile obtained in column studies using free and im-mobilized NNLP adsorbents with untreated effluent profile (Initial pH: 6; bed height: 2.5 cm; flow rate: 1 mL min-1; inlet adsorbate concentration: 215 mg L^{-1}; temperature: 301 K), Table S1: Effect of NNLP adsorbent composition in various polymeric matrices for CR dye adsorption (Initial pH: 6; initial dye concentration: 250 mg L^{-1}; free NNLP adsorbent particle size: 94 μm; immobilized NNLP ad-sorbent dosage: 6 g L^{-1}; agitation speed: 150 rpm; temperature: 301 K; contact time: 24 h), Table S2: Desorption studies for the removal of CR dye from sodium silicate gel-immobilized NNLP adsorbent in various runs, Table S3: Reusability of sodium silicate gel-immobilized NNLP adsorbent for the decolorization of CR dye in various runs.

Author Contributions: Conceptualization, V.P.; methodology, V.P., G.J. and N.R.; software, V.P. and G.J.; validation, V.P. and N.R.; formal analysis, V.P., G.J. and N.R.; investigation, V.P. and G.J.; resources, G.J. and N.R.; data curation, V.P., G.J. and N.R.; writing—original draft preparation, V.P.; writing—review and editing, G.J. and N.R.; visualization, V.P. and G.J.; supervision, V.P. All authors have read and agreed to the published version of the manuscript.

Funding: This research received no external funding.

Institutional Review Board Statement: Not applicable.

Informed Consent Statement: Not applicable.

Data Availability Statement: The data used to support the findings of this study are available from the corresponding author upon request.

Acknowledgments: The authors would like to thank the Department of Chemical engineering, Manipal Institute of Technology (MIT), Manipal Academy of Higher Education (MAHE) for providing all the facilities to perform the research work.

Conflicts of Interest: The authors declare no conflict of interest.

References

1. Taher, T.; Rohendi, D.; Mohadi, R.; Lesbani, A. Congo red dye removal from aqueous solution by acid-activated bentonite from sarolangun: Kinetic, equilibrium, and thermodynamic studies. *Arab J. Basic Appl. Sci.* **2019**, *26*, 125–136. [CrossRef]
2. Jhilirani, M.; Banashree, D.; Soumen, D. Highly porous iron-zirconium binary oxide for efficient removal of Congo red from water. *Desalin. Water Treat.* **2020**, *189*, 227–242. [CrossRef]

3. Vairavel, P.; Murty, V.R. Residence time distribution studies and modeling of rotating biological contactor reactor for decolorization of Congo red from synthetic dye wastewater. *Desalin. Water Treat.* **2021**, *220*, 380–391. [CrossRef]
4. Rahimi, R.; Kerdari, H.; Rabbani, M.; Shafiee, M. Synthesis, characterization and adsorbing properties of hollow $Zn-Fe_2O_4$ nanospheres on removal of Congo red from aqueous solution. *Desalination* **2011**, *280*, 412–418. [CrossRef]
5. Islam, M.T.; Aimone, F.; Ferri, A.; Rovero, G. Use of N-methylformanilide as swelling agent for meta-aramid fibers dyeing: Kinetics and equilibrium adsorption of Basic blue 41. *Dyes Pigment.* **2015**, *113*, 554–561. [CrossRef]
6. Vairavel, P.; Murty, V.R. Decolorization of Congo red dye in a continuously operated rotating biological contactor reactor. *Desalin. Water Treat.* **2020**, *196*, 299–314. [CrossRef]
7. Zahir, A.; Aslam, Z.; Kamal, M.S.; Ahmad, W.; Abbas, A.; Shawabkeh, R.A. Development of novel cross-linked chitosan for the removal of anionic Congo red dye. *J. Mol. Liq.* **2017**, *244*, 211–218. [CrossRef]
8. Vahidhabanu, S.; Swathika, I.; Adeogun, A.I.; Roshni, R.; Babu, B.R. Biomass-derived magnetically tuned carbon modified Sepiolite for effective removal of Congo red from aqueous solution. *Desalin. Water Treat.* **2020**, *184*, 326–339. [CrossRef]
9. Srilakshmi, C.; Saraf, R. Ag-doped hydroxyapatite as efficient adsorbent for removal of Congo red dye from aqueous solutions: Synthesis, kinetic and equilibrium adsorption isotherm analysis. *Microporous Mesoporous Mater.* **2016**, *219*, 134–144. [CrossRef]
10. Jiao, C.; Liu, D.; Wei, N.; Gao, J.; Fu, F.; Liu, T.; Wang, J. Efficient Congo red removal using porous cellulose/gelatin/sepiolite gel Beads: Assembly, characterization, and adsorption mechanism. *Polymers* **2021**, *13*, 3890. [CrossRef]
11. Dawood, S.; Sen, T.K. Removal of anionic dye Congo red from aqueous solution by raw pine and acid-treated pine cone powder as adsorbent: Equilibrium, thermodynamic, kinetics, mechanism and process design. *Water Res.* **2012**, *46*, 1933–1946. [CrossRef]
12. Kittappa, S.; Jais, F.M.; Ramalingam, M.; Mohd, N.S.; Ibrahim, S. Functionalized magnetic mesoporous palm shell activated carbon for enhanced removal of azo dyes. *J. Environ. Chem. Eng.* **2020**, *8*, 104081. [CrossRef]
13. Lakshmipathy, R.; Sarada, N.C. Methylene blue adsorption onto native watermelon rind: Batch and fixed bed column studies. *Desalin. Water Treat.* **2016**, *57*, 10632–10645. [CrossRef]
14. Bulgariu, L.; Escudero, L.B.; Bello, O.S.; Iqbal, M.; Nisar, J.; Adegoke, K.A.; Alakhras, F.; Kornaros, M.; Anastopoulos, I. The utilization of leaf-based adsorbents for dyes removal: A review. *J. Mol. Liq.* **2019**, *276*, 728–747. [CrossRef]
15. Lim, L.B.L.; Priyantha, N.; Latip, S.A.A.; Lu, Y.C.; Mahadi, A.H. Converting *Hylocereus undatus* (white dragon fruit) peel waste into a useful potential adsorbent for the removal of toxic Congo red dye. *Desalin. Water Treat.* **2020**, *185*, 307–317. [CrossRef]
16. Dai, H.; Huang, Y.; Zhang, H.; Ma, L.; Huang, H.; Wu, J.; Zhang, Y. Direct fabrication of hierarchically processed pineapple peel hydrogels for efficient Congo red adsorption. *Carbohydr. Polym.* **2020**, *230*, 115599. [CrossRef] [PubMed]
17. Vairavel, P.; Murty, V.R.; Nethaji, S. Removal of Congo red dye from aqueous solutions by adsorption onto a dual adsorbent (*Neurospora crassa* dead biomass and wheat bran): Optimization, isotherm, and kinetics studies. *Desalin. Water Treat.* **2017**, *68*, 274–292. [CrossRef]
18. Garba, Z.N.; Bello, I.; Galadima, A.; Lawal, A.Y. Optimization of adsorption conditions using central composite design for the removal of copper(II) and lead(II) by defatted papaya seed. *Karbala Int. J. Mod. Sci.* **2016**, *2*, 20–28. [CrossRef]
19. Song, J.Y.; Zou, W.H.; Bian, Y.Y.; Su, F.Y.; Han, R.P. Adsorption characteristics of methylene blue by peanut husk in batch and column modes. *Desalination* **2011**, *265*, 119–125. [CrossRef]
20. Vairavel, P.; Rampal, N.; Jeppu, G. Adsorption of toxic Congo red dye from aqueous solution using untreated coffee husks: Kinetics, equilibrium, thermodynamics and desorption study. *Int. J. Environ. Anal. Chem.* **2021**, in press. [CrossRef]
21. Debamita, C.; Nakul, R.; Gautham, J.P.; Vairavel, P. Process optimization, isotherm, kinetics, and thermodynamics studies for removal of Remazol Brilliant Blue—R dye from contaminated water using adsorption on guava leaf powder. *Desalin. Water Treat.* **2020**, *185*, 318–343. [CrossRef]
22. Divya, J.M.; Palak, K.; Vairavel, P. Optimization, kinetics, equilibrium isotherms, and thermodynamics studies of Coomassie violet dye adsorption using *Azadirachta indica* (neem) leaf adsorbent. *Desalin. Water Treat.* **2020**, *190*, 353–382. [CrossRef]
23. Romzia, A.A.; Lim, L.B.L.; Chan, C.M.; Priyantha, N. Application of *Dimocarpus longan ssp. malesianus* leaves in the sequestration of toxic brilliant green dye. *Desalin. Water Treat.* **2020**, *189*, 428–439. [CrossRef]
24. Meghana, C.; Juhi, B.; Rampal, N.; Vairavel, P. Isotherm, kinetics, process optimization and thermodynamics studies for removal of Congo red dye from aqueous solutions using *Nelumbo nucifera* (lotus) leaf adsorbent. *Desalin. Water Treat.* **2020**, *207*, 373–397. [CrossRef]
25. Aksu, Z.; Gonen, F. Biosorption of phenol by immobilized activated sludge in a continuous packed bed: Prediction of breakthrough curves. *Process Biochem.* **2004**, *39*, 599–613. [CrossRef]
26. Bai, R.S.; Abraham, T.E. Studies on chromium(VI) adsorption-desorption using immobilized fungal biomass. *Bioresour. Technol.* **2003**, *87*, 17–26. [CrossRef] [PubMed]
27. Vairavel, P.; Murty, V.R. Continuous fixed-bed adsorption of Congo red dye by dual adsorbent (*Neurospora crassa* dead fungal biomass and wheat bran): Experimental and theoretical breakthrough curves, immobilization and reusability studies. *Desalin. Water Treat.* **2017**, *98*, 276–293. [CrossRef]
28. Yang, Y.; Lin, E.; Tao, X.; Hu, K. High efficiency removal of Pb(II) by modified spent compost of *Hypsizygus marmoreus* in a fixed-bed column. *Desalin. Water Treat.* **2018**, *102*, 220–228. [CrossRef]
29. Aksu, Z.; Cagatay, S.S.; Gonen, F. Continuous fixed bed biosorption of reactive dyes by dried *Rhizopus arrhizus*: Determination of column capacity. *J. Hazard. Mater.* **2007**, *143*, 362–371. [CrossRef]

30. Han, X.; Yuan, J.; Ma, X. Adsorption of malachite green from aqueous solutions onto lotus leaf: Equilibrium, kinetic, and thermodynamic studies. *Desalin. Water Treat.* **2014**, *52*, 5563–5574. [CrossRef]
31. Vairavel, P.; Murty, V.R. Optimization of batch process parameters for Congo red color removal by *Neurospora crassa* live fungal biomass with wheat bran dual adsorbent using response surface methodology. *Desalin. Water Treat.* **2018**, *103*, 84–101. [CrossRef]
32. Boyaci, E.; Eroglu, A.E.; Shahwan, T. Sorption of As(V) from waters using chitosan and chitosan-immobilized sodium silicate prior to atomic spectrometric determination. *Talanta* **2010**, *80*, 1452–1460. [CrossRef]
33. Rangsayatorn, N.; Pokethitiyook, P.; Upatham, E.S.; Lanza, G.R. Cadmium biosorption by cells of *Spirulina platensis* TISTR 8217 immobilized in alginate and silica gel. *Environ. Int.* **2004**, *30*, 57–63. [CrossRef]
34. Vairavel, P.; Murty, V.R. Optimization, kinetics, equilibrium isotherms and thermodynamics studies for Congo red dye adsorption using calcium alginate beads immobilized with dual adsorbent (*Neurospora crassa* dead fungal biomass and wheat bran). *Desalin. Water Treat.* **2017**, *97*, 338–362. [CrossRef]
35. Vairavel, P.; Rampal, N. Continuous fixed-bed column study for removal of Congo red dye from aqueous solutions using *Nelumbo nucifera* leaf adsorbent. *Int. J. Environ. Anal. Chem.* **2021**, in press. [CrossRef]
36. Padmesh, T.V.N.; Vijayaraghavan, K.; Sekaran, G.; Velan, M. Biosorption of acid blue 15 using water macroalga *Azolla filiculoides*: Batch and column studies. *Dyes Pigment.* **2006**, *71*, 77–82. [CrossRef]
37. Lim, A.P.; Aris, A.Z. Continuous fixed-bed column study and adsorption modeling: Removal of cadmium (II) and lead (II) ions in aqueous solution by dead calcareous skeletons. *Biochem. Eng. J.* **2014**, *87*, 50–61. [CrossRef]
38. Alardhi, S.M.; Albayati, T.M.; Alrubaye, J.M. Adsorption of the methyl green dye pollutant from aqueous solution using mesoporous materials MCM-41 in a fixed-bed column. *Heliyon* **2020**, *6*, e03253. [CrossRef]
39. Sadaf, S.; Bhatti, H.N. Batch and fixed bed column studies for the removal of Indosol Yellow BG dye by peanut husk. *J. Taiwan Inst. Chem. Eng.* **2014**, *45*, 541–553. [CrossRef]
40. Padmesh, T.V.N.; Vijayaraghavan, K.; Sekaran, G.; Velan, M. Batch and column studies on biosorption of acid dyes on fresh water macro alga *Azolla filiculoides*. *J. Hazard. Mater.* **2005**, *B125*, 121–129. [CrossRef]
41. Vishali, S.; Karthikeyan, R.; Prabhakar, S. Utilization of seafood processing waste as an adsorbent in the treatment of paint industry effluent in a fixed-bed column. *Desalin. Water Treat.* **2017**, *66*, 149–157. [CrossRef]
42. Adhami, L.; Mirzaei, M. Removal of copper (II) from aqueous solution using granular sodium alginate/activated carbon hydrogel in a fixed-bed column. *Desalin. Water Treat.* **2018**, *103*, 208–215. [CrossRef]
43. Kishor, B.; Rawal, N. Column adsorption studies on copper(II) ion removal from aqueous solution using natural biogenic iron oxide. *Desalin. Water Treat.* **2019**, *153*, 216–225. [CrossRef]
44. Jain, S.N.; Gogate, P.R. Fixed bed column study for the removal of Acid blue 25 dye using NaOH-treated fallen leaves of *Ficus racemose*. *Desalin. Water Treat.* **2017**, *85*, 215–225. [CrossRef]
45. Olivares, J.C.; Alonso, C.P.; Diaz, C.B.; Nunez, F.U.; Mercado, M.C.C.; Bilyeue, B. Modeling of lead (II) biosorption by residue of allspice in a fixed-bed column. *Chem. Eng. J.* **2013**, *228*, 21–27. [CrossRef]
46. Omar, W.; Dwairi, R.A.; Hamatteh, Z.S.A.; Jabarin, N. Investigation of natural Jordanian zeolite tuff (JZT) as adsorbent for TOC removal from industrial wastewater in a continuous fixed bed column: Study of the influence of particle size. *Desalin. Water Treat.* **2019**, *152*, 26–32. [CrossRef]
47. APHA. *Standard Methods for Examination of Water and Wastewater*, 22nd ed.; American Public Health Association: Washington DC, USA, 2012.
48. Tchobanoglous, G.; Burton, F.L.; Stensel, H.D. *Metcalf & Eddy Wastewater Engineering: Treatment and Reuse*, 4th ed.; Tata McGraw-Hill Publishing Company Limited: New York, NY, USA, 2003.
49. ASTM International. *Annual Book of ASTM Standards*; v. 11.0; Water and Environmental Technology: Montgomery, PA, USA, 2003.
50. APHA. *Standard Methods for Examination of Water and Wastewater*, 19th ed.; American Public Health Association: Washington DC, USA, 1995.
51. Trivedy, R.K.; Goel, P.K. *Chemical and Biological Methods for Water Pollution Studies*; Environmental Publication: Varanasi, India, 1984.
52. Spinti, M.; Zhuang, H.; Trujillo, E.M. Evaluation of immobilized biomass beads for removing heavy metals from waste water. *Water Environ. Res.* **1995**, *67*, 943–952. [CrossRef]
53. Patel, R.; Suresh, S. Kinetic and equilibrium studies on the biosorption of Reactive black 5 dye by *Aspergillus foetidus*. *Bioresour. Technol.* **2008**, *99*, 51–58. [CrossRef]
54. Purkait, M.K.; Maiti, A.; Gupta, S.D.; De, S. Removal of Congo red using activated carbon and its regeneration. *J. Hazard. Mater.* **2007**, *145*, 287–295. [CrossRef] [PubMed]
55. Jain, S.N.; Gogate, P.R. Adsorptive removal of azo dye in a continuous column operation using biosorbent based on NaOH and surfactant activation of *Prunus dulcis* leaves. *Desalin. Water Treat.* **2019**, *141*, 331–341. [CrossRef]
56. Nethaji, S.; Sivasamy, A.; Vimal Kumar, R.; Mandal, A.B. Preparation of char from lotus seed biomass and the exploration of its dye removal capacity through batch and column adsorption studies. *Environ. Sci. Pollut. Res.* **2013**, *20*, 3670–3678. [CrossRef]
57. Vijayaraghavan, K.; Lee, M.W.; Yun, Y.S. A new approach to study the decolorization of complex reactive dye bath effluent by biosorption technique. *Bioresour. Technol.* **2008**, *99*, 5778–5785. [CrossRef]
58. Gupta, V.K.; Mittal, A.; Krishnan, L.; Gajbe, V. Adsorption kinetics and column operations for the removal and recovery of Malachite green from wastewater using bottom ash. *Sep. Purif. Technol.* **2004**, *40*, 87–96. [CrossRef]

59. Vieira, M.L.G.; Martinez, M.S.; Santos, G.B.; Dotto, G.L.; Pinto, L.A.A. Azo dyes adsorption in fixed bed column packed with different deacetylation degrees chitosan coated glass beads. *J. Environ. Chem. Eng.* **2018**, *6*, 3233–3241. [CrossRef]
60. Yoon, Y.H.; Nelson, J.H. Application of gas adsorption kinetics I. A theoretical model for respirator cartridge service life. *Am. Ind. Hyg. Assoc. J.* **1984**, *45*, 509–516. [CrossRef]
61. Malkoc, E.; Nuhoglu, Y.; Abali, Y. Cr(VI) adsorption by waste acron of *Querus ithaburensis* in fixed beds: Prediction of breakthrough curves. *Chem. Eng. J.* **2006**, *119*, 61–68. [CrossRef]
62. Yaseen, D.A.; Scholz, M. Textile dye wastewater characteristics and constituents of synthetic effluents: A critical review. *Int. J. Environ. Sci. Technol.* **2019**, *16*, 1193–1226. [CrossRef]
63. Koprivanac, N.; Bosanac, G.; Grabaric, Z.; Papic, S. Treatment of wastewaters from dye industry. *Environ. Technol.* **1993**, *14*, 385–390. [CrossRef]
64. Holkar, C.R.; Jadhav, A.J.; Pinjari, D.V.; Mahamuni, N.M.; Pandit, A.B. A critical review on textile wastewater treatments: Possible approaches. *J. Environ. Manag.* **2016**, *182*, 351–366. [CrossRef]
65. Yusuff, R.O.; Sonibare, J.A. Characterization of textile industries effluents in Kaduna, Nigeria and pollution implications. *Glob. Nest Int. J.* **2004**, *6*, 212–221. [CrossRef]

Article

Kinetics, Isotherm and Thermodynamic Studies for Efficient Adsorption of Congo Red Dye from Aqueous Solution onto Novel Cyanoguanidine-Modified Chitosan Adsorbent

Nouf F. Al-Harby [1,*], Ebtehal F. Albahly [1] and Nadia A. Mohamed [1,2]

[1] Department of Chemistry, College of Science, Qassim University, P.O. Box 6644, Buraydah 51452, Saudi Arabia; e.albahly@qu.edu.sa (E.F.A.); NA.AHMED@qu.edu.sa (N.A.M.)
[2] Department of Chemistry, Faculty of Science, Cairo University, Giza 12613, Egypt
* Correspondence: hrbien@qu.edu.sa

Abstract: Novel Cyanoguanidine-modified chitosan (CCs) adsorbent was successfully prepared via a four-step procedure; first by protection of the amino groups of chitosan, second by insertion of epoxide rings, third by opening the latter with cyanoguanidine, and fourth by restoring the amino groups through elimination of the protection. Its structure and morphology were checked using Fourier-transform infrared spectroscopy (FTIR), X-ray diffraction (XRD) and scanning electron microscopy (SEM) techniques. The adsorption capacity of CCs for Congo Red (CR) dye was studied under various conditions. It decreased significantly with the increase in the solution pH value and dye concentration, while it increased with increasing temperature. The adsorption fitted to the pseudo-second order kinetic model and Elovich model. The intraparticle diffusion model showed that the adsorption involved a multi-step process. The isotherm of CR dye adsorption by CCs conforms to the Langmuir isotherm model, indicating the monolayer nature of adsorption. The maximum monolayer coverage capacity, q_{max}, was 666.67 mg g^{-1}. Studying the thermodynamic showed that the adsorption was endothermic as illustrated from the positive value of enthalpy (34.49 kJ mol^{-1}). According to the values of $\Delta G°$, the adsorption process was spontaneous at all selected temperatures. The value of $\Delta S°$ showed an increase in randomness for the adsorption process. The value of activation energy was 2.47 kJ mol^{-1}. The desorption percentage reached to 58% after 5 cycles. This proved that CCs is an efficient and a promising adsorbent for the removal of CR dye from its aqueous solution.

Keywords: cyanoguanidine-modified chitosan adsorbent; Congo red dye; adsorption kinetics; adsorption isotherms; adsorption thermodynamic

Citation: Al-Harby, N.F.; Albahly, E.F.; Mohamed, N.A. Kinetics, Isotherm and Thermodynamic Studies for Efficient Adsorption of Congo Red Dye from Aqueous Solution onto Novel Cyanoguanidine-Modified Chitosan Adsorbent. *Polymers* **2021**, *13*, 4446. https://doi.org/10.3390/polym13244446

Academic Editors: Irene S. Fahim, Ahmed K. Badawi and Hossam E. Emam

Received: 12 November 2021
Accepted: 13 December 2021
Published: 18 December 2021

Publisher's Note: MDPI stays neutral with regard to jurisdictional claims in published maps and institutional affiliations.

Copyright: © 2021 by the authors. Licensee MDPI, Basel, Switzerland. This article is an open access article distributed under the terms and conditions of the Creative Commons Attribution (CC BY) license (https://creativecommons.org/licenses/by/4.0/).

1. Introduction

In recent years, many pollutants, especially synthetic dyes, have flowed into the water environment. A small quantity of these dyes in water is very clear and dangerous to aquatic life. In addition, dyes can lead to severe damage to the liver, digestive and central nervous systems of humans. [1–3]. Due to the synthetic origin and the complex structure of these dyes, they are stable in light and heat and are not bio-degradable [4]. Many techniques have been applied to remove dyes, including biodegradation, membrane filtration, coagulation-aggregation, electrochemical and oxidation processes. These techniques have a considerable efficiency, but they face some issues, including high energy consumption, a large cost, a large amount of toxic residues and poor performance at low dye concentrations [1–4].

Adsorption methods have been proven to be a highly effective technique for the removal of dyes due to their simplicity, easy operation, low initial cost, and the adsorbent abundance with no undesirable secondary product, in addition to easy restoring of the adsorbents for reuse [1–5]. The most important adsorbent is the carbon because of its high efficiency to remove the various contaminants, but it is highly expensive and is difficult in regeneration [6]. Therefore, it is necessary to find other effective and low-cost adsorbents. The use of adsorbents based on natural biopolymers has attracted considerable attention, especially chitosan and its derivatives.

Chitosan, a natural linear polysaccharide, is a non-toxic, eco-friendly, antimicrobial, biocompatible and biodegradable material [7–9]. Chitosan is one of the most promising adsorbents due to its low initial cost, ease of operation and efficient removal of dyes compared with activated carbon and other adsorbents. Because of its unique polycationic nature, it has a high capacity to remove anionic dyes through protonation processes of its amino groups in acidic medium. However, the use of chitosan as an adsorbent is limited due to its low surface area, the formation of colloid in water and its deterioration by chemical and microbial procedures. In addition, the high dissolution ability of chitosan in acidic media is considered to be its greatest shortcoming, especially in the case of its usage as an adsorbent for the removal of dyes since the effluent is usually acidic. Thus, modification of chitosan via grafting [2,10–13], blending [14–16] and cross linking [17–22] is used to improve chitosan properties, such as an increase of its functionality, and a lowering of its solubility, inhibiting its degradation rate and increasing the life span of its products in different media.

In this study, to avoid the aforementioned drawbacks of chitosan and to improve its adsorption capacity for anionic dyes, chitosan was chemically modified via prior protection of its primary amino groups using benzaldehyde. Thus, the reaction of epichlorohydrin was confined to the primary hydroxyl groups on C6 of the produced chitosan Schiff's base derivative, to incorporate epoxy moieties which can be readily opened by compounds possessing a free electron pair as cyanoguanidine. Afterwards, benzaldehyde was eliminated in an acidic medium and the primary amino groups on chitosan were regained to obtain cyanoguanidine-modified chitosan. It would be expected that the inclusion of nitrogen-rich cyanoguanidine moieties into the repeating units of chitosan, in addition to the regained amino groups of chitosan, will increase the basic sites available for adsorption the anionic dyes as CR dye. The adsorption kinetic was studied under the influence of different variables, such as initial dye concentration, temperature, time period and pH of the adsorption medium to reach the ideal conditions of adsorption and to determine the optimum adsorption capacity. The thermodynamic parameters, such as enthalpy, entropy and the free energy of adsorption, were determined. The possibility of regeneration of the modified biomaterial for reuse was also studied.

2. Materials and Methods
2.1. Materials

Chitosan (1.0–3.0×10^5 g mol^{-1} and 98% deacetylation degree) was obtained from Acros Organics (Newark, NJ, USA). Benzaldehyde and epichlorohydrin were purchased from PanReac. AppliChem- ITW Reagent (Darmstadt, Germany). Cyanoguanidine was supplied by Sigma-Aldrich (Munich, Germany). CR dye was supplied by Winlab (Leicestershire, UK). The other chemicals and solvents were obtained from Sigma-Aldrich (Munich, Germany).

2.2. Preparation of Novel Cyanoguanidine-Modified Chitosan (CCs) Adsorbent

First step: chitosan Schiff's base should be prepared via reaction of chitosan with benzaldehyde to direct the chemical modification to the primary -OH groups on chitosan. The suspended chitosan (5 g) that swollen in 50 mL methanol was well stirred with 20 mL of benzaldehyde at room temperature for 24 h. The produced chitosan Schiff's base has been filtered, rinsed repeatedly with MeOH and dried at 50 °C to constant weight [21].

Second step: 4 g of chitosan Schiff's base were suspended and well stirred in aqueous sodium hydroxide solution (120 mL, 0.001 mol L^{-1}) at room temperature for 15 min for swelling. To the suspended solution, 10 mL of epichlorohydrin were slowly added with continuous stirring for a further 6 h. The yielded epoxy chitosan Schiff's base has been filtered, rinsed repeatedly with H$_2$O and dried at 50 °C to constant weight [22].

Third step: 2 g of cyanoguanidine were dissolved in 25 mL of water, then were slowly mixed with a solution of epoxy chitosan Schiff's base (2 g) that swelled and was suspended in 60 mL of aqueous sodium hydroxide solution (0.001 mol L^{-1}) and stirred at room temperature overnight. The produced cyanoguanidine chitosan Schiff's base was obtained by filtration, washing frequently with methanol as well as acetone and drying at 50 °C to constant weight.

Forth step: to remove the protection from the NH$_2$ groups, cyanoguanidine chitosan Schiff's base (2 g) was treated with 60 mL of hydrochloric acid in ethanol (0.24 mol L^{-1} HCl) and stirred at room temperature for 24 hrs. The formed cyanoguanidine-modified chitosan (CCs) adsorbent was neutralized with aqueous sodium carbonate solution (1 wt%) till pH 7, filtered, washed frequently with ethanol and dried at 50 °C to constant weight (Scheme 1).

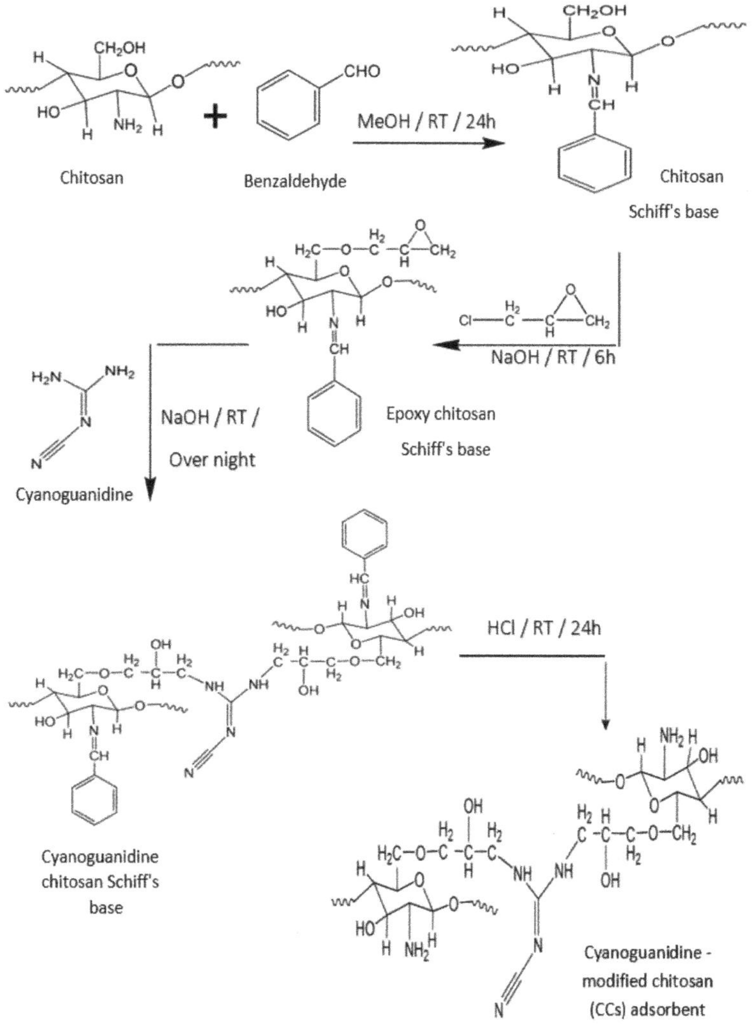

Scheme 1. Preparation of novel CCs adsorbent.

2.3. Measurements

2.3.1. FTIR Spectroscopy

A Thermo Scientific Nicolet 6700 FTIR spectrometer (Tokyo, Japan) was utilized to record FTIR spectra of the modified chitosan derivatives using KBr pellets in the wave number range of 4000–500 cm^{-1}.

2.3.2. X-ray Diffractometry

A Rigaku Ultima-IV wide-angle X-ray diffractometer (Tokyo, Japan) was used to study the morphology of the modified chitosan derivatives at diffraction angles (2θ) at a range between 5 and 80° with a speed of 5° min^{-1}.

2.3.3. Scanning Electron Microscopy

A field emission scanning electron microscope JSM-7610F (Freising, Germany) was used to photograph the surface topography of the modified chitosan derivatives after coating with a thin layer of gold at an accelerating voltage of 15 kV and at a magnification of 8000×.

2.4. Adsorption Studies

2.4.1. Standard Curve of CR Dye

Anionic CR dye was chosen due to its extensive applications. Its stock solution was prepared by dissolution of CR dye in double distilled water of a concentration of 1000 mg L^{-1}. The prepared stock solution was then diluted into 10 different dye concentrations using double distilled water. The absorbance of each dye solution was measured using UV–vis spectrophotometer (Shimadzu UV/Vis 1601 spectrophotometer, Kyoto, Japan) at λ_{max} = 497 nm. A calibration curve was plotted between concentration and absorbance using the predetermined concentrations of CR dye. The molar absorptivity was determined using Beer–Lambert law Equation (1).

$$A = \varepsilon l c \qquad (1)$$

where A is the absorbance, ε is the molar absorptivity (L mol^{-1} cm^{-1}), l is the path length of the cuvette that containing the sample (cm) and c is the concentration of dye in solution (mg L^{-1}). The molar absorptivity of CR dye was 0.045 L mol^{-1} cm^{-1} (Figure 1).

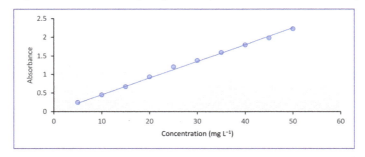

Figure 1. Absorption coefficient of CR dye.

2.4.2. Adsorption of CR Dye Using CCs Adsorbent

The removal of CR dye by CCs adsorbent was studied using batch experiment. 50 mL of dye solution was added into an Erlenmeyer flask with 50 mg of the absorbent. The solution was then shaken in a water bath shaker (70 rpm) until the equilibrium was attained. The residual dye concentration was analyzed spectrophotometrically at λ_{max} = 497 nm.

The amount of adsorbed dye onto adsorbent at certain time, q_t (mg g^{-1}), and at equilibrium, q_e (mg g^{-1}), were calculated using Equations (2) and (3), respectively.

$$q_t = \frac{(C_o - C_t)\,V}{W} \qquad (2)$$

$$q_e = \frac{(C_o - C_e)\,V}{W} \qquad (3)$$

where q_t and q_e are the amount of adsorbed dye (mg g^{-1}) at time t and at equilibrium, respectively, C_o is the initial concentration of the dye (mg L^{-1}), C_t is the dye concentration at time t (mg L^{-1}), V is the volume of the dye solution (L) and W is the weight of the adsorbent (g).

The removal efficiency (R.E.) of dye can be calculated by Equation (4).

$$\%\ R.E. = \frac{(C_o - C_e)}{C_o} \times 100 \qquad (4)$$

where C_o is the initial concentration of the dye (mg L^{-1}) and C_e is the equilibrium concentration of the dye (mg L^{-1}).

Batch experiments were performed in stoppered Erlenmeyer flasks at various pH values (4, 7 and 9) using 50 mL of CR dye solution (600 mg L^{-1}) and 50 mg of adsorbent. The solutions were continuously shaken at 70 rpm in a water bath shaker at 55 °C. For pH adjustment, HCl and NaOH solutions were utilized.

For studying the effect of temperature, 50 mg of CCs adsorbent was added to 50 mL of CR dye solution (600 mg L^{-1}) at different temperatures (25, 35, 45 and 55 °C) at pH 9.

Studying the effect of CR dye concentration is carried out by adding 50 mg of CCs adsorbent to 50 mL of dye solution of various concentrations (400, 500, 600 and 1000 mg L^{-1}) at pH 4 and 55 °C.

2.4.3. Kinetic Studies

Studying the kinetics of the adsorption process is very significant for understanding the adsorption rate onto the particle surface, since adsorption kinetics show the influence of different conditions on the speed of the process by using models that could describe this reaction. In addition, adsorption kinetics determine the mechanism of dye adsorption onto the adsorbent material.

The kinetic data of CR dye were modeled using four different kinetic models: pseudo-first order, pseudo-second order, Elovich and intraparticle diffusion models.

Pseudo-First-Order Model (Lagergren Model)

This kinetic model determines the relationship between the change in time and the adsorption capacity with order of one. It is expressed by Equation (5), where q_t and q_e are adsorption capacity at time t and at equilibrium time, respectively, and k_1 is the rate constant.

$$\frac{dq_t}{dt} = k_1(q_e - q_t) \qquad (5)$$

Integrating Equation (5) with boundary conditions of $q_t = 0$ at $t = 0$ and $q_t = q_t$ at $t = t$ gives a linear equation of pseudo-first order, Equation (6).

$$\log(q_e - q_t) = \log q_e - \frac{k_1}{2.303}t \qquad (6)$$

where q_e and q_t are the adsorption capacity at equilibrium time and time t (mg g^{-1}), respectively, k_1 is the pseudo first order rate constant (min^{-1}) and t is the time (min). The values of q_e and k_1 can be determined from the intercept and the slope of the linear plot of $\log(q_e - q_t)$ versus t.

Pseudo-Second-Order Model (Ho and Mckay Model)

The pseudo-second-order kinetic model is expressed by Equation (7), which shows the relationship of the adsorption capacity and concentration with second order. This model describes the adsorption of the dissolved ions of the dye, via a cation exchange or chemical sharing, onto the adsorbent surface, assigning that a chemical process is involved during the adsorpt.

$$\frac{dq_t}{dt} = k_2(q_e - q_t)^2 \tag{7}$$

The integrated form of Equation (7) gives the linear form of pseudo-second order equation, Equation (8).

$$\frac{t}{q_t} = \frac{1}{k_2 q_e^2} + \frac{t}{q_e} \tag{8}$$

where q_e and q_t are the amount of adsorbed dye onto adsorbent at equilibrium time and time t (mg g^{-1}), respectively, k_2 is the pseudo-second order constant (g mg^{-1} min^{-1}). The slope and intercept of the linear plot of t/q against t yielded the values of q_e and k_2, respectively [23].

Elovich Model

This model is an interesting one for describing the activated chemisorption process, since it is generally applicable for chemisorption kinetics. It can cover a large range of slow adsorption processes. It is valid for the heterogenous adsorbent surfaces and expressed by Equation (9).

$$q_t = \frac{1}{\beta}\ln(\alpha\beta) + \frac{1}{\beta}\ln t \tag{9}$$

where α is the initial adsorption rate constant (mg g^{-1} min^{-1}), β is the desorption constant related to the chemisorption activation energy and the surface coverage (g mg^{-1}), and q_t is the adsorbed dye (mg g^{-1}) at time t (min). By plotting q_t versus ln t, a straight line is obtained where values of α and β can be obtained.

Intraparticle Diffusion Model (Webber and Morris Model)

This model is used to determine the rate-controlling for the adsorption process and is expressed by Equation (10).

$$q_t = (k_{int} t^{1/2}) + C \tag{10}$$

where k_{int} is the intraparticle diffusion constant (mg g^{-1} min$^{-1/2}$), C is a constant (mg g^{-1}) that is directly proportional to the boundary layer thickness. Both values of k_{int} and C can be obtained by calculations from slope and intercept resulted from the linear curve of q_t versus $t^{1/2}$, respectively [24].

Kinetic Validation

To determine the most applicable model for describing the adsorption process, the normalized standard deviation Δq_e (%) is used and calculated using Equation (11) [25].

$$\Delta q_e\ (\%) = 100 \times \sqrt{\frac{\left[\left(q_{t,exp} - q_{t,cal}\right)/q_{t,exp}\right]^2}{N-1}} \tag{11}$$

where $q_{t,exp}$ is the experimental adsorption capacity (mg g^{-1}), $q_{t,cal}$ is the calculated adsorption capacity for pseudo-first and pseudo-second models (mg g^{-1}) and N is the number of data points.

2.4.4. Adsorption Isotherm for CR Dye

Studying the adsorption isotherm is important for describing the ability of adsorbate molecules to distribute between the liquid and the solid phases at equilibrium state of the adsorption process. For studying adsorption isotherm, 10 mL of dye solution (400–1000 mg L^{-1}) was shaken with 10 mg of adsorbent in a shaking water bath (70 rpm) at 55 °C and pH 9. After reaching the equilibrium, the concentrations of un-adsorbed dye were determined by measuring absorbance of dye solution using UV-vis spectrophotometer at λ_{max} = 497 nm. The adsorption isotherm was studied using four models; Langmuir, Freundlich, Temkin and Dubinin–Radushkevich models.

Langmuir Isotherm Model

Langmuir model supposes the formation of a monolayer coverage of adsorbate molecules onto a homogenous surface of adsorbent without interaction between the adsorbate molecules. This isotherm can be expressed by Equation (12).

$$\frac{C_e}{q_e} = \frac{1}{(q_{max} \cdot K_L)} + \frac{C_e}{(q_{max})} \tag{12}$$

where q_e is the amount of adsorbed dye at equilibrium (mg g^{-1}), C_e is the equilibrium concentration of dye in solution (mg L^{-1}), q_{max} is the maximum monolayer coverage adsorption capacity (mg g^{-1}), and K_L is a Langmuir coefficient that is concerned with adsorption energy (L mg^{-1}). By plotting C_e/q_e versus c_e, results in a linear relationship, the values of both q_{max} and K_L can be obtained from the slope ($1/q_{max}$) and the intercept ($1/(q_{max} K_L)$), respectively. There is an important parameter for Langmuir model which is the separation factor (R_L), Equation (13).

$$R_L = \frac{1}{(1 + K_L C_o)} \tag{13}$$

where K_L is the Langmuir constant and C_o is the initial concentration of the adsorbate in solution. This parameter indicates whether the process is irreversible, favorable, unfavorable and linear if $R_L = 0$, $0 < R_L < 1$, $R_L > 1$ and $R_L = 1$, respectively.

Freundlich Isotherm Model

A Freundlich isotherm model is used for describing the adsorption onto heterogeneous surfaces. It possesses the formation of multilayer adsorption with interaction between the adsorbate molecules. Moreover, the adsorption active sites are non-identical for the occupation of adsorbate due to differences in adsorption energy. Its linear form is given by Equation (14).

$$\ln q_e = \ln K_F + \frac{1}{n} \ln C_e \tag{14}$$

where q_e is the adsorbed dye amount at equilibrium (mg g^{-1}) and C_e is the equilibrium concentration of dye in the solution at a constant temperature (mg L^{-1}). K_F is the adsorption capacity ((mg g^{-1}) (L mg^{-1})$^{1/n}$), while $1/n$ shows an indication of the favorability and feasibility of the adsorption process and the surface heterogeneity, which is related to adsorption intensity. The value of $1/n$ indicates the type of isotherm; irreversible ($1/n = 0$); favorable ($0 < 1/n < 1$), and unfavorable ($1/n > 1$). By plotting ln q_e against ln C_e, the constant K_F and the indictor $1/n$ can be obtained.

Temkin Isotherm Model

The Temkin model assumes the decrease in heat uptake for dye molecules with surface saturation under the effect of continuous interaction between the studied adsorbent and dye molecules. The Temkin model can be expressed by Equation (15).

$$q_e = B \ln k_T + B \ln C_e \tag{15}$$

where B is the Temkin constant that is controlled by the adjusted uptake temperature (J mol^{-1}), K_T is the binding constant of Temkin isotherm (L g^{-1}), q_e is the amount of dye adsorbed at equilibrium (mg g^{-1}), and C_e is the dye equilibrium concentration at a constant temperature (mg L^{-1}). By plotting q_e against ln C_e, the constants B and K_T can be obtained from the slope and the intercept, respectively.

Dubinin–Radushkevich (D–R) Isotherm

The Dubinin–Radushkevich model is used for estimating the free energy of adsorption and the characteristic porosity. This model can successfully help in determining the type of the adsorption process whether it is chemisorption or physisorption. The (D–R) isotherm model does not assume that the surface is homogenous; so, it is more general than the Langmuir isotherm. The D–R isotherm is given by Equation (16).

$$\ln q_e = \ln (X_m) - \beta \varepsilon^2 \tag{16}$$

where q_e is the amount of dye adsorbed at equilibrium (mg g^{-1}), X_m is monolayer saturation capacity (mg g^{-1}), β is the activity coefficient that is related to adsorption mean free energy (mol^2 j^{-2}) and ε is the Polanyi potential which is determined by Equation (17).

$$\varepsilon = RT \ln \left(1 + \frac{1}{C_e}\right) \tag{17}$$

where R is universal gas constant (8.314 J mol^{-1}K^{-1}) and T is the absolute temperature (K).

By plotting ln q_e versus ε^2 using Equation (16), a straight line is obtained with a slope and intercept β and ln X_m, respectively. The mean free energy, E, of adsorption (kJ mol^{-1}) can be determined by Equation (18):

$$E = \frac{1}{(2\beta)^{0.5}} \tag{18}$$

This means that the free energy can determine the type of adsorption; if 8 < E < 16 kJ mol^{-1} the process is chemisorption, while if it is less than 8 kJ mol^{-1} the process is physisorption.

2.4.5. Thermodynamic Studies

Thermodynamic parameters provide in-depth information on inherent energetic changes associated with adsorption; therefore, these parameters should be accurately evaluated. Studying thermodynamic parameters give a better understanding of the adsorptive behavior for the dye towards adsorbents. Gibbs free energy change, $\Delta G°$ (kJ mol^{-1}), enthalpy change, $\Delta H°$ (kJ mol^{-1}) and entropy change, $\Delta S°$ (J mol^{-1} K^{-1}) were calculated for the adsorption of CR dye using four different temperatures (298, 308, 318 and 328 K) according to Equations (19) and (20).

$$\Delta G° = -RT \ln K_c \tag{19}$$

$$\ln k_c = \frac{\Delta S°}{R} - \frac{\Delta H°}{RT} \tag{20}$$

where R is the universal gas constant (8.314 J mol^{-1} K^{-1}), T is the absolute temperature (K) and K_c is the distribution coefficient (q_e/C_e). By plotting ln K_c versus 1/T, the thermodynamic parameters can be obtained.

2.4.6. Activation Energy

The activation energy (E_a) is determined by applying Arrhenius equation, which refers to the minimum energy for proceeding the reaction as shown in Equation (21).

$$\ln k = \ln A - \frac{E_a}{RT} \tag{21}$$

where A is the Arrhenius factor, R is the universal gas constant (8.314 J mol^{-1} K^{-1}), E_a is the activation energy (kJ mol^{-1}) and T is the absolute temperature (K). By plotting ln k of the pseudo-second order constant versus inverse temperature (1/T), a straight line is obtained with slope of $-E_a/R$, where E_a can be determined.

2.4.7. Desorption Study

The dye desorption from adsorbent was performed by its washing with distilled water to remove any un-adsorbed dye molecules. Afterwards, the adsorbent (0.01 g) was immersed in the desorption medium (10 mL of ethanol, methanol, acetone or aqueous NaOH solution (0.1 N)) at 25 °C for 24 h. The amount of the desorbed dye can be calculated using Equation (22).

$$\% \text{ Dye desorption} = q_d/q_a \times 100 \tag{22}$$

where q_d is the amount of desorbed dye from the adsorbent surface (mg g^{-1}) and q_a is the amount of the dye adsorbed onto the adsorbent (mg g^{-1}) [24].

3. Results and Discussion

3.1. Synthesis of Novel CCs Adsorbent

The CCs adsorbent (Scheme 1) was prepared via a four-step procedure since the primary amine groups in chitosan were firstly protected by reaction with benzaldehyde for achieving chitosan Schiff's base, in which the primary hydroxyl groups on C6 reacted with epichlorohydrin for generating epoxy chitosan Schiff's base, followed by reaction of the epoxy rings with cyanoguanidine for attaining cyanoguanidine chitosan Schiff's base, which was finally hydrolyzed in acidic medium to eliminate the benzaldehyde moieties and retrieve the amino groups to get CCs adsorbent. The amino and hydroxyl groups on chitosan in addition to the basic functional groups of cyanoguanidine incorporated into chitosan can potentially remove the acidic pollutants such as acidic dyes.

3.2. Characterization of Novel CCs Adsorbent

3.2.1. FTIR Spectra of CCs Adsorbent

FTIR spectra of chitosan and its modified derivatives were demonstrated in Figure 2. In the spectrum of chitosan, the existence of the saccharide moieties was confirmed by the appearance of four absorption peaks at 1158, 1074, 1029, and 894 cm^{-1}. A dense broad absorption peak at around 3700 to 3000 cm^{-1} appeared, relating to the stretching vibration of -OH groups overlapped with that for -NH$_2$ and their hydrogen bonds. The symmetric absorption peak corresponded to -CH and -CH$_2$ groups in the pyranose rings appeared at 2924 and 2864 cm^{-1}, respectively. The high extent of deacetylation of chitosan was confirmed by the appearance of two weak absorption peaks at 1658 and 1593 cm^{-1} assigning to amide I and amide II, respectively. The overlapping between the amino groups deforming vibration at 1600 cm^{-1} and the stretching vibration peak of amide I at 1658 cm^{-1} resulted in an intensive peak [22,26].

The spectrum of chitosan Schiff's base displayed similar absorption peaks of chitosan in addition to some new peaks as follows: (1) at 3052 and 3027 cm^{-1} indicated to C-H groups in aromatic ring, (2) at 1691 cm^{-1} corresponded to C=N groups, (3) at 1600, 1579, 1493 and 1454 cm^{-1} related to C=C bond in aromatic rings, and (4) at 757 and 692 cm^{-1} (strong) due to mono-substituted benzene rings [16,21].

Epoxy chitosan Schiff's base spectrum, in addition to the afore-mentioned peaks, displayed a new peak at 1250 cm^{-1} due to the epoxide moieties [27]. The spectrum of the cyanoguanidine chitosan Schiff's base showed that the disappearance of the peak corresponded to the epoxide linkages at 1250 cm^{-1} and the appearance of two new peaks at 2207 and 2162 cm^{-1} related to C≡N group of the cyanoguanidine moiety [28], indicating occurrence of the interaction between the epoxide rings and the NH$_2$ groups of cyanoguanidine. Moreover, the stretching vibration peak corresponded to C=N group of the cyanoguanidine moiety appeared at 1641 cm^{-1}.

Removal of benzaldehyde moieties to obtain CCs adsorbent was confirmed by the disappearance of the absorption bands of mono-substituted benzene ring at 757 and 692 cm^{-1}.

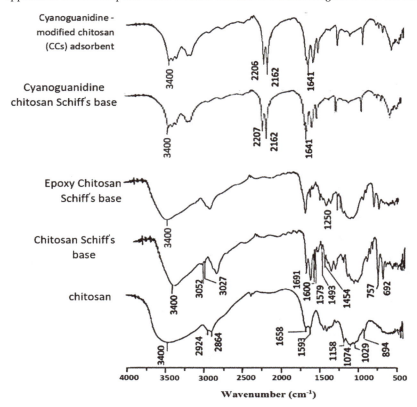

Figure 2. FTIR spectra of the modified chitosan derivatives.

3.2.2. Powder X-ray Diffraction of CCs Adsorbent

X-ray diffractometry was used to explore the inner structures of the modified chitosan derivatives and their X-ray diffraction patterns were shown in Figure 3. There were two broad peaks in chitosan appeared near to $2\theta = 10°$ and $20°$ which were attributable to its amorphous and crystalline regions, respectively [29]. This can be ascribed to the formation of the hydrogen bonds along its chains due to its possession of a great number of hydroxyl and amino groups. On the other hand, the modified chitosan derivatives were less crystalline than chitosan. This was illustrated by a disappearance of the peak at $2\theta = 10°$ and a decrease of the intensity of the peak at $2\theta = 20°$. After modification of chitosan, its functionality greatly changed with a reduction of the hydrogen bonds due to consumption of its polar –NH$_2$ and/or –OH groups and incorporation of the modifiers' moieties that separated the chitosan chains away from each other. This led to an increase of the amorphous region and a decrease of the crystalline region.

3.2.3. SEM Analysis of CCs Adsorbent

SEM micrographs of surface topography of the modified chitosan derivatives were shown in Figure 4. The original chitosan showed a smooth surface; however, its derivatives showed a raucous surface, containing lumps of diverse sizes because the incorporated substituent groups have assorted sizes. It can be noted that the lumps were homogeneously distributed over each derivative, suggesting that every stage for the chitosan modification process was successfully completed. The inserted substituent groups separated chitosan chains away from each other, reduced the formation of their hydrogen bonds and created a porous matrix with a large surface area.

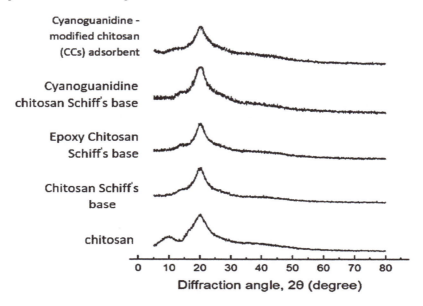

Figure 3. X-ray diffraction patterns of the modified chitosan derivatives.

3.3. Adsorption of CR Dye Using CCs Adsorbent

CCs was used in the present work for the removal of CR dye from water. Its adsorption capacity came from the incorporation of functional chelating cyanoguanidine moieties, which possessed more binding centers for CR dye. Various adsorption factors were studied, in addition to studying kinetics, isotherm and thermodynamics for understanding the adsorption mechanism.

3.3.1. Optimization of the Adsorption Conditions
Effect of pH

One of the main parameters that has a major role on adsorption of dyes is the pH of the solution due to some mechanisms such as protonation processes between dyes and adsorbent surface.

The effect of pH on the adsorption of CR dye onto CCs adsorbent was studied at different pH values of 4, 7 and 9, and the results were presented in Figure 5. It can be noted that the adsorption of CR dye onto CCs decreased with increasing pH values. From the results, the percentages of CR dye removal were 92.55, 85.59 and 82.40% at pH 4, 7 and 9, respectively. This is in agreement with the study for the removal of CR dye using chitosan-coated quartz sand [30].

At low pH, an electrostatic attraction between the positively charged active sites on CCs adsorbent and dye anions takes place. The incorporation of cyanoguanidine moieties improved the adsorption performance of CCs, because they introduced more positive charges onto the CCs surface [31].

At a high pH, a strong competition between OH^- groups and dye anions for the positively charged sites of CCs occurred. Since the OH^- groups have a smaller size than the dye anions, binding of OH^- groups with CCs predominated, decreasing the adsorption performance of CCs.

Interestingly, the cyanoguanidine moieties in CCs have an important role in enhancing and improving its adsorptive capacity. The % R.E. values of CR dye (adsorbent dose = 50 mg, dye solution (50 mL, 600 mg L^{-1}) pH = 9 and temperature = 55 °C) by chitosan Schiff's base, epoxy chitosan Schiff's base and cyanoguanidine chitosan Schiff's base were 20.66, 23.70 and 41.48%, respectively, which were lower than that of CCs (82.40%).

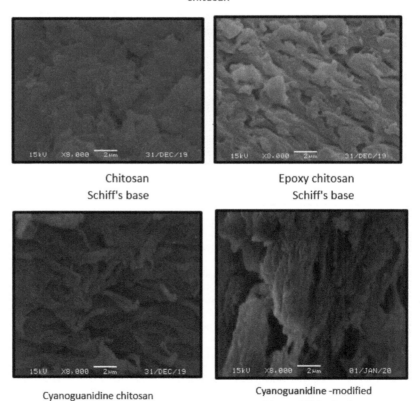

Figure 4. SEM images of the modified chitosan derivatives.

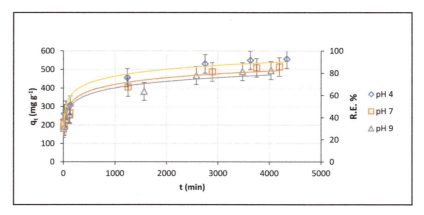

Figure 5. Effect of pH on CR dye adsorption onto CCs (Experimental conditions: 50 mg of adsorbent, 50 mL of dye solution (600 mg L^{-1}) and temperature 55 °C).

Effect of Temperature

Temperature is an important parameter which can change the adsorption capacity of adsorbent and highly affect the adsorption of dyes from aqueous solution. The adsorption of CR dye onto CCs was studied at four different temperatures (25, 35, 45 and 55 °C) and the results were illustrated in Figure 6. It was obvious that increasing the temperature was accompanied by increasing the adsorption capacity of the adsorbent towards CR dye molecules. The % R.E. values were 55.51, 72.22, 77.0 and 82.40% at temperatures 25, 35, 45 and 55 °C, respectively. This indicated that a high temperature can facilitate the adsorption of CR dye onto CCs. This might be attributed to the swelling effect of the internal surface of an adsorbent, which contributed for the penetration of dye molecules into the interlayer of the adsorbent. This indicated that the adsorption of CR dye onto CCs was endothermic in nature [32]. Additionally, the gradual increase of temperature was followed by an increase in dye diffusivity and an increase in the dimensions of adsorbent pores. This consequently led to a reduction in the contribution of intraparticle resistance and decreased the effect of the boundary-layer [30]. The maximum adsorption capacity was achieved at 55 °C, which was used as the optimum temperature for batch experiments.

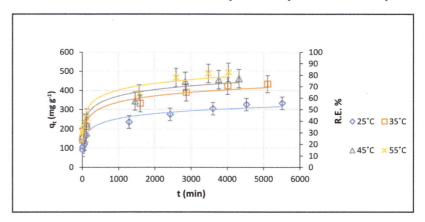

Figure 6. Impact of temperature on adsorption of CR dye onto CCs (Experimental conditions: 50 mg of adsorbent, 50 mL of dye solution (600 mg L^{-1}) and pH = 9).

Impact of Initial Concentrations of the Dye

The adsorbed dye amount is strongly dependent on the initial dye concentration. This finding comes from the relationship between the initial dye concentration and the available active sites onto the adsorbent surface. In the present study, the use of 150 mg L^{-1} dye solution resulted in a complete adsorption, which reached 100% of CR dye removal (Experimental condition: pH = 9, temperature = 55 °C and equilibrium time = 6 h).

The adsorption of CR dye using CCs was studied at different concentrations (400, 500, 600 and 1000 mg L^{-1}) and the results were illustrated in Figure 7. It could be noted that increasing the initial concentration of the dye was accompanied by a decrease in the removal percentage of CR dye. It decreased gradually to 99.5, 94.9, 92.55 and 76% at concentrations 400, 500, 600 and 1000 mg L^{-1}, respectively. This can be attributed to the limited obtainable reactive centers available for adsorption process at higher concentrations. While the adsorption capability increased with increasing initial dye concentration, this is attributed to the high driving force of the concentration gradient at higher initial CR dye concentrations [30]. The maximum removal rate of CR dye was achieved at initial dye concentration of 400 mg L^{-1}, which was considered the optimum concentration.

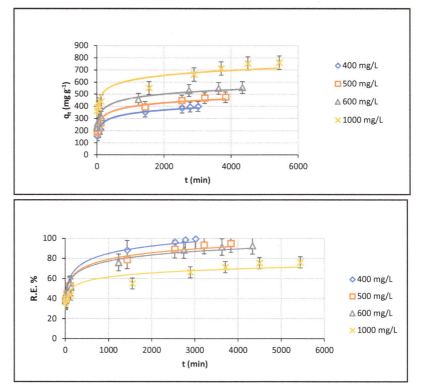

Figure 7. Impact of concentration of CR dye on adsorption onto CCs (Experimental conditions: 50 mg of adsorbent, 50 mL of dye solution, temperature 55 °C and pH = 4).

3.3.2. Adsorption Kinetics

Studying adsorption kinetics is essential for suggesting the mechanism of adsorption, determining the optimum parameters for reaching the maximum removal percentage. and explaining the rate of dye uptake. The effect of time on the adsorption of CR dye using CCs was studied at different pHs, temperatures and initial dye concentrations using three kinetic models: pseudo-first-order, Equation (6), pseudo-second-order, Equation (8), and Elovich model, Equation (9), in order to understand the adsorption mechanism.

At a Different pH

The results of kinetic data of the adsorption of CR dye onto CCs are summarized in Table 1. The plots of the adsorption for the three kinetic models at pH 4, 7 and 9 were illustrated in Figure 8.

Table 1. Kinetic model constants and correlation coefficients for the adsorption of CR dye onto CCs at different pH.

Kinetic Models	Parameters	PH		
		4	7	9
pseudo-first-order	$q_{e.exp}$ (mg g^{-1})	555.33	513.56	494.44
	R^2	0.989	0.969	0.958
	$q_{e.cal}$ (mg g^{-1})	310.38	321.07	322.18
	K_1 (10^{-4}) (min^{-1})	9.21	9.21	11.15
	Δq_e (%)	16.67	14.17	13.17
pseudo-second-order	R^2	0.997	0.996	0.991
	$q_{e,cal}$ (mg g^{-1})	555.56	526.32	500
	K_2 (10^{-5}) (g·mg^{-1}·min^{-1})	2.12	1.86	1.81
	Δq_e (%)	0.02	0.94	0.42
Elovich	R^2	0.961	0.951	0.94
	β (g·mg^{-1})	0.019	0.019	0.02
	α (mg g^{-1} min^{-1})	325.64	168.8	148.17

The pseudo-second-order model showed the highest value of correlation coefficient (R^2) compared with the other models. It ranged from 0.997 to 0.991 at pH values from 4 to 9. This indicated that the pseudo-second order model has a perfect fit to describe the CR dye adsorption by CCs surface. In addition, there is a good agreement between the values of experimental and calculated q_e for the pseudo-second-order, which is equal to 555.33 and 555.56 mg g^{-1}, respectively, at pH = 4. This reflected and confirmed the excellent fitting of the pseudo-second-order kinetic model for the adsorption process of CR dye onto CCs.

On the other hand, the values of R^2 obtained the for pseudo-first order showed relatively lower values than that obtained by the pseudo-second order model, since they ranged from 0.989 to 0.958 at pH values from 4 to 9. Additionally, the theoretical equilibrium adsorption capacities, q_e, for pseudo-first-order were significantly different from that of the experimental ones which equal to 555.33 and 310.38 mg g^{-1}, respectively, suggesting the inadequacy of the pseudo-first-order model for describing the adsorption kinetics of CR dye onto CCs.

The normalized standard deviation, Δq_e, (%) was applied to check the validity of the pseudo-first-order and pseudo-second-order models quantitatively. Based on the values of Δq_e (%) shown in Table 1, it is clearly noted that the pseudo-second-order model showed the best fit since it yielded lower values of Δq_e (%) than those obtained from pseudo-first-order at different pH values. These results confirmed that the adsorption kinetic studies of CR dye onto CCs would be more favorable when applied by pseudo-second-order. Comparable data were published in a previous study to remove CR dye using chitosan-coated quartz sand [30].

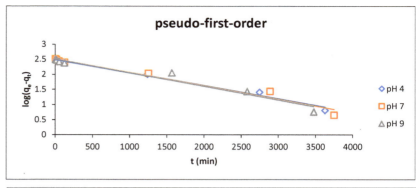
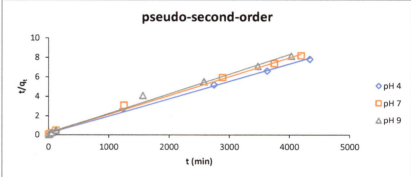
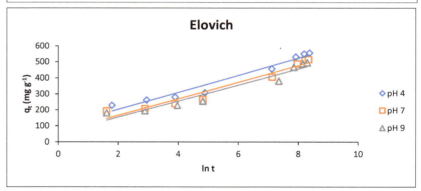

Figure 8. Linear regressions of kinetic plots at different pH for the adsorption of CR dye onto CCs.

The fitness of the pseudo-second order kinetic model for the adsorption of CR dye onto CCs indicated that the adsorption process was proceeded by various mechanisms, including chemical interaction and electrostatic attraction of anions of the dye with the binding reactive centers onto the surface of CCs [30].

On the other hand, it is noted that the values of rate constant, k_2, decreased gradually with the increase in the pH from 4 to 9, suggesting that pH 4 was the optimum value for adsorption. At pH 4, an enhancement in the electrostatic attraction force between the protonated sites of CCs surface and the negatively charged sulfonate groups ($-SO_3^-$) of CR dye molecules took place. This led to an increase in the adsorption process at a low pH.

The Elovich model is one of the most interesting kinetic models that has a wide applicability to describe the adsorption systems related to the chemical nature. This model occurs when the adsorption involves the chemisorption reaction onto adsorbent surface, and the adsorption speed decreases as time passes due to covering of the adsorbent surface with an adsorbate [33].

Figure 8 showed the linear plots of the Elovich kinetic model, which was obtained by applying Equation (9) at pH 4, 7 and 9. The values of α, β and the respective R^2 were summarized in Table 1. The obtained correlation coefficients were relatively high, which confirmed that this model fitted fully to the experimental results of the adsorbent. The values of R^2 ranged from 0.96 and 0.94 at pH from 4 to 9. The good fitness of the experimental data for Elovich confirmed that the adsorption of CR dye onto CCs was controlled by a chemisorption process.

On the other hand, it was obvious that the values of the initial adsorption rate, α, decreased with increasing pH values. They ranged from 325.64 to 148.17 mg g^{-1} min^{-1} at pH values from 4 to 9 for the adsorption of CR dye onto CCs. Whereas, the values of the extent of surface coverage, β, increased with increasing pH; they ranged from 0.019 to 0.020 g mg^{-1} at pH values from 4 to 9.

At Different Temperatures

Temperature is a crucial parameter which highly affects both adsorption performance and adsorbent behavior. The experimental results of kinetic data were summarized in Table 2 for the adsorption of CR dye onto CCs. The plots showing the three kinetic models for CR dye adsorption by CCs, at 25 to 55 °C were illustrated in Figure 9.

Table 2. Kinetic model constants and correlation coefficients for the adsorption of CR dye onto CCs at different temperatures.

Kinetic Models	Parameters	Temperatures			
		25 °C	35 °C	45 °C	55 °C
	$q_{e.exp}$ (mg g^{-1})	333.11	433.33	462	494.44
pseudo-first-order	R^2	0.957	0.973	0.978	0.958
	$q_{e.cal}$ (mg g^{-1})	230.57	288.07	313.47	322.18
	K_1 (10^{-4}) (min^{-1})	6.909	6.909	6.909	11.151
	Δq_e (%)	10.88	12.67	12.15	13.17
pseudo-second-order	R^2	0.99	0.993	0.992	0.991
	$q_{e,cal}$ (mg g^{-1})	333.33	434.78	476.19	500
	K_2 (10^{-5}) (g·mg^{-1}·min^{-1})	1.64	1.7	1.73	1.81
	Δq_e (%)	0.02	0.13	1.16	0.42
Elovich	R^2	0.963	0.963	0.937	0.94
	β (g·mg^{-1})	0.027	0.021	0.02	0.02
	α (mg g^{-1} min^{-1})	36.24	64.72	96.36	148.17

Based on the high values of correlation coefficient (R^2) and the low values of the normalized standard deviation, Δq_e (%), it was clearly apparent that the pseudo-second-order model showed fully fit for the obtained results compared with pseudo-first-order model. R^2 values obtained from pseudo-second-order adsorption model ranged from 0.990 to 0.991 at 25 to 55 °C. In addition, the q_e for pseudo-second-order model was similar to that of the experimental one, since the values of experimental and calculated q_e for pseudo-second-order model at 55 °C equal to 494.44 and 500 mg g^{-1}, respectively. On the other hand, the correlation coefficient obtained from the pseudo-first order model showed lower values than those obtained from pseudo-second order, since they ranged from 0.957 and 0.958 at temperatures from 25 and 55 °C for the adsorption of CR dye onto CCs. In addition, the experimental and calculated capacities showed different values and equalled 494.44 and 322.18 mg g^{-1} at 55 °C, respectively. This indicated the unsuitability of the pseudo-first order model for describing the adsorption process.

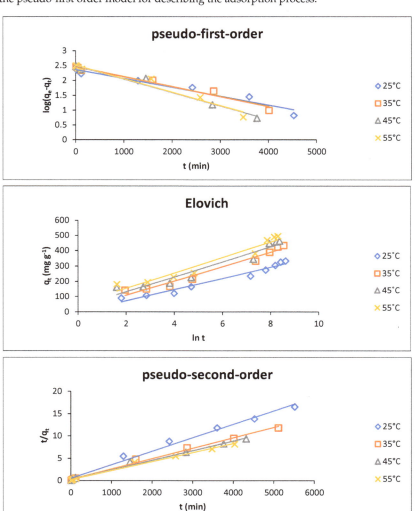

Figure 9. Linear regressions of kinetic plots at different temperatures for the adsorption of CR dye onto CCs.

Additionally, the value of rate constant, k$_2$, increased gradually by increasing the temperature, as it increased from 1.64×10^{-5} to 1.81×10^{-5} g mg^{-1} min^{-1} when the temperature increased from 25 to 55 °C. This can be attributed to the low viscosity of the dye solution, which subsequently entrapped more dye molecules onto the adsorbent surface [34]. Furthermore, this might be due to the increase in collision probability between the active sites of adsorbent and the adsorbate and the decrease in the boundary layer thickness of adsorbent at an elevated temperature [35]. In addition, this might be due to the effect of swelling of the adsorbent internal structure, resulting from increasing temperature. This led to a greater diffusion capability for dye molecules into the adsorbent surface, reaching to its internal structure, and finally increasing the rate of adsorption [32].

Figure 9 showed the linear form of the Elovich model (Equation (9)). The values of the initial adsorption rate constant (α) and the desorption constant (β) at different temperatures (25, 35, 45 and 55 °C) were summarized in Table 2. Obviously, the experimental data for the adsorption of CR dye onto CCs agreed with the Elovich model. This result was reflected by R^2 values which ranged from 0.963 to 0.940 at temperatures from 25 to 55 °C. The Elovich model assumes that the adsorbent active sites were heterogenous, and exhibited various energies for chemical adsorption [33]. This model is used for describing the kinetics of ion exchange systems. Hence, it is concluded that the adsorption of CR dye onto CCs was chemisorption. This finding is in agreement with the above results observed throughout the present study, which concluded that pseudo-second order is the best fit model. The obtained results also suggested that the chemisorption is the rate-determining step for the adsorption of CR dye onto CCs.

On the other hand, based on the results in Table 2, the values of initial adsorption rate constant, α, showed an increase with the increasing temperature. It ranged from 36.24 to 148.17 mg g^{-1} min^{-1} when the temperature increased from 25 to 55 °C for the adsorption onto CCs, whereas the desorption constant, β, showed a decrease with increasing temperature; the values of β ranged from 0.027 to 0.020 g mg^{-1} when the temperature increased from 25 to 55 °C for the adsorption onto CCs. This is attributed to the low number of available binding sites for the adsorption at an elevated temperature [36].

At Different Initial Dye Concentrations

The experimental kinetic data were summarized in Table 3. Figure 10 showed the graphical linear forms of the studied three models of kinetics for adsorbing CR dye onto CCs at different initial dye concentrations (400, 500, 600 and 1000 mg L^{-1}).

Table 3. Kinetic model constants and correlation coefficients for the adsorption of CR dye onto CCs at different dye concentrations.

Kinetic Models	Parameters	Dye Concentrations			
		400 mg L^{-1}	500 mg L^{-1}	600 mg L^{-1}	1000 mg L^{-1}
	$q_{e.exp}$ (mg g^{-1})	398	474.67	555.33	760.44
pseudo-first-order	R^2	0.972	0.975	0.989	0.879
	$q_{e.cal}$ (mg g^{-1})	240.99	279.58	310.38	429.64
	K_1 (10^{-4}) (min^{-1})	13.82	9.21	9.21	6.91
	Δq_e (%)	13.95	14.53	16.67	15.38
pseudo-second-order	R^2	0.998	0.996	0.997	0.99
	$q_{e,cal}$ (mg g^{-1})	400	476.19	555.56	769.23
	K_2 (10^{-5}) (g·mg^{-1}·min^{-1})	3.97	2.58	2.12	0.88
	Δq_e (%)	0.18	0.11	0.02	0.41
Elovich	R^2	0.963	0.959	0.961	0.913
	β (g mg^{-1})	0.023	0.021	0.019	0.017
	α (mg g^{-1} min^{-1})	111.97	207.29	325.64	1874.6

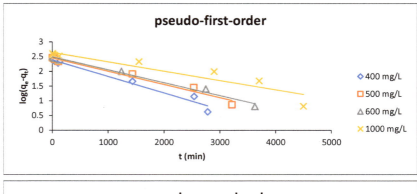
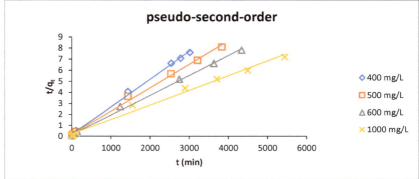
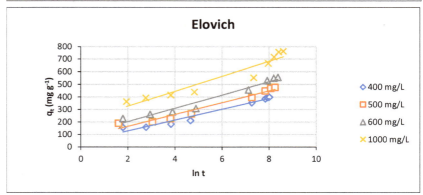

Figure 10. Linear regressions of kinetic plots for the adsorption of CR dye onto CCs.

The adsorption of CR dye onto the adsorbent showed excellent compliance with the pseudo-second-order, better than the pseudo-first-order, according to their respective correlation coefficient (R^2) values and the values of the normalized standard deviation, Δq_e (%). The R^2 values obtained for pseudo-second-order ranged from 0.998–0.990, which were higher than those obtained from the pseudo-first-order (0.972–0.879) at concentrations which ranged from 400–1000 mg L^{-1}. On the other hand, the values of the normalized standard deviation, Δq_e (%), showed lower values for the pseudo-second-order than the pseudo-first-order. In addition, the calculated equilibrium capacities were close to the experimental ones for the pseudo-second-order. These findings indicated that the adsorption kinetics of CR dye by CCs perfectly followed the pseudo-second-order model, implying that the overall rate of the adsorption process was controlled by chemisorption [35]. Similar results were obtained for adsorption of CR dye onto trimellitic anhydride isothiocyanate-cross-linked chitosan hydrogels [23].

Regarding the values of rate constants for pseudo-second-order model, k_2, there was a significant decrease for k_2 values with increasing dye concentrations, as it decreased from 3.97×10^{-5} to 0.88×10^{-5} g mg^{-1} min^{-1} when the dye concentration increased from 400 to 1000 mg L^{-1}. The same results are in agreement with another study [36]. The obtained results confirmed that the optimum concentration value was 400 mg L^{-1}. This might be ascribed to the decrease in the unoccupied available reactive centers with time, which led to a decrease in the adsorption rate [35].

The linear form of Elovich model, represented in Equation (9), was illustrated in Figure 10 at different dye concentrations (400–1000 mg L^{-1}) for the adsorption onto CCs. The values of the initial adsorption rate, α, and the extent of surface coverage, β, were summarized in Table 3. It is noted that the experimental results agreed adequately with the Elovich model, since R^2 ranged from 0.963 and 0.913 at a dye concentration from 400 and 1000 mg L^{-1}. The good applicability of the Elovich model indicated that the process was governed by chemisorption. On the other hand, the values of the initial rate of adsorption increased with increasing concentrations, and it increased from 111.97 to 1874.60 mg g^{-1} min^{-1} by increasing concentrations from 400 to 1000 mg L^{-1}. This result might be ascribed to the higher concentration gradient, whereas a decrease in the extent of coverage (β) with increasing concentration was observed and decreased from 0.23 to 0.017 g mg^{-1} with increasing concentrations from 400 to 1000 mg L^{-1}. Since at a higher concentration the gradient forces the particles of CR dye to show more adsorption towards the surface of adsorbent, this might negatively affect the desorption process [37].

3.3.3. Mechanism of CR Dye Adsorption onto CCs

Mainly, in the aqueous solution, the CR dye (CR-SO$_3$Na) would dissolve and dissociate giving the dye anions (CR-SO$_3^-$) as shown in Equation (23).

$$CR\text{-}SO_3Na \rightarrow CR\text{-}SO_3^- + Na^+ \qquad (23)$$

CCs is characterized by polar functional groups (–NH$_2$ and –OH) on its surface. At a low pH, the protonation of these groups took place, leading to the formation of NH$_3^+$ and -OH$_2^+$, respectively. Thus, an electrostatic interaction occurred between these two positively charged functional groups with the negatively charged dye anions, as shown in Equations (24) and (25).

$$-OH \xrightarrow{H^+} -OH_2^+ \xrightarrow{CR-SO_3^-} CR-SO_3^- OH_2^+ - \qquad (24)$$

$$-NH_2 \xrightarrow{H^+} -NH_3^+ \xrightarrow{CR-SO_3^-} CR-SO_3^- NH_3^+ - \qquad (25)$$

Additionally, the surface of CCs contains several heteroatoms, comprising oxygen and nitrogen. Therefore, binding of CR dye on the CC surface via hydrogen bonding and Van der Waals forces cannot be ruled out [30]. Thus, the possible mechanism involves a combination of adsorbent–sorbate electrostatic interactions, between the positively charged protonated groups and the negatively charged CR dye ions, in addition to H-bonding interactions and other physical forces such as π–π stacking and van der Waals forces [38]. Whereas, the electrostatic interaction might be the main mechanism for the removal behavior (Scheme 2).

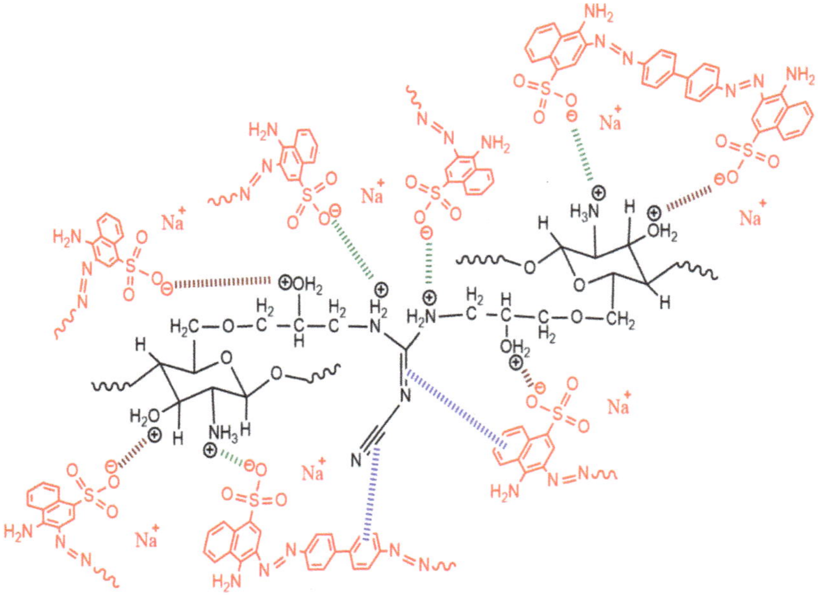

Scheme 2. Schematic illustration of the interaction of CR dye with CCs at a low pH.

To identify and explore the diffusion mechanism for the adsorption of the CR dye, the intraparticle diffusion model was applied. By studying the intraparticle diffusion, the adsorption process involves a multi-step process: (i) bulk diffusion of dye molecules, (ii) film diffusion; dye molecules can diffuse to the surface of adsorbent through the boundary layer, (iii) intraparticle diffusion; dye molecules are transported from the surface into adsorbent pores, and (iv) dyes are adsorbed onto the adsorbent active sites through chemical or physical reaction. Due to the continuous stirring of the batch system, ignoring diffusion by bulk could be suggested and the step that determines the rate would be the intraparticle diffusion in biosorption [39].

The Intraparticle Diffusion

The intraparticle diffusion model was studied for the determination of the rate determining step. When the plot of qt versus $t^{1/2}$ (Equation (10)) gives a straight-line passing via the origin, this implies that the process of adsorption is governed by diffusion, and the particle diffusion would be the rate controlling step. However, if the plot shows two or more linear regions, this indicates that the adsorption process takes place by a multistage adsorption [36].

The intraparticle diffusion results for the adsorption of CR dye onto CCs were analyzed at different pH values (from 4 to 9), temperatures (from 25 to 55 °C) and initial dye concentrations (from 400 to 1000 mg L^{-1}), and are illustrated in Figures 11–13, respectively. The obtained plots did not pass through the origin. This confirmed that the intraparticle diffusion model was not the only rate-limiting step, suggesting other kinetic factors controlling and contributing the rate and the mechanism of adsorption of CR dye onto CCs. The multilinearity of the curves confirmed that the adsorption of CR dye molecules into the adsorbent surface took place in three stages. Firstly, the dye molecule diffused into the external surface of adsorbent by film diffusion, then in a slower step it diffused into the internal pores of the adsorbent throughout the intraparticle diffusion. The second stage was more stable than the first stage, indicating that it was the step for determining the rate, and the final step corresponded to the stage for final equilibrium [40].

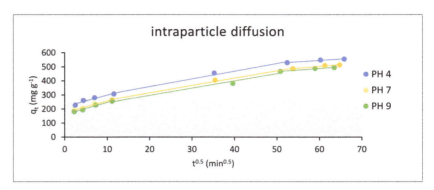

Figure 11. The intraparticle diffusion plots for adsorption of CR dye onto CCs using different pH values.

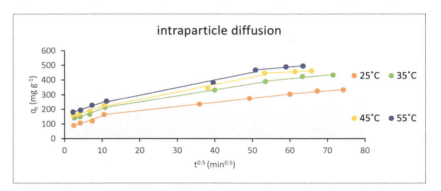

Figure 12. The intraparticle diffusion plots for adsorption of CR dye onto CCs at different temperatures.

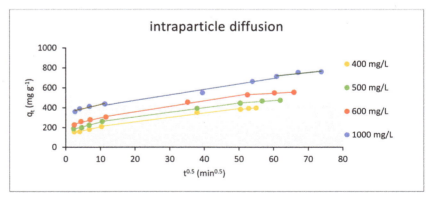

Figure 13. The intraparticle diffusion plots for adsorption of CR dye onto CCs using different initial concentrations.

The first slope ($k_{int.1}$) represented the fast adsorption of dye molecules onto adsorbent surface by electrostatic interactions, while the second slope ($k_{int.2}$) revealed to the diffusion of dye molecules into the internal pores of adsorbent. The third slope ($k_{int.3}$) corresponded to the saturation of all active sites on adsorbent surface (equilibrium stage). These findings suggested that CR dye adsorbed onto the surface of adsorbent and in its internal structure [38].

Studying the intraparticle diffusion with the variation of pH was carried out using pH values of 4, 7 and 9. The values of intraparticle diffusion rate (k_{int}) and correlation coefficients (R^2) were summarized in Table 4 for the adsorption of CR dye onto CCs. It is noted that the values of $k_{int.2}$, decreased with increasing pH values; they equal 5.50, 5.24 and 5.17 at pH 4, 7 and 9, respectively. A similar result was illustrated in a previous study for adsorption of CR dye onto pine bark [41].

Table 4. Parameters of intraparticle diffusion model using different pH for adsorption of CR dye onto CCs.

	Parameters	pH		
		4	7	9
Whole processes	R^2	0.979	0.988	0.989
	k_i	5.19	5.19	5.18
1st	$K_{i.1}$	11.26	9.72	9.54
2nd	$K_{i.2}$	5.5	5.24	5.17
3rd	$K_{i.3}$	1.15	1.26	1.27

The intraparticle diffusion at 25, 35, 45 and 55 °C was studied, and k_{int} and R^2 values were summarized in Table 5. The results showed an increase in the diffusion rate constant with the increasing temperature. For the adsorption of CR dye onto CCs, the diffusion rate constant ($k_{int.2}$) increased from 2.84 to 5.17 mg g^{-1} min^{-1} when the temperature increased from 25 to 55 °C. All values of R^2 are higher than 0.9, suggesting that the adsorption process took place by a combination mechanism.

Table 5. Parameters of intraparticle diffusion model for adsorption of CR dye onto CCs at different temperatures.

	Parameters	Temperatures			
		25 °C	35 °C	45 °C	55 °C
Whole processes	R^2	0.976	0.984	0.987	0.989
	k_i	3.33	4.37	4.96	5.18
1st	$K_{i.1}$	5.9	6.25	6.43	9.54
2nd	$K_{i.2}$	2.84	4.1	5.06	5.17
3rd	$K_{i.3}$	0.95	1.2	1.21	1.27

The intraparticle diffusion was studied also at various initial dye concentrations (400, 500, 600 and 1000 mg L^{-1}) and the values of intraparticle diffusion rate (k_{int}) and correlation coefficients (R^2) for adsorbing CR dye onto CCs were summarized in Table 6. It was found that the values of k_{int} increased with the increase in the concentration, since the values of $k_{int.2}$ for the second stage equal to 4.46 and 5.51 mg g^{-1} min^{-1} at concentrations of 400 and 1000 mg L^{-1}, respectively. These findings indicated that the diffusion of CR dye molecules into the interior pores of adsorbent surface increased with increasing the initial dye concentration, and it is in agreement with some previous studies [40].

3.3.4. Adsorption Isotherm

The adsorption isotherm is considered one of the main important fundamentals for describing the mechanism of dye adsorption onto the adsorbent surface. These isotherms efficiently show the way of interaction between the dye and the sorbent surface.

The adsorption isotherms for CR dye removal by CCs were represented in Figure 14 for Langmuir, Freundlich, Temkin and Dubinin–Radushkevich (D–R) isotherm models. Additionally, Table 7 listed the results of the four models and their corresponding fitting correlation coefficients (R^2). The suitability of the model would be determined by the value of R^2 which is the closest to unity [36].

Table 6. Parameters of intraparticle diffusion model for adsorption onto CCs using different CR dye concentrations.

	Parameters	Dye Concentrations (mg L^{-1})			
		400	500	600	1000
whole processes	R^2	0.988	0.986	0.979	0.989
	k_i	4.7	4.88	5.19	5.55
1st	$K_{i.1}$	6.59	7.9	11.26	11.99
2nd	$K_{i.2}$	4.46	4.63	5.5	5.51
3rd	$K_{i.3}$	1.91	1.41	1.15	0.96

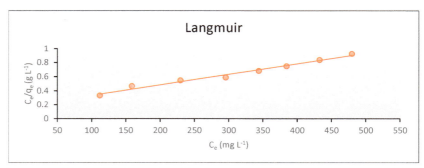

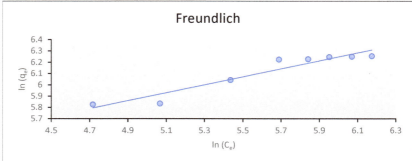

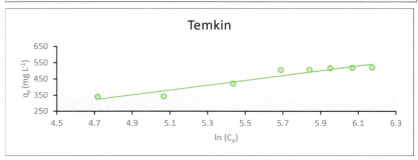

Figure 14. *Cont.*

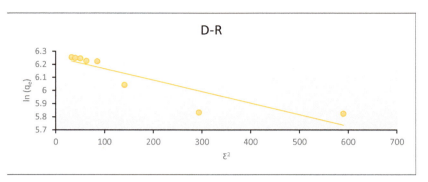

Figure 14. Adsorption isotherms for CR dye onto CCs at 55 °C and pH 9.

Table 7. Adsorption isotherm constants of CR dye by CCs.

Isotherm Model	Parameters	
Langmuir	R^2	0.98
	K_L (L mg^{-1})	0.008
	q_{max} (mg g^{-1})	666.67
	R_L	0.217–0.111
Freundlich	R^2	0.917
	K_f (L mg^{-1})	62.11
	$1/n$	0.35
Temkin	R^2	0.92
	K_T (L mg^{-1})	0.08
	B (J mol^{-1})	149.74
D-R	R^2	0.807
	X_{max} (mg g^{-1})	520.04
	β (mol^2 J^{-2})	0.0009
	E (kJ mol^{-1})	0.023

According to R^2 values, the adsorptive systems were better modeled by Langmuir adsorption isotherms than Freundlich adsorption isotherms, over the concentration ranges studied, since the value of R^2 for Langmuir was 0.980 and for Freundlich was 0.917.

The good applicability of Langmuir adsorption isotherm for the removal of CR dye by the adsorbent indicated that the adsorption process occurred in a monolayer coverage manner with no interaction between adsorbed dye molecules with each other. In addition, the adsorption active sites were homogenously distributed on the adsorbent surface and they were identical for all dye molecules. Thus, each active site binds only with one dye molecule [39]. The maximum monolayer coverage capacity, q_{max}, at 55 °C was 666.67 mg g^{-1} for the removal of CR dye using CCs. Previous studies also reported the Langmuir model as the best fit for the adsorption of dyes using chitosan [42]. The obtained R_L values for adsorbing CR dyes onto CCs ranged from 0.217 to 0.111, which indicates that the adsorption of CR dye onto CCs was a favorable process.

The Temkin model took into consideration the effect of the indirect interaction between the molecules of adsorbate. Assuming the linear decrease of adsorption heat of all the molecules in the layer due to the interactions between adsorbent and adsorbate, the adsorption process is characterized with a uniform distribution of binding energies [40].

Temkin constants, B_T and K_T, were listed in Table 7. The obtained plot provided a curve which indicated that the adsorption process did not follow this model.

The Dubinin–Radushkevich (D-R) isotherm model is used to describe the nature of the adsorption process. It is similar to Langmuir, while it is more general because it rejects the homogenous surfaces [40]. The linear form of D-R isotherm (Equation (16)) was obtained by plotting $\ln q_e$ versus ε^2, as illustrated in Figure 14. The values of β and $\ln X_m$ were calculated from the slope and intercept, respectively, and were listed in Table 7. Based on the value of R^2, the D-R isotherm cannot be used to describe the adsorption of CR dye onto CCs, as it equals 0.807 for the adsorption onto CCs.

From the analysis of the experimental results showed in Table 7, it is concluded that the Langmuir isotherm model was the best model can fit the experimental results due to its highest R^2. Freudlich and Temkin models did not show high accuracy as their correlation coefficients were not as high as the Langmuir isotherm, while the D-R model could not describe the adsorption process using CCs due to its low value of R^2. These findings reflected the homogenous nature of the adsorbent. Therefore, it is assumed that the adsorption of CR dye by CCs took place uniformly onto its active sites [40].

3.3.5. Adsorption Thermodynamics

The adsorption process of CR dye onto the adsorbent was studied using various temperatures (298, 308, 318 and 328 K). The values of thermodynamic parameters $\Delta G°$ were calculated according to Equation (19), while $\Delta H°$ and $\Delta S°$ were calculated from the Van't Hoff linear plot (Equation (20)).

The results were illustrated in Figure 15 and Table 8 for the adsorption CR dye using CCs. The values of $\Delta H°$ was 34.49 kJ mol^{-1}, this positive value suggested that the interaction of CR dye adsorbed onto CCs was endothermic in nature [23].

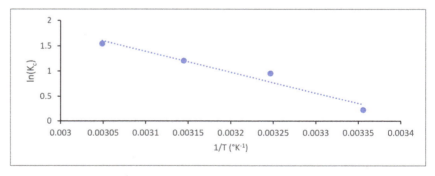

Figure 15. Adsorption thermodynamic of CR dye onto CCs.

Table 8. Parameters of thermodynamic of adsorbing CR dye by CCs.

Thermodynamic Parameters	Temperature (K)	
$\Delta G°$ (KJ mol^{-1})	298	−0.55
	308	−2.45
	318	−3.19
	328	−4.21
$\Delta H°$ (kJ mol^{-1})	34.49	
$\Delta S°$ (J K^{-1} mol)	118.48	

The negative values of $\Delta G°$ at all selected temperatures indicated that the adsorption of CR dye onto CCs was a spontaneous process. It decreased with increasing temperature, and this indicated the feasibility and favorability of the process at elevated temperatures [32].

The positive value of ΔS° (118.48 J mol^{-1}K^{-1}) indicated that the randomness increased at the interface between adsorbent and adsorbate solution during CR dye adsorption by CCs [42]. It can be explained as follows: throughout the adsorption process, the dye molecule received more entropy due to the displacement of adsorbed solvent molecule which resulting finally in an increasing randomness [43]. This actually occurred due to increasing the number of free ions of adsorbate that were found in order form near adsorbent surface than the adsorbed ions before adsorption process. Hence, the distribution of rotational energy increased, which consequently increased randomness at the interface between solid and liquid, and finally resulted in an increasing adsorption [35]. These findings are in a good agreement with previous studies [42].

3.3.6. Activation Energy

The type of adsorption process, whether it is physisorption or chemisorption, can be determined by the magnitude of activation energy. When the activation energy is found in the range 0–40 kJ mol^{-1}, this indicates that the reaction is physisorption, while if it is in the range 40–800 kJ mol^{-1}, the reaction is chemisorption [24]. The value of E_a for the adsorption of CR dye onto CCs was 2.47 kJ mol^{-1}, as illustrated in Figure 16. This value indicated that the adsorption was a physisorption in nature and was ascribed to a low potential barrier [44]. This result agrees with the adsorption of CR dye onto activated Moringa oleifera seed [45] and onto guava leaf-based activated carbon [36].

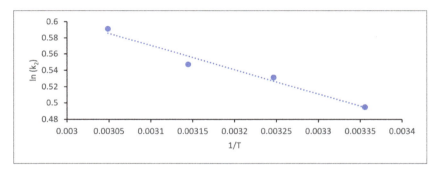

Figure 16. Arrhenius plot for adsorption of CR dye onto CCs.

3.3.7. Comparison between CCs and Other Adsorbents for CR Dye Removal

To evaluate the affinity and efficiency of the CCs adsorbent, it became of interest to compare its adsorption maximum capacity for removal of CR dye with other adsorbents in previous reported studies as shown in Table 9. It is clearly observed that CCs possesses the highest removal capacities for CR dye than other reported adsorbents. It could efficiently adsorb CR dye at 55 °C with q_{max} of 760.44 mg g^{-1}. This proved that CCs is an efficient and a promising adsorbent for the removal of CR dye from aqueous solution.

Table 9. Comparison of adsorption capacity of CCs with other adsorbents for the removal of CR dye.

Adsorbent	Q_{max} (mg g^{-1})	Temperature °C	Dye Concentration mg L^{-1}	Adsorbent Dose (g)	pH	Ref.
Tunics of the corm of the saffron	6.2	25	50–500	0.5	10	[46]
Quaternized chitosan/chitosan cationic polyelectrolyte microsphere	1500	25	0–1000	0.075	5	[47]
Activated carbon coffee waste	90.90	25	50 mg	0.1	3	[48]
xanthated chitosan/cellulose sponges	289.855	30	6.25–200	0.05	6	[49]
Chitosan and Laponite based nanocomposite	390.3	30	500	1	6	[50]
CCs	760.44	55	1000	0.05	4	Present study

3.3.8. Desorption Studies

The reusability is considered one of the most important aspects to minimize the cost of adsorption by regeneration of adsorbent [51].

The desorption of CR dye was applied using Equation (22). No desorption was obtained by using ethanol, methanol and acetone as desorption medium, while using 0.1 N NaOH solution showed that the desorption percentage reached 58% after 5 cycles. These findings confirmed that CCs can be efficiently reused for the adsorption of CR dye from the aqueous solution.

4. Conclusions

Modification of chitosan was successfully done via a four-step procedure using cyanoguanidine at the final step. The produced cyanoguanidine-modified chitosan (CCs) possessed free primary amino and hydroxyl active centers in its main chains, in addition to nitrogen-rich cyanoguanidine as crosslinking linkages. Its structure was characterized by different techniques; FTIR, XRD and SEM. Its active basic groups acted as efficiently binding sites for the anionic CR dye from its aqueous solution. The adsorption capacity for CR dye onto CCs increased with increasing temperature reaching its maximum at 55 °C. The adsorption capacity decreased with increasing pH achieving its optimum value at pH 4. The removal percentage of CR dye decreased with the increase in the initial concentration of the dye. The pseudo-second-order model showed a perfect fit to describe the adsorption process, indicating the chemisorption process. The values of rate constant, k_2, decreased gradually with the increasing pH from 4 to 9 and initial concentration of dye solution from 400 to 1000 mg L^{-1}, while it increased with increasing temperature from 25 °C to 55 °C. The relatively high correlation coefficient R^2 obtained from the Elovich model confirmed that this model fitted fully to the experimental results for the adsorbent at the studied parameters. This indicated that the adsorption of CR dye onto CCs was controlled by a chemisorption process. According to intraparticle diffusion, the transfer of CR dye molecules into the adsorbent surface took place in three stages. The plots of intraparticle diffusion showed that it was not the only rate-limiting step. The good applicability of the Langmuir adsorption isotherm indicated that the adsorption process occurred in a monolayer coverage manner with no interaction between adsorbed dye molecules with each other. The maximum monolayer coverage capacity, q_{max}, at 55 °C was 666.67 mg g^{-1}. The Freudlich and Temkin models did not show a high accuracy as their correlation coefficients were not as high as the Langmuir isotherm, whereas, the experimental data confirmed the unsuitability of the Dubinin–Radushkevich isotherm model. The positive value of $\Delta H°$ (34.49 kJ mol^{-1}) suggests that the adsorption was endothermic in nature.

The negative values of ΔG° at all selected temperatures reflected that the adsorption was a spontaneous process. The positive value of ΔS° indicated the randomness elevation at the interface between adsorbent and adsorbate solution. The value of activation energy E_a was 2.47 kJ mol^{-1}, assigning that the adsorption was physisorption. Thus, it is concluded that the adsorption of CR dye CCs involved physisorption and chemisorption processes.

Author Contributions: Supervision, N.A.M., N.F.A.-H.; Conceptualization, N.A.M. and N.F.A.-H.; Methodology, N.A.M., N.F.A.-H.; Investigation, E.F.A.; Formal analysis, N.A.M. and N.F.A.-H.; Writing—original draft, E.F.A.; Review and editing, N.A.M., N.F.A.-H. All authors have read and agreed to the published version of the manuscript.

Funding: This research received no external funding.

Institutional Review Board Statement: Not Applicable.

Informed Consent Statement: Not Applicable.

Data Availability Statement: The data presented in this study are available on request from the corresponding author.

Conflicts of Interest: The authors declare no conflict of interest.

References

1. Shariatinia, Z.; Jalali, A.M. Chitosan-based hydrogels: Preparation, properties and applications. *Int. J. Biol. Macromol.* **2018**, *115*, 194–220. [CrossRef] [PubMed]
2. Vakili, M.; Rafatullah, M.; Salamatinia, B.; Abdullah, A.Z.; Ibrahim, M.H.; Tan, K.B.; Gholami, Z.; Amouzgar, P. Application of chitosan and its derivatives as adsorbents for dye removal from water and wastewater: A review. *Carbohydr. Polym.* **2014**, *113*, 115–130. [CrossRef] [PubMed]
3. Mohamed, N.A.; Al-Harby, N.F.; Almarshed, M.S. Effective removal of Basic Red 12 dye by novel antimicrobial trimellitic anhydride isothiocyanate-cross linked chitosan hydrogels. *Polym. Polym. Comp.* **2021**, *29*, S274–S287. [CrossRef]
4. Buthelezi, S.P.; Olaniran, A.O.; Pillay, B. Textile dye removal from wastewater effluents using bioflocculants produced by indigenous bacterial isolates. *Molecules* **2012**, *17*, 14260–14274. [CrossRef]
5. Crini, G. Non-conventional low-cost adsorbents for dye removal: A review. *Bioresour. Technol.* **2006**, *97*, 1061–1085. [CrossRef] [PubMed]
6. Kadirvelu, K.; Kavipriya, M.; Karthika, C.; Radhika, M.; Vennilamani, N.; Pattabhi, S. Utilization of various agricultural wastes for activated carbon preparation and application for the removal of dyes and metal ions from aqueous solutions. *Bioresour. Technol.* **2003**, *87*, 129–132. [CrossRef]
7. Annu, S.; Ahmed, S.; Ikram, S. *Chitin and Chitosan: History, Composition and Properties, Chitosan*; Ahmed, S., Ikram, S., Eds.; Scrivener & Wiley: Hoboken, NJ, USA, 2017; pp. 3–24.
8. Ahmed, S.; Ikram, S. Chitosan & its derivatives: A review in recent innovations. *Int. J. Pharm. Sci. Res.* **2015**, *6*, 14–30.
9. Elmehbad, N.Y.; Mohamed, N.A. Synthesis, characterization, and antimicrobial activity of novel N-acetyl, N'-chitosanacetohydrazide and its metal complexes. *Int. J. Polym. Mater. Polym. Biomater.* **2021**. [CrossRef]
10. Sabaa, M.W.; Mohamed, N.A.; Mohamed, R.R.; Khalil, N.M.; Abd El Latif, S.M. Synthesis, characterization and antimicrobial activity of poly (N-vinyl imidazole) grafted carboxymethyl chitosan. *Carbohydr. Polym.* **2010**, *79*, 998–1005. [CrossRef]
11. Sabaa, M.W.; Mohamed, N.A.; Ali, R.; Abd El Latif, S.M. Chemically induced graft copolymerization of acrylonitrile onto carboxymethyl chitosan and its modification to amidoxime derivative. *Polym. Plast. Technol. Eng.* **2010**, *49*, 1055–1064. [CrossRef]
12. Sabaa, M.W.; Mohamed, N.A.; Mohamed, R.R.; Abd El Latif, S.M. Chemically induced graft copolymerization of 4-vinyl pyridine onto carboxymethyl chitosan. *Polym. Bull.* **2011**, *67*, 693–707. [CrossRef]
13. Mohamed, N.A.; Abd El-Ghany, N.A. Synthesis, characterization, and antimicrobial activity of carboxymethyl chitosan-graft-poly(N-acryloyl,N'-cyanoacetohydrazide) copolymers. *J. Carbohydr. Chem.* **2012**, *31*, 220–240. [CrossRef]
14. Sabaa, M.W.; Abdallah, H.M.; Mohamed, N.A.; Mohamed, R.R. Synthesis, characterization and application of biodegradable crosslinked carboxymethyl chitosan/poly(vinyl alcohol) clay nanocomposites. *Mater. Sci. Eng. C* **2015**, *56*, 363–373. [CrossRef] [PubMed]
15. Abraham, A.; Soloman, A.; Rejin, V.O. Preparation of chitosan-polyvinyl alcohol blends and studies on thermal and mechanical properties. *Procedia Technol.* **2016**, *24*, 741–748. [CrossRef]
16. Bahrami, S.B.; Kordestani, S.S.; Mirzadeh, H.; Mansoori, P. Poly (vinyl alcohol)-chitosan blends: Preparation, mechanical and physical properties. *Iran. Polym. J.* **2003**, *12*, 139–146.
17. Mohamed, N.A.; El-Ghany, N.A.A.; Abdel-Aziz, M.M. Synthesis, characterization, anti-inflammatory and anti-Helicobacter pylori activities of novel benzophenone tetracarboxylimide benzoyl thiourea cross-linked chitosan hydrogels. *Int. J. Biol. Macromol.* **2021**, *181*, 956–965. [CrossRef] [PubMed]

18. Elmehbad, N.Y.; Mohamed, N.A. Terephthalohydrazido cross-linked chitosan hydrogels: Synthesis, characterization and applications. *Int. J. Polym. Mater. Polym. Biomater.* **2021**. [CrossRef]
19. Mohamed, N.A.; Abd El-Ghany, N.A.; Fahmy, M.M. Novel antimicrobial superporous cross-linked chitosan/pyromellitimide benzoyl thiourea hydrogels. *Int. J. Biol. Macromol.* **2016**, *82*, 589–598. [CrossRef]
20. Elsayed, N.H.; Monier, M.; Youssef, I. Fabrication of photo-active trans -3-(4-pyridyl)acrylic acid modified chitosan. *Carbohydr. Polym.* **2017**, *172*, 1–10. [CrossRef] [PubMed]
21. Mohamed, N.A.; Abd El-Ghany, N.A. Novel aminohydrazide cross-linked chitosan filled with multi-walled carbon nanotubes as antimicrobial agents. *Int. J. Biol. Macromol.* **2018**, *115*, 651–662. [CrossRef]
22. Mohamed, N.A.; Abd El-Ghany, N.A. Synthesis, characterization and antimicrobial activity of novel aminosalicylhydrazide cross linked chitosan modified with multi-walled carbon nanotubes. *Cellulose* **2019**, *26*, 1141–1156. [CrossRef]
23. Mohamed, N.A.; Al-Harby, N.F.; Almarshed, M.S. Enhancement of adsorption of Congo red dye onto novel antimicrobial trimellitic anhydride isothiocyanate-cross-linked chitosan hydrogels. *Polym. Bull.* **2020**, *77*, 6135–6160. [CrossRef]
24. Alharby, N.F.; Almutairi, R.S.; Mohamed, N.A. Adsorption Behavior of Methylene Blue Dye by Novel Crosslinked O-CM-Chitosan Hydrogel in Aqueous Solution: Kinetics, Isotherm and Thermodynamics. *Polymers* **2021**, *13*, 3659. [CrossRef] [PubMed]
25. Surikumaran, H.; Mohamad, S.; Muhamad Sarih, N. Molecular Imprinted Polymer of Methacrylic Functionalised β-Cyclodextrin for Selective Removal of 2,4-Dichlorophenol. *Int. J. Mol. Sci.* **2014**, *15*, 6111–6136. [CrossRef]
26. Mohamed, N.A.; Abd El-Ghany, N.A. Synthesis, characterization, and antimicrobial activity of chitosan hydrazide derivative. *Int. J. Polym. Mater. Polym. Biomater.* **2017**, *66*, 410–415. [CrossRef]
27. Elmehbad, N.Y.; Mohamed, N.A. Designing, preparation and evaluation of the antimicrobial activity of biomaterials based on chitosan modified with silver nanoparticles. *Int. J. Biol. Macromol.* **2020**, *151*, 92–103. [CrossRef] [PubMed]
28. Gonçalves, J.O.; Dotto, G.L.; Pinto, L.A.A. Cyanoguanidine-crosslinked chitosan to adsorption of food dyes in the aqueous binary system. *J. Mol. Liq.* **2015**, *211*, 425–430. [CrossRef]
29. Burkhanova, N.; Yugai, S.; Pulatova, K.P.; Nikononvich, G.; Milusheva, R.Y.; Voropaeva, N.; Rashidova, S.S. Structural investigations of chitin and its deacetylation products. *Chem. Nat. Compd.* **2000**, *36*, 352–355. [CrossRef]
30. Feng, T.; Zhang, F.; Wang, J.; Wang, L. Application of Chitosan-Coated Quartz Sand for Congo Red Adsorption from Aqueous Solution. *J. Appl. Polym. Sci.* **2013**, *125*, 1766–1772. [CrossRef]
31. Salama, H.E.; Saad, G.R.; Sabaa, M.W. Synthesis, characterization, and biological activity of cross-linked chitosan biguanidine loaded with silver nanoparticles. *J. Biomater. Sci. Polym. Ed.* **2016**, *27*, 1880–1898. [CrossRef]
32. El-Harby, N.F.; Ibrahim, S.M.A.; Mohamed, N.A. Adsorption of Congo red dye onto antimicrobial terephthaloyl thiourea cross-linked chitosan hydrogels. *Water Sci. Technol.* **2017**, *76*, 2719–2732. [CrossRef] [PubMed]
33. Guarın, J.R.; Moreno-Pirajan, J.C.; Giraldo, L. Kinetic Study of the Bioadsorption of Methylene Blue on the Surface of the Biomass Obtained from the Algae D. antarctica. *J. Chem.* **2018**, *2018*, 2124845. [CrossRef]
34. Swan, N.B.; Abbas, M.; Zaini, A. Adsrption of Malachite green and Congo red dyes from water: Recent progress and future outlook. *Ecol. Chem. Eng.* **2019**, *26*, 119–132.
35. Banerjee, S.; Chattopadhyaya, M.C. Adsorption characteristics for the removal of a toxic dye, tartrazine from aqueous solutions by a low cost agricultural by-product. *Arab. J. Chem.* **2017**, *10*, S1629–S1638. [CrossRef]
36. Ojedokun, A.T.; Bello, O.S. Kinetic modeling of liquid-phase adsorption of Congo red dye using guava leaf-based activated carbon. *Appl. Water Sci.* **2017**, *7*, 1965–1977. [CrossRef]
37. Md Ahsanul Haque, A.N.; Remadevi, R.; Wang, X.; Naebe, M. Adsorption of anionic Acid Blue 25 on chitosan-modified cotton gin trash film. *Cellulose* **2020**, *27*, 9437–9456. [CrossRef]
38. Kumar, R.; Ansari, S.A.; Barakat, M.A.; Aljaafari, A.; Cho, M.H. A polyaniline@MoS2-based organic–inorganic nanohybrid for the removal of Congo red: Adsorption kinetic, thermodynamic and isotherm studies. *New J. Chem.* **2018**, *42*, 18802–18809. [CrossRef]
39. Lin, C.; Li, S.; Chen, M.; Jiang, R. Removal of Congo red dye by gemini surfactant C12-4-C12 · 2Br-modified chitosan hydrogel beads. *J. Dispers. Sci. Technol.* **2017**, *38*, 46–57. [CrossRef]
40. Titi Ojedokun, A.; Solomon Bell, O. Liquid phase adsorption of Congo red dye on functionalized corn cobs. *J. Dispers. Sci. Technol.* **2017**, *38*, 1285–1294. [CrossRef]
41. Litefti, K.; Freire, M.S.; Stitou, M.; González-Álvarez, J. Adsorption of an anionic dye (Congo red) from aqueous solutions by pine bark. *Sci. Rep.* **2019**, *9*, 16530. [CrossRef] [PubMed]
42. Tahira, I.; Aslam, Z.; Abbas, A.; Monim-ul-Mehboob, M.; Ali, S.; Asghar, A. Adsorptive removal of acidic dye onto grafted chitosan: A plausible grafting and adsorption mechanism. *Int. J. Biol. Macromol.* **2019**, *136*, 1209–1218. [CrossRef] [PubMed]
43. Zahir, A.; Aslam, Z.; Kamal, M.S.; Ahmad, W.; Abbas, A.; Shawabkeh, R.A. Development of novel cross-linked chitosan for the removal of anionic Congo red dye. *J. Mol. Liq.* **2017**, *244*, 211–218. [CrossRef]
44. Bulut, Y.; Karaer, H. Adsorption of Methylene Blue from Aqueous Solution by Crosslinked Chitosan/Bentonite Composite. *J. Dispers. Sci. Technol.* **2015**, *36*, 61–67. [CrossRef]
45. Jabar, J.M.; Odusote, Y.A.; Alabi, K.A.; Ahmed, I.B. Kinetics and mechanisms of congo-red dye removal from aqueous solution using activated Moringa oleifera seed coat as adsorbent. *Appl. Water Sci.* **2020**, *10*, 136. [CrossRef]
46. Dbik, A.; Bentahar, S.; el Khomri, M.; el Messaoudi, N.; Lacherai, A. Adsorption of Congo red dye from aqueous solutions using tunics of the corm of the saffron. *Mater. Today* **2020**, *22*, 134–139. [CrossRef]

47. Cai, L.; Ying, D.; Liang, X.; Zhu, M.; Lin, X.; Xu, Q.; Cai, Z.; Xu, X.; Zhang, L. A novel cationic polyelectrolyte microsphere for ultrafast and ultra-efficient removal of heavy metal ions and dyes. *Chem. Eng. J.* **2021**, *410*, 128404. [CrossRef]
48. Lafi, R.; Montasser, I.; Hafiane, A. Adsorption of Congo red dye from aqueous solutions by prepared activated carbon with oxygen-containing functional groups and its regeneration. *Adsorpt. Sci. Technol.* **2019**, *37*, 160–181. [CrossRef]
49. Xu, X.; Yu, J.; Liu, C.; Yang, G.; Shi, L.; Zhuang, X. Xanthated chitosan/cellulose sponges for the efficient removal of anionic and cationic dyes. *React. Funct. Polym.* **2021**, *160*, 104840. [CrossRef]
50. Xu, G.; Zhu, Y.; Wang, X.; Wang, S.; Cheng, T.; Ping, R.; Cao, J.; Lv, K. Novel chitosan and Laponite based nanocomposite for fast removal of Cd(II), methylene blue and Congo red from aqueous solution. *e-Polymers* **2019**, *19*, 244–256. [CrossRef]
51. Wang, P.; Yan, T.; Wang, L. Removal of Congo Red from Aqueous Solution Using Magnetic Chitosan Composite Microparticles. *BioResources* **2013**, *8*, 6026–6043. [CrossRef]

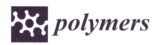

Review

Advanced Polymeric Nanocomposites for Water Treatment Applications: A Holistic Perspective

Adedapo Oluwasanu Adeola [1,2,3,*] and Philiswa Nosizo Nomngongo [2,3,*]

1. Department of Chemical Sciences, Adekunle Ajasin University, Akungba Akoko 001, Ondo State, Nigeria
2. Department of Chemical Sciences, Doornfontein Campus, University of Johannesburg, Doornfontein, Johannesburg 2028, South Africa
3. Department of Science and Innovation-National Research Foundation South African Research Chair Initiative (DSI-NRF SARChI) in Nanotechnology for Water, University of Johannesburg, Doornfontein, Johannesburg 2028, South Africa
* Correspondence: adedapo.adeola@aaua.edu.ng (A.O.A.); pnnomngongo@uj.ac.za (P.N.N.)

Citation: Adeola, A.O.; Nomngongo, P.N. Advanced Polymeric Nanocomposites for Water Treatment Applications: A Holistic Perspective. *Polymers* 2022, *14*, 2462. https://doi.org/10.3390/polym14122462

Academic Editors: Irene S. Fahim, Hossam E. Emam and Ahmed K. Badawi

Received: 20 May 2022
Accepted: 14 June 2022
Published: 16 June 2022

Publisher's Note: MDPI stays neutral with regard to jurisdictional claims in published maps and institutional affiliations.

Copyright: © 2022 by the authors. Licensee MDPI, Basel, Switzerland. This article is an open access article distributed under the terms and conditions of the Creative Commons Attribution (CC BY) license (https://creativecommons.org/licenses/by/4.0/).

Abstract: Water pollution remains one of the greatest challenges in the modern era, and water treatment strategies have continually been improved to meet the increasing demand for safe water. In the last few decades, tremendous research has been carried out toward developing selective and efficient polymeric adsorbents and membranes. However, developing non-toxic, biocompatible, cost-effective, and efficient polymeric nanocomposites is still being explored. In polymer nanocomposites, nanofillers and/or nanoparticles are dispersed in polymeric matrices such as dendrimer, cellulose, resins, etc., to improve their mechanical, thermophysical, and physicochemical properties. Several techniques can be used to develop polymer nanocomposites, and the most prevalent methods include mixing, melt-mixing, in-situ polymerization, electrospinning, and selective laser sintering techniques. Emerging technologies for polymer nanocomposite development include selective laser sintering and microwave-assisted techniques, proffering solutions to aggregation challenges and other morphological defects. Available and emerging techniques aim to produce efficient, durable, and cost-effective polymer nanocomposites with uniform dispersion and minimal defects. Polymer nanocomposites are utilized as filtering membranes and adsorbents to remove chemical contaminants from aqueous media. This study covers the synthesis and usage of various polymeric nanocomposites in water treatment, as well as the major criteria that influence their performance, and highlights challenges and considerations for future research.

Keywords: fabrication techniques; inorganic contaminants; organic pollutants; polymer nanocomposites; water treatment

1. Introduction

Water plays a major role in the evolution of human civilization and industrialization. The population of the world is predicted to increase to 9 billion people, which will result in an increase in demand for freshwater and necessitate wastewater treatment and reuse [1–3]. In recent decades, human activities have exacerbated major environmental and conservation challenges [4–6]. The environmental challenges we currently face that pose a threat to sustainable life on land (15th Sustainable Development Goal) include water and air pollution, poor waste management, fallen groundwater tables, loss of biodiversity, land/soil debasement, global warming/climate change, depletion of natural resources, etc. [7–9].

Nanomaterials could be a viable and effective technique for overcoming significant obstacles in the development of efficient remedial technologies and environmental protection [10,11]. However, within the last century, the volume of industrial chemicals produced has increased dramatically, from 1 to 400 million tons in the year 2000 [12]. Furthermore, between 2000 and 2017, the worldwide production capacity of industries increased from

1.2 billion tons to 2.3 billion tons [13]. The advances in science and the phenomenal growth in production capacity in the 21st century have also resulted in the emergence of troubling realities, such as the exponential increase in the spectrum of different classes of emerging pollutants detected in our water systems. As a result, new materials are urgently needed to clean up the polluted environment.

Nanostructured sorbents have a high capacity for the treatment of polluted water and may be tailored to target specific pollutants [10,11]. There have been recent improvements in the development of polymer nanocomposites (PNC), which have enhanced their novel applications in pollution remediation. Many chemical pollutants, such as dyes, heavy metals, and hydrocarbons, have been removed from wastewater using polymer nanocomposites [14,15]. Natural and synthetic polymers are both accessible. Natural polymers are those that are found in nature and can be extracted for usage. Water-based natural polymers include silk, wool, DNA, cellulose, and proteins [16]. Synthetic polymers, on the other hand, include nylon, polyethylene, polyester, Teflon, and epoxy, etc. Polymer nanocomposites possess unique synergistic features that are not possible to achieve with individual components functioning alone [17]. Inorganic nanofillers such as nanoclays, metal-oxide nanoparticles, carbon nanomaterials, and metal nanoparticles, can be introduced into a polymer matrix to create a PNC with better properties for a specific application [18,19]. Polymer nanocomposites have gained scientific prominence due to their broad range of applications in environmental remediation and treating various environmental challenges [16,20].

The aim of this review is to comprehensively examine the current state of polymeric nanocomposites as water remediation materials around the world, discussing various classifications of polymer nanocomposites, state-of-the-art synthetic methods, applications, merits, limitations, and potentials. The search keywords for this review are polymeric material, polymer nanocomposite, adsorbent, photocatalyst, organic pollutant, heavy metal, photocatalytic degradation, adsorption, and water treatment, and the literature scope is based on SCOPUS and Web of Science published papers and books.

2. Synthetic Methods and Remedial Application of Polymeric Nanocomposites

The choice of design, precursors, and synthetic methods for polymer nanocomposite development are germane. It entails choosing monomers, fillers, and other composite materials, as well as synthesis methods from a vast variety in order to produce PNC with the required property [21]. The determination of optimal processing strategy, considering the application intended for the PNC, and lastly, fabricating the product yield of the composite. This emphasizes the significance of the design and synthesis steps in the manufacture of PNC [18]. The ultrasonication-assisted mixing, shear mixing, microwave-assisted synthesis, roll milling, in-situ polymerization, ball milling, selective laser sintering, double-screw extrusion, and 3D printing (additive manufacturing) are among the most often utilized synthesis procedures [18,22]. Generally, there are two methods for making polymer nanocomposites: direct compounding and in situ synthesis [23]. The choice of synthetic methods demonstrates researchers' trend toward simple, scalable, and ultimately commercially viable and reproducible PNC [24].

Atomic layer deposition and plasma-assisted mechanochemistry have been reported to solve nanoparticle aggregation challenges in melt-mixing techniques. In addition, high-frequency sonication plays a similar role in the disaggregation of nanoparticles in mixing procedures. In-situ polymerization allows for the creation of thermodynamically stable nanocomposites, while electrospinning is an efficient way of creating porous objects. Furthermore, fabricating nanocomposites using selective laser sintering offers obvious advantages in terms of overcoming the aggregation problem [25]. Due to the unique features and various suitable applications of nanofillers, they are increasingly being used in the fabrication of PNC. One-dimensional (1D) nanofillers are fillers with one dimension smaller than 100 nanometers [18]. They normally come in sheets ranging in thickness from a few nanometers to hundreds of thousands of nanometers in length. One-dimensional nanofillers include montmorillonite clay and graphene nano-platelets. Both academics and industry are interested in polymer nanocomposites made from different nanofillers and polymers [2]. Due to its consistent volumetric heating and the huge increase in reaction rate, microwave-assisted synthesis is an emerging technology used in numerous domains of materials science and chemistry. Microwave-assisted fabrication has also been used to create functionalized polymer-based nanoadsorbents [26].

2.1. Dendritic Polymers

Dendrimers are branched structures with a specific form, size, and molecular weight [27,28]. Dendrimers, which are made up of monomers that radiate from a central core, are becoming a popular type of polymer. Different classes of dendritic polymers are depicted in Figure 1. Due to the large number of functional groups in the core, on the surface, and the periphery and pocket, dendrimers are a high-capacity nanoscaled multidentate chelating agent/ligand, suitable for ion separation technology. The microenvironment within the dendrimer scaffold, on the other hand, may be an appropriate host for diverse contaminants. Dendrimers' potential for environmental clean-up has also been explored. Dendrimers' value for removing inorganic contaminants as well as organic pollutants has been extensively demonstrated in this regard [29]. Dendrimers are also attractive chemicals for the preparation of electrochemical sensors [30].

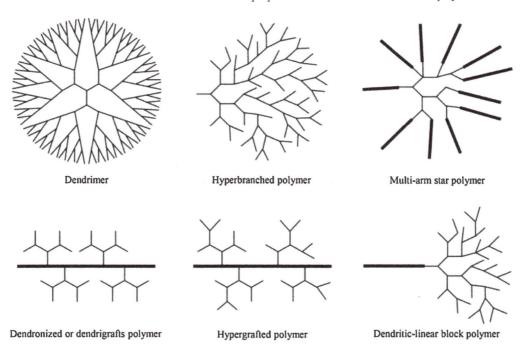

Figure 1. Six subclasses of dendritic polymeric material. Adapted with permission from Ma et al. [28]. Copyright (2016) Ivyspring International Publisher.

Poly(amidoamine) (PAMAM) dendrimers are extensively utilized dendrimers for the adsorption of pollutants in aqueous media. They are made up of ethylenediamine core, repeating units, and terminal units that make up the three fundamental units [29]. Their synthesis is performed by serially repeating two reactions: amino group addition to the double bond of methyl acrylate, accompanied by amidation of the resulting methyl ester with ethylenediamine. Each reaction step results in the production of a new dendrimer

(Figure 2). The addition-amidation reaction increases the diameter of PAMAM dendrimers by increasing the repeating units. This is a 1 nm per generation increase [29,31,32].

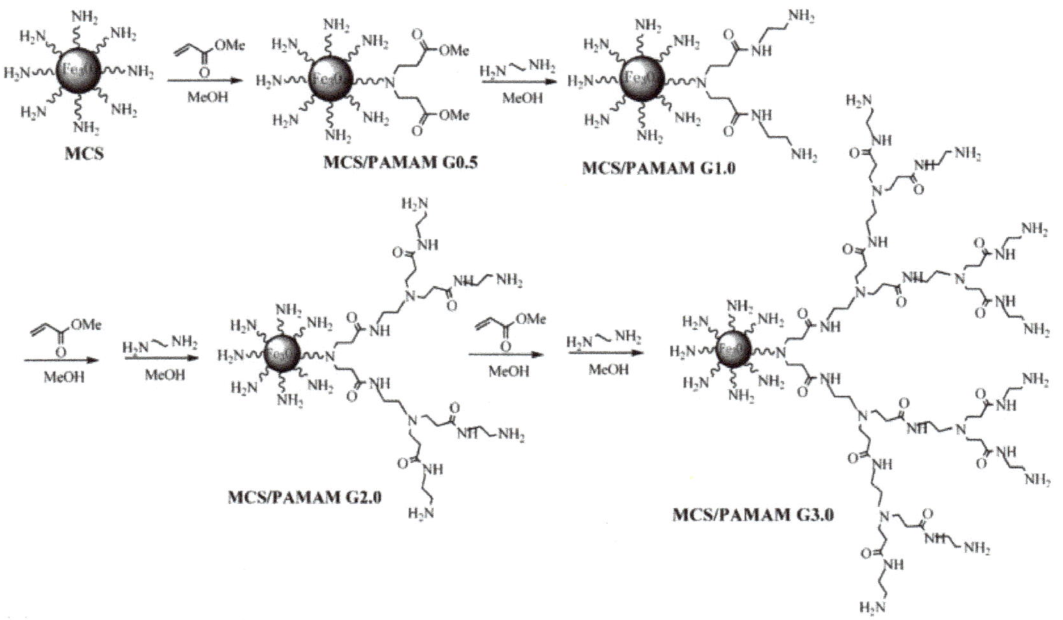

Figure 2. PAMAM growth on magnetic chitosan nanoparticles. Adapted with minor modifications with permission from Wang et al. [32]. Copyright (2015) Elsevier.

Most polymers are synthesized via single step; however, dendrimers are made in a series of steps, giving them well-tuned structures and narrow polydispersity. Divergent or convergent techniques, or a combination of both, can be used to synthesize dendrimers [33,34]. The divergent strategy entails the preparation of the dendrimer from a multifunctional core that is stretched outward by repeated reaction sequence (Figure 3A). On the other hand, a convergent method involves a bottom-up approach where dendrimers are constructed with tiny molecules that build-up on dendrimer surface and get connected to a central core through a sequence of inward-oriented interactions (Figure 3B). To combine the advantages of divergent and convergent synthesis, a combined divergent/convergent strategy, also known as the double-stage convergent approach (Figure 3C), has been further developed and employed for dendrimer synthesis [31,35].

However, structural flaws are a major problem in the synthesis of high-generation dendrimers because they are often caused by incomplete reactions or side reactions that take place as steric hindrance increases. Furthermore, due to their comparable chemical compositions and physical properties, dendrimers with structural flaws are frequently difficult to distinguish from intact dendrimers [33]. The convergent technique gives better control of the preparation of dendritic polymers, thus limiting structural flaws and impurities can be gotten rid of easily, as they conspicuously differ in morphology from the synthesized dendrimers. However, steric congestion is a challenge in the convergent approach, thus they are only suitable for small-scale dendrimers [36–38]. In the combination approach, a divergent technique is used to synthesize building blocks, which is then followed by convergent dendrimer assembly. Furthermore, when compared to either divergent or convergent synthesis, the sequence of reactions necessary for dendrimer preparation and purification can be minimized, thus ensuring more efficient production of higher generation [31,39].

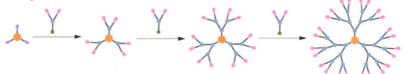

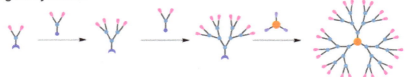

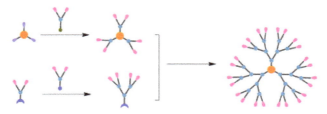

Figure 3. Covalent dendrimer synthesis using: (**A**) divergent synthetic route, (**B**) divergent synthetic route, (**C**) Combined approach (Adapted with permission from Lyu et al. [31]. Copyright (2019) Elsevier).

Environmental Remediation Using Dendritic Polymers

Researchers are interested in dendritic polymer-based nanocomposites because of their tunable architectures and characteristics. Using appropriate functional moieties, dendritic polymers can be tailored to the target contaminants (Table 1). Dendritic PNC has high specific surface areas and pore volumes for trapping chemical contaminants. They can be combined with a variety of supports and other materials. A challenge to field/large scale applications is the multi-step synthesis, which requires profound expertise in polymer development. This challenge is solvable by collaborative research between organic chemists and separation science professionals.

Dendrimers have shown good potential for remediation of inorganic and organic pollutants (Table 1) in aqueous matrices based on the following reasons:

i. They are a unique sort of macromolecules with a highly branched structure, high porosity, and a three-dimensional functionalized structure.
ii. Dendrimers can be grafted onto large supports, leading to increased selectivity through size exclusion by modified cavities and/or via selective binding to contaminants due to well-tailored support/substrate. Large support may also enhance surface area of the nanocomposite, leading to higher adsorption/separation capacity.
iii. They have large external and interior regions, as well as a large network of peripheral functional moieties, which permits the capture of large amount of contaminants.
iv. Adjustment of the physicochemical parameters of the core, interior cells, and outer end groups plays a major role in their adsorption capacity.
v. The existence of a high number of required peripheral functional groups ensures good selectivity. More intriguingly, the character of functional groupings of the nanocomposite can be tailored to target pollutants.

Heavy metal decontamination of polluted water with aid of dendritic polymeric materials is influenced by pH due to its impacts on the moieties on the surface of dendritic polymer. Furthermore, solution pH plays a crucial role in the adsorption performance of dendritic polymeric materials by altering the speciation and shape of metal ions [29,40]. The mercury (Hg) adsorption onto silica gel-supported salicylaldehyde-modified PAMAM dendrimers was investigated using the density functional theory (DFT) approach in another study [41]. Hg ions may bind with dendritic polymers by chelation due to the presence of oxygen and nitrogen species. The results of the Dubinin–Radushkevich (D–R) isotherm model indicate that mercury was removed via chemisorption onto sorbent pores, while thermodynamic tests revealed the endothermic and spontaneous nature of the sorption process, similar to what was reported in a recent study [42].

There are hundreds of polycyclic aromatic hydrocarbons (PAHs), which are essentially hydrocarbons containing two or more fused aromatic rings. Partial combustion or burning of carbon-based materials generates hazardous PAHs [43]. They can cause endocrine disruption and are carcinogenic [44]. Arkas et al. used hyperbranched poly(ethylene imine) in combination with silicic acid to remove pyrene and phenanthrene from an aqueous solution [45]. Sol–gel processes were used to make the dendrimer–silica nanoparticle composite. It was discovered that adding dendrimer to the silica nanospheres increased the adsorption of PAHs. The development of complexes with transferable charges between the tertiary amino groups of the dendrimer and PAHs resulted in a greater water treatment performance of PEI–silica nanoparticles.

Textile dyes, particularly acid dyes, are frequently utilized and they are poisonous, causing nausea, sleepiness, diarrhea, blood clots, and breathing difficulties [37,46,47]. Textile production is critical to commerce and industry. Unexhausted dye is being thrown into natural streams without being adequately treated by some businesses due to non-compliance with environmental requirements and wastewater treatment before discharge [24]. Various colors have been removed from aqueous solutions using dendritic polymers (Table 1). Hydrogen bonding, Vander Waals forces, electrostatic attractions, and dye trapping in dendrimers facilitate dye adsorption on dendrimer-based hybrid complexes [48]. A dendritic polymer was used by Hayati et al. to remove Acid Blue 7 (AB7), Direct Red 23 (DR23), Acid Green 25 (AG25), and Direct Red 80 (DR80) aqueous solution [49]. The adsorption process was discovered to be affected by pH, dye concentration and adsorbent dosage. The Langmuir (monolayer) adsorption isotherm model was shown to be the best fit for all dye adsorption by PPI dendrimers. The dye removal rate decreases as the pH rise, with the highest adsorption capacity occurring at pH 2 as a result of substantial electrostatic interaction between the anionic dye and the positively charged dendrimer surface. Removal of dye anions reduces when pH rises as a result of the decline in the number of positively charged sites.

Dendrimers enhance the activity of the photocatalysts for dye degradation when used as supports. PAMAM dendrimer, for example, was used to modify the photocatalytic activity of a polyoxometalate (POM) cluster for methyl red (MR) degradation by blocking POM-MR aggregation. After 20 min of irradiation under the same reaction conditions, POM-dendrimer showed an increased dye degradation of 83% compared to 11% for POM [48,50]. Dendritic polymers and composites have demonstrated extraordinary potential in the field of water remediation. Dendrimers can be utilized on their own or with other materials. In dynamic systems, they can be immobilized over membranes. Dendrimer production, solution pH, contact time, and other factors influence their performance [37,51]. They are less toxic and biocompatible, which bodes well for their future commercialization and applications in water treatment.

Table 1. Various dendritic polymers used for the treatment of contaminated water.

Dendritic Nanocomposites	Target Pollutant	Remediation Approach	Removal Capacity	References
PAMAM/Graphene oxide	**Heavy metals**: Pb, Cd, Cu, MnCd, Cu, Mn	Adsorption	568.18, 253.81, 68.68, 18.29 253.81, 68.68, 18.29 (mg/g)	[52,53]
Dendrimer-clay nanocomposite	Cr	Adsorption	6–10 (mg/g)	[54]
Polystyrene PAMAMiminodiacetic acid	Ni	Adsorption	24.09 (mg/g)	[55]
PAMAM-grafted cellulosenanofibril	Cr	Adsorption	377.36 (mg/g)	[56]
Hyperbranched PAMAM/polysulfone membrane	Cd	Ultrafiltration	27.29 µg/cm^2	[57]
Dendrimer/titania	Pb	Adsorption	400 (mg/g)	[58]
PAMAM-grafted core-shellmagnetic silica nanoparticles	Hg	Adsorption	134.6 (mg/g)	[59]
PAMAM dendrimers withethylenediamine (EDA) core	Cu	Ultrafiltration	451 (mg/g)	[60]
Amine terminated-Magneticcored dendrimer	Pb, Cd	Adsorption	170.42, 75.15 (mg/g)	[61]
Carbon nanotube-dendrimer	Pb, Cu	Adsorption	3333–4320 (mg/g)	[62]
Polyacrylonitrile/PAMAM composite nanofibers,	**Dyes**: Direct red 80, Direct red 23	Adsorption	2000 (mg/g)	[63]
Magnetic Chitosan/PAMAM	Reactive blue 21	Adsorption	555.56 (mg/g)	[32]
PPI-grafted cotton fabrics	Direct red 80, Disperse yellow 42, Basic blue 9	Adsorption	143.3, 104.8, 105.8 (mg/g)	[64]
PPI dendrimer	Direct red 80, Acid green 25, Acid blue 7, Direct red 23	Adsorption	33,333–50,000 (mg/g)	[49]
Graphene oxide-PPI dendrimer	Acid red 14, Acid blue 92	Adsorption	434.78, 196.08 (mg/g)	[65]
PAMAM-titaniananohybrid	Phenol	Adsorption	77 (mg/g)	[61]
PPI dendrimers functionalized with long aliphatic chains	**PAHs**: Fluoranthene, Phenanthrene, Pyrene	Adsorption	19, 67, 57 (mg/g)	[66]
Alkylated hyperbranched polymers	Fluoranthene, Phenanthrene, Pyrene	Adsorption	6–54 (mg/g)	[45]

2.2. Polymeric Aerogels and Hydrogels

Polymeric aerogels are unique types of porous material with interesting physicochemical features, including ultra-low thermal conductivity, low density, high porosity, large specific surface area, and controllable surface chemistry [67]. Aerogels are polymeric nanoparticle networks that are expanded by a gas across their whole volume, while hydrogels are cross-linked polymeric networks that may contain water within the interstitial spaces between chains [68,69]. The feasibility of this porous material in various applications has been thoroughly studied, thanks to recent breakthroughs in the synthesis of several forms of aerogels [70]. Aerogels and hydrogels are used as adsorbents for removing a variety of pollutants that are hazardous to the environment and human health. Figure 4 describes the general approach for the synthesis of aerogels, which involves the mixing of starting materials, gelling and a crucial step of removing solvents from pores of wet gels without distortion of the structure or network of molecules [71]. The ability to regulate the construction of a tunable aerogel network is enabled by the flexibility of the process conditions (change of the synthesis parameters, composition, etc.) [72]. The drying stage is critical for retaining porosity and integrity. The supercritical drying approach is the best method so far for achieving well-defined structures among the many drying processes, although freeze-drying is considerably easier, less expensive, and ecofriendly [73–75].

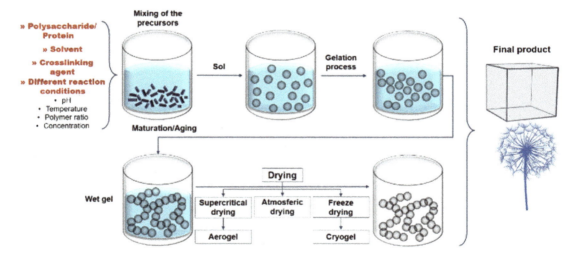

Figure 4. A general method for the preparation of aerogels. Adapted from Nita et al. [71].

On the other hand, hydrogel materials are frequently made by polymerizing acrylic monomers using a free radical initiator [76]. The resins can be made in an aqueous medium with solution polymerization or in a hydrocarbon medium with well-dispersed monomers (Figures 5 and 6). Hydrogel has gained a lot of attention in recent years due to its exceptional mechanical properties, swellability, biocompatibility, etc., but there are still a few things that can be done to improve it. A simple approach was reportedly used to make a novel alginate/graphene double-network (GAD) hydrogel, and its mechanical characteristics, stability, and adsorption performance were compared with alginate/graphene single-network hydrogel (GAS) [77]. It was discovered that GAD has a smaller swelling power than GAS, resulting in better gel stability in highly concentrated alkali/salt solution. The GAD beads have a considerably greater adsorption capacity for heavy metal ions and dye than GAS beads.

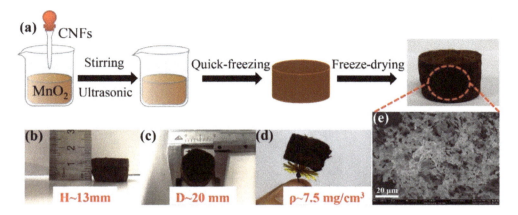

Figure 5. Preparation of macroporous MnO$_2$-based aerogel crosslinked with cellulose nanofibers. (**a**) scheme for the synthesis of cellulose nanofibers/MnO$_2$ hybrid aerogels; (**b–d**) the height, diameter and density of cellulose nanofibers/MnO$_2$ hybrid aerogel; (**e**) the SEM image of cellulose nanofibers/MnO$_2$ hybrid aerogel. Adapted with permission from Cao et al. [75]. Copyright (2021) Elsevier.

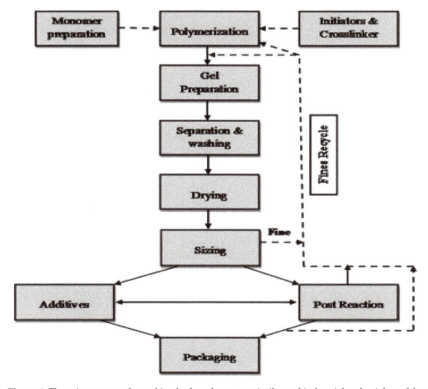

Figure 6. The primary steps for making hydrogels on a semi-pilot and industrial scale. Adapted from Ahmed et al. [76].

Apart from the adsorption of aqueous- and gas-phase pollutants, several aerogel composites are acknowledged as useful tools for photocatalytic remediation by harnessing the (photo)catalytic moieties present in their framework [78]. Although aerogels and hydrogels are good remedial tools, there is still the challenge of sophisticated drying technology requirements, structural fragility and instability, and high processing costs, which should be subject to advanced research [78,79]. To strengthen the structure of aerogels and hydrogels, for improved performance and wider applications, aerogel and hydrogel-based nanocomposites have been developed (Table 2).

A starch-graft-poly(acrylamide)/graphene oxide/hydroxyapatite nanocomposite hydrogel was prepared by free radical cross-linking copolymerization [80]. It is worth noting that the n-HAp nanoparticles served as cross-linkers, allowing for ionic interactions and hydrogen bond formation between phosphates and the polymer network's grafted acrylic groups. It also demonstrated that many hydrogen bonds and potential covalent connections developed between electrons of graphene oxide (GO) sheets and acrylamide (AM) monomers grafted on a starch framework (Figure 7). The bioadsorbent was used for malachite green dye adsorption. The results revealed that the adsorptive interaction was viable, spontaneous, and endothermic. The pseudo-second-order model was used to characterize the malachite green (MG) sorption rates. The sorption data were best-fit by the Langmuir model with a maximum adsorption capacity of 297 mg/g. The hydrogel-based adsorbent demonstrated good regeneration capacity for up to five cycles [80]. The polymeric nanocomposite could be an environmentally benign and promising adsorbent for water treatment applications.

Similarly, a starch/MnO_2/cotton hydrogel nanocomposite has been synthesized [81]. 0.008 M potassium permanganate, 0.7 g starch, and 0.6 M sodium hydroxide were used to make the optimal starch hydrogel nanocomposite at 50–55 °C. Potassium permanganate was used as a strong and cheap oxidizing agent to prepare manganese dioxide nanoparticles and cross-link the starch molecular chains to cellulose molecular chains. Because of its simple one-step synthesis approach, in-situ preparation of nanoparticles, cost-effectiveness, and desirable activity such as photocatalysis, biocompatibility, antibacterial properties of 93 percent against *S. aureus*, the starch hydrogel nanocomposite is a good material for various applications such as agriculture, medical, textile engineering, and water treatment [81].

A selective adsorbent was developed for the remediation of mercury (Hg) contaminants in aqueous solution [82]. The functionalized silica-gelatin hybrid aerogel with 24 wt.% gelatin content is ideal for the high efficiency and selective adsorption of aqueous Hg(II). By exposing cultures of *Paramecium caudatum* to Hg(II) and observing the model cultures with time-lapse video microscopy, the remediation efficacy of this adsorbent was assessed under realistic aquatic settings. With Hg(II) concentration, the viability of *Paramecium* shows a definite exposure-response relationship. When the Hg(II) concentration is greater than 125 g/L, viability diminishes. Only at Hg(II) concentrations greater than 500 ppm do the cells lose viability in the presence of 0.1 mg/mL aerogel adsorbent. The importance of the provided quasi-realistic aquatic toxicity model system in meeting the needs of environmental and chemical engineering technology is highlighted by the need for actual testing during adsorbent development [82].

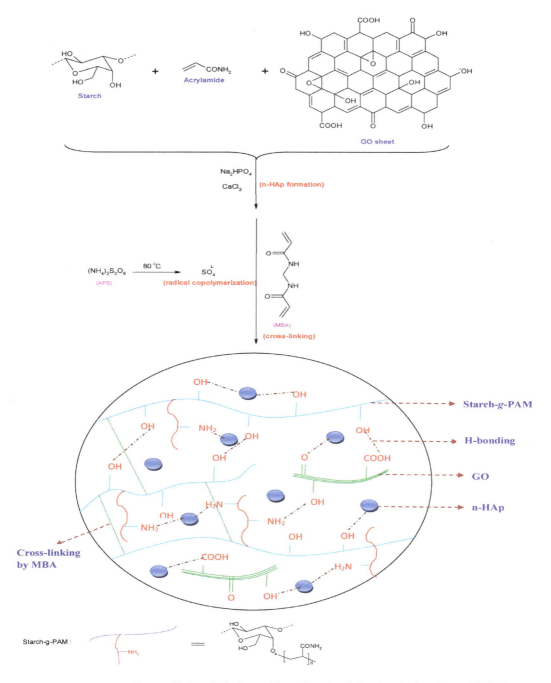

Figure 7. Hydrogel adsorbent with starch-graft-poly(acrylamide)/graphene oxide/hydroxyapatite nanocomposite as constituent. Adapted from Hosseinzadeh and Ramin [80].

Table 2. Selected hydrogel/aerogel nanocomposites for water treatment applications.

Material Description	Core Findings	Reference
MnO_2 coated cellulose nanofibers	Oxidation occurred at acidic pH. Over 99.8% removal of methylene blue dye	[83]
MnO_2/graphene aerogel (GMA)	GMA had 100% adsorption of rhodamine B and 89.02% COD, compared to 73.80% and 59.65% for SMA (silica wool-MnO_2 deposition)	[81]
Poly(acrylic acid)/starch hydrogel	Adsorption of cadmium was best described by Langmuir (monolayer) adsorption model with a maximum adsorption capacity of 588 mg/g	[84]
3D MnO_2 modified biochar-based porous hydrogels	Cd(II) and Pb(II) removal from aquatic and soil systems could be possible uses. Reusable and highly stable	[85]
Cassava starch-based double network hydrogel	The high adsorption capacity of about 417 mg/g and adsorption performance of 70% after regeneration five times. Physically and mechanically stable.	[86]
Chitosan-Gelatin based hydrogel	CH-GEL/ZSPNC (MW) eliminated 99% of cationic dye from the solution. The adsorption capacity of about 10.5 mg/g	[87]
CdS amended nano-ZnO/chitosan hydrogel	For 5.0 mg/L, 95 percent of Congo Red was removed in 1 min. Pollutant removal is quick, with high apparent rate constants and good reusability.	[88]
MnO_2 NWs/chitosan hydrogels	Abundant sunlight absorption (94%). The conversion efficiency of sunlight to thermal energy (90.6%)	[89]

2.3. Polymeric Membrane and Biopolymers

The need for freshwater will continue to rise over time, necessitating the development of cost-effective, eco-friendly water treatment technology [90,91]. Membrane processes such as reverse osmosis, membrane distillation, pervaporation, and others are among the existing technologies for drinking water management [10,92]. The mechanical and thermal stability of the present polymer membranes employed in water treatment procedures has various drawbacks [93].

Simple and cost-effective water filtration technologies based on recyclable biobased natural polymers such as chitosan, cellulose, and carbohydrate polymer modified with nanoparticles are needed [94]. These novel chitosan-based nanomaterials have been shown to effectively remove a range of pollutants from wastewater to permissible levels [95]. Other biobased polymer and polymeric support includes porous resins, polyaniline, polyacrylamides, cellulose acetate, cellulose or carboxymethyl cellulose, chitosan, alginate, eggshells, nanofibers and cellulose nanofillers (CNFs) [19,96]. CNFs potentially hold an important role in the advancement of the development of polymer nanocomposites. However, the challenge of extracting, isolating, and refining them, as well as surface improvements, can be developed systematically. It is safe to assume that CNF exploits will first find use in high-end applications, since their utilization offers significant benefits in limited quantities.

Membrane-based separation technologies are regarded as the most advanced separation technology due to various merits: they can meet a wide range of separation requirements; change of phase is not required, which saves energy; they are highly selective; they can retrieve trace amounts of contaminant; and they are simple, adaptable, and versatile [91,97]. Cellulose nanofillers have found applications in membranes/filters, injury dressings, dental implants, sophisticated adhesives, and materials requiring high transparency and mechanical qualities [98]. Findings suggest that when the choice of the polymer matrix is properly done, this enhances mechanical properties and lead to a high degree of optical transparency in the finished product [99]. The greatest barrier to their industrial applications remains the efficiency of recovery and facile strategies for morphological modifications as integrated with a system. It is worth noting that some recent research suggests that CNFs have increased barrier qualities as a function of cellulose crystal content, implying that they could potentially be used in more environmentally friendly packaging materials [100,101].

Adsorption and rejection are the two strategies for extracting solutes from wastewater using adsorptive membranes. Molecular sieving and filtration rejects any solutes larger than the membrane's pore size when water-containing solutes come into contact with the active layer of the membrane (Figure 8). Smaller solutes will pass through the active layer and into the support layer, which acts as an adsorption microsphere [19,102]. These solutes then react/bind to form a tight internal spherical complex, resulting in filtered water permeating from the adsorptive membrane that meets the requisite standards [103]. The nanofiltration (NF) and reverse osmosis (RO) membranes are widely utilized due to their high-water permeability, low-pressure need, and low cost (Table 3). The improvement of the membrane's functional adsorption sites has been reported to enhance its performance [104]. Therefore, using hydrophilic nanoparticles to boost the clean treated water flux can be quite successful, as long as the number of nanoparticles introduced is low. Having less than 6% adsorbent in the membrane matrix is typically recommended to avoid clogging membranes and counterproductive effects [105,106].

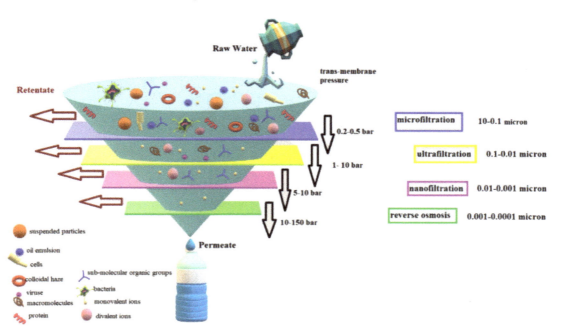

Figure 8. Capabilities of different separation/adsorptive membranes using microfiltration (MF), ultrafiltration (UF), nanofiltration (NF) and reverse osmosis (RO) processes. Adapted with permission from Hmtshirazi et al. [19]. Copyright (2022) Elsevier.

Recently, a carbon molecular sieve (CMS) with different concentrations was used to make asymmetric polyethersulfone (PES-15 wt. percent) mixed-matrix membranes via the phase inversion method [107]. Loading 1 wt.% of CMS, the membranes improved in hydrophilicity, heterogeneity, porosity, net surface charge and mean pore diameter. Furthermore, 1 wt.% CMS addition to PES, led to increased pure water flux from 55.77 to 75.05 L/m^2.hr. This polymeric nanocomposite showed improved dissolved ions-removal efficiency, compared to the unmodified PES membrane [107]. Table 3 presents various polymeric membranes that have been used to remove several types of pollutants from water, and the process parameters as well as the treatment technology adopted by various researchers.

Polymeric adsorptive membranes are generally considered potent pollution remediation technologies and have various applications in wastewater treatment plants and household water polishing. Cellulosic and other polymeric membranes have the capacity to remove various classes of persistent and emerging chemical pollutants from wastewater that are recalcitrant to conventional methods. The combined advantage of adsorption and filtration mechanisms, as well as the variety of forms and configurations, make these imprinted membranes (or membranes in general) and biopolymers appealing (Figure 9). Adsorptive membranes address fouling, minimize operational cost, adsorbent reusability, and enhances adsorption capacity, membrane permeability, rejection rates, and selectivity.

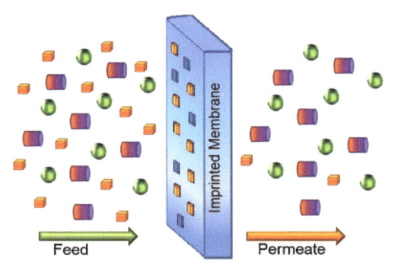

Figure 9. Removal of multiple contaminants selectively via an adsorptive membrane. Adapted from Salehi et al. [106].

Table 3. Selected polymeric membrane nanocomposites for water treatment applications.

Polymeric Membrane	Treatment Technology	Target Pollutants	Core Process Conditions	Reference
ES-10- polyamide, NTR-729HF-polyvinyl alcohol	Reverse osmosis (RO)	As, Sb	As(V) and Sb(V) removals are substantially higher than As(III) and Sb(V) removals at pH 3–10.	[108]
ES-10 and HS5110/HR3155	Nanofiltration (NF)/RO	As	NF: pressure 0.2–0.7 MPa/RO: pressure 4 MPa	[109]
NF90-4040	NF	Cr, As	pH = 9, temp. 45 °C, pressure 3.1 MPa	[110]
UiO-66 (Zr-MOF)/TFN	NF	Se, As	1.15 L/m^2·h/MPa	[111]
The P[MPC-co-AEMA] co-polymer	NF	Se, As	0.85 L/m^2·h/MPa	[112]
PVDF with melanin nanoparticles from the marine bacterium *Pseudomonas stutzeri*	Vacuum filtration (VF)	Hg, Cu, Cr, Pb	45 °C; pH = 3 for Cr and pH = 5 for other metals; flow rate of 0.5 mL/min	[113]
M-I	Micellar enhanced filtration (MEF)	Cu, Pb, Cd	Operating pressure 0.025 MPa; the flux 63,579 L/m^2·h	[114]
PAN- Polyacrylonitrile—Osmonic 100 kDa	Electro-ultrafiltration (EUF)	As	an averaged crossflow velocity of 0.1 m/s; pressure 0.098 MPa	[115]
Desal AG-2540 RO, TFC-ULP-2540 RO, and TFC-SR2-2540 NF	NF/RO	Sr	Applied pressure 0.10–0.15 MPa, pH = 3–6	[116]
Polyelectrolyte multilayer membrane	NF	Mg, Sr, Ca, Ba	low ionic strength conditions (e.g., <50 mM NaCl as a background electrolyte); 0.345 MPa; crossflow velocity 21.4 cm/s; 25 °C.	[117]
tubular Kerasep® ceramic membrane	Hybrid: Oxidation	Fe	Oxidation: 0.07 MPa; 20–22 °C; MF: tangential velocity 3.2 m/s; trans-membrane pressure 0.06–0.3 MPa; pH = 6.8–7.2; 20–22 °C	[118]
PPSU—sulfonated polyphenylenesulfone polymer; TBF—triangle-shape tri-bore hollow fiber membranes	UF	Oil	Transmembrane pressure of 0.1 MPa; a flow rate of 300 mL/min along the lumen side; a velocity range of 2.58–2.81 m/s	[119]

Table 3. Cont.

Polymeric Membrane	Treatment Technology	Target Pollutants	Core Process Conditions	Reference
NiCo-LDH—nickel cobalt layered double hydroxide; PVDF—the polydopamine modified polyvinylidenefluoride membrane	Gravity-driven filtration	Soybean oil, petroleum ether, 1,2-dichloroethane, n-hexadecane	Glass sand core filter device; water-in-oil emulsions—the volume ratio of 1:99	[120]
APTES—3-aminopropyltriethoxysilane; ATPR—atomic transfer radical polymerization/Graphene oxide	Filtration	Oil	Polymerization with ATRP; a volume ratio of organics and water: 1:99; the pressure of 0.05 MPa; complex environments, such as 2 M HCl, 2 M NaOH and saturated NaCl; permeation flux $10{,}000 \pm 440$ L/m^2·h·MPa	[121]
Nanofibrous PVDF membrane	Gravity-driven filtration	Oil	Permeability 88 1660 $\pm$ 6520 L/m^2·h·MPa; water-in-oil emulsions (chloroform, toluene, dichloromethane, and high viscosity oils: D4 and D5)	[122]
TiO$_2$-Nanoparticles/PVDF—polydopamine modified polyvinylidenefluoride membrane/TrFE—trifluoro ethylene	Photoreactor	Oily industrial wastewater	The flow rate 100.8 L/h; pH = 4–5.5	[123]
SiO$_2$-NPs/PVDF	Separation	Oil	The pressure of 0.09 MPa; fluxes of over 10,000 L/m^2 h	[124]
PVDF—polydopamine modified polyvinylidenefluoride membrane	RO	Oil	The cross-flow velocity 2 m/s; operating pressure 6 MPa; crossflow membrane sequencing batch reactor inoculated with isolated tropical halophilic microorganisms	[125]
Chitosan–SiO$_2$–glutaraldehyde composite/PVDF-polydopamine modified polyvinylidenefluoride membrane	VDF system	Oil	Separation area ~1.6 cm^2; the pressure 0.03 MPa.	[126]
TiO$_2$-NP/polydopamine modified polyvinylidenefluoride membrane	Separation	Petroleum ether; n-hexadecane; 1,3,5-trimethylbenzene; diesel oil	Pressure difference of 0.09 MPa; separation area 1.77 cm^2; permeation flux for SDS/oil/H$_2$O emulsion: 428 L/m^2·h, 605 L/m^2·h, 524 L/m^2·h, 382 L/m^2·h respectively	[127]

3. Considerations for Future Research

Pollutants in aqueous media have been remedied using a variety of polymeric nanocomposites adsorbents [128]. The toxicity of the adsorbent, when used for water filtration, is a major concern. Cellulose and other carbonaceous adsorbents (i.e., graphene, carbon nanotubes) have received significant attention for water purification due to their large surface area and affinity for inorganic and organic pollutants, as well as microbial contaminants [129,130]. Their toxicity, on the other hand, is relatively unknown and this requires more evaluation. Carbonaceous compounds and organic-based nanomaterials with reactive moieties may be hazardous to human health [131]. Therefore, only rational chemical design combined with a comprehensive understanding of potential biological interaction and risks can result in the fabrication of polymeric nanocomposites that is both safe and effective [18].

In recent research, polymers have been combined with materials such as carbon nanotubes, graphene oxide, and magnetic nanoparticles, in order to improve physicochemical and morphological properties [51,53]. Therefore, it makes sense to evaluate the toxicity of the entire composite system and not just the polymers. It is a prevalent fallacy that certain metals are only harmful in their pure form and functionalizing with a polymeric framework lessens toxicity. However, such statements should be backed up by scientific evidence.

Polymeric membrane technology has some concerns and challenges that need to be addressed by the scientific community. Examples include, but are not limited to:

(1) Fouling has long been a severe issue encountered during polymeric membrane applications in water treatment. Antifouling nanoparticles and surface functionalization are some of the ways to address this challenge (post or pre-treatment) [132]. Future studies should concentrate on inhibiting the growth of microbial colonies on the surface of the membrane, as well as minimizing filler leaching.

(2) In real-world applications, polymer nanocomposites' availability, reusability, cost, stability, agglomeration, and reactivity are all major concerns. As a result, developing novel, inexpensive, and effective nanofillers and polymeric nanocomposites for adsorptive membrane technology still requires attention.

(3) It is difficult to ensure that the adsorptive material combined with the polymeric membrane is safe and harmless. Some composite materials are hazardous because their application in water purification generates secondary pollution. Environmental health and human safety can be achieved by carrying out comprehensive post-treatment evaluation to determine the quality of the water, its suitability for human consumption, and/or its safety for release into the water bodies.

(4) The development of new materials for polymeric nanocomposites remains a major issue, as most materials have been limited to laboratory-scale testing and advanced field trials are needed. Because many innovative materials are not marketable yet due to high pricing or time-consuming synthesis procedures. There is a need for continuous material science research for sustainable and cost-effective polymeric membranes.

(5) In addition to identifying the necessary steps for scaling up new membranes for large-scale industrial applications, there is a need for the development of facile synthesis methods capable of producing defect-free polymeric membranes, without compromising water treatment efficiency.

(6) Furthermore, models for the prediction of the lifespan of the polymeric membrane, regenerability, and reusability are required. To forecast membrane performance and economic viability, models that take into account the morphology and specific characteristics of the polymeric nanocomposite must be developed and validated.

4. Conclusions

Polymer nanocomposites' development and application in water and wastewater treatment holds immense potential due to the added benefits of the synergism between adsorption and membrane filtration approaches. Polymeric membranes have shown remarkable performance in removing both established and emerging chemical pollutants from water. In the synthesis of polymer nanocomposites, water, atomic layer deposition, and plasma-assisted mechanochemistry have been identified to solve the problem of nanoparticle aggregation in melt-mixing techniques. Mixing techniques also benefit from sonication at high frequencies. In-situ polymerization allows for the creation of thermodynamically stable polymeric nanocomposites. Electrospinning is an efficient way for developing porous materials. In addition, fabricating nanocomposites with selective laser sintering offers obvious advantages in terms of avoiding aggregation.

Furthermore, the diversity in the types of materials, i.e., dendritic, biopolymeric, cellulosic and various polymeric composites, available for adsorptive membranes application is advantageous. However, the use of polymeric nanocomposites as adsorbents or as membranes (involving both filtration and adsorption) have various challenges associated with their preparation, human and environmental health, and as well as scalability and adaptability for industrial and field applications. As a result, future research should focus on overcoming these obstacles in order to effectively and sustainably use polymeric nanocomposite adsorbents and membranes in water treatment sectors.

Author Contributions: A.O.A.: conceptualization; investigation; data curation; formal analysis; writing—original draft preparation; writing—review and editing. P.N.N.: fund acquisition, project administration; supervision; validation; writing—review and editing. All authors have read and agreed to the published version of the manuscript.

Funding: This research was funded by Department of Science and Innovation-National Research Foundation South African Research Chairs Initiative (DSI-NRF SARChI) (grant no. 91230).

Acknowledgments: This work was supported by the Department of Science and Innovation-National Research Foundation South African Research Chairs Initiative (DSI-NRF SARChI) (grant no. 91230) and Global Excellence Stature (GES 4.0) funding received from the University of Johannesburg.

Conflicts of Interest: The authors declare no conflict of interest.

References

1. Bushra, R.; Shahadat, M.; Ahmad, A.; Nabi, S.A.; Umar, K.; Oves, M. Synthesis, characterization, antimicrobial activity and applications of polyanilineTi(IV)arsenophosphate adsorbent for the analysis of organic and inorganic pollutants. *J. Hazard. Mater.* **2014**, *264*, 481–489. [CrossRef] [PubMed]
2. Bushra, R. 11—Nanoadsorbents-based polymer nanocomposite for environmental remediation. In *New Polymer Nanocomposites for Environmental Remediation*; Hussain, C.M., Mishra, A.K., Eds.; Elsevier: Amsterdam, The Netherlands, 2018; pp. 243–260.
3. UNESCO. The United Nations World Water Development Report 2021: Valuing Water. UNESCO World Water Assessment Programme. Available online: https://unesdocunescoorg/ark:/48223/pf00003757242021 (accessed on 3 May 2022).
4. Manisalidis, I.; Stavropoulou, E.; Stavropoulos, A.; Bezirtzoglou, E. Environmental and Health Impacts of Air Pollution: A Review. *Front. Public Health* **2020**, *8*, 14. [CrossRef] [PubMed]
5. Patel, A.K.; Singhania, R.R.; Albarico, F.P.J.B.; Pandey, A.; Chen, C.-W.; Dong, C.-D. Organic wastes bioremediation and its changing prospects. *Sci. Total Environ.* **2022**, *824*, 153889. [CrossRef] [PubMed]
6. Adeola, A.O.; Akingboye, A.S.; Ore, O.T.; Oluwajana, O.A. Adewole, A.H.; Olawade, D.B. Crude oil exploration in Africa: So-cio-economic implications, environmental impacts, and mitigation strategies. *Environ. Syst. Decis.* **2021**, *42*, 26–50. [CrossRef] [PubMed]
7. Adeola, A.O.; Ore, O.T.; Fapohunda, O.; Adewole, A.H.; Akerele, D.D.; Akingboye, A.S.; Oloye, F.F. Psychotropic Drugs of Emerging Concerns in Aquatic Systems: Ecotoxicology and Remediation Approaches. *Chem. Afr.* **2022**, *5*, 481–508. [CrossRef]
8. Selwe, K.P.; Thorn, J.P.R.; Desrousseaux, A.O.S.; Dessent, C.E.H.; Sallach, J.B. Emerging contaminant exposure to aquatic systems in the Southern African Development Community. *Environ. Toxicol. Chem.* **2022**, *41*, 382–395. [CrossRef]
9. Hannah, L. Chapter 21—Mitigation: Reducing Greenhouse Gas Emissions, Sinks, and Solutions. In *Climate Change Biology*, 3rd ed.; Hannah, L., Ed.; Academic Press: Cambridge, MA, USA, 2022; pp. 439–472.
10. Adeola, A.; Forbes, P.B.C. Advances in water treatment technologies for removal of polycyclic aromatic hydrocarbons: Existing concepts, emerging trends, and future prospects. *Water Environ. Res.* **2020**, *93*, 343–359. [CrossRef]
11. Singh, V.V. Chapter 2—Green nanotechnology for environmental remediation. In *Sustainable Nanotechnology for Environmental Remediation*; Koduru, J.R., Karri, R.R., Mubarak, N.M., Bandala, E.R., Eds.; Elsevier: Amsterdam, The Netherlands, 2022; pp. 31–61.
12. UNEP—United Nations Environment Programme. *Global Chemicals Outlook II: From Legacies to Innovative Solutions—Implementing the 2030 Agenda for Sustainable Development*; The Global Chemicals Outlook; UNEP: Geneva, Switzerland, 2019.

13. Sanganyado, E. Policies and regulations for the emerging pollutants in freshwater ecosystems: Challenges and opportunities. *Emerg. Freshw. Pollut.* **2022**, *2022*, 361–372. [CrossRef]
14. Ambika, S.P.P. 10—Environmental remediation by nanoadsorbents-based polymer nanocomposite. In *New Polymer Nanocomposites for Environmental Remediation*; Hussain, C.M., Mishra, A.K., Eds.; Elsevier: Amsterdam, The Netherlands, 2018; pp. 223–241.
15. Maurya, A.K.; Gogoi, R.; Manik, G. Polymer-Based Nanocomposites for Removal of Pollutants from Different Environment Using Catalytic Degradation. In *Advances in Nanocomposite Materials for Environmental and Energy Harvesting Applications*; Shalan, A.E., Hamdy, M.A.S., Lanceros-Méndez, S., Eds.; Springer International Publishing: Cham, Switzerland, 2022; pp. 331–368.
16. Darwish, M.S.A.; Mostafa, M.H.; Al-Harbi, L.M. Polymeric Nanocomposites for Environmental and Industrial Applications. *Int. J. Mol. Sci.* **2022**, *23*, 1023. [CrossRef]
17. Fu, S.-Y.; Sun, Z.; Huang, P.; Li, Y.-Q.; Hu, N. Some basic aspects of polymer nanocomposites: A critical review. *Nano Mater. Sci.* **2019**, *1*, 2–30. [CrossRef]
18. Akpan, E.I.; Shen, X.; Wetzel, B.; Friedrich, K. 2—Design and Synthesis of Polymer Nanocomposites. In *Polymer Composites with Functionalized Nanoparticles*; Pielichowski, K., Majka, T.M., Eds.; Elsevier: Amsterdam, The Netherlands, 2019; pp. 47–83.
19. Hmtshirazi, R.; Mohammadi, T.; Asadi, A.A.; Tofighy, M.A. Electrospun nanofiber affinity membranes for water treatment applications: A review. *J. Water Process Eng.* **2022**, *47*, 102795. [CrossRef]
20. Dhillon, S.K.; Kundu, P.P. Polyaniline interweaved iron embedded in urea–formaldehyde resin-based carbon as a cost-effective catalyst for power generation in microbial fuel cell. *Chem. Eng. J.* **2021**, *431*, 133341. [CrossRef]
21. Bustamante-Torres, M.; Romero-Fierro, D.; Arcentales-Vera, B.; Pardo, S.; Bucio, E. Interaction between Filler and Polymeric Matrix in Nanocomposites: Magnetic Approach and Applications. *Polymers* **2021**, *13*, 2998. [CrossRef] [PubMed]
22. Xu, W.; Jambhulkar, S.; Zhu, Y.; Ravichandran, D.; Kakarla, M.; Vernon, B.; Lott, D.G.; Cornella, J.L.; Shefi, O.; Miquelard-Garnier, G.; et al. 3D printing for polymer/particle-based processing: A review. *Compos. B Eng.* **2021**, *223*, 109102. [CrossRef]
23. Ucankus, G.; Ercan, M.; Uzunoglu, D.; Culha, M. 1—Methods for preparation of nanocomposites in environmental remediation. In *New Polymer Nanocomposites for Environmental Remediation*; Hussain, C.M., Mishra, A.K., Eds.; Elsevier: Amsterdam, The Netherlands, 2018; pp. 1–28.
24. Boikanyo, D.; Masheane, M.L.; Nthunya, L.N.; Mishra, S.B.; Mhlanga, S.D. 5—Carbon-supported photocatalysts for organic dye photodegradation. In *New Polymer Nanocomposites for Environmental Remediation*; Hussain, C.M., Mishra, A.K., Eds.; Elsevier: Amsterdam, The Netherlands, 2018; pp. 99–138.
25. Kamal, A.; Ashmawy, M.S.S.; Algazzar, A.M.; Elsheikh, A.H. Fabrication techniques of polymeric nanocomposites: A comprehensive review. *Proc. Inst. Mech. Eng. C—J. Mech. Eng. Sci.* **2021**, *236*, 009544062211055662.
26. Oladipo, A.A. 14—Microwave-assisted synthesis of high-performance polymer-based nanoadsorbents for pollution control. In *New Polymer Nanocomposites for Environmental Remediation*; Hussain, C.M., Mishra, A.K., Eds.; Elsevier: Amsterdam, The Netherlands, 2018; pp. 337–359.
27. Bizzarri, B.M.; Fanelli, A.; Botta, L.; Zippilli, C.; Cesarini, S.; Saladino, R. Dendrimeric Structures in the Synthesis of Fine Chemicals. *Materials* **2021**, *14*, 5318. [CrossRef] [PubMed]
28. Ma, Y.; Mou, Q.; Wang, D.; Zhu, X.; Yan, D. Dendritic Polymers for Theranostics. *Theranostics* **2016**, *6*, 930–947. [CrossRef]
29. Sajid, M.; Nazal, M.K.; Ihsanullah; Baig, N.; Osman, A.M. Removal of heavy metals and organic pollutants from water using dendritic polymers based adsorbents: A critical review. *Sep. Purif. Technol.* **2018**, *191*, 400–423. [CrossRef]
30. Sadjadi, S. 13—Dendritic polymers for environmental remediation. In *New Polymer Nanocomposites for Environmental Remediation*; Hussain, C.M., Mishra, A.K., Eds.; Elsevier: Amsterdam, The Netherlands, 2018; pp. 279–335.
31. Lyu, Z.; Ding, L.; Huang, A.-T.; Kao, C.-L.; Peng, L. Poly(amidoamine) dendrimers: Covalent and supramolecular synthesis. *Mater. Today Chem.* **2019**, *13*, 34–48. [CrossRef]
32. Wang, P.; Ma, Q.; Hu, D.; Wang, L. Removal of Reactive Blue 21 onto magnetic chitosan microparticles functionalized with polyamidoamine dendrimers. *React. Funct. Polym.* **2015**, *91–92*, 43–50. [CrossRef]
33. Karatas, O.; Keyikoglu, R.; Gengec, N.A.; Vatanpour, V.; Khataee, A. A review on dendrimers in preparation and modification of membranes: Progress, applications, and challenges. *Mater. Today Chem.* **2021**, *23*, 100683. [CrossRef]
34. Eghbali, P.; Gürbüz, M.U.; Ertürk, A.S.; Metin, Ö. In situ synthesis of dendrimer-encapsulated palladium(0) nanoparticles as catalysts for hydrogen production from the methanolysis of ammonia borane. *Int. J. Hydrog. Energy* **2020**, *45*, 26274–26285. [CrossRef]
35. Walter, M.V.; Malkoch, M. Simplifying the synthesis of dendrimers: Accelerated approaches. *Chem. Soc. Rev.* **2012**, *41*, 4593–4609. [CrossRef] [PubMed]
36. Ambekar, R.S.; Choudhary, M.; Kandasubramanian, B. Recent advances in dendrimer-based nanoplatform for cancer treatment: A review. *Eur. Polym. J.* **2020**, *126*, 109546. [CrossRef]
37. Viltres, H.; López, Y.C.; Leyva, C.; Gupta, N.K.; Naranjo, A.G.; Acevedo–Peña, P. Polyamidoamine dendrimer-based materials for environmental applications: A review. *J. Mol. Liq.* **2021**, *334*, 116017. [CrossRef]
38. Arkas, M.; Anastopoulos, I.; Giannakoudakis, D.A.; Pashalidis, I.; Katsika, T.; Nikoli, E. Catalytic Neutralization of Water Pol-lutants Mediated by Dendritic Polymers. *Nanomaterials* **2022**, *12*, 445. [CrossRef] [PubMed]
39. Lakshmi, K.; Rangasamy, R. Synthetic modification of silica coated magnetite cored PAMAM dendrimer to enrich branched Amine groups and peripheral carboxyl groups for environmental remediation. *J. Mol. Struct.* **2020**, *1224*, 129081. [CrossRef]

40. Wang, Q.; Zhu, S.; Xi, C.; Zhang, F. A Review: Adsorption and Removal of Heavy Metals Based on Polyamide-amines Composites. *Front. Chem.* **2022**, *10*, 814643. [CrossRef]
41. Niu, Y.; Qu, R.; Chen, H.; Mu, L.; Liu, X.; Wang, T. Synthesis of silica gel supported salicylaldehyde modified PAMAM dendrimers for the effective removal of Hg(II) from aqueous solution. *J. Hazard. Mater.* **2014**, *278*, 267–278. [CrossRef]
42. Lin, Z.; Pan, Z.; Zhao, Y.; Qian, L.; Shen, J.; Xia, K. Removal of Hg^{2+} with Polypyrrole-Functionalized Fe_3O_4/Kaolin: Synthesis, Performance and Optimization with Response Surface Methodology. *Nanomaterials* **2020**, *10*, 1370. [CrossRef]
43. Munyeza, C.F.; Osano, A.; Maghanga, J.K.; Forbes, P.B. Polycyclic Aromatic Hydrocarbon Gaseous Emissions from Household Cooking Devices: A Kenyan Case Study. *Environ. Toxicol. Chem.* **2019**, *39*, 538–547. [CrossRef] [PubMed]
44. Hardonnière, K.; Saunier, E.; Lemarié, A.; Fernier, M.; Gallais, I.; Héliès-Toussaint, C.; Mograbi, B.; Antonio, S.; Bénit, P.; Rustin, P.; et al. The environmental car-cinogen benzo[a]pyrene induces a Warburg-like metabolic reprogramming dependent on NHE1 and associated with cell survival. *Sci. Rep.* **2016**, *6*, 30776. [CrossRef] [PubMed]
45. Arkas, M.; Eleades, L.; Paleos, C.M.; Tsiourvas, D. Alkylated hyperbranched polymers as molecular nanosponges for the purification of water from polycyclic aromatic hydrocarbons. *J. Appl. Polym. Sci.* **2005**, *97*, 2299–2305. [CrossRef]
46. Igwegbe, C.A.; Onukwuli, O.D.; Ighalo, J.O.; Okoye, P.U. Adsorption of Cationic Dyes on Dacryodes edulis Seeds Activated Carbon Modified Using Phosphoric Acid and Sodium Chloride. *Environ. Process.* **2020**, *7*, 1151–1171. [CrossRef]
47. Mudhoo, A.; Gautam, R.K.; Ncibi, M.C.; Zhao, F.; Garg, V.K.; Sillanpää, A. Green synthesis, activation and functionalization of adsorbents for dye sequestration. *Environ. Chem. Lett.* **2019**, *17*, 157–193. [CrossRef]
48. Wazir, M.B.; Daud, M.; Ali, F.; Al-Harthi, M.A. Dendrimer assisted dye-removal: A critical review of adsorption and catalytic degradation for wastewater treatment. *J. Mol. Liq.* **2020**, *315*, 113775. [CrossRef]
49. Hayati, B.; Mahmoodi, N.M.; Arami, M.; Mazaheri, F. Dye Removal from Colored Textile Wastewater by Poly(propylene imine) Dendrimer: Operational Parameters and Isotherm Studies. *CLEAN—Soil Air Water* **2011**, *39*, 673–679. [CrossRef]
50. Kutz, A.; Mariani, G.; Schweins, R.; Streb, C.; Gröhn, F. Self-assembled polyoxometalate–dendrimer structures for selective photo-catalysis. *Nanoscale* **2018**, *10*, 914–920. [CrossRef]
51. Cui, C.; Xie, Y.-D.; Niu, J.-J.; Hu, H.-L.; Lin, S. Poly(Amidoamine) Dendrimer Modified Superparamagnetic Nanoparticles as an Efficient Adsorbent for Cr(VI) Removal: Effect of High-Generation Dendrimer on Adsorption Performance. *J. Inorg. Organomet. Polym. Mater.* **2022**, *32*, 840–853. [CrossRef]
52. Zhang, F.; Wang, B.; He, S.; Man, R. Preparation of Graphene-Oxide/Polyamidoamine Dendrimers and Their Adsorption Properties toward Some Heavy Metal Ions. *J. Chem. Eng. Data* **2014**, *59*, 1719–1726. [CrossRef]
53. Lotfi, Z.; Mousavi, H.Z.; Sajjadi, S.M. Covalently bonded dithiocarbamate-terminated hyperbranched polyamidoamine polymer on magnetic graphene oxide nanosheets as an efficient sorbent for preconcentration and separation of trace levels of some heavy metal ions in food samples. *J. Food Meas. Charact.* **2019**, *14*, 293–302. [CrossRef]
54. Beraa, A.; Hajjaji, M.; Laurent, R.; Delavaux-Nicot, B.; Caminade, A.-M. Removal of chromate from aqueous solutions by dendrimers-clay nanocomposites. *Desalin. Water Treat.* **2016**, *57*, 14290–14303. [CrossRef]
55. Liu, Y.-C.; Li, X.-N.; Wang, C.-Z.; Kong, X.; Zhoug, L.-Z. Poly(styrene-co-divinylbenzene)-PAMAM-IDA chelating resin: Synthesis, characterization and application for Ni(II) removal in aqueous. *J. Cent. South Univ.* **2014**, *21*, 3479–3484. [CrossRef]
56. Zhao, J.; Zhang, X.; He, X.; Xiao, M.; Zhang, W.; Lu, C. A super biosorbent from dendrimer poly(amidoamine)-grafted cellulose nanofibril aerogels for effective removal of Cr(vi). *J. Mater. Chem. A* **2015**, *3*, 14703–14711. [CrossRef]
57. Han, K.N.; Yu, B.Y.; Kwak, S.-Y. Hyperbranched poly(amidoamine)/polysulfone composite membranes for Cd(II) removal from water. *J. Membr. Sci.* **2012**, *396*, 83–91. [CrossRef]
58. Barakat, M.; Ramadan, M.; Kuhn, J.; Woodcock, H. Equilibrium and kinetics of Pb^{2+} adsorption from aqueous solution by dendrimer/titania composites. *Desalin. Water Treat.* **2013**, *52*, 5869–5875. [CrossRef]
59. Liang, X.; Ge, Y.; Wu, Z.; Qin, W. DNA fragments assembled on polyamidoamine-grafted core-shell magnetic silica nanoparticles for removal of mercury(II) and methylmercury(I). *J. Chem. Technol. Biotechnol.* **2016**, *92*, 819–826. [CrossRef]
60. Diallo, M.S.; Christie, S.; Swaminathan, P.; Johnson, J.H.; Goddard, W.A. Dendrimer Enhanced Ultrafiltration. 1. Recovery of Cu(II) from Aqueous Solutions Using PAMAM Dendrimers with Ethylene Diamine Core and Terminal NH2 Groups. *Environ. Sci. Technol.* **2005**, *39*, 1366–1377. [CrossRef]
61. Hayati, B.; Arami, M.; Maleki, A.; Pajootan, E. Application of dendrimer/titania nanohybrid for the removal of phenol from contaminated wastewater. *Desalin.Water Treat.* **2016**, *57*, 6809–6819. [CrossRef]
62. Hayati, B.; Maleki, A.; Najafi, F.; Daraei, H.; Gharibi, F.; McKay, G. Super high removal capacities of heavy metals (Pb^{2+} and Cu^{2+}) using CNT dendrimer. *J. Hazard. Mater.* **2017**, *336*, 146–157. [CrossRef]
63. Almasian, A.; Olya, M.E.; Mahmoodi, N.M. Synthesis of polyacrylonitrile/polyamidoamine composite nanofibers using electrospinning technique and their dye removal capacity. *J. Taiwan Inst. Chem. Eng.* **2015**, *49*, 119–128. [CrossRef]
64. Salimpour Abkenar, S.; Malek, R.M.A.; Mazaheri, F. Dye adsorption of cotton fabric grafted with PPI dendrimers: Isotherm and kinetic studies. *J. Environ. Manag.* **2015**, *163*, 53–61. [CrossRef] [PubMed]
65. Ghasempour, A.; Pajootan, E.; Bahrami, H.; Arami, M. Introduction of amine terminated dendritic structure to graphene oxide using Poly(propylene Imine) dendrimer to evaluate its organic contaminant removal. *J. Taiwan Inst. Chem. Eng.* **2017**, *71*, 285–297. [CrossRef]
66. Arkas, M.; Tsiourvas, D.; Paleos, C.M. Functional Dendrimeric "Nanosponges" for the Removal of Polycyclic Aromatic Hydrocarbons from Water. *Chem. Mater.* **2003**, *15*, 2844–2847. [CrossRef]

67. Azum, N.; Rub, M.A.; Khan, A.; Khan, A.A.P.; Asiri, A.M. Chapter 19—Aerogel applications and future aspects. In *Advances in Aerogel Composites for Environmental Remediation*; Khan, A.A.P., Ansari, M.O., Khan, A., Asiri, A.M., Eds.; Elsevier: Amsterdam, The Netherlands, 2021; pp. 357–367.
68. Paraskevopoulou, P.; Chriti, D.; Raptopoulos, G.; Anyfantis, G.C. Synthetic Polymer Aerogels in Particulate Form. *Materials* 2019, *12*, 1543. [CrossRef]
69. Pal, K.; Banthia, A.K.; Majumdar, D.K. Polymeric Hydrogels: Characterization and Biomedical Applications. *Des. Monomers Polym.* 2009, *12*, 197–220. [CrossRef]
70. Wang, Y.; Su, Y.; Wang, W.; Fang, Y.; Riffat, S.B.; Jiang, F. The advances of polysaccharide-based aerogels: Preparation and potential application. *Carbohydr. Polym.* 2019, *226*, 115242. [CrossRef]
71. Nita, L.E.; Ghilan, A.; Rusu, A.G.; Neamtu, I.; Chiriac, A.P. New Trends in Bio-Based Aerogels. *Pharmaceutics* 2020, *12*, 449. [CrossRef]
72. Mekonnen, B.T.; Ding, W.; Liu, H.; Guo, S.; Pang, X.; Ding, Z.; Seid, M.H. Preparation of aerogel and its application progress in coatings: A mini overview. *J. Leather Sci. Eng.* 2021, *3*, 25. [CrossRef]
73. García-González, C.; Camino-Rey, M.; Alnaief, M.; Zetzl, C.; Smirnova, I. Supercritical drying of aerogels using CO_2: Effect of extraction time on the end material textural properties. *J. Supercrit. Fluids* 2012, *66*, 297–306. [CrossRef]
74. Guastaferro, M.; Reverchon, E.; Baldino, L. Polysaccharide-Based Aerogel Production for Biomedical Applications: A Comparative Review. *Materials* 2021, *14*, 1631. [CrossRef] [PubMed]
75. Cao, R.; Li, L.; Zhang, P. Macroporous MnO_2-based aerogel crosslinked with cellulose nanofibers for efficient ozone removal under humid condition. *J. Hazard. Mater.* 2020, *407*, 124793. [CrossRef] [PubMed]
76. Ahmed, E.M. Hydrogel: Preparation, characterization, and applications: A review. *J. Adv. Res.* 2015, *6*, 105–121. [CrossRef] [PubMed]
77. Yu, F.; Li, Y.; Ma, J. Environmental application and design of alginate/graphene double-network nanocomposite beads. *New Polym. Nanocompos. Environ. Remediat.* 2018, *2018*, 47–76. [CrossRef]
78. Maleki, H.; Hüsing, N. 16—Aerogels as promising materials for environmental remediation—A broad insight into the environmental pollutants removal through adsorption and (photo)catalytic processes. In *New Polymer Nanocomposites for Environmental Remediation*; Hussain, C.M., Mishra, A.K., Eds.; Elsevier: Amsterdam, The Netherlands, 2018; pp. 389–436.
79. Chhetri, K.; Subedi, S.; Muthurasu, A.; Ko, T.H.; Dahal, B.; Kim, H.Y. A review on nanofiber reinforced aerogels for energy storage and conversion applications. *J. Energy Storage* 2022, *46*, 103927. [CrossRef]
80. Hosseinzadeh, H.; Ramin, S. Fabrication of starch-graft-poly(acrylamide)/graphene oxide/hydroxyapatite nanocomposite hydrogel adsorbent for removal of malachite green dye from aqueous solution. *Int. J. Biol. Macromol.* 2018, *106*, 101–115. [CrossRef]
81. Saraf, P.; Abdollahi Movaghar, M.; Montazer, M.; Mahmoudi Rad, M. Bio and photoactive starch/MnO_2 and starch/MnO_2/cotton hydrogel nanocomposite. *Int. J. Biol. Macromol.* 2021, *193*, 681–692. [CrossRef]
82. Herman, P.; Kiss, A.; Fábián, I.; Kalmár, J.; Nagy, G. In situ remediation efficacy of hybrid aerogel adsorbent in model aquatic culture of Paramecium caudatum exposed to Hg(II). *Chemosphere* 2021, *275*, 130019. [CrossRef]
83. Wang, Y.; Zhang, X.; He, X.; Zhang, W.; Zhang, X.; Lu, C. In situ synthesis of MnO_2 coated cellulose nanofibers hybrid for effective removal of methylene blue. *Carbohydr. Polym.* 2014, *110*, 302–308. [CrossRef]
84. Abdel-Halim, E.S.; Al-Deyab, S.S. Preparation of poly(acrylic acid)/starch hydrogel and its application for cadmium ion removal from aqueous solutions. *React. Funct. Polym.* 2014, *75*, 1–8. [CrossRef]
85. Wu, Z.; Chen, X.; Yuan, B.; Fu, M.-L. A facile foaming-polymerization strategy to prepare 3D MnO_2 modified biochar-based porous hydrogels for efficient removal of Cd(II) and Pb(II). *Chemosphere* 2019, *239*, 124745. [CrossRef] [PubMed]
86. Arayaphan, J.; Maijan, P.; Boonsuk, P.; Chantarak, S. Synthesis of photodegradable cassava starch-based double network hydrogel with high mechanical stability for effective removal of methylene blue. *Int. J. Biol. Macromol.* 2020, *168*, 875–886. [CrossRef] [PubMed]
87. Kaur, K.; Jindal, R. Comparative study on the behaviour of Chitosan-Gelatin based Hydrogel and nanocomposite ion exchanger synthesized under microwave conditions towards photocatalytic removal of cationic dyes. *Carbohydr. Polym.* 2018, *207*, 398–410. [CrossRef] [PubMed]
88. Jiang, R.; Zhu, H.-Y.; Fu, Y.-Q.; Jiang, S.-T.; Zong, E.-M.; Zhu, J.-Q. Colloidal CdS sensitized nano-ZnO/chitosan hydrogel with fast and efficient photocatalytic removal of congo red under solar light irradiation. *Int. J. Biol. Macromol.* 2021, *174*, 52–60. [CrossRef] [PubMed]
89. Irshad, M.S.; Wang, X.; Abbasi, M.S.; Arshad, N.; Chen, Z.; Guo, Z. Semiconductive, Flexible MnO_2 NWs/Chitosan Hydrogels for Efficient Solar Steam Generation. *ACS Sustain. Chem. Eng.* 2021, *9*, 3887–3900. [CrossRef]
90. Adeola, A.O.; Abiodun, B.A.; Adenuga, D.O.; Nomngongo, P.N. Adsorptive and photocatalytic remediation of hazardous organic chemical pollutants in aqueous medium: A review. *J. Contam. Hydrol.* 2022, *248*, 104019. [CrossRef]
91. Ray, P.; Singh, P.S.; Polisetti, V. 2—Synthetic polymeric membranes for the removal of toxic pollutants and other harmful contaminants from water. In *Removal of Toxic Pollutants through Microbiological and Tertiary Treatment*; Shah, M.P., Ed.; Elsevier: Amsterdam, The Netherlands, 2020; pp. 49–93.
92. Mukherjee, M.; Roy, S.; Bhowmick, K.; Majumdar, S.; Prihatiningtyas, I.; Van der Bruggen, B.; Mondal, P. Development of high performance pervaporation desalination membranes: A brief review. *Process Saf. Environ. Prot.* 2022, *159*, 1092–1104. [CrossRef]

93. Wilson, R.; George, G.; Jose, A.J. Polymer membranes reinforced with carbon-based nanomaterials for water purification. *New Polym. Nanocompos. Environ. Remediat.* **2018**, *2018*, 457–468. [CrossRef]
94. Potara, M.; Focsan, M.; Craciun, A.-M.; Botiz, I.; Astilean, S. 15—Polymer-coated plasmonic nanoparticles for environmental reme-diation: Synthesis, functionalization, and properties. In *New Polymer Nanocomposites for Environmental Remediation*; Hussain, C.M., Mishra, A.K., Eds.; Elsevier: Amsterdam, The Netherlands, 2018; pp. 361–387.
95. Leudjo Taka, A.; Klink, M.J.; Yangkou Mbianda, X.; Naidoo, E.B. Chitosan nanocomposites for water treatment by fixed-bed con-tinuous flow column adsorption: A review. *Carbohydr. Polym.* **2021**, *255*, 117398. [CrossRef]
96. Sapna, K.D. 12—Biodegradable polymer-based nanoadsorbents for environmental remediation. In *New Polymer Nanocomposites for Environmental Remediation*; Hussain, C.M., Mishra, A.K., Eds.; Elsevier: Amsterdam, The Netherlands, 2018; pp. 261–278.
97. Semyonov, O.; Kogolev, D.; Mamontov, G.; Kolobova, E.; Trelin, A.; Yusubov, M.S.; Guselnikova, O.; Postnikov, P.S. Synergetic effect of UiO-66 and plasmonic AgNPs on PET waste support towards degradation of nerve agent simulant. *Chem. Eng. J.* **2021**, *431*, 133450. [CrossRef]
98. Fahma, F.; Febiyanti, I.; Lisdayana, N.; Arnata, I.; Sartika, D. Nanocellulose as a new sustainable material for various applications: A review. *Arch. Mater. Sci. Eng.* **2021**, *2*, 49–64. [CrossRef]
99. Iba, H.; Chang, T.; Kagawa, Y. Optically transparent continuous glass fibre-reinforced epoxy matrix composite: Fabrication, optical and mechanical properties. *Compos. Sci. Technol.* **2002**, *62*, 2043–2052. [CrossRef]
100. Rol, F.; Belgacem, M.N.; Gandini, A.; Bras, J. Recent advances in surface-modified cellulose nanofibrils. *Prog. Polym. Sci.* **2018**, *88*, 241–264. [CrossRef]
101. Eichhorn, S.J.; Etale, A.; Wang, J.; Berglund, L.A.; Li, Y.; Cai, Y.; Chen, C.; Cranston, E.D.; Johns, M.A.; Fang, Z.; et al. Current international research into cellulose as a functional nanomaterial for advanced applications. *J. Mater. Sci.* **2022**, *57*, 5697–5767. [CrossRef]
102. Adam, M.R.; Hubadillah, S.K.; Esham, M.I.M.; Othman, M.H.D.; Rahman, M.A.; Ismail, A.F. Chapter 12—Adsorptive Mem-branes for Heavy Metals Removal From Water. In *Membrane Separation Principles and Applications*; Ismail, A.F., Rahman, M.A., Othman, M.H.D., Matsuura, T., Eds.; Elsevier: Amsterdam, The Netherlands, 2019; pp. 361–400.
103. Zhang, X.; Fang, X.; Li, J.; Pan, S.; Sun, X.; Shen, J.; Han, W.; Wang, L.; Zhao, S. Developing new adsorptive membrane by modification of support layer with iron oxide microspheres for arsenic removal. *J. Colloid Interface Sci.* **2018**, *514*, 760–768. [CrossRef]
104. Tabrizi, S.H.; Tanhaei, B.; Ayati, A.; Ranjbari, S. Substantial improvement in the adsorption behavior of montmorillonite toward Tartrazine through hexadecylamine impregnation. *Environ. Res.* **2021**, *204*, 111965. [CrossRef] [PubMed]
105. Qalyoubi, L.; Al-Othman, A.; Al-Asheh, S. Recent progress and challenges on adsorptive membranes for the removal of pollutants from wastewater. Part I: Fundamentals and classification of membranes. *Case Stud. Chem. Environ. Eng.* **2021**, *3*, 100086. [CrossRef]
106. Salehi, E.; Daraei, P.; Shamsabadi, A.A. A review on chitosan-based adsorptive membranes. *Carbohydr. Polym.* **2016**, *152*, 419–432. [CrossRef]
107. Qadir, D.; Nasir, R.; Mukhtar, H.B.; Keong, L.K. Synthesis, characterization, and performance analysis of carbon molecular sieve-embedded polyethersulfone mixed-matrix membranes for the removal of dissolved ions. *Water Environ. Res.* **2020**, *92*, 1306–1324. [CrossRef]
108. Kang, M.; Kawasaki, M.; Tamada, S.; Kamei, T.; Magara, Y. Effect of pH on the removal of arsenic and antimony using reverse osmosis membranes. *Desalination* **2000**, *131*, 293–298. [CrossRef]
109. Oh, J.; Yamamoto, K.; Kitawaki, H.; Nakao, S.; Sugawara, T.; Rahman, M. Application of low-pressure nanofiltration coupled with a bicycle pump for the treatment of arsenic-contaminated groundwater. *Desalination* **2000**, *132*, 307–314. [CrossRef]
110. Mojarrad, M.; Noroozi, A.; Zeinivand, A.; Kazemzadeh, P. Response surface methodology for optimization of simultaneous Cr (VI) and as (V) removal from contaminated water by nanofiltration process. *Environ. Prog. Sustain. Energy* **2017**, *37*, 434–443. [CrossRef]
111. He, Y.; Tang, Y.P.; Ma, D.; Chung, T.-S. UiO-66 incorporated thin-film nanocomposite membranes for efficient selenium and arsenic removal. *J. Membr. Sci.* **2017**, *541*, 262–270. [CrossRef]
112. He, Y.; Liu, J.; Han, G.; Chung, N.T.-S. Novel thin-film composite nanofiltration membranes consisting of a zwitterionic co-polymer for selenium and arsenic removal. *J. Membr. Sci.* **2018**, *555*, 299–306. [CrossRef]
113. Manirethan, V.; Gupta, N.; Balakrishnan, R.M.; Raval, K. Batch and continuous studies on the removal of heavy metals from aqueous solution using biosynthesised melanin-coated PVDF membranes. *Environ. Sci. Pollut. Res.* **2019**, *27*, 24723–24737. [CrossRef] [PubMed]
114. Yoo, H.; Kwak, S.-Y. Surface functionalization of PTFE membranes with hyperbranched poly(amidoamine) for the removal of Cu^{2+} ions from aqueous solution. *J. Membr. Sci.* **2013**, *448*, 125–134. [CrossRef]
115. Weng, Y.-H.; Chaung-Hsieh, L.H.; Lee, H.-H.; Li, K.-C.; Huang, C.P. Removal of arsenic and humic substances (HSs) by elec-tro-ultrafiltration (EUF). *J. Hazard. Mater.* **2005**, *122*, 171–176. [CrossRef]
116. Cai, Y.-H.; Yang, X.J.; Schäfer, A.I. Removal of Naturally Occurring Strontium by Nanofiltration/Reverse Osmosis from Ground-water. *Membranes* **2020**, *10*, 321. [CrossRef]
117. Cheng, W.; Liu, C.; Tong, T.; Epsztein, R.; Sun, M.; Verduzco, R. Selective removal of divalent cations by polyelectrolyte mul-tilayer nanofiltration membrane: Role of polyelectrolyte charge, ion size, and ionic strength. *J. Membr. Sci.* **2018**, *559*, 98–106. [CrossRef]

118. Fakhfekh, R.; Chabanon, E.; Mangin, D.; Ben Amar, R.; Charcosset, C. Removal of iron using an oxidation and ceramic microfiltration hybrid process for drinking water treatment. *Desalin. Water Treat.* **2017**, *66*, 210–220. [CrossRef]
119. Luo, L.; Han, G.; Chung, T.-S.; Weber, M.; Staudt, C.; Maletzko, C. Oil/water separation via ultrafiltration by novel triangle-shape tri-bore hollow fiber membranes from sulfonated polyphenylenesulfone. *J. Membr. Sci.* **2014**, *476*, 162–170. [CrossRef]
120. Cui, J.; Zhou, Z.; Xie, A.; Wang, Q.; Liu, S.; Lang, J. Facile preparation of grass-like structured NiCo-LDH/PVDF composite membrane for efficient oil–water emulsion separation. *J. Membr. Sci.* **2019**, *573*, 226–233. [CrossRef]
121. Cui, J.; Xie, A.; Zhou, S.; Liu, S.; Wang, Q.; Wu, Y.; Meng, M.; Lang, J.; Zhou, Z.; Yan, Y. Development of composite membranes with irregular rod-like structure via atom transfer radical polymerization for efficient oil-water emulsion separation. *J. Colloid Interface Sci.* **2018**, *533*, 278–286. [CrossRef] [PubMed]
122. Wu, J.; Ding, Y.; Wang, J.; Li, T.; Lin, H.; Wang, J.; Liu, F. Facile fabrication of nanofiber- and micro/nanosphere-coordinated PVDF membrane with ultrahigh permeability of viscous water-in-oil emulsions. *J. Mater. Chem. A* **2018**, *6*, 7014–7020. [CrossRef]
123. Zioui, D.; Salazar, H.; Aoudjit, L.; Martins, P.M.; Lanceros-Méndez, S. Polymer-Based Membranes for Oily Wastewater Remediation. *Polymers* **2020**, *12*, 42. [CrossRef]
124. Wei, C.; Dai, F.; Lin, L.; An, Z.; He, Y.; Chen, X.; Chen, L.; Zhao, Y. Simplified and robust adhesive-free superhydrophobic SiO_2-decorated PVDF membranes for efficient oil/water separation. *J. Membr. Sci.* **2018**, *555*, 220–228. [CrossRef]
125. Fakhru'l-Razi, A.; Pendashteh, A.; Abidin, Z.Z.; Abdullah, L.C.; Biak, D.R.A.; Madaeni, S.S. Application of membrane-coupled se-quencing batch reactor for oilfield produced water recycle and beneficial re-use. *Bioresour. Technol.* **2010**, *101*, 6942–6949. [CrossRef]
126. Liu, J.; Li, P.; Chen, L.; Feng, Y.; He, W.; Lv, X. Modified superhydrophilic and underwater superoleophobic PVDF membrane with ultralow oil-adhesion for highly efficient oil/water emulsion separation. *Mater. Lett.* **2016**, *185*, 169–172. [CrossRef]
127. Shi, H.; He, Y.; Pan, Y.; Di, F.; Zeng, G.; Zhang, L. A modified mussel-inspired method to fabricate TiO_2 decorated superhydrophilic PVDF membrane for oil/water separation. *J. Membr. Sci.* **2016**, *506*, 60–70. [CrossRef]
128. Ighalo, J.O.; Yap, P.-S.; Iwuozor, K.O.; Aniagor, C.O.; Liu, T.; Dulta, K.; Iwuchukwu, F.U.; Rangabhashiyam, S. Adsorption of persistent organic pollutants (POPs) from the aqueous environment by nano-adsorbents: A review. *Environ. Res.* **2022**, *212*, 113123. [CrossRef]
129. Peng, B.L.; Yao, Z.L.; Wang, X.C.; Crombeen, M.; Sweeney, D.G.; Tam, K.C. Cellulose-based materials in wastewater treatment of petroleum industry. *Green Energy Environ.* **2020**, *5*, 37–49. [CrossRef]
130. Adeola, A.O.; Kubheka, G.; Chirwa, E.M.N.; Forbes, P.B.C. Facile synthesis of graphene wool doped with oleylamine-capped silver nanoparticles (GW-αAgNPs) for water treatment applications. *Appl. Water Sci.* **2021**, *11*, 172. [CrossRef]
131. Sajid, M.; Ilyas, M.; Basheer, C.; Tariq, M.; Daud, M.; Baig, N.; Shehzad, F. Impact of nanoparticles on human and environment: Review of toxicity factors, exposures, control strategies, and future prospects. *Environ. Sci. Pollut. Res.* **2014**, *22*, 4122–4143. [CrossRef] [PubMed]
132. Bhoj, Y.; Tharmavaram, M.; Rawtani, D. A comprehensive approach to antifouling strategies in desalination, marine environment, and wastewater treatment. *Chem. Phys. Impact* **2020**, *2*, 100008. [CrossRef]

Review

Functionalized Nanomembranes and Plasma Technologies for Produced Water Treatment: A Review

Anton Manakhov [1,*], Maxim Orlov [1], Vyacheslav Grokhovsky [1], Fahd I. AlGhunaimi [2] and Subhash Ayirala [2]

1. Aramco Innovations LLC, Aramco Research Center, 119234 Moscow, Russia; maxim.orlov@aramcoinnovations.com (M.O.); vyacheslav.grokhovsky@aramcoinnovations.com (V.G.)
2. EXPEC Advanced Research Center, Saudi Aramco, Dhahran 31311, Saudi Arabia; fahad.ghunaimi@aramco.com (F.I.A.); nethajisubhash.ayirala@aramco.com (S.A.)
* Correspondence: anton.manakhov@aramcoinnovations.com; Tel.: +7-9158-494-059

Abstract: The treatment of produced water, associated with oil & gas production, is envisioned to gain more significant attention in the coming years due to increasing energy demand and growing interests to promote sustainable developments. This review presents innovative practical solutions for oil/water separation, desalination, and purification of polluted water sources using a combination of porous membranes and plasma treatment technologies. Both these technologies can be used to treat produced water separately, but their combination results in a significant synergistic impact. The membranes functionalized by plasma show a remarkable increase in their efficiency characterized by enhanced oil rejection capability and reusability, while plasma treatment of water combined with membranes and/or adsorbents could be used to soften water and achieve high purity.

Keywords: plasma; produced water; membranes; desalination; gliding arc; nanofiltration; nanofibers

Citation: Manakhov, A.; Orlov, M.; Grokhovsky, V.; AlGhunaimi, F.I.; Ayirala, S. Functionalized Nanomembranes and Plasma Technologies for Produced Water Treatment: A Review. *Polymers* **2022**, *14*, 1785. https://doi.org/10.3390/polym14091785

Academic Editors: Irene S. Fahim, Ahmed K. Badawi and Hossam E. Emam

Received: 7 April 2022
Accepted: 24 April 2022
Published: 27 April 2022

Publisher's Note: MDPI stays neutral with regard to jurisdictional claims in published maps and institutional affiliations.

Copyright: © 2022 by the authors. Licensee MDPI, Basel, Switzerland. This article is an open access article distributed under the terms and conditions of the Creative Commons Attribution (CC BY) license (https://creativecommons.org/licenses/by/4.0/).

1. Introduction

The sustainable development of modern society is increasing the energy demand, leading to a higher power being produced [1]. At the same time, the importance of all environmental concerns is growing substantially every year [2]. While attention to renewable sources is great for multiple reasons, the improvements and optimization of the technologies in the current format of the energy sector are also highly essential [3]. This necessitates the need to revise processes in the oil & gas industry towards more sustainable strategies in terms of energy efficiency, process-water reusability, and recycling of chemicals [4].

The water resources have a significant impact on the efficiency of the oil & gas industry, as the energy sector is 6th in the most water-intensive industries with 52,000,000,000 m³ of freshwater consumed annually for global energy production [5]. The reuse of wastewater streams became an important research topic with more than 130 000 publications in this field. The number of publications focused on the produced water treatment is also rapidly growing and will be even more significant in the coming years (Figure 1a).

In the oil & gas sector, the produced water is the most significant waste stream, especially for the wells with high maturity [6]. Production activities generate increasing loads of produced water up to thousands of tons each day or more, depending on the geographic area, formation depth, oil production techniques, and age of oil supply wells. The regulations related to the discharge of the produced water are stringent, although they might vary for different countries. For example, in Denmark, it must be below 30 ppm [7]. The Regional Organization for Protection of the Marine Environment (ROPME) issued several regional agreements controlling drilling operations in Arabian Gulf. According to the ROPME protocol concerning Marine Pollution resulting from Exploration and Exploitation of the Continental Shelf, the effluent discharge should contain less than 15 ppm of oil, and and no drilling waste above 40 ppm [8]. The water is produced from different oil fields containing different chemical compositions. Currently, produced water is also known as industrial wastewater containing heavy metals that are toxic to humans and the environment, requiring special processing so that they can be properly disposed [9]. The ever-evolving and increasingly stringent regulatory standards for discharging produced water pose major environmental and economic implications [6]. Hence, the treatment of produced water creates a significant challenge that new materials and technologies must be capable of solving.

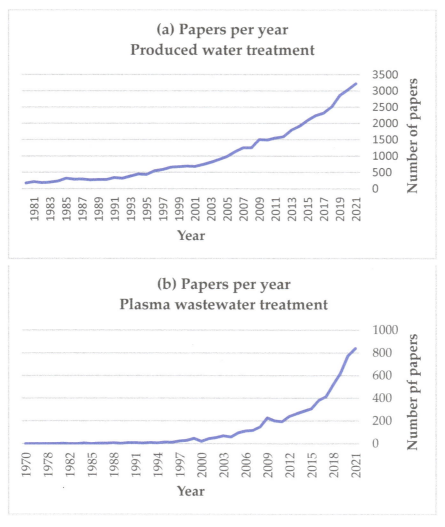

Figure 1. The evolution of publications number (per annum) related to produced water treatment (**a**) and plasma wastewater treatment (**b**). The data was acquired from Scopus database.

The importance of this topic is also supported by the decisive engagement/increase of R&D funds, among international organizations and private companies, with a total number of awarded research grants of about 328 (determined according to the literature analysis using Scopus database). For example, the efforts to enhance the economic efficiency of sustainable produced water treatment processes were supported by different initiatives, including the U.S. Department of Energy (DOE). These initiatives actually aim to develop advanced technologies that can improve oil and gas production while protecting water resources, reducing water use [10].

The topic of produced water treatment was reviewed several times in the last decade, e.g., in [11–13]. The water treatment methods are generally based on chemicals, thermal processes, electrodialysis, membrane distillation, etc. However, chemical water treatment and thermal treatment have certain limitations, including economic efficiency and environmental impact (production of chemical, energy-intensive processes). In contrast, membrane technologies, plasma treatment, or combined technologies have excellent potential for oil/water separation, desalination, and heavy metals/toxic chemical removal. The use of plasma technologies for wastewater treatment also became rapidly growing topic with several hundred publications per annum, as shown in Figure 1b.

This review is focused on hybrid strategies and technologies based on nanomaterials and plasma processes with a significant focus on the sustainability of each proposed methodology. Indeed, as the plasma process is an environmentally friendly technology enabling new materials, it has excellent potential to enhance the water/oil separation process, induce new reactive cites to improve the desalination, and, finally, destroy toxic organic chemicals or convert heavy metals into sediments.

2. Plasma Modified Membranes Used for Oil/Water Separation

Plasma modification of different materials enables enhanced properties by adjusting surface chemistry and roughness without damaging the bulk properties [14]. The nonthermal plasma is generally used to modify the surface of materials as it will not damage the surface polymeric materials, including temperature-sensitive materials [15]. Nonthermal plasma can be ignited at different pressures by applying the high voltage to the electrodes separated by the gap (filled by noble gases, air or a mixture of precursors) [16]. To obtain homogenous plasma at atmospheric pressure, the surface of the electrodes could be covered by a dielectric layer (to prevent arcing). At the same time, the frequency of the applied voltage can be maintained in the range from a few kHz up to several MHz. It was often used to increase the superhydrophobicity or superhydrophilicity of nanomaterials to develop highly efficient nanofiltration membranes [17,18].

Nanofiltration (NF) is a perspective technology that provides a combined solution to reject both organic and inorganic pollutants, thereby making it more beneficial than ultrafiltration (UF). However, compared to reverse osmosis (RO), the salt and metal ion rejection percentages are typically lower (50–80% vs. 99%). The rejection of organic molecules is also somewhat lower. In comparison, ultrafiltration membranes allow passing almost all components with MW < 10,000 g/mol, while the pore size of nanofiltration membrane is much smaller, allowing many smaller organic molecules to be rejected. As a result, nanofiltration can be classified between reverse osmosis and ultrafiltration on the filtration spectrum [19].

Nanofiltration has a strong potential for oil/water separation, as it allows to obtain cleaner water from oily solutions, including the possibility for in-house recycling [20]. If the discharge of the produced water is preferred, nanofiltration permeate can also be clean enough to meet the stringent discharge requirements [21]. However, different challenges, including economic viability, robustness, and modest efficiency, hinder the large-scale nanofiltration application. This section summarizes the most advanced eco-friendly oil/water separation solutions based on the modified nanofiltration membranes.

A novel environment-friendly method to prepare nanofiltration membranes for highly efficient separation of water-in-oil emulsions was reported by Yang et al. [22], where durable biphilic (oleophilic and hydrophilic) surface was prepared by atmospheric air plasma treatment of cellulose filter paper followed by various grafting procedures. The hydrophobic surface with a microscale hierarchical structure was constructed using long-chain alkyl silane grafting. The authors utilized silanization by acting aminopropyltrimethoxy silane

(APTMS), glycidyl propyl trimethoxy silane (GPTMS) and hexadecyl trimethoxy silane (HDTMS) onto plasma-activated SiO$_2$ nanoparticles using surface chemistry approach as shown in Figure 2. The samples exhibited excellent separation performance for surfactant-stabilized water-in-oil emulsions with the filtration efficiency of up to 99% and the oil recovery purity > 99.98 wt%. As shown in Figure 2, the water-amine attraction (for APTMS functionalized layers) and oil-alkyl attraction (for the GPTMS side) synergistically destabilize the emulsion and separate the two phases. The formatted coating was very durable against ultrasonication in acetone and can be stored for 12 months without degradation upon aging. The coating also exhibited excellent recyclability (over 20 cycles), although the multi-step process is too complicated and not cost-effective for large-scale application.

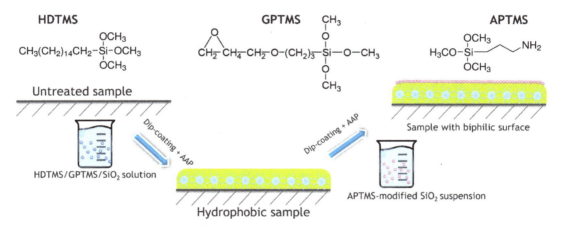

Figure 2. Scheme for the nanofiltration membranes preparation using atmospheric air plasma and modifications by dip-coating.

Various materials, including fluoropolymers, treated by plasma exhibited enhanced oil-water separation properties. For example, the traditional polytetrafluoroethylene (PTFE) membrane was pretreated using the atmospheric pressure glow discharge (APGD) plasma. Then, the surface of the plasma-treated PTFE membrane was modified by the photo-induced graft polymerization of 2-acrylamido-2-methyl-propyl-sulfoacid (AMPS). The obtained surface demonstrated a highly negative charge, ensuring substantial oleophobic property of the PTFE membrane and leading to improved anti-fouling performance compared to the unmodified sample [23].

A superhydrophilic PVDF membrane obtained by acrylic acid plasma polymerization followed by TiO$_2$ nanoparticles self-assembly was employed for oily produced water treatment [24]. The TiO$_2$ nanoparticles were immobilized onto the membrane surface via the coordination of Ti^{4+} with a carboxylic group from plasma layers. It is worth noting that the immobilization step did not affect the valence state of Ti^{4+} (Figure 3). The TiO$_2$-decorated PVDF membranes exhibited high uniformity and dramatically improved surface hydrophilicity. After modification, the permeation flux was increased more than four times, and the oil rejection rate was higher than 92%.

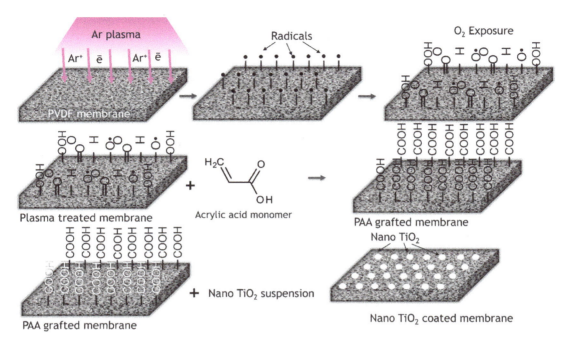

Figure 3. Preparation of nanofiltration membrane by TiO_2 self-assembling on a PVDF surface.

Lin and co-workers developed Janus membrane composed of the PVDF substrate functionalized with a special silicification layer inducing the asymmetric wetting selectivity [25]. The hydrophobicity of the membrane was facilitated by grafting the hydrophobic perfluorodecyltriethoxysilane, while unilateral plasma etching treatment of modified membrane allowed to induce the hydrophilicity. By tuning the surface chemistry of hierarchical nanosphere-like architecture of a robust membrane with superhydrophobicity/superoleophilicity (perfluorodecyltriethoxysilane modified side), and underwater superoleophobicity/superhydrophilicity (plasma-modified side) were obtained. The membranes showed an extremely opposing surface wettability with a high water/underwater oil contact angle (CA) difference up to 150° due to the exceptionally asymmetric wetting selectivity incorporated with the 15 nm ultrathin silicification layer. As a result, the water-oil separation efficiencies in surfactant-stabilized oil-in-water and water-in-oil emulsions reached 99.8% and 99.0%, respectively. In addition, the authors demonstrated nearly 100% recovery ratio of permeate flux after several cycles of oil-in-water and water-in-oil emulsion filtration tests.

It is worth noting that the multi-step functionalization of PVDF membranes has a particular drawback: this process requires several reactors and a relatively long treatment time. As a result, it increases the production cost. Therefore, many research works were focused on utilizing cheaper materials and improving the cost-efficiency of the modification to one-step processes.

Complimentary to the PVDF, polyethersulfone (PES) ultra- and nanofiltration membranes are attractive materials due to their high efficiency and stability [26]. Sadeghi and co-workers prepared efficient PES ultrafiltration membranes using a corona plasma treatment to reduce membrane fouling during the separation of oil/water emulsions [21]. These PES membranes were fabricated by the phase inversion method and thoroughly analyzed. The authors investigated the effects of solvent in casting solution and corona treatment conditions on membrane surface properties, morphology, and separation performance. The corona treatment induced the grafting of hydrophilic OH groups leading to the higher water flux in all configurations of membranes without significant changes in oil rejection. Finally, the rejection rate up to 99.5% was observed at the oil/water permeability range from 34 to 70 (L m^{-2} h^{-1} bar^{-1}). Similar approach was exploited by Adib et al. [27] to explore the potential of corona plasma-treated PES membranes for water-oil separation. By treating the membrane in corona discharge at 360 W for 6 min, the authors achieved their most promising sample with the rejection rate of 92%. Membranes exhibited mean pore size ~30 nm at a porosity of 85%. To enhance the grafting of hydrophilic COOH groups and, additionally, to utilize CO_2 to improve the sustainability of the approach further, carbon dioxide can be admixed to the plasma-forming gas mixture. CO_2 plasmas were used to modify hydrophobic polysulfone ultrafiltration membranes to create hydrophilic surfaces throughout the membrane structure [28]. The water contact angle (WCA) of the upstream side of the membrane (facing the plasma) decreased to zero after treatment and remained at the same condition even after several months of aging.

Fedotova and co-workers employed radio-frequency (RF) glow discharge to treat polysulfonamide membranes (pore size of 0.01 µm) [29]. Low-pressure RF discharge in an argon and nitrogen atmosphere at an anode voltage of U_a = 1.5 kV and contact time τ = 1.5 min led to a decrease in the contact angle from 59.6° to 47.9°, and grafting of oxygen groups. Experiments on the membrane separation of a 3% oil-in-water emulsion have been conducted. The results have shown that the use of plasma-treated polysulfonamide membranes leads to the intensification of the oil/water separation process.

In order to enhance the wettability or superhydrophobicity, many researchers prefer to use the materials with advanced nanoporous structures, e.g., electrospun nanofibers. Indeed, the production of nanofibrous membranes has approached the commercial scale processes with many products, including air-filters and masks, showing high cost efficiency [30–32]. Thus, the implementation of nanofibrous membranes for water treatment offered significant potential.

Nanofibrous membranes with superhydrophilic/superoleophobic surfaces were constructed by grafting acrylic acid onto low-pressure nitrogen plasma treated electrospun polystyrene/polyacrylonitrile (PS/PAN) nanofibrous foils as shown in Figure 4 [33]. The WCAs of the PS/PAN membranes decreased to 0° after grafting treatment of acrylic acid, thereby proving that the modification improved the surface hydrophilicity of the membranes due to the introduction of hydrophilic groups. Oil/water permeation tests confirmed a high oil/water separation potential of plasma-treated PS/PAN membranes. The results showed that these membranes effectively separated the layered oil/water mixture with permeate flux up to 57,509 L m^{-2} h^{-1}, while high fluxes of 1390–6460 L m^{-2} h^{-1} were observed for the separation of different oil-in-water emulsions. The prepared membranes exhibited superhydrophilic and underwater superoleophobic surfaces, which could prevent oil droplets from adhering to the surface, thereby contributing to membrane anti-fouling. In the separation step, the resultant membranes, solely driven by gravity, had high separation efficiency as well as one to two orders of magnitude higher fluxes than that of traditional polymeric filtration membranes with similar permeation properties [33].

Notably, the membranes maintained high flux and efficiency even after several separation cycles to demonstrate excellent recyclability. Therefore, such flexibility in reusing these membranes is expected to provide potential cost-effective materials for oil/water treatment.

The strategy shown above for PS/PAN membrane modification by acrylic acid grafting using plasma activation required a wet chemical step that is preferably switched to a gas-phase process or a single-step plasma modification for better cost-efficiency. Of course, the one-step atmospheric plasma method would be the most desirable, but it is pretty challenging at the same time [34,35]. Moreover, the wet chemical steps are often eliminated by low or atmospheric pressure plasmas to improve the economic viability.

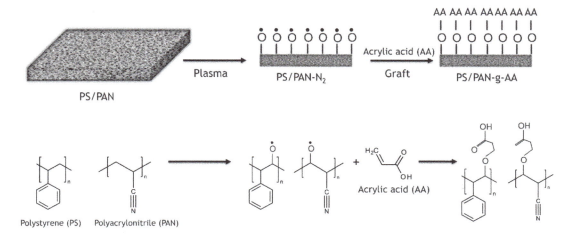

Figure 4. PS/PAN membrane modification by acrylic acid grafting using plasma activation by grafting of radical Oxygen species (O*).

Several researchers demonstrated the capability of atmospheric plasma for modification of membranes to stimulate the oil/water separation. Chen et al. proposed the environment-friendly oil/water separation methodology based on atmospheric plasma-treated nylon meshes [36]. Nylon fibers were exposed to atmospheric pressure plasma for surface modification, leading to micro/nanostructures and oxygen-containing groups. Consequently, the functionalized meshes possess superhydrophilicity in air and thus superoleophobicity underwater. The obtained materials exhibited an efficiency above 97.5% for various oil/water mixtures. Moreover, the authors showed that the functionalized nylon meshes have excellent recyclability and durability in oil/water separation. Fedotova et al. optimized the plasma modification of polysulfonamide membranes (with a pore size of 0.01 µm) for the separation of model "oil-in-water" emulsions containing 3% of industrial oil [37]. The atmospheric pressure plasma torch was ignited for short time periods from 1 to 7 min at the anode voltage ranging from 1.5 to 7.5 kV. The highest performance was observed in the case of membranes treated with plasma for 7 min, leading to increased membrane efficiency from 90 to 99% owing to the surface hydrophilization. Indeed, the membrane exhibited a water contact angle 59.6° before treatment, and after a 4-min plasma treatment at 7.5 kV it decreased to 19.5°. The authors also confirmed that the plasma treatment leads to a change in the surface structure of the membranes, namely, a decrease in their roughness. The membrane internal microstructure also undergoes changes that increase their crystallinity.

You et al. employed $He/CH_4/C_4F_8$ atmospheric plasma deposition to modify the surface of stainless steel meshes [38] and applied them for oil/water separation. The fine-tuning of the plasma process parameters enabled the selective functionalization of each side of the membranes, leading to superhydrophobic-superoleophilic and superhydrophobic-oleophobic sides. The authors demonstrated the successful separation in a simple test, where a 50 mL oil-water solution in 6 min was treated. According to the FTIR analysis, approximately 88% of the hydrocarbons in the oil-water mixture were separated in 6 min without any external forces, and no water was detected in the collected oil. The separation efficiency of up to 99% has been achieved. A similar concept of oil/water separation using oxygen plasma treatment of Cu coated meshes to induce hydrophobic and oleophilic behaviors is presented by Agarwal et al. [39]. This concept's advantage is the possibility of using cheaper and more robust steel grades for separating meshes.

Sometimes, produced water may contain special chemicals, e.g., hydrolyzed polyacrylamide (HPAM), which is added to the displacing fluid to enhance the mobility and extraction of the oil phase. Optimized ultrafiltration, using ceramic membranes with a surface pore size of 15 kDa, to effectively separate HPAM from produced water was demonstrated by Ricceri et al. [40]. The precipitation and ultrafiltration may be used in sequence as they complement each other in several ways. Moreover, ultrafiltration membranes with 15 kDa cut-off provided the best combination of productivity and HPAM removal (92%). However, the authors highlighted the problem of membrane fouling under severe conditions.

3. Nanofiltration Membranes for Desalination

The rejection of salts from the produced water is very challenging task. However, a sustainable nanofiltration approach using modified nanomaterials significantly progressed to solve this problem by using various advanced 2D nanomaterials, including graphene and boron nitride [41,42]. An exciting approach based on membrane crystallization was demonstrated by Ali et al. [43]. The authors employed membrane crystallization for desalination and salt recovery from produced water streams at a semi-pilot scale using convetional polypropylene or in-lab prepared PVDF hollow fiber membranes. The experiments were carried out in lab scale and semi-pilot scale. It is worth noting that the salts may also be in high demand, and the recovery of ions can be highly interesting if economically viable. Unlike conventional crystallizers, in membrane crystallization, well-controlled nucleation and crystal growth are achieved through a uniform evaporation rate through the membrane pores. The crystals recovered by membrane crystallization have higher purity and narrow size distribution. The recovered crystals were sodium chloride with high purity (>99.9%). It was demonstrated experimentally that at a recovery factor of 37%, 16.4 kg NaCl per cubic meter of produced water could be recovered.

Another method to extract salts from produced water is the airgap membrane distillation. Woo et al. modified the PVDF membrane by using electrospinning of PVDF, modified the surface of the nanofibrous foil, and evaluated the membrane's performance using real reverse osmosis brine with a produced water as a feed [44]. The results suggest the successful chemical modification of the membrane by plasma treatment without significantly altering the morphology and its physical properties, providing increased fluorination primarily through the formation of CF_3 and CF_2-CF_2 bonds, with the treatment duration. The optimal 15 min plasma treatment induced the omniphobic property of the membrane with the highest hydrophobicity and high wetting resistance to low surface tension liquids such as methanol, mineral oil, ethylene glycol, and consequently high liquid entry pressure (187 kPa). These improved properties translate to high flux (15.28 L/m^2h) and salt rejection (~100%) performances of membrane even with the addition of up to 0.7 mM sodium dodecyl sulfate in the reverse osmosis brine feed during air gap membrane distillation.

4. Plasma-Aided Technologies for Produced Water Treatment

The plasma-aided water treatment is generally achieved by activating highly efficient adsorbents or direct plasma treatment of water by atmospheric discharges. Each technique has its own advantages and drawbacks, and the most noticeable results are reviewed below.

4.1. Plasma-Activated Adsorbents

The economic aspects for the oil/water separation play a crucial role in the evolution of produced water treatment technology. Often cost-effective reusable materials with lower performance may have gained economic benefits against polymeric membranes based on PVDF, fluoropolymers, polysulfone, or other expensive materials. Hence cheaper feedstock material can play a crucial role in reducing costs and enhancing the business value for membrane oil/water separation. In this regard, the two-step surface modification of polyamide meshes and nonwoven fabrics for oil/water separation investigated by Zhao et al. attracted extreme attention from the scientific community [45]. Their methodology was based on pre-etching the polyamide surface using plasma treatment and coating of a pre-etched surface by eco-friendly polydopamine (PDA)/cellulose, as shown in Figure 5. The pre-etching increased the surface roughness, which further improved the underwater superoleophobicity of the coating. Therefore, the modified polyamide separated various oil/water mixtures and showed a higher intrusion pressure than the original sample and the samples, which were only etched or coated. The grooves on the surface that resulted from the pre-etching prevented the coating from peeling off. In durability tests, after six repeated uses, the modified nonwoven sample lost its underwater oleophobicity due to severe oil fouling, coming to a complete failure in oil/water separation. After 19 cycles, the modified mesh could still separate a certain amount of oil/water but showed reduced intrusion pressure because of slight oil contamination. Filters with different structures, like meshes with one layer of pores and nonwoven fabrics with complex three-dimensional pores, had different oil fouling levels that affected oil/water separation. The recoverability of filters from oil contamination should be considered for practical applications.

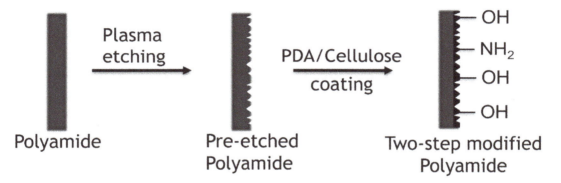

Figure 5. A fabrication of hydrophilic polyamide fabrics for produced water treatment.

Anupriyanka et al. presented the efficient and cost-effective sustainable oil/water separation by fabricating innovative oil-recovery materials (fluorine and silane-free) using an environment-friendly route [46]. Researchers produced superhydrophobic PET (polyethylene terephthalate) fabric by a DC glow discharge oxygen plasma treatment followed by the incorporation of the superhydrophobic agents. The WCA of this fabric was 163°. The prepared fabric possesses excellent repellency to water while absorbing oil and hydrocarbons. Indeed, PET is a significantly cheaper material, and even disposed fabrics can be used as a feedstock to fabricate such adsorbents.

A superhydrophobic cotton nonwoven fabric aimed at oil-water separation was prepared by a two-step strategy based on (1) atmospheric-pressure N_2/O_2 plasma treatment and (2) graft polymerization of siloxane [47]. Different process conditions (plasma power, composition of the plasma gas) influenced the contact angle, stability, surface morphology of the hydrophobic coating, and the growth of silica nanoparticles. The water contact angle of treated nonwoven was up to 155°. The superhydrophobic nonwovens treatment showed excellent stability toward severe acid and alkaline conditions and subsequently the resulting fabrics may be used under harsh environmental conditions. The modified cotton nonwoven exhibited an oil/water separation efficiency higher than 97%, and the result was repeated at least ten times for each sample.

4.2. Direct Plasma Treatment of Water

Atmospheric pressure discharges can directly decompose the organic pollutants. The decomposition of model organic pollutants: chloroform, benzene, and methanol through the gliding arc (GA) plasma technology was reviewed by Gong et al. [48]. The conventional GA reactor is depicted in Figure 6. It is equipped with an AC power supply and flat electrodes contributing to the dispersion of heat and restriction of the gas flow. When a high voltage is applied, the gas flow initiates the GA plasma.

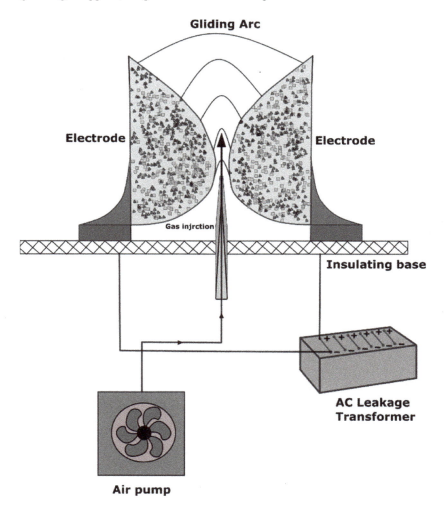

Figure 6. A typical gliding arc setup scheme.

GA plasma devices can be directly used to treat the organic waste gas derived from industrial emissions and be coupled with other technologies or devices, such as catalysts, washing towers, combustion chambers, and activated carbon. GA plasma has extensive adaptability for the types and concentrations of various pollutants, including linear hydrocarbons, aromatic hydrocarbons, halogenated compounds, ethers, and alcohols. Nevertheless, although the gliding arc requires less energy for organic waste gas treatment than other nonthermal technologies, it is still a little higher considering the practical engineering applications. It is worth noting that the plasma treatment can be used to treat produced water and emulsions of water in heavy oils, allowing the separation of oil from water and improving oil properties, as shown elsewhere [49].

Plasma discharges were also used to soften produced water, and the investigation of such a process was funded by US Department of Energy (US DOE) [50]. It was shown that plasma treatment enables the decrease of bicarbonate ions down to 100 ppm.

US DOE has also investigated novel processes for water softening [51]. Specifically, in the plasma discharge project, as mentioned above, the research was focused on reducing the "temporary hardness" (caused by the presence of calcium and magnesium bicarbonate). In this DOE-supported research by Drexel University, plasma arc discharges were used to dissociate and remove bicarbonate ions. In turn, this effect helps prevent fouling from the calcium carbonate scale providing a more prudent strategy for fouling prevention.

Later, Cho et al. investigated stretched arc plasma to increase the volume of produced water treated by plasma, increasing the practical efficiency of the process [52]. Stretching of an arc discharge in produced water was accomplished using a ground electrode and two high-voltage electrodes: one positioned close to the ground electrode, and the other placed farther away from the ground (Figure 7). The contact between the arc and water significantly increased, resulting in twice more efficient removal of bicarbonate ions from produced water when compared to plasma with and without stretching. The usage of plasma treatment for removing bicarbonate ions was further studied by Kim [53]. Although the results were quite promising, the energy cost for this method to treat the produced water was too expensive, and there were no recent publications after 2016.

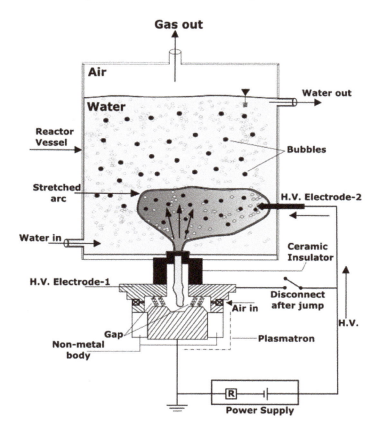

Figure 7. The scheme of a stretch arc plasma applied for produced water treatment.

Choi et al. analyzed the decomposition of water-insoluble 1-decanol by a water plasma system operated at atmospheric pressure [54]. 1-decanol was dispersed in water by a surfactant generating an oil-in-water emulsion. The emulsion was used as the water plasma jet's feeding liquid and plasma-forming gas (Figure 8). A high decomposition rate of over 99.9999% was achieved by converting 1-decanol emulsion into H_2, CO, CO_2, CH_4, condensed liquid, and solid-state carbon despite relatively low input power (<1 kW). The main organic substance in the treated liquid of 1-decanol emulsion was methanediol produced by the hydration of formaldehyde. The authors concluded that the gas conversion rate of carbon in 1-decanol and the removal rate of total organic carbon concentration were increased by increasing the arc current due to enhanced O radicals in the high temperature of the water plasma jet.

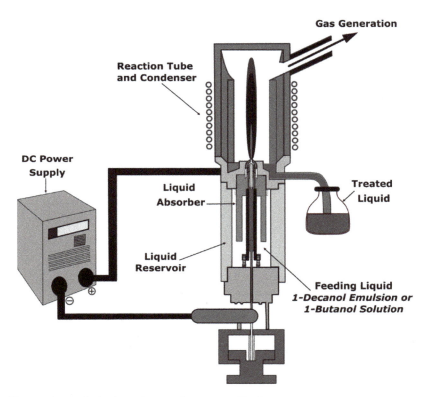

Figure 8. Atmospheric plasma jet setup for water purification.

The decomposition of organic pollutants (dyes as a model substance) in water under plasma micro-discharges was studied by Wright [55]. The low-temperature plasma contacting with aqueous solutions, including organic dyes, was analyzed. Using atmospheric plasma for cleaning water was demonstrated by relatively good energy efficiency: 1.3–12 g of removed pollutants per 1 kWh.

Millie et al. and Lemont et al. designed a plasma tool ELIPSE to decompose organic impurities in water using an atmospheric plasma torch [56,57]. The ELIPSE process is a novel technology of organic liquid destruction involving a thermal plasma working under a water column, ensuring the cooling, filtration, and scrubbing of the gases coming from the degradation (Figure 9). The ELIPSE demonstrated the ability to destroy the pure organic liquids and then eliminate the organic compounds remaining in the aqueous solution through plasma's thermal or radiative properties. Preliminary tests have shown how efficient the process is to destroy the organic liquids when they are directly fed in the plasma. By applying plasma for 60 min, the concentration of organic contaminants decreased from 1200 to 400 ppm.

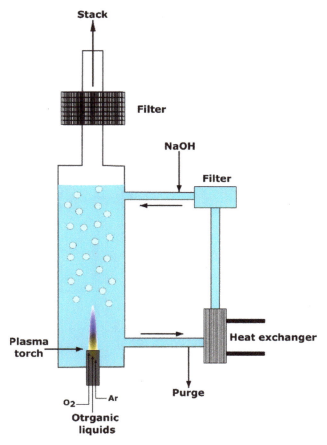

Figure 9. Atmospheric plasma torch employed for the destruction of organic pollutions in water.

Yu et al. studied the decomposition of different polycyclic aromatic hydrocarbons (acenaphthene, fluorene, anthracene, and pyrene) by DC gliding arc discharge. The results indicated that the highest destruction rate was achieved with oxygen as a carrier gas and the external resistance of 50 kΩ independently of the type of hydrocarbons. Furthermore, experimental results suggest that the destruction energy efficiency of gliding arc plasma can be improved by treating higher concentration pollutants [58].

The efficient decomposition of toluene in a gliding arc was shown by Du et al. [59]. The toluene removal efficiency increased with the inlet gas temperature, while the presence of water vapors accelerated the toluene decomposition in the plasma. The energy efficiency was 29.46 g per 1 kWh at a relative humidity of 50% and a specific energy input of 0.26 kWh/m^3, which is higher than other types of nonthermal plasmas. The primary gas-phase decomposition products were determined using FT-IR analysis of the gas components: CO, CO_2, H_2O, and NO_2. Some small deposits of benzaldehyde, benzoic acid, quinine, and nitrophenol were found in the reactor. The authors concluded that these are the minor products from the reaction of toluene with radicals.

5. Combined Techniques for Efficient Water Treatment

The increase of the water quality purification related to the final use of the treated water might be extremely challenging to achieve in an economically viable scenario if a single technology is used (e.g., only sorption or filtration). However, the researchers recently implemented an exciting approach based on a combination of various techniques: plasma and sorption of membranes and photocatalysis. These combination approaches allow separating or adsorb parts of the pollutants and decomposing the rest of the contaminants. For example, Gushchin et al. developed a combined process involving the sorption of oil ($C_{22}H_{38}$) on a sorbent (diatomite) followed by regeneration of the sorbent by plasma-oxidative destruction in an atmospheric dielectric barrier discharge (DBD) [60]. The reactor depicted in Figure 10 was composed of a glass tube diameter of 22 mm, which served as a dielectric barrier; the outer electrode was made from the aluminum foil and located outside the glass tube, and the internal electrode, a 15 mm diameter aluminum rod. The sorbent was placed inside the reactor along the length of the discharge zone.

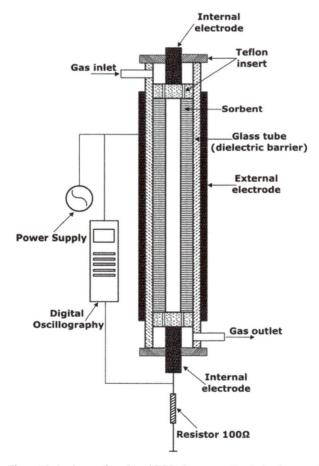

Figure 10. A scheme of combined DBD plasma-sorption technology used for water treatment.

The energy efficiency of the decomposition was 0.169 molecules of oil per 100 eV of input energy. It was shown that the complete deterioration of oil on the diatomite surface reached after 5 min DBD treatment, while the decomposition products were water-soluble and non-toxic carboxylic acids, aldehydes, and CO_2. The complete removal of acids and aldehydes requires a time of about 40 min [60].

Another emerging sustainable approach for decontamination of water is a photocatalytic membrane treatment. Photocatalytic membrane reactor (PMR) is an emerging green technology for removing organic pollutants, photoreduction of heavy metals, photoinactivation of bacteria, and resource recovery.

The combination of photocatalysis and membrane separation improves the removal of contaminants and alleviates membrane fouling. However, the turbidity and color of produced water reduce the efficient light transmission. Efficient pretreatment improves the resiliency for produced water treatment and minerals recovery using PMR processes [61].

Recently, Butman and co-workers compared photocatalytic, plasma, and combined plasma–photocatalytic water treatment processes using a model pollutant, Rhodamine B dye, solutions with a concentration of 40 mg/L [62]. The TiO_2-pillared montmorillonite was used as a photocatalyst, while plasma was ignited in dielectric barrier discharge (DBD) plasma. By using the combined DBD and photocatalysis process, a significant increase of degradation efficiency has been observed: combined approach (100%, 8 sec), plasmolysis (94%), and UV photolysis (92%, 100 min of UV irradiation). In contrast to photolysis, destructive processes are more profound and lead to the formation of simple organic compounds such as carboxylic acids. The plasma–catalytic method enhances by 20% the energetic efficiency of the destruction of Rhodamine B compared to simple DBD plasma. The efficiency of dye destruction with the plasma–catalytic method increases with specific surface area and total pore volume, and the size of the TiO_2 crystallites. This approach may bring significant added value for produced water treatment if the methodology is further optimized.

Implementing the oxidation to the produced water treatment system may provide additional advantages in terms of continuous process reliability. For example, the advanced oxidation processes based on reactions with ozone and Fenton's reagent efficiently clean the polluted water streams, as revealed by Simões et al. in their bibliometric study [63]. They indicated the increase of publications dedicated to this approach in the last few years, although the first work appeared only in 1995. However, the selected works revealed the limitation of these processes due to the high costs with energy consumption.

Finally, the problem of low water flux attributed to commercial ceramic membranes applied in the treatment of produced water was also studied using complex combined technologies [64]. For minimizing this problem, the titanium dioxide (TiO_2) nanocomposites, synthesized via a sol-gel method, were deposited on the active layer of the hydrolyzed bentonite membrane [65]. The grafting time of TiO_2 nanocomposite improved the performance of the coated bentonite membranes. The pure waterpermeability performance showed an increment from 262.3 L h^{-1} m^{-2} bar^{-1} (pristine bentonite membrane) to 337.1 L $h^{-1} m^{-2} bar^{-1}$ (bentonite membrane with TiO_2 grafted for 30 min) and 438.3 L $h^{-1} m^{-2}$ bar^{-1} (TiO_2 grafted for 60 min). The oil rejection performance also revealed an increase in the oil rejection performance from 95 to 99%. These findings can be an excellent example to further investigate and exploit the advantages of modified ceramic membranes in produced water treatment.

6. Discussion

The produced water treatment solutions based on plasma-aided technologies and nanomembranes show serious potential for industrial applications. However, it is obvious that not every kind of treatment would reach the stage of commercialization. To sum up the advantages and limitations of all of the reviewed technologies, the oil/water separation and desalination technologies are listed in Tables 1 and 2, respectively.

Table 1. Summary of the oil/water separation technologies.

Technology	Oil Rejection Rate/Decomposition Rate	Water Flux $L \cdot m^{-2} \cdot h^{-1}$	Advantage	Limitation
Superhydrophilic PVDF membrane with TiO_2 nanoparticles [24]	92%	63,492. Four times higher than pristine membrane	Durability, high permeability.	Multi-step process involving expensive materials.
RF plasma modified polysulfonamide membrane for oil/water emulsion separation [37]	Up to 99%	Up to 30	Simple technology	Low productivity
Superhydrophilic plasma modified PS/PAN nanofibrous membranes for layered oil separation [33]	Up to 99.8%	57,509	Very high oil rejection rate and water permeability	Poor durability: 5–10 cycles only.
Superhydrophilic PS/PAN nanofibrous membranes for water/oil emulsions separation [33]	99.5%	Up to 6460	Very high oil rejection rate and water permeability	Poor durability: 5–10 cycles only.
Superhydrophobic plasma modified steel meshes [38]	Up to 99%	-	Quick separation of water/oil mixtures, simple methodology	No available data for emulsion tests
Superhydrophobic nonwovens	97%	-	Simple method	The industrial implementation is difficult
GA plasma treatment of produced water [56]	66.7%	-	Decomposing organic chemical contaminations and salts	High remaining concentration of organic contaminations in the discharged effluent, high power consumption.
Combined plasma-adsorbent approach [62]	94%	-	High decomposition rate	Slow process
DC gliding Arc [58]	Up to 98.5%	-	Robustness, high decomposition rate	High power consumption

Table 2. Summary of the desalination technologies.

Technology	Salt Rejection Property	Water Flux $L \cdot m^{-2} \cdot h^{-1}$	Advantage	Limitation
PVDF hollow fiber membrane [43,66]	~100%	15.28	Straightforward technique, clear pathway for commercialization	Questions about cost efficiency and durability.
Stretched Arc Plasma treatment [52]	Removal of bicarbonate ions as low as 100 ppm	-	Robust technique	High power consumption
Distillation by a PVDF nanofibrous mebrane [67]	>99.9%	-	Robust technique with high productivity	Unclear recycling pathway and cost-efficiency, due to a high cost of the PVDF nanofibers

Proposed technologies can provide the oil/water separation process with high efficiency. The applicability of the proposed methods would greatly depend on the pretreatment of an effluent and the remaining oil contaminants. In this review, the water/oil separation by most of the membranes allowed to achieve the oil rejection rate ranging from 92% to 99.8%. Hence, knowing the ROPME protocol standard value of 15 ppm (oil in water), it is quite easy to estimate the range of maximum oil concentrations passing membranes. The very well-known formula for oil rejection rate available elsewhere [68] and is presented below:

$$Rejection\ Rate = \left(1 - \frac{C_{permeate}}{C_{feed}}\right) \times 100\%$$

$C_{permeate}$ and C_{feed} are the oil in water concentrations in the permeate and the feed, respectively. Hence, for the technologies exhibiting a rejection rate of 92%, the maximum concentration in the feed can be 187 ppm to ensure the oil concentration in the permeate is below 15 ppm. For the technologies exhibiting a rejection rate above 99%, the C_{feed} may exceed 1500 ppm. Therefore, the majority of reviewed technologies may satisfy the requirements for the effluent discharge composition according to ROPME protocol.

The implementation of any new water treatment technologies is often depending on the economic, environmental, and technological advantages proposed by the innovative approach. The reviewed technologies are still being tested in a lab or at the semi-pilot scale setups. Hence, the discussion related to the potential economic effects are generally based on a very rough estimations and have low accuracy. Nevertheless, it is worth noting that a great economic potential of some techniques can be predicted based on techno-economic assessments available in the literature. For example the techno-economic effect of the implementation of nanofiltration membranes has been shown by Wenzlick and Siefert [69]. They have modeled the base (without nanofiltration) and advanced (with nanofiltration) processes to model the desalination of produced water. The cost related to NF process was estimated as 0.18 ± 0.07 \$/m^3. The usage of NF process allowed to save the cost of chemicals and turn the economically negative case (-0.98 \$/m^3) to the positive one with a revenue of +0.6 \$/m^3. Hence, the implementation of nanofiltration may add very significant economic advantage to the produced water treatment business case. Furthermore, the costs related to the synthesis of nanofiltration membrane can be further reduced by the means of implementing cheaper recyclable materials as a feedstock for nanomembranes. The techno-economic assessment of using a direct plasma treatment to purify produced water is not available in the literature, most probably due to a very complicated capital costs projection. However, the operational costs can be roughly estimated on the basis of the power consumption estimated for lab-scale setups. Due et al. indicated the power 29.46 g/kWh of organic contaminants is decomposing in their plasma setup [59]. This means that to treat 1 m^3 of produced water containing 400 ppm of organic contaminants, the required energy will be 400/29.46 = 13.58 kWh. Taking into account energy price at the level 50 to 100 \$/MWh, the energy part for operational costs would be 0.68 to 1.36 \$/m^3. The energy consumption for combined plasma-adsorbent technology [60] calculated by a similar procedure would be 21.0 kWh/m^3. The price for consumed electricity would be in the range from 1.05 to 2.10 \$/m^3 if the same range for electricity pricing is taken into account.

The estimated produced water treatment levelized costs for membrane distillation and evaporation/crystallization processes are 8.1 and 22.1 \$/m^3, respectively [70]. Therefore, the nanofiltration and plasma-based water treatment processes have evident potential to compete with many water treatment technologies in terms of business efficiency.

It is worth noting that presented methodologies would require pre-treatment of the produced water preliminary to the nanofiltration or plasma processing. The variety of the pre-treatment methods, required for the upscaled process cannot be considered at a present technology readiness level as it would be dependent on the properties of the effluent. The pre-treatment of produced water may include de-emulsification, mechanical treatment, or chemical treatments. Generally, the nanomembranes or plasma technologies cannot be considered as a stand-alone one step solution. The produced water treatment should be composed of a combination of several techniques and processes in order to fulfil the requirements for the water quality and cost-efficiency.

7. Conclusions & Outlook

This work reviewed multiple techniques for produced water treatment based on various environment-friendly methods and summarized their advantages and drawbacks. It is crucial to summarize the most recent research advancements and suggest the perspectives for further development of this topic. The solutions for produced water treatment would require the proposed technologies to have several essential attributes, such as:

(1) Being reliable and cost efficient (the solution should not deteriorate EBITDA)
(2) Capable of yielding high separation/recovery efficiency in compliance with the Zero Discharge Liquid (ZLD) approach
(3) Provision of a positive impact on the environment (the solution cannot rely on highly toxic chemicals or high CO_2 emissions due to low energy efficiency)
(4) Priority usage of the sustainable materials and technologies employed to realize the proposed solution (the utilization of renewable feedstock or recycled materials are preferable)

It is highly challenging to comply with all of these attributes, and all techniques reviewed in this work need to be optimized. Yet, nanomaterials, surface treatment techniques, and plasma discharges have great potential to become reliable, cost-effective, and environment-friendly solutions for produced water treatment if the following challenges are overcome:

(1) Production of the nanomaterials (nanofibers, nanoparticles) used for the membranes or adsorbents must be optimized and upscaled to reduce the current cost. This may happen naturally similar to the witnessed continuous decrease in carbon nanotubes cost.
(2) Optimizing the plasma setup to decrease the power consumption and prevent the extreme energy loss for heating. The modeling of efficient plasma setups using computation fluid dynamics and plasma chemistry kinetics must be performed to reach good power efficiency.
(3) Combining and upscaling of technologies to accumulate quantitative data for the calculation of energy efficiency and evaluate the economic viability for large scale practical applications.

From these perspectives, it is recommended that multi-step modular approaches based on synergetic effects achieved by membrane distillation, adsorption, and plasma or photocatalytic treatment for regeneration of active materials/membranes have excellent potential. They are likely going to gain increasing attention from researchers and engineers in the coming years.

Author Contributions: Conceptualization, A.M. and M.O.; methodology, S.A. and V.G., F.I.A.; writing—original draft preparation, A.M., M.O. and V.G.; project administration, F.I.A. All authors have read and agreed to the published version of the manuscript.

Funding: This research received no external funding.

Institutional Review Board Statement: Not applicable.

Informed Consent Statement: Not applicable.

Data Availability Statement: Not applicable.

Conflicts of Interest: The authors declare no conflict of interest.

References

1. Global Electricity Demand is Growing Faster Than Renewables, Driving Strong Increase in Generation from Fossil Fuels. Available online: https://www.iea.org/news/global-electricity-demand-is-growing-faster-than-renewables-driving-strong-increase-in-generation-from-fossil-fuels (accessed on 8 February 2022).
2. Talal Rafi Why Corporate Strategies Should Be Focused On Sustainability. Available online: https://www.forbes.com/sites/forbesbusinesscouncil/2021/02/10/why-corporate-strategies-should-be-focused-on-sustainability/?sh=2a3079577e9f (accessed on 8 February 2022).

3. IEA Report. A Sustainable Recovery Plan for the Energy Sector. Available online: https://www.iea.org/reports/sustainable-recovery/a-sustainable-recovery-plan-for-the-energy-sector (accessed on 1 April 2022).
4. Anderson, J. The environmental benefits of water recycling and reuse. *Water Supply* **2003**, *3*, 1–10. [CrossRef]
5. Ghurye, G.L. Evaluation of a minimum liquid discharge (Mld) desalination approach for management of unconventional oil and gas produced waters with a focus on waste minimization. *Water* **2021**, *13*, 2912. [CrossRef]
6. Alzahrani, S.; Mohammad, A.W. Challenges and trends in membrane technology implementation for produced water treatment: A review. *J. Water Process Eng.* **2014**, *4*, 107–133. [CrossRef]
7. Jepsen, K.L.; Bram, M.V.; Hansen, L.; Yang, Z.; Lauridsen, S.M.Ø. Online backwash optimization of membrane filtration for produced water treatment. *Membranes* **2019**, *9*, 68. [CrossRef]
8. Abu Khamsin, S.A. Environmental regulations for drilling operations in Saudi Arabia. In Proceedings of the SPE/IADC Middle East Drilling Technology Conference, Manama, Bahrain, 23–25 November 1997; SPE: Manama, Bahrain, 1997; pp. 119–123.
9. Hammod, N.M.; Al-Janabi, K.W.S.; Hasan, S.A.; Al-Garawi, Z.S. High pollutant levels of produced water around Al-Ahdab oil field in Wasit governorate (Iraq). *J. Phys. Conf. Ser.* **2021**, *1853*, 012004. [CrossRef]
10. Folio, E.; Ogunsola, O.; Melchert, E.; Frye, E. Produced water treatment R&d: Developing advanced, cost-effective treatment technologies. In Proceedings of the SPE/AAPG/SEG Unconventional Resources Technology Conference, Houston, TX, USA, 23–25 July 2018; OnePetro: Houston, TX, USA, 2018; pp. 1–8.
11. Igunnu, E.T.; Chen, G.Z. Produced water treatment technologies. *Int. J. Low-Carbon Technol.* **2014**, *9*, 157–177. [CrossRef]
12. Jiménez, S.; Micó, M.M.; Arnaldos, M.; Medina, F.; Contreras, S. State of the art of produced water treatment. *Chemosphere* **2018**, *192*, 186–208. [CrossRef]
13. Dawoud, H.D.; Saleem, H.; Alnuaimi, N.A.; Zaidi, S.J. Characterization and Treatment Technologies Applied for Produced Water in Qatar. *Water* **2021**, *13*, 3573. [CrossRef]
14. Manakhov, A.; Kedroňová, E.; Medalová, J.; Černochová, P.; Obrusník, A.; Michlíček, M.; Shtansky, D.V.; Zajíčková, L. Carboxyl-anhydride and amine plasma coating of PCL nanofibers to improve their bioactivity. *Mater. Des.* **2017**, *132*, 257–265. [CrossRef]
15. Manakhov, A.; Nečas, D.; Čechal, J.; Pavliňák, D.; Eliáš, M.; Zajíčková, L. Deposition of stable amine coating onto polycaprolactone nanofibers by low pressure cyclopropylamine plasma polymerization. *Thin Solid Films* **2015**, *581*, 7–13. [CrossRef]
16. Manakhov, A.; Zajíčková, L.; Eliáš, M.; Čechal, J.; Polčák, J.; Hnilica, J.; Bittnerová, Š.; Nečas, D. Optimization of Cyclopropylamine Plasma Polymerization toward Enhanced Layer Stability in Contact with Water. *Plasma Process. Polym.* **2014**, *11*, 532–544. [CrossRef]
17. Mozaffari, A.; Gashti, M.P.; Mirjalili, M.; Parsania, M. Argon and Argon–Oxygen Plasma Surface Modification of Gelatin Nanofibers for Tissue Engineering Applications. *Membranes* **2021**, *11*, 31. [CrossRef]
18. Essa, W.; Yasin, S.; Saeed, I.; Ali, G. Nanofiber-Based Face Masks and Respirators as COVID-19 Protection: A Review. *Membranes* **2021**, *11*, 250. [CrossRef]
19. Park, E.; Barnett, S.M. Oil/water separation using nanofiltration membrane technology. *Sep. Sci. Technol.* **2001**, *36*, 1527–1542. [CrossRef]
20. Papaioannou, E.H.; Mazzei, R.; Bazzarelli, F.; Piacentini, E.; Giannakopoulos, V.; Roberts, M.R.; Giorno, L. Agri-Food Industry Waste as Resource of Chemicals: The Role of Membrane Technology in Their Sustainable Recycling. *Sustainability* **2022**, *14*, 1483. [CrossRef]
21. Sadeghi, I.; Aroujalian, A.; Raisi, A.; Dabir, B.; Fathizadeh, M. Surface modification of polyethersulfone ultrafiltration membranes by corona air plasma for separation of oil/water emulsions. *J. Memb. Sci.* **2013**, *430*, 24–36. [CrossRef]
22. Yang, X.; Liu, S.; Zhao, Z.; He, Z.; Lin, T.; Zhao, Y.; Li, G.; Qu, J.; Huang, L.; Peng, X.; et al. A facile, clean construction of biphilic surface on filter paper via atmospheric air plasma for highly efficient separation of water-in-oil emulsions. *Sep. Purif. Technol.* **2021**, *255*, 117672. [CrossRef]
23. Lin, A.; Shao, S.; Li, H.; Yang, D.; Kong, Y. Preparation and characterization of a new negatively charged polytetrafluoroethylene membrane for treating oilfield wastewater. *J. Memb. Sci.* **2011**, *371*, 286–292. [CrossRef]
24. Chen, X.; Huang, G.; An, C.; Feng, R.; Yao, Y.; Zhao, S.; Huang, C.; Wu, Y. Plasma-induced poly(acrylic acid)-TiO2 coated polyvinylidene fluoride membrane for produced water treatment: Synchrotron X-ray, optimization, and insight studies. *J. Clean. Prod.* **2019**, *227*, 772–783. [CrossRef]
25. Lin, Y.; Salem, M.S.; Zhang, L.; Shen, Q.; El-shazly, A.H.; Nady, N.; Matsuyama, H. Development of Janus membrane with controllable asymmetric wettability for highly-efficient oil/water emulsions separation. *J. Memb. Sci.* **2020**, *606*, 118141. [CrossRef]
26. Zinadini, S.; Zinatizadeh, A.A.; Derakhshan, A.A. Preparation and characterization of high permeance functionalized nanofiltration membranes with antifouling properties by using diazotization route and potential application for licorice wastewater treatment. *Sep. Purif. Technol.* **2022**, *280*, 119639.
27. Adib, H.; Raisi, A. Surface modification of a PES membrane by corona air plasma-assisted grafting of HB-PEG for separation of oil-in-water emulsions. *RSC Adv.* **2020**, *10*, 17143–17153. [CrossRef]
28. Wavhal, D.S.; Fisher, E.R. Modification of polysulfone ultrafiltration membranes by CO_2 plasma treatment. *Desalination* **2005**, *172*, 189–205. [CrossRef]
29. Fedotova, A.V.; Dryakhlov, V.O.; Shaikhiev, I.G.; Nizameev, I.R.; Garaeva, G.F. Effect of Radiofrequency Plasma Treatment on the Characteristics of Polysulfonamide Membranes and the Intensity of Separation of Oil-in-Water Emulsions. *Surf. Eng. Appl. Electrochem.* **2018**, *54*, 174–179. [CrossRef]

30. Liu, X.; Ruan, W.; Wang, W.; Zhang, X.; Liu, Y.; Liu, J. The Perspective and Challenge of Nanomaterials in Oil and Gas Wastewater Treatment. *Molecules* **2021**, *26*, 3945. [CrossRef] [PubMed]
31. Algieri, C.; Chakraborty, S.; Candamano, S. A Way to Membrane-Based Environmental Remediation for Heavy Metal Removal. *Environments* **2021**, *8*, 52. [CrossRef]
32. Liu, Y.; Zhang, G.; Zhuang, X.; Li, S.; Shi, L.; Kang, W.; Cheng, B.; Xu, X. Solution Blown Nylon 6 Nanofibrous Membrane as Scaffold for Nanofiltration. *Polymers* **2019**, *11*, 364. [CrossRef]
33. Yi, Y.; Tu, H.; Zhou, X.; Liu, R.; Wu, Y.; Li, D.; Wang, Q.; Shi, X.; Deng, H. Acrylic acid-grafted pre-plasma nanofibers for efficient removal of oil pollution from aquatic environment. *J. Hazard. Mater.* **2019**, *371*, 165–174. [CrossRef]
34. Michlíček, M.; Manakhov, A.; Dvořáková, E.; Zajíčková, L. Homogeneity and penetration depth of atmospheric pressure plasma polymerization onto electrospun nanofibrous mats. *Appl. Surf. Sci.* **2019**, *471*, 835–841. [CrossRef]
35. Manakhov, A.; Moreno-Couranjou, M.; Choquet, P.; Boscher, N.D.; Pireaux, J.-J. Diene functionalisation of atmospheric plasma copolymer thin films. *Surf. Coatings Technol.* **2011**, *205*, S466–S469. [CrossRef]
36. Chen, F.; Song, J.; Liu, Z.; Liu, J.; Zheng, H.; Huang, S.; Sun, J.; Xu, W.; Liu, X. Atmospheric Pressure Plasma Functionalized Polymer Mesh: An Environmentally Friendly and Efficient Tool for Oil/Water Separation. *ACS Sustain. Chem. Eng.* **2016**, *4*, 6828–6837. [CrossRef]
37. Fedotova, A.V.; Shaikhiev, I.G.; Dryakhlov, V.O.; Nizameev, I.R.; Abdullin, I.S. Intensification of separation of oil-in-water emulsions using polysulfonamide membranes modified with low-pressure radiofrequency plasma. *Pet. Chem.* **2017**, *57*, 159–164. [CrossRef]
38. You, Y.S.; Kang, S.; Mauchauffé, R.; Moon, S.Y. Rapid and selective surface functionalization of the membrane for high efficiency oil-water separation via an atmospheric pressure plasma process. *Sci. Rep.* **2017**, *7*, 15345. [CrossRef]
39. Agarwal, A.; Manna, S.; Nath, S.; Sharma, K.; Chaudhury, P.; Bora, T.; Solomon, I.; Sarma, A. Controlling oil/water separation using oleophillic and hydrophobic coatings based on plasma technology. *Mater. Res. Express* **2020**, *7*, 036411. [CrossRef]
40. Ricceri, F.; Farinelli, G.; Giagnorio, M.; Zamboi, A.; Tiraferri, A. Optimization of physico-chemical and membrane filtration processes to remove high molecular weight polymers from produced water in enhanced oil recovery operations. *J. Environ. Manag.* **2022**, *302*, 114015. [CrossRef]
41. Thi, H.Y.N.; Nguyen, B.T.D.; Kim, J.F. Sustainable Fabrication of Organic Solvent Nanofiltration Membranes. *Membranes* **2020**, *11*, 19.
42. Doménech, N.G.; Purcell-Milton, F.; Arjona, A.S.; García, M.-L.C.; Ward, M.; Cabré, M.B.; Rafferty, A.; McKelvey, K.; Dunne, P.; Gun'ko, Y.K. High-Performance Boron Nitride-Based Membranes for Water Purification. *Nanomaterials* **2022**, *12*, 473.
43. Ali, A.; Quist-Jensen, C.A.; Macedonio, F.; Drioli, E. Application of membrane crystallization for minerals' recovery from produced water. *Membranes* **2015**, *5*, 772–792. [CrossRef]
44. Woo, Y.C.; Chen, Y.; Tijing, L.D.; Phuntsho, S.; He, T.; Choi, J.S.; Kim, S.H.; Shon, H.K. CF4 plasma-modified omniphobic electrospun nanofiber membrane for produced water brine treatment by membrane distillation. *J. Memb. Sci.* **2017**, *529*, 234–242.
45. Zhao, P.; Qin, N.; Ren, C.L.; Wen, J.Z. Surface modification of polyamide meshes and nonwoven fabrics by plasma etching and a PDA/cellulose coating for oil/water separation. *Appl. Surf. Sci.* **2019**, *481*, 883–891. [CrossRef]
46. Anupriyanka, T.; Shanmugavelayutham, G.; Sarma, B.; Mariammal, M. A single step approach of fabricating superhydrophobic PET fabric by using low pressure plasma for oil-water separation. *Colloids Surf. A Physicochem. Eng. Asp.* **2020**, *600*, 124949. [CrossRef]
47. Yang, J.; Pu, Y.; He, H.; Cao, R.; Miao, D.; Ning, X. Superhydrophobic cotton nonwoven fabrics through atmospheric plasma treatment for applications in self-cleaning and oil–water separation. *Cellulose* **2019**, *26*, 7507–7522. [CrossRef]
48. Gong, X.; Lin, Y.; Li, X.; Wu, A.; Zhang, H.; Yan, J.; Du, C. Decomposition of volatile organic compounds using gliding arc discharge plasma. *J. Air Waste Manag. Assoc.* **2020**, *70*, 138–157. [CrossRef]
49. Honorato, H.D.A.; Silva, R.C.; Piumbini, C.K.; Zucolotto, C.G.; De Souza, A.A.; Cunha, A.G.; Emmerich, F.G.; Lacerda, V.; De Castro, E.V.R.; Bonagamba, T.J.; et al. 1H low-and high-field NMR study of the effects of plasma treatment on the oil and water fractions in crude heavy oil. *Fuel* **2012**, *92*, 62–68. [CrossRef]
50. Cho, Y.I.; Fridman, A.; Rabinovich, A.; Cho, D.J. Pre-Treating Water with Non-Thermal Plasma. U.S. Patent No. 9,695,068, 4 July 2017.
51. Wright, K. Study of Plasma Treatment of Produced Water from Oil and Gas Exploration. Ph.D. Thesis, Drexel University, Philadelphia, PA, USA, 2015.
52. Cho, Y.I.; Wright, K.C.; Kim, H.S.; Cho, D.J.; Rabinovich, A.; Fridman, A. Stretched arc discharge in produced water. *Rev. Sci. Instrum.* **2015**, *86*, 013501. [CrossRef]
53. Kim, H.S. Plasma Discharges in Produced Water and Its Applications to Large Scale Flow. Ph.D. Thesis, Drexel University, Philadelphia, PA, USA, 2016.
54. Choi, S.; Watanabe, T. Decomposition of 1-decanol emulsion by water thermal plasma jet. *IEEE Trans. Plasma Sci.* **2012**, *40*, 2831–2836. [CrossRef]
55. Wright, K. Plasma Water Treatment and Oxidation of Organic Matter in Water. *IEEE Int. Pulsed Power Conf.* **2019**, *2019*, 35–38.
56. Milelli, D.; Lemont, F.; Ruffel, L.; Barral, T.; Marchand, M. Thermo- and photo-oxidation reaction scheme in a treatment system using submerged plasma. *Chem. Eng. J.* **2017**, *317*, 1083–1091. [CrossRef]
57. Lemont, F.; Marchand, M.; Mabrouk, M.; Milelli, D.; Baronnet, J.M. ELIPSE: An Innovative Technology for the Treatment of Radioactive Organic Liquids. *Nucl. Technol.* **2017**, *198*, 53–63. [CrossRef]

58. Yu, L.; Tu, X.; Li, X.; Wang, Y.; Chi, Y.; Yan, J. Destruction of acenaphthene, fluorene, anthracene and pyrene by a dc gliding arc plasma reactor. *J. Hazard. Mater.* **2010**, *180*, 449–455. [CrossRef]
59. Du, C.M.; Yan, J.H.; Cheron, B. Decomposition of toluene in a gliding arc discharge plasma reactor. *Plasma Sources Sci. Technol.* **2007**, *16*, 791–797. [CrossRef]
60. Gushchin, A.A.; Grinevich, V.I.; Gusev, G.I.; Kvitkova, E.Y.; Rybkin, V.V. Removal of Oil Products from Water Using a Combined Process of Sorption and Plasma Exposure to DBD. *Plasma Chem. Plasma Process.* **2018**, *38*, 1021–1033. [CrossRef]
61. Chen, L.; Xu, P.; Wang, H. Photocatalytic membrane reactors for produced water treatment and reuse: Fundamentals, affecting factors, rational design, and evaluation metrics. *J. Hazard. Mater.* **2022**, *424*, 127493. [CrossRef] [PubMed]
62. Butman, M.F.; Gushchin, A.A.; Ovchinnikov, N.L.; Gusev, G.I.; Zinenko, N.V.; Karamysheva, S.P.; Krämer, K.W. Synergistic Effect of Dielectric Barrier Discharge Plasma and TiO_2-Pillared Montmorillonite on the Degradation of Rhodamine B in an Aqueous Solution. *Catalysts* **2020**, *10*, 359. [CrossRef]
63. Simões, A.J.A.; Macêdo-Júnior, R.O.; Santos, B.L.P.; Silva, D.P.; Ruzene, D.S. A Bibliometric Study on the Application of Advanced Oxidation Processes for Produced Water Treatment. *Water, Air Soil Pollut.* **2021**, *232*, 297. [CrossRef]
64. Santos, E.N.; Fazekas, Á.; Hodúr, C.; László, Z.; Beszédes, S.; Firak, D.S.; Gyulavári, T.; Hernádi, K.; Arthanareeswaran, G.; Veréb, G. Statistical Analysis of Synthesis Parameters to Fabricate PVDF/PVP/TiO_2 Membranes via Phase-Inversion with Enhanced Filtration Performance and Photocatalytic Properties. *Polymers* **2021**, *14*, 113.
65. Mohamad Esham, M.I.; Ahmad, A.L.; Othman, M.H.D. Fabrication, Optimization, and Performance of a TiO2 Coated Bentonite Membrane for Produced Water Treatment: Effect of Grafting Time. *Membranes* **2021**, *11*, 739. [CrossRef]
66. Woo, Y.C.; Tijing, L.D.; Park, M.J.; Yao, M.; Choi, J.S.; Lee, S.; Kim, S.H.; An, K.J.; Shon, H.K. Electrospun dual-layer nonwoven membrane for desalination by air gap membrane distillation. *Desalination* **2017**, *403*, 187–198. [CrossRef]
67. Chiao, Y.H.; Ang, M.B.M.Y.; Huang, Y.X.; Depaz, S.S.; Chang, Y.; Almodovar, J.; Wickramasinghe, S.R. A "graft to" electrospun zwitterionic bilayer membrane for the separation of hydraulic fracturing-produced water via membrane distillation. *Membranes* **2020**, *10*, 402. [CrossRef]
68. Elshorafa, R.; Saththasivam, J.; Liu, Z.; Ahzi, S. Efficient oil/saltwater separation using a highly permeable and fouling-resistant all-inorganic nanocomposite membrane. *Environ. Sci. Pollut. Res.* **2020**, *27*, 15488–15497. [CrossRef]
69. Wenzlick, M.; Siefert, N. Techno-economic analysis of converting oil & gas produced water into valuable resources. *Desalination* **2020**, *481*, 114381.
70. Ghalavand, A.; Hatamipour, M.S.; Ghalavand, Y. Clean treatment of rejected brine by zero liquid discharge thermal desalination in Persian Gulf countries. *Clean Technol. Environ. Policy* **2021**, *23*, 2683–2696. [CrossRef]

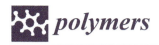

Review

Recent Developments in the Application of Bio-Waste-Derived Adsorbents for the Removal of Methylene Blue from Wastewater: A Review

Hamad Noori Hamad and Syazwani Idrus *

Department of Civil Engineering, Faculty of Engineering, Universiti Putra Malaysia, Serdang 43400, Malaysia; gs59432@student.upm.edu.my
* Correspondence: syazwani@upm.edu.my; Tel.: +60-13-692-2301

Abstract: Over the last few years, various industries have released wastewater containing high concentrations of dyes straight into the ecological system, which has become a major environmental problem (i.e., soil, groundwater, surface water pollution, etc.). The rapid growth of textile industries has created an alarming situation in which further deterioration to the environment has been caused due to substances being left in treated wastewater, including dyes. The application of activated carbon has recently been demonstrated to be a highly efficient technology in terms of removing methylene blue (MB) from wastewater. Agricultural waste, as well as animal-based and wood products, are excellent sources of bio-waste for MB remediation since they are extremely efficient, have high sorption capacities, and are renewable sources. Despite the fact that commercial activated carbon is a favored adsorbent for dye elimination, its extensive application is restricted because of its comparatively high cost, which has prompted researchers to investigate alternative sources of adsorbents that are non-conventional and more economical. The goal of this review article was to critically evaluate the accessible information on the characteristics of bio-waste-derived adsorbents for MB's removal, as well as related parameters influencing the performance of this process. The review also highlighted the processing methods developed in previous studies. Regeneration processes, economic challenges, and the valorization of post-sorption materials were also discussed. This review is beneficial in terms of understanding recent advances in the status of biowaste-derived adsorbents, highlighting the accelerating need for the development of low-cost adsorbents and functioning as a precursor for large-scale system optimization.

Keywords: methylene blue; activated carbon; agro-waste; wastewater; adsorption; cationic dyes; low-cost adsorbents; bio-waste

1. Introduction

The pervasiveness of pollutants in the ecosystem is often linked to population growth and anthropogenic activity [1]. Water resource contamination is an extremely contentious issue on a worldwide scale, as it has long-term or even lethal consequences for living creatures [2]. Dyes in effluents are a severe issue since they harm many sorts of life [3]. Toxicological and aesthetic issues are intertwined with regard to color dye pollution [4]. According to recent data, approximately 100 thousand commercially dyed products with a total 7×10^5 tons of yearly production of dyestuff (about 10% of dyes used in industrial applications) have been released into the aquatic environment [5–7]. The water pollution issue was first caused by the textile industry, followed by the printing industries, as well as paper, paint, and leather production companies [8,9]. The amount of textile wastewater generated per year in the United States, United Kingdom, and China was estimated to be around 12.4, 1, and 26 million tons, respectively. This is equivalent to 1–10 million liters of textile wastewater being produced per day [10].

Over a third of the world's renewable freshwater resources are used for industrial, residential, and agricultural purposes, and the majority of these activities pollute water with a wide range of geogenic and synthetic substances, including dyes, pesticides, fertilizers, radionuclides, and heavy metals. [11,12]. As a result, it is not surprising that water poisoning induced by a variety of human activities has created alarm regarding public health problem on a global scale. Dye-induced water pollution is one of the most serious pollutants since it alters water. Even at extremely low quantities, water retains its natural look [13,14]. These industries consume a vast proportion of the coloration and produce dye-laden effluent that is eventually released straight into the environment, posing significant environmental problems due to the dyes' toxic and unpleasant properties [15].

MB is much more commonly used dye and is a heterocyclic molecule with the chemical formula $C_{16}H_{18}N_3SC_1$. Initially, it was manufactured as a synthetic aniline dye for textiles in 1876 by Heinrich Caro of Badische Aniline and Soda Fabrik. Its utility in staining and inactivating species of microbes was also revealed [16]. Additionally, it was identified in 1932 to be a cyanide and carbon monoxide antidote [17]. The ingredient is a dark green powder that causes water to turn blue at room temperature. It absorbs the most visible light at around 665 nm. MB is known to be an extensively explored dye because of its favorable and negative qualities. Its application has a wide range, with it being used in the pharmaceutical and textile industries as a coloring, as well as in the plastic, tannery, cosmetics, paper, food, and medicinal industries, and it is also used as a staining agent for the classification of microorganisms [18,19]. On the other hand, MB has garnered considerable attention due to its antagonistic nature, which has a detrimental effect on human health and the environment. This dye's adverse effects include skin irritation, as well as mouth, throat, and stomach irritation; in addition, esophagus irritation, nausea, gastrointestinal pain, headache, diarrhea, vomiting, fever, dizziness, and high blood pressure are all common side effects of this dye [20]. The discharge of colored waste without sufficient treatment can cause severe environmental effects, including an increase in toxicity via an increase in water bodies' chemical-oxygen demand (COD) [21]. Due to the fact that synthetic dyes in wastewater cannot be effectively decolored using currently available technologies as a result of their synthetic roots and predominantly aromatic structures, which are not biodegradable, the need to remove color from waste effluents has grown in importance. Several strategies for removing MB from waste water have been studied, including enzymatic procedures, photodegradation reactions, electrochemical extraction, membrane filtration, physical adsorption, and chemical coagulation [22,23].

Adsorption as a physico-chemical treatment has been identified as one of the most appropriate methods and has been extensively explored for MB elimination, with its total use cases more than doubling in the last decade. The adsorption approach employed a straightforward procedure with a cheap and plentiful adsorbent, and it was also capable of achieving a high removal efficiency of MB [24,25]. Additionally, adsorption prevents the formation of secondary contaminants due to the reactions of the oxidation or degradation processes of MB [26,27]. As a result, the findings have attracted the interest of numerous researchers over the last decade.

Most of the recent studies on adsorbent development focus on the application of carbon-based adsorbents, including magsorbents [28], nano catalyst applications [29], and the function of all types of carbon-based adsorbents [30] for MB's removal from wastewater. To the best of our knowledge, no recent literature has addressed the removal of MB through the extensive use of bio-waste-derived adsorbents and compared the bio-waste-derived adsorbents' characteristics as well as related parameters that influence the performance of the process. Aiming at the further evaluation of current advances and methods developed in previous studies, this review also highlights regeneration processes, economic challenges, and the valorization of post-sorption materials. This article provides new perspectives for the development of adsorbents, serving as a precursor for large-scale and low-cost adsorbent applications. Figure 1 depicts the trends in the research on the removal of MB from wastewater using carbon-based adsorbent and sources of activated carbon published between 2008 and 2021.

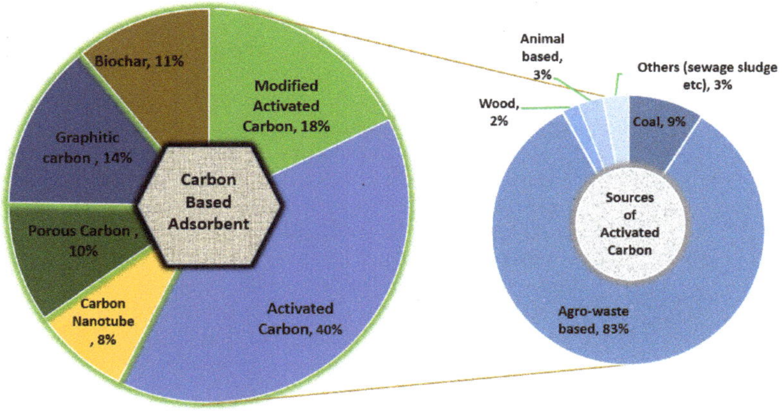

Figure 1. Research on carbon-based adsorbents and sources of activated carbon for methylene blue elimination from 2008 to 2021.

2. Carbon Structural Characteristics and Their Relationship to Adsorption Capacity

Carbon's adsorbent quality is determined by its sorption capacity. The characteristics of the adsorbent are considered to be the most critical factors that can affect MB's adsorption, and include the surface area, pore structure, carbon particle size, surface acidity, and functionality [31–33]. As illustrated in Table 1, carbon adsorbents can be classified as superior (adsorption capacity >1000 mg/g), excellent (500–1000 mg/g), moderate (100–500 mg/g), and weak (adsorption capacity 100 mg/g) based on their MB adsorption capacities. The surface area of carbon adsorbent was reported to be positively correlated with its adsorption capacity. Nonetheless, not all carbon adsorbents follow this trend, as some have low adsorption capabilities due to having excessive surface areas. The highest MB adsorption capacity, exceeding 800 mg/g, was found in adsorbents with large surface areas but small pore diameters. The MB dimensions of $0.400 \times 0.793 \times 1.634$ nm were reported in water. In terms of facilitating MB's diffusion via the adsorbent's pores, the pore opening size is critical. At its maximum, carbon was found to have an adsorption capacity of 1791 mg/g, a surface area of 2138 m^2/g, and a pore diameter of 3.33 nm [34]. Interestingly, pores with dimensions of greater than 6 nm, with total surface areas of 500 m^2/g, were reported to have less adsorption capability than other adsorbents.

Table 1. Structural characteristics and adsorption capacity of adsorbent in relation to the efficiency of the elimination of MB within the 2008 to 2020 period.

No	Adsorbents	Surface Area (m^2/g)	Diameter, ϕ (nm)	Q_{max} (mg/g)	Sources
1	Activated charcoal	4.445–2854	1.0–15.9	0.71–1030	[35–41]
2	Biochar	2.05–2054.49	2.29–20.57	2.06–1282.6	[42–44]
3	Modified activated carbon and modified biochar	4.02–1229	1.038–7.477	9.72–986.8	[45–47]
4	Carbon graphics and modifications	32–295.56	2–50	41.67–847	[45,47–49]
5	Porous Carbon	21–3496	0.74–5.45	8.96–1791	[50–53]
6	Carbon Nanotube	140–558.7	2.19–25	33.4–1189	[49,54–59]

3. Wastewater Treatment Methods for MB's Removal

Dye users, industrial entities, and the government should take all appropriate steps in the treatment of dye effluents in order to improve public health and protect the environment. In general, industrial wastewater treatment technologies are divided into several stages, including pre-primary, primary, secondary, and tertiary processes [60]. The initial one is a preliminary process that is applied for the removal of contaminants (such as papers, grits, wood, plastics, cloths, etc.) with minimal effort, as well as the comminution and screening of floating, suspended particles, and oil and grease traps. The following process is the primary treatment, which includes skimming to remove frothy solids and flotation and sedimentation to remove settleable inorganic and organic impurities. Secondary wastewater treatment involves the microbial breakdown of dissolved organic and colloidal materials, which maintains the waste's stability [61]. Biological agents are used in advanced and tertiary treatment (i.e., anoxic, aerobic and anaerobic, facultative, or a mix of these), chemical (ozonation, fenton reagents, chemical precipitation, ion exchange, photocatalysis, ultrasound, and solar-driven processes) or physical (sedimentation, membrane filtration, coagulation and flocculation, ultrafiltration, nanofiltration, adsorption, and reverse osmosis) strategies for treating effluents that are incapable of being removed during secondary treatment [62–64]. Likewise, during treatment of effluent-containing dye, there could be substances left in treated wastewater which require post treatment including the application of bio-waste-derived adsorbent. Previous studies reported on the disadvantages of various wastewater treatment, including lower efficiency, greater capital or operating costs, a large amount of sludge production, and high costs of maintenance, that make these technologies inappropriate for economic application [65,66]. In contrast, adsorption technology offers a wide range of techniques due to its cost efficiency, ease of operation, low energy consumption, simple set up, toughness towards harmful contaminants, capacity to eliminate all dyes, and great efficiency [67,68]. Furthermore, no harmful materials are generated as a consequence of using this treatment method. Figure 2 depicts tertiary treatment and adsorption technology as an alternative for MB's removal from wastewater.

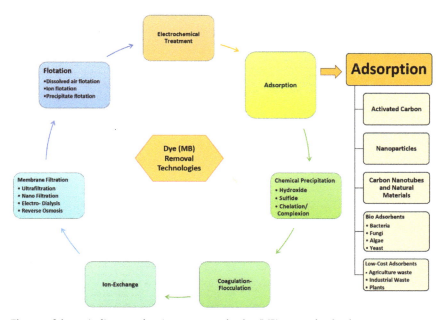

Figure 2. Schematic diagram of tertiary treatment for dye (MB)-removal technologies.

Current color removal treatment approaches involve chemical, physical and biological processes. There are two sorts of dye molecules: chromophores, which provide colors, and auxochromes, which not only act as a substitute for the chromophore but also increase the solubility of dye in water, thus increasing its affinity (ability to join) to fibers [69]. Chemical, physical, and biological remediations are the most often used ways for treating colored wastewater. These technologies, however, have both advantages and disadvantages. Most of these traditional procedures are inapplicable on a broad scale because of the high expense and disposal issues associated with the significant the quantity of sludge produced in the final treatment process [70].

3.1. Physical Techniques

Membrane filtration, reverse osmosis, electrolysis, and adsorption technology are classified as physical treatment methods. The main disadvantage of the membrane technique, in particular, is the short life due to fouling, and thus, frequent maintenance is needed. As a result, costs associated with periodic chemical cleaning and replacement have to be considered during the evaluation of its viability economically. The adsorption procedure is considered to be the most effective way for water purification among all physical treatments [71]. Adsorption is acknowledged as a potential strategy with substantial significance in the decolorization process, due to its simplicity in operation and comparably cheap application. From the point of view of its commercial scale potential, activated carbon is an extraordinary substance that is sustainable in treating polluted groundwater and industrial contaminants such as colored effluents. These natural adsorbents have been studied extensively to recover undesired hazardous chemicals at a relatively low cost from polluted water [72]. Nevertheless, the application of activated carbon is limited due to its expensive cost; thus, improvement in terms of development and regeneration is indispensable. Numerous non-traditional low-cost adsorbents have also been proposed, including zeolites, clay materials, agricultural wastes, siliceous material, and industrial waste products, in an attempt to develop more affordable and effective adsorbents [73,74].

3.2. Chemical Techniques

Coagulants and flocculants are the primary agents used in the treatment of dye wastewater chemically [75]. It is accomplished by adding chemicals to the influent, such as ferric ion aluminum and calcium, to produce flocs [76]. Moreover, the utilization of various chemical agents, for instance, ferric sulphate, polyaluminium chloride, and several organic synthetic polymers, in chemical treatment was previously reported [77,78]. The combination of more than one coagulant or flocculant could be applied for improving the removal rates, as suggested by Shi et al. [75]. In a nutshell, the chemical technique is generally economical and efficient, but the main disadvantage is that chemical cost is high, and prices fluctuate in the market due to the demand and manufacturing cost. Furthermore, despite its efficiency, major drawback of this technique is the formation of large sludge volume, which causes disposal issues including higher operating costs and pH dependence, thus limiting its application as a biofertilizer [79].

3.3. Biological Techniques

Biological treatment is the most cost-effective treatment method as compared with physical and chemical treatments. In the treatment of industrial effluents, biodegradation technologies including the use of adsorbents as alternatives for filter media to promote microbial population, have gained attention for treating bio-waste in fungal decolorization processes. Microorganisms such as algae, yeasts, fungi, and bacteria can accumulate and decompose various contaminants; however, their applications are frequently limited due to technical limits. Aerobic and anaerobic biological treatments are both possible [80]. Conversely, the main disadvantage is that it requires a large area of land and is restricted by sensitivity to diurnal change as well as chemical toxicity [73]. Furthermore, contrary findings were published in a study of existing technologies [81], which reported that the biological remediation process is incapable of achieving good color eradication while utilizing present conventional technologies. Furthermore, due to their complicated chemical structure, synthetic organic origin, and xenobiotic character, azo dyes are not easily biodegradable. Table 2 summarizes the benefits and drawbacks of different approaches for treating dye-contaminated water.

Table 2. Benefits and drawbacks of various wastewater treatment technologies for MB's removal.

Technologies	Benefits	Drawbacks	Reference
Advanced oxidation process	At normal atmospheric pressure and temperature, the dyes are degraded efficiently, and organic contaminants are transformed into carbon dioxide.	Significant operating and maintenance expenses; inflexibility	[69,73]
Chemical precipitation	Simple; low-cost; can manage high pollutant loads; is easy to use; has an integrated physio-chemical process; and results in a significant reduction in COD.	Contains a huge amount of chemicals and generates a lot of sludge	[82]
Ion exchange	Absence of sludge; requires less time; water of superior purity is generated; and an effective decolorization procedure is used. No adsorbent loss during regeneration	pH has a significant effect on performance; not suitable for all colors; costly in terms of recharging and the formation of significant amounts of sludge	[73,81]
Electrochemical	Chemicals are either unnecessary or are limited; the process is quick; suited to both insoluble and soluble dyes, with a lower COD.	High operating expenses; rising electricity prices; sludge formation; contamination from chlorinated organics and heavy metals as a result of indirect oxidation	[65,69]
Oxidation	Dyes are completely degraded, and the reaction time is minimal.	pH maintenance; catalyst required for optimal treatment; high cost	[69,83]
Ozonation	Disinfection that is quick and effective, as well as equipment installation that is simple; no volume growth in the gas phase	A relatively brief half-life; costly process; hazardous by-products and intermediates in manufacturing; strict pH control of effluent	[81,84]
Hydrogen peroxide	Oxidation causes water-insoluble colors to decolorate; reduction in COD; and non-toxic by-products of manufacturing	Increased reaction time; increased need for space; more costly	[65]
Fenton reagents	Removal of both soluble and insoluble dyes with effective decolorization	Sludge production	[63]
Sodium hypochloride	Cleavage of azo bonds develops and accelerates	Production of aromatic amines	[63]
Electrochemical destruction	The breakdown products are not dangerous.	Electricity is costly	[63]

Table 2. Cont.

Technologies	Benefits	Drawbacks	Reference
Coagulation–Flocculation	A wide range of physiochemical approaches used for color elimination; the coagulating agent entirely removes dyes from remediated wastewater; it is effective and simple to operate, and as a result decolorization occurs completely.	Recycling high-priced chemicals is impractical; not suited for very water-soluble colors; generates colorful coagulated solid waste; produces hazardous sludge; raises TDS in treated wastewater; is not ecologically sustainable.	[65,82]
Ultrafiltration and Nanofiltration	Effective with all types of dyes	Extreme operational pressure, significant energy consumption, high price of membrane, limited lifespan, and concentrated production of sludge	[83,85,86]
Reverse osmosis	The most efficient decolorizing and desalting technology, with maximal salt removal, and high-quality water	Extreme pressure and operating costs, as well as membrane clogging, are involved on a frequent basis.	[83,86]
Biological techniques (aerobic and anaerobic)	Low-cost, environmentally friendly, and non-dangerous product; is fully mineralized.	Dye biodegradability is lower, extremely dependent on reaction circumstances, design and operation inflexibility, requires a vast land area, and the requires a longer period for decolorization	[69]
Adsorption technique	Highly efficient and easy; simple and adaptable to a wide variety of pollutants; excellent capacity to remove a wide variety of impurities; economical; adsorbents can be made from wastes; potential regeneration of the adsorbent	Adsorbents' compositions influence their efficacy; their chemical modification is necessary to boost their adsorption capacity; certain adsorbents are highly expensive.	[83,86]

Despite significant advances in dye wastewater treatment methods, achieving commercially viable, cost-effective, and short-retention-time treatment remains a challenge. A previous study concentrated on an adsorption technique for dye treatment from wastewater [87]. This approach is capable of handling relatively high flow rates while creating high-quality effluent that does not develop hazardous chemicals such as free radicals and ozone [88]. Furthermore, it can eliminate or reduce a variety of contaminants, giving it a broader range of applications in the controlling of pollution. Adsorption is thus acknowledged as the most adaptable technique employed in less developed countries, and it is now widely used for the removal of organic pollutants from aquatic environments [89].

4. Adsorption

The adsorption process is an efficient, affordable, and widely utilized color removal approach [90]. Biomass is commonly used as a low-cost activated carbon in wastewater remediation for the removal of impurities. Several non-conventional and cost-effective biomass-derived adsorbents have been studied in relation to the treatment of dye-containing wastewater, as shown in Figure 3.

Figure 3. Numerous inexpensive adsorbents' capacities for dye (MB) elimination.

Environmentally friendly sorbents, which include organic waste compounds (compounds from leaves, barks, and peels) and microbial biomass (fungus bio-sorbents, green algal, and bacterial biomasses), are gaining popularity as types of commercial activated carbon (CAC). Likewise, carbon nanomaterials (graphene, carbon nanotube, and derivatives) have also been employed for decolorization [91]. Zeolite, as an inorganic adsorbent and activated carbon, can be categorized as a type of carbon compound with high oscillation and internal surfaces. Special techniques for producing them in the form of granular, powdered, and spherical activated carbons have been devised. Activated carbon is made by pyrolyzing carbon or carbon-containing plant materials such as coal, bamboo wood, charcoal, kernels, or fruit shells, for example, coconut shells [92]. Carbon can be activated by steam, carbon dioxide, or chemical means, thus making it an ideal material for dye binding. Steam activation is the most eco-friendly and cost-effective approach, whereas chemical activation leads to the highest porosity and surface area. Following the activation process, carbon can be easily rinsed and dried to eliminate the chemicals used (including acid) [92]. In terms of the sorption capacity of carbon groups, the highest theoretical adsorption capacities were recorded at 348, 527, and 394 mg/g at 25° C for Norit Darco 12–20 (DARCO).

Charcoal-derived activated carbon was revealed to be the most superior adsorbent with an efficiency of 99.8%, and it can handle different types of dyes. Researchers discovered that MB performed better as an adsorbate as compared to Rhodamine B in wastewater [93]. At a pH of 2 and a temperature of 25 °C, the highest capabilities of microalgae and CAC in the adsorption of dye were 482.2 mg/g and 267.2 mg/g, respectively. Dye was removed at a rate of 93.6–97.7% using AC and at a rate of 94.4–99.0% with microalgae. In another investigation, CAC outperformed olive stone activated carbon in the adsorption of Remazol Red [94]. The replacement of CAC via the development of alternatives requires comprehensive research on activation methods and adsorbent characteristics. The initial dye concentration, pH, temperature, adsorbent dose and type, and contact duration are the parameters that determine the dye-adsorption ability. Effective adsorbents should have the capacity for high adsorption amounts and quick adsorption rates, be effective against a range of dyes or pollutants, and be easily regenerable and reusable to ensure efficient treatment [95].

Despite the good functioning of activated carbon, which has successfully removed dyes from industrial wastewater effluents, it has drawbacks such as high capital costs, high energy consumption, and sorption–desorption cycles. For color and heavy metal elimination, bio adsorbents made from bacteria or fungi are promising ecologically acceptable adsorbents [90].

5. Adsorption Mechanism

Functional groupings such as the aromatic ring, —C=O, —C—O—C-, —OH, —NH$_2$, —C=S, —C=N, and —S=O on the carbon surface also play important roles in improving the adsorption capacity in terms of MB's disconnection from water [96,97]. MB is a positively charged chemical. It has a six-carbon aromatic ring, sulfur, and nitrogen in its chemical structure. Figure 4 shows that the electron dispersion forces between the carbon surface functional groups and MB molecules induce, via electrostatic contact, hydrogen bridge generation, electron donor–acceptor relationships, and π—π forces after MB's adsorption on carbon [57,59]. Most commonly, thermal activation involves the annealing of carbon adsorbent at high temperatures with nitrogen gas (N$_2$) flowing through it. Furthermore, MB's adsorption capacity can be maximized by increasing the carbon's porosity and surface area. This technique is known as the addition of carboxyl group numbers (—COOH) [40,41]. Another technique to improve carbon surface functionality is to use compounds that contain the functional groups required for the chemical activation of the composites. Carbon from bio-waste is treated with propylene diamine, ethylene diamine, aniline, and ethylene amine to form amino radical (NH$_2$) groups. Additionally, poly (sodium 4-sterenesulfonate) can be used to enclose carbon nanotubes to graft sulphur trioxide (SO$_3$) groups [57]. This occurs via reactions with cysteamine, on the nanocarbon surface, with carboxylic groups to form imidogen (NH) and —sodium hypochlorite (SH) functional groups [58]. Another method that can be applied to increase the MB adsorption capability involves coating charcoal with sodium dodecyl sulfate (SO$_3$) groups [98]. For charcoal and chitosan groups, the improvement of MB's adsorption can be obtained through enhanced numbers of —C=O, —OH, and —NH$_2$ [99].

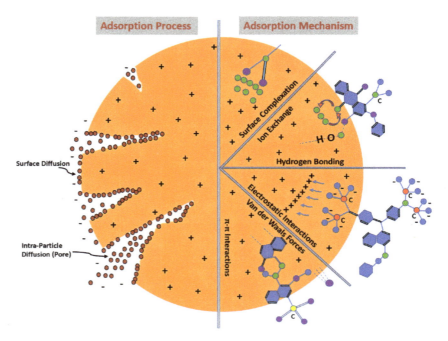

Figure 4. Mechanism and adsorption process for the elimination of dye (MB).

6. Characterization and Formation of Carbon-Derived Adsorbents

Adsorption processes are influenced by adsorbent structures, fluid characteristics, the nature of contaminant structures, operating circumstances, and system design features. The adsorbents used for removing impurities from wastewater include biochar, activated carbon, clays, silica gel, composites, zeolites, agro-wastes, and biological and polymeric materials [100]. Most of the pollutants are easily absorbed by carbon-based materials, including hazardous metal ions, medicines, insecticides, metalloids, and other inorganic and organic compounds [101]. The role of adsorbents in water or wastewater treatment is to concentrate and transfer contaminants, thus improving the performance of the process. The reaction also depends on adsorbate–adsorbent interactions. pH, ionic strength, and temperature are the factors that influence the adsorption capability of carbon-based adsorbents [102]. The forces involved in the removal process are hydrogen, van der Waals bonds, covalent and electrostatic interactions, and the hydrophobic effect. Meanwhile, donor–acceptor forces are responsible for the binding and accumulation of chemical compounds on the surfaces of adsorbents [102,103]. These reactions occur in all carbon-based adsorbents including carbon aerogels, carbon nanotubes, carbon nanofibers, and graphene (CAs). The carbon-based materials (CBMs) utilized in adsorption are shown in Figure 5. The advantages and disadvantages of CBMs are tabulated in Table 3.

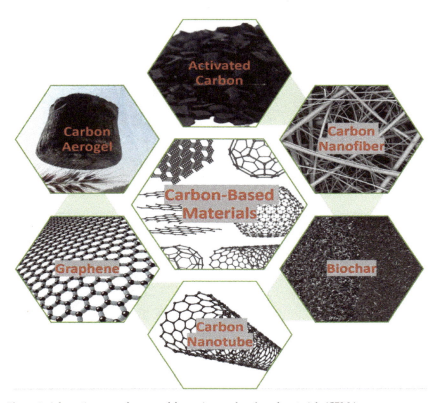

Figure 5. Adsorption procedures used for various carbon-based materials (CBMs).

Table 3. Classification of various carbon compounds and their associated benefits and drawbacks.

Classifications	Adsorbents	Formation	Benefits	Drawbacks	Sources
Composition of carbon	Activated carbon	Carbonized and activated (e.g., lignite, coal, peat, wood)	Large and specific chemical functional groups; large surface area; large pore volume	Hygroscopicity; pore resistance; flammability; incomplete desorption; high permeability	[104]
	Biochar	Formed under moderate pyrolysis conditions in an inert environment	Abundant resources; highly efficient; affordable; low energy usage	Plug hole; flammability; hygroscopicity; gas release	[105]
	Carbon fiber, activated	A microfilament fiber	Hydrophobic and efficient	Expensive	[106]
	Graphene	2D graphene is made up of carbon sheets hexagonal that portion three extra carbon atoms' sp2 hybridized orbitals	Superior electrical conductivity; a large amount of physical specific surface area; great mechanical strength	Synthesis is difficult and dangerous	[107]
	Carbon nanotubes	The cylindrical structure is composed of carbon atoms that have undergone sp2 hybridization.	Strong thermal stability; good electrical conductivity; wide surface area; inherent hydrophobicity	Serious aggregation	[108]

Table 3. Cont.

Classifications	Adsorbents	Formation	Benefits	Drawbacks	Sources
Materials containing oxygen	Zeolite	Zeolite is composed of an endless (3D) arrangement of TO4 tetrahedra in a crystalline aluminosilicate frame (T is Al or Si)	High adsorption capacity; huge surface area; tunable porosity; incombustibility; hydrothermal and chemical stability; good hydrophobicity	The synthetic technique is intricate, lengthy, and costly	[109]
	Frameworks of metal organic	Metal ions or coordination clusters containing organic ligands are created in a single-, two-, or three-dimensional manners.	Extremely large surface area; outstanding thermal stability; oxidizable porous structure; simplicity of functionalization	A large vacuum space; a weak dispersion force; an unsuitable environment for coordination; an inadequate number of active metal catalyst areas; expensive preparation costs	[110]
	Clay	Clay is a layered aluminosilicate mineral that contains water and is found in rocks and soils	Strong thermal stability; excessive heat resistance; great surface area; a special micro-porous medium; inexpensive cost	Because of its underdeveloped pore structure, clay's adsorption affinity for carbon-based gases is restricted	[111]
	Silica gel	Silica gel is a three-dimensional tetrahedral inorganic substance with silicol groups on its surface	Low density; substantial porous surface area; multiple functional groupings; excellent mechanical, thermal, and chemical stabilities	Hygroscopicity	[112]
Organic polymers	Macroporous and hyper cross-linked polymers	Other known porous materials have a higher density than organic polymers made of light nonmetallic components such as C, H, O, N, and B	Large specific surface area; excellent porosity; low weight; excellent thermal stability, repeatability, and hydrophobicity	Complex synthesis	[113]

Activated Carbons

Recently, activated carbon has been reported to be useful in the remediation of heavy effluents and dye. Activated carbons are generated from commercially available wood, animal-based sources, or coal, and are all natural materials. However, practically any carbonaceous substance can be employed as a precursor in the synthesis of carbon adsorbents [114]. Coal is a widely utilized precursor for activated carbon generation due to its accessibility and low cost [115,116]. Various carbon and mineral combinations emerge from the decomposition of plants to form coal. The sorption qualities and the characteristics of coal are established as a result of the nature, source, and scope of the physical and chemical changes that happen upon deposition. Karaka et al. [117] investigated coal's use as a dye sorbent. Furthermore, the irregular surface of coal can influence its sorption properties. Peanut shell, [118], bael shell carbon [119], powdered pine cones (both raw and acid-treated) [120], Calotropis procera [121], neem leaf [122], coconut shell [123], and polyvinyl acetate (PVA) alginate super paramagnetic microspheres [124] were successful in reducing the contaminant concentrations of wastewater. Their sorption capacities increased as their adsorbent dosages increased [125].

7. Low-Cost Adsorbents

Many variables influence the characteristics of low-cost adsorbents. The precursor should be easily accessible, cheap, and non-toxic. Recent research has focused on natural solids that can remove contaminants from polluted water at cheap cost. Cost is a crucial factor when comparing sorbents. Generally, a sorbent is considered "low cost" if it needs minimal processing, is plentiful in nature, or is a by-product of another business. Many low-cost adsorbents have been employed to remove dyes including agricultural waste, natural materials, and bio-sorbents. Their efficacy in dye removal has been thoroughly investigated. Trash-derived adsorbents have been identified as the most challenging field since they can treat wastewater and reduce waste.

7.1. Natural Adsorbent

7.1.1. Clay

Clay is a layered natural adsorbent; with layers including vermiculite, smectites (saponite and montmorillonite), pyrophyllite (talc), mica (illite), kaolinite, serpentine, and sepiolite, clay minerals are accessible [126]. Adsorption occurs as a result of the minerals' net-negative charge, and this negative charge allows the clay substance to absorb positively charged ions. Their high surface area and porosity account for the majority of their sorption properties [127].

7.1.2. Siliceous

Siliceous is one of the most common materials and reasonably priced adsorbents. It contains glasses, silica beads, alunite, dolomite, and perlite. These minerals were utilized on the basis of the hydrophilic surface's chemical reactiveness and stability, which was due to a silanol group's presence. However, special consideration was given to the use of silica beads as adsorbents due to their low resistance to the application of alkaline solutions, limiting their use to media with pH values of less than 8 [73,128–130].

7.1.3. Zeolites

Zeolites are aluminosilicate porous materials that naturally form porous aluminosilicates with a variety of cavity configurations linked together by common oxygen atoms. There are numerous species of zeolite [131]. The natural species include chabazite and clinoptilolite. Conversely, clinoptilolite, a heulandite mineral, is the most common investigated substance due to its strong selectivity for specific pollutants. Zeolite has a special characteristic, namely a cage-like structure that is perfect for the elimination of trace pollutants including phenols and heavy metal ions. [132,133].

7.2. Bio Adsorbents

Different technologies can be used for the treatment of wastewater that contains dyes. Biological adsorbents that use nonliving biomass have been identified as the most promising approach due to their environmentally safe treatment capability [134]. The effective removal of dyes from the effluent depends on the unique surface chemistry with the presence of different functional groups in the cell wall of microorganisms, such as alcohol, aldehydes, ketones, carboxylic, ether, and phenolic compounds, which make the bio-sorbents have a high affinity toward dye and are attractive materials for dye removal [135]. Biological materials including chitin, peat, chitosan, yeast, and fungi biomass are frequently used in the sorption of dye from the solution through the mechanism of chelation and complexion [136]. A good adsorbent used in the removal of dye must have several desirable properties, including a large surface area, high adsorption capacity, large porosity, easy availability, stability, feasibility, compatibility, eco-friendliness, and ease of regeneration, as well as being highly selective in terms of removing different varieties of dyes [137]. The pore volume of the bio adsorbents and the functional groups of dyes are the deciding factors in the achievement of high dye adsorption. The presence of a large pore volume allows the binding of the highest number of dye molecules to the adsorbent [92]. Higher surface area, higher porosity, and low ash content lead to high adsorption capacity. Functional groups (hydroxyl, carboxyl, etc.) on the surface of biomass-based adsorbents are important properties determining the hydrophobicity or hydrophilicity of biochar as well as their adsorptive mechanism [138]. Likewise, the diversity of microbes consisting of different species of bacteria, fungi, yeast, and algae was studied in relation to the removal of dye molecules [139]. Besides the high sorption capacity toward dye, the dye removal performance can be improved by combining the biosorption process with the biodegradation processes using living cells [140]. The pH, bio-sorbent dose, initial dye concentration, temperature, and contact time are the influencing factors for the biosorption capacities of biomass [141].

7.2.1. Bacterial

Bacteria can play a role in bioremediation processes by adsorbing pollutants from aqueous media through a variety of methods, including dead biomasses [142]. Due to their tiny size, widespread distribution, and capacity to grow in a variety of environmental circumstances, they make excellent adsorbents [143]. Bacterial species were identified to successfully adsorb reactive dyes from wastewater under optimal environmental conditions [144]. The rates of bacterial dye decolorization vary according to the bacterium type, dye reactivity, and operational factors such as temperature, pH, co-substrate, electron donor, and dissolved oxygen content. It is possible to successfully treat textile dyes using extremophiles. According to the Langmuir adsorption isotherm, the maximum solubility capacity of basic blue dye is 139.74 mg/g. Carboxyl and phosphonate groups that are present on adsorbent surfaces may operate as possible surface functional groups, which are capable of binding cationic contaminants [145]. Numerous functional groups on the surface of the Penaeus indicus biomass were probably involved in the binding of the Acid Blue 25 dye, although the amino groups and alpha-chitin were by far the most significant [146]. Bacillus subtilis was immobilized on a calcium alginate bead and then used in batch and continuous reactors to remove MB. The kinetic analysis of the batch and continuous contactors revealed a removal rate of more than 90% [147]. Additionally, bacteria were adapted for MB's removal using electro-spun nanofibrous-encapsulated cells (Sarioglu et al., 2017a). Due to their variable cell wall compositions, biosorption fidelity is dependent not only on the group of ions but also on the type of bacteria.

7.2.2. Fungal

Fungal biomasses include sugars, proteins, and lipids, as well as functional groups (alcohols, carboxyls, and alkanes), which provide them with specific qualities and uses in wastewater treatment [148]. The biotreatment of dye-containing wastewater effluent by fungal cells was reported to be cost-effective, simple to implement, environmentally benign, and devoid of nutritional requirements [149]. Numerous fungi have been applied as effective candidates for the removal of a variety of dyes from effluents, including *Trichoderma* sp. [149], *Sarocladium* sp. [150], growing *Rhizopus arrhizus* [151], and several varieties of white-rot fungi [152]. It was shown that the removal rate of anionic dyes increases whereas the removal rate of cationic dyes decreases in low-pH solutions. In contrast, a high-pH solution enhances the removal of cationic dyes and results in a low proportion of anionic colors being removed [148]. The point of zero charges (pHpzc) is a critical metric for understanding the adsorption mechanism and its favorability. The pHpzc value provides information on the active sites and adsorption capacity of adsorbents. When the pH is larger than the pHpzc, cationic dye adsorption is more advantageous owing to the presence of functional groups (OH^-, COO^-), but anionic dye adsorption is more favorable when the pH is less than the pHpzc due to the positively charged surfaces of the adsorbents [95]. In general, the use of fungal biomass as a dye decolorizer and adsorbent is a viable alternative to existing technology. Along with the regulation of environmental factors, it is critical to consider the genotype and preparation of the biomass in order to ensure optimum dye-adsorption performance.

7.2.3. Algae

Algae are one of the best sources of bio-sorbents since they have high biosorption ability and are readily accessible [153]. The algal cell wall is composed of polysaccharides, including xylan, mannan, alginic acid, and chitin. In addition to proteins, these components may include amino, amine, hydroxyl and imidazole, phosphate, and sulfate groups [143]. Pretreatments such as encapsulation and surface modification may improve the sorption capacity of algae. The adsorption ability of citric acid-functionalized brown algae for textile dye (crystal violet) removal in aqueous solutions was investigated. It was found to improve the uptake capacity by up to 279.14 mg/g [154]. This process was also due to electrostatic interactions.

The adsorption of five water-soluble dyes was performed using magnetically sensitive brown algae (Sargassum horneri). Using microwave-synthesized iron oxide nano- and microparticles, the magnetic modification allowed for quick and selective separation [155]. After 2 h contact time, the sorbent demonstrated maximal acridine orange sorption capacity (193.8 mg/g) but not malachite green sorption capacity (110.4 mg/g). Sargassum macroalgae are frequently investigated for their ability to remove colors. MB is a popular dye that is removed by dye species. Anionic dyes are eliminated in acid, and cationic dyes (e.g., MB) are eliminated in alkaline. This is because hydrogen (H+) ions are involved in the biomass–pollutant interaction mechanism. To reduce the quantity of absorbed dye, the adsorbent's surface might be charged positively to compete with the dye's cations. At increased pH, carboxyl groups have a negative charge, resulting in the electrostatic binding of cationic dyes. Other criteria that influence the biosorption efficiency include the processing of the biomass into the adsorbent, the starting contaminant concentration, and the biomass dose, temperature, and contact duration.

7.2.4. Yeast

Yeast is a single-celled organism that has numerous advantages over filamentous fungus in terms of the adsorption and accumulation of pollutants, as well as its growth rate, decolorization rate, and ability to live in harsh settings [156]. The carboxyl hydroxide, polymer, amino, and phosphate functional groups on the yeast surface alter the pH of the tested solution [157]. Yeast biomass has been shown to bio-adsorb several types of colors. The bio-sorption process is affected by the pH, pollutant concentration, yeast mass, temperature, and contact duration [158]. Reactive Blue 19 (RB 19) and Red 141 (RR 141) were studied in Antarctic yeast (Debaryomyces hansenii F39A). At pH 6.0, with 100 mg/L as the initial dye concentration, and a 2 g/L biomass dose, 90% of RR 141 and 50% of RB 19 were adsorbed. However, at a 6 g/L biomass dose, 90% of RB 19 was adsorbed. The Langmuir isotherm was defined as the pseudo-second-order kinetics for each dye system, and the Langmuir isotherm was the best-matched model [159].

The removal of Reactive Blue 160 dye using residual yeast and diatomaceous earth (RB 160) was also investigated. The dye removal capability of the two bio-sorbents was 8.66 mg/g and 7.96 mg/g at pH 2 [160]. The biomass functional group's positive charge interacted with the anionic dye. The yeast biosorption data were better fitted to the Freundlich isotherm model, whereas the diatomaceous earth data were better fitted to the Langmuir isotherm. Another study used brewer's yeast biomass that was able to adsorb the basic dyes safranin O (SO), MB, and malachite green (MG)) from aqueous solutions within 1 h. This study also reported that MB's and MG's adsorption kinetics were pseudo-second-order, whereas those of SO were pseudo-first-order. Yeast was also found to adsorb hydroxyl, cyano, and other functional groups [161].

7.3. Agricultural and Industrial Materials' Adsorbents

7.3.1. Agricultural Waste and Plant Adsorbents

The use of agricultural wastes and plants to adsorb organic and inorganic contaminants is considered to be a viable alternative to standard wastewater treatment procedures [162]. Numerous investigations on the elimination of MB have recently been conducted, which involved employing dead or living agro-waste, algae, fungi, and a variety of naturally occurring and low-cost agro-waste sources as adsorbents, including fruit peels, seeds, leaves, straw, sawdust, bark, sludge, and ash [163].

Numerous studies demonstrated that the dye-adsorption properties of certain biomasses are highly dependent on the kind of dyes used, and the processing procedures used were successful in reducing the contaminant concentrations of wastewater. This group of biological compounds of agro-waste-derived adsorbents was capable of collecting and concentrating dyes in aqueous solutions. Due to the non-selective nature of these biomaterials, all pollutants, both target and non-target, became concentrated on the adsorbent's surface, providing significant removal for the purpose of pollution control. The technique allows the adsorption of only those ions for which it has a particular affinity. In comparison to other methods, bioadsorption is rated as preferable due to its low cost, simplified design, great efficiency, and capacity to separate a wide variety of contaminants [164].

7.3.2. Industrial Products

Fly ash, metal hydroxide sludge, bio solids, red mud, and waste slurry are examples of industrial products that may be employed as dye adsorbents since they are low-cost and readily available. Adsorbents made from industrial waste may be used instead of more expensive traditional adsorbents [165].

Fly Ash

Fly ash is a type of industrial waste that may be used to adsorb dyes. Fly ash is generated in enormous quantities during combustion operations and may include certain harmful chemicals, such as heavy metals [166]. However, bagasse fly ash, created in the sugar industry, is devoid of hazardous metals and is often employed for color adsorption. Its qualities are very variable and are dependent on its source. Adsorption investigations were conducted on congo red and MB textile dye solutions and it was discovered that the monolayer development on the adsorbent surface and the adsorption process are exothermic in nature. Fly ash from thermal power plants may be efficiently utilized as an adsorbent to remove colors from dyeing industry effluents [167]. The removal of methylene blue, using fly ash as an absorbent, was investigated and a maximum removal of 58.24 percent was reported at pH 6.75 and 900 mg/L adsorbent for an initial methylene blue dye concentration of 65 mg/L. At various beginning conditions, fly ash could remove 95–99 percent of the dye from the solution, and the Langmuir constant q_m was 1.91 mg/g and the K_a value was 48.94 L/mg with a liner regression coefficient of 0.999 [168].

Metal Hydroxide Sludge

Sludge made from metal hydroxide is used to clean up azo dyes. It has insoluble metal hydroxides and salts. Researchers discovered that at 30 °C and pH 8–9, electroplating industrial hydroxide sludge had maximal adsorption capacities of 45.87 and 61.73 mg/g for Reactive Red 120 and Reactive Red 2, respectively. The pH also influenced the adsorption and development of dye–metal complexes. Sludge of metal hydroxide was used as an adsorbent and it was found to have a maximum adsorption capacity of 270.8 mg/g at 30 °C and an initial pH of 10.4. Metal hydroxide, as a low-cost adsorbent for the removal of the Remazol Brilliant Blue reactive dye from a solution, was reported to have a 91.0 mg/g monolayer adsorption capacity at 25 °C and pH 7 [169].

Red Mud

Red mud is another industrial byproduct and bauxite manufacturing waste product used to make alumina. The capacity of discarded red mud as an adsorbent for the removal of dye from its solution was examined, and it was found to be effective. It was found that the greatest dye removal via adsorption occurred at pH 2, and this was followed by the Freundlich isotherm [170]. Red mud was used as an adsorbent to remove a basic dye, methylene blue, from its aqueous solution. The adsorption capacity of red mud was determined to be 7.8×10^{-6} mol/g. The use of discarded red mud as an adsorbent was shown in order to extract congo red from aqueous solution. The dye-adsorption capability of the red mud was determined to be 4.05 mg/g. Using acid-activated red mud, the adsorption of congo red from wastewater was examined [171]. The Langmuir isotherm provided the greatest match to the experimental data. Using red mud, the removal of methylene blue, quick green, and rhodamine B from wastewater was investigated. Fast green, Methylene blue, and rhodamine B were removed with red mud at percentages of 75.0, 94.0, and 92.5, respectively; the adsorption process followed both the Langmuir and Freundlich isotherms and was exothermic in nature [172].

7.4. Activated Carbon-Based Adsorbent Derived from Low-Cost Waste

Agricultural wastes are rich in hemicellulose, cellulose, and lignin. Their surfaces are covered with a variety of active groups, including carboxyl, hydroxyl, methyl, and amino [95]. These functional groups may adsorb dyes in a variety of ways, including via complexation, hydrogen bonding, and ion exchange [152]. Tables 4–13 highlight different agricultural and forest waste types, their biosorption capacity, and the activation reagents required. Numerous acids have been utilized to activate biosorbents to increase their binding sites, aqueous solution chemistry, specific surface area, and porosity. Phosphoric acid increases the bond-breaking process in agricultural waste biomass, thereby boosting its carbon output [173]. Sodium hydroxide (NaOH), sulphuric acid (H_2SO_4), and potassium hydroxide (KOH) are often utilized as activators in the manufacturing of agricultural waste-based bioadsorbents (Tables 4–13). Numerous environmental factors, including the adsorbent dosage, temperature, contact time, solution pH, particle size of the plant-based adsorbent, agitation, and initial dye concentration, all have a significant influence on the biosorption process. The pH of the solution, the particle size of the plant-based adsorbent, the rate of agitation, and the initial dye concentration all have a substantial effect on the biosorption process. The pH of a solution has an effect on both the aqueous solution's chemistry and the binding sites on the surfaces of the adsorbents [174]. Due to the abundance of low-cost products, they constitute excellent raw materials for the manufacturing of activated carbon. Tables 4–13 summarize the different types of activated carbon derived from biomass and their maximal adsorption capacities for MB elimination. A schematic clarification of bio-waste-derived adsorbents is shown in Figure 6.

Isotherm Equilibrium and Sorption Capacity of Biowaste-Derived Adsorbents

Tables 4–13 show the outstanding capabilities and operating conditions of bio-waste-derived adsorbents with high sorption capacities that have been established over the last decade. Furthermore, this review sought to enclose a broad range of recent research on unconventional adsorbents to educate researchers about the design parameters and sorption capacities for the adsorption of various bio-waste materials. Phosphoric acid improved dye biosorption by grafting phosphate functions onto the biomass and enhancing the acid functions involved in dye fixation

Previous studies addressed equilibrium isotherms and kinetic features by employing models ranging from Henry's law to the Langmuir (monolayer), Redlich–Peterson, Sips, and Freundlich models for fitness analyses. The kinetic and isotherm models are useful predictive tools for adsorbent system regeneration, design parameter optimization, and adsorption and desorption capacity maximization, and can, by these means, optimize waste disposal. Additionally, most of the previous studies were conducted in batch mode, which enables more cost-efficient and effective treatments for the design of continuous systems. The adsorption capacity of an adsorbent can be determined using equilibrium isotherms. Equilibrium isotherms link the equilibrium concentration of the adsorbate (Ce) to the quantity of the adsorbent (qe). Furthermore, the adsorbate characteristics and adsorbent surfaces can be studied in detail using liquid–solid isotherms.

Tables 4–13 illustrate the operating conditions, sorption capacities, and appropriate kinetic isotherms for adsorbents derived from bio-waste over the last decade. Furthermore, this review sought to enclose a broad range of recent research on unconventional adsorbents in order to educate researchers about the design parameters and sorption capacities related to the adsorption of various bio-waste materials [175–177].

Figure 6. Schematic clarification of activated carbon derived from bio-waste and its potential uses.

Table 4. Summary of bio-waste-derived adsorbent studies in 2012.

Biosorbents	Q_{max} (mg/g)	Most Appropriate Model	pH	Temperature (°C)	Time (min)	Reference
Pink Guava leaf	250	L-K2	NA	30	300	[178]
Malted sorghum mash	357.1	L	7.3	33	18	[179]
Rice husk	8.13	L-K2	5.2	25	NA	[180]
Water Hyacinth	8.04	L-K2	8	25	80	[181]
Date stones	398.19	S-K2	7	30	270	[182]
Oil palm shell	133.13	NA	NA	30	10	[183]
Swede rape straw	246.4	L	NA	25	NA	[184]
Pyrolysis of wheat	12.03	S	8–9	20	50	[185]

Table 5. Summary of bio-waste-derived adsorbent studies in 2013.

Biosorbents	Q_{max} (mg/g)	Most Appropriate Model	pH	Temperature (°C)	Time (min)	Reference
Pea shells	246.91	L	2–11.5	25	180	[185]
Coconut fiber	500	L-K2	7.8	30	30	[186]
Papaya leaves	231.65	L	2–10	30	300	[187]
Untreated Alfa grass	200	L-K2	12	20	180	[188]
Neem leaf Powder	401.6, 352.6	F-K2	7	87	60	[189]
Corn husk	662.25	F	4	25	120	[190]
Lagerstroemia microcarpa	229.8	L-K2	NA	30	360	[191]
watermelon (Citrullus lanatus)	489.80	L-K2	NA	30	30	[192]
Sugarcane bagasse	95.19%	NA	8.76	25	193	[193]

Table 6. Summary of bio-waste-derived adsorbent studies in 2014.

Biosorbents	Q_{max} (mg/g)	Most Appropriate Model	pH	Temperature (°C)	Time (min)	Reference
Iron oxide-modified montmorillonite	69.11	L-K2	8	35	240	[194]
Magnetic NaY Zeolite	2.046	L	10.3	50	45	[195]
Fe_3O_4 graphene/MWCNTs	65.79	L-K2	7	10	30	[196]
Water hyacinth	111.1	L	8-10	30	300	[197]
Lantana camara stem	19.84	F-K2	3-11	20	60	[198]
Natural peach gum (PG)	298	L-K2	6-10	25	30	[199]
Activated fly ash (AFSH)	14.28	F-K2	3.0-10.0	20	100	[200]

Table 7. Summary of bio-waste-derived adsorbent studies in 2015.

Biosorbents	Q_{max} (mg/g)	Most Appropriate Model	pH	Temperature (°C)	Time (min)	Reference
Magnetic biochar derived from empty fruit bunch	31.25	L-K2	2-10	25	120	[201]
Magnetic adsorbent derived from corncob	163.93	L-K2	NA	25	500	[202]
Fe_3O_4 bentonite	NA	K2	7	NA	20	[203]
Magnetic chitosan/organic rectorite	24.69	L-K2	6	25	60	[204]
Poly acrylic acid/$MnFe_2O_4$	NA	K2	8.3	25	NA	[205]
Fe_3O_4 xylan/poly acrylic acid	438.6	L-K2	8	25	NA	[206]
Fe_3O_4 modified graphene sponge	526	L-K2	6	NA	NA	[207]
Xanthate/Fe_3O_4 graphene oxide	714.3	L-K2	5.5	25	120	[208]
Magnetic carbonate hydroxyapatite/graphene oxide	405.4	L-K2	9.1	25	90	[209]

Table 8. Summary of bio-waste-derived adsorbent studies in 2016.

Biosorbents	Q_{max} (mg/g)	Most Appropriate Model	pH	Temperature (°C)	Time (min)	Reference
Palm shell	163.3	F-K2	NA	25	NA	[210]
Fe_3O_4-activated montmorillonite	106.38	L-K2	7.37	20	25	[211]
Clay (montmorillonite and vermaculti)/polyaniline/Fe_3O_4	184.5	L-K2	6.3	25	30	[212]
Magnetic chitosan/active charcoal	200	L-K2	7.73	25	200	[99]
Fe_3O_4/poly acrylic acid	73.8	L-K2	NA	45	NA	[213]
Magnetized graphene oxide	306.5	L-K2	9	25	360	[214]
Corn straw	267.38	F-K2	8	25	20	[215]
Magnetic chitosan and graphene oxide	243.31	K2-L	12	60	60	[216]

Table 9. Summary of bio-waste-derived adsorbent studies in 2017.

Biosorbents	Q_{max} (mg/g)	Most Appropriate Model	pH	Temperature (°C)	Time(min)	Reference
Corn shell	357.1	L	4	25	30	[217]
Magnetic activated carbon	2.046	F-K2	10	25	120	[218]
Magnetic halloysite nanotube nano-hybrid	689.66	L-K2	10	25	180	[219]
Magnetic polyvinyl alcohol/laponite RD	251	L-K2	5.5	25	60	[220]
Aegle marmelos leaves	500	F-K2	6	25	120	[221]
Oak-acorn peel	109.43	L-K2	7	24	120	[222]
Geopolymers	15.95–20.22	S-K2	4-12	25	80	[223]
Ouricuri fiber	31.7	S-K2	5.5	25	5	[224]

Table 10. Summary of bio-waste-derived adsorbent studies in 2018.

Biosorbents	Q_{max} (mg/g)	Most Appropriate Model	pH	Temperature (°C)	Time (min)	Reference
Carboxymethyl/cellulose/Fe_3O_4/SiO_2	31.02	L-K1	11	NA	60	[225]
Cellulose-grafted	7.5	L	8		5.5	[226]
$NiFe_2O_4Ca$/alginate	1243	R-K1	6.5	25	180	[227]
Magnetic alginate	161	L	7	20	120	[228]
Magnetic hydrogel Nanocomposite of poly acrylic acid	507.7	L-K1	7	25	120	[229]
Magnetized graphene oxide	232.56	L-K2	9	30	10	[230]
Soursop	55.397	R-K2	5.5	25	300	[231]
Sugarcane Bagasse	17.434	S-K2	5.5	25	300	[231]
Palm sawdust	53.476	F-K2	8	25	120	[232]
Eucalyptus sawdust	99.009	F-K2	6	20	60	[232]

Table 11. Summary of bio-waste-derived adsorbent studies in 2019.

Biosorbents	Q_{max} (mg/g)	Most Appropriate Model	pH	Temperature (°C)	Time (min)	Reference
Fir bark	330.00	F-K2	NA	25	40	[233]
Pumpkin peel	198.15	L-K2	7	50	180	[234]
Rice husk	608	L	7	25	60	[235]
date stones	163.67	F-K2	10	25	360	[236]
Seaweed	1279.00	L-K2	4	25	50	[237]
Moroccan cactus	14.04	L	5	25	60	[238]
Syagrus oleracea	893.78	L-K2	7	25	20	[239]
Mentha plant	588.24	L	10	25	30	[240]
Palm leaf	500	L	2	30-60	30	[241]

Table 12. Summary of bio-waste-derived adsorbent studies in 2020.

Biosorbents	Q_{max} (mg/g)	Most Appropriate Model	pH	Temperature (°C)	Time (min)	Reference
Kendu fruit peel	144.90	L-K2	6	25	100	[242]
Magnesium oxide nanoparticles	163.87	L-K2	7.3	25	70	[243]
Fava bean peel	140.00	L	5.8	27	NA	[244]
Dicarboxymethyl cellulose	887.60	L-K2	3	25	60	[245]
Alginate-based beads	400.00	L-K1	7	25	NA	[246]
Black cumin seeds	16.85	F-K2	4.8	25	20	[247]
Dragon fruit peels	195.2	L-K1	3-10	50	60	[248]
Litsea glutinosa seeds	29.03	L-K2	9	40	600	[249]
Moringa oleifera leaf	136.99	F-K2	7	25	90	[250]

Table 13. Summary of bio-waste-derived adsorbent studies in 2021.

Biosorbents	Q_{max} (mg/g)	Most Appropriate Model	pH	Temperature (°C)	Time (min)	Reference
Grass waste	364.2	L	10	45	15	[251]
Mangosteen peel	871.49	L-K2	10	25	60	[252]
Coconut shell	156.25	F-K2	4.9	25	360	[253]
Core shell	34.3	L-K2	7	25	120	[254]
Banana stem	101.01	F-K2	7	25	90	[255]
Alginate beads	769	L-K2	8	30	NA	[256]
Ulva lactuca	344.83	L-K2	11	25	NA	[257]
Cassava Stem	384.61	L-K2	9.2	25	60	[258]
Corncob	864.58	L-K2	5	25	360	[259]

General equation

$$Q_{max} = \frac{(C_0 - C_e)V}{W} \quad (1)$$

where V is the solution volume (L) and W is the adsorbent mass (mg/L) (g), C_0 and C_e are the initial and equilibrium dye concentration in mg/L, respectively.

Langmuir (L) isotherm model:

$$qe = \frac{Q^\circ \cdot K \cdot C_e}{1 + K \cdot C_e} \quad (2)$$

where qe is the adsorbate quantity per unit of adsorbent (mg/mg), C_e is the equilibrium concentration of the adsorbate (mg), K is the Langmuir adsorption coefficient (mg/g) (L/mg)$^{1/n}$.

Freundlich (F) isotherm model:

$$qe = Kf\, C_e^{1/n} \quad (3)$$

where qe is the quantity of adsorbates per unit of adsorbent (mg/g), C_e is the adsorbate equilibrium concentration in the solution (mg), n is the empirical coefficient, Kf is the Freundlich adsorption coefficient (mg/g) (L/mg)$^{1/n}$.

Redlich–Peterson (R) isotherm model:

$$qe = \frac{K_R C_e}{1 + a_R C_e^g} \quad (4)$$

where qe is the adsorbate quantity per unit of adsorbent (mg/mg), K_R (L g^{-1}) and a_R (L^g·mg^{-g}) are constants, C_e is the equilibrium concentration of the adsorbate (mg), g is the exponent ($0 \leq g \leq 1$).

Sips (S) isotherm model:

$$qe = \frac{q_{ms} K_s C_e^{n_s}}{1 + K_s C_e^{n_s}} \quad (5)$$

where q_{ms} is the maximum adsorbed amount (mg/g), K_s (L^{n_s}·mg$^{-n_s}$) and n_s are the Sips constants, C_e is the equilibrium concentration of adsorbate (mg), qe is the quantity of adsorbate per unit of adsorbent (mg/mg).

Modeling adsorption kinetics:

Adsorption kinetics were used to explore the pace and mechanisms of adsorption, which may occur due to physical and chemical events, and to compare these with experimental data.

Pseudo-First-Order Kinetics (K1):

$$\ln(Q_e - Q_t) = \ln Q_t - k_1 t \quad (6)$$

where Q_t is the adsorbed amount at time t, Q_e is the equilibrium amount, t is the time in minutes, and k_1 is the rate constant.

Pseudo-Second-Order Kinetics (K2):

$$\frac{t}{Qt} = \frac{1}{k_2 Q_e^2} + \left(\frac{1}{Q_e}\right)t \quad (7)$$

where Q_t is the adsorbed amount at time t, Q_e is the equilibrium amount, t is the time in minutes, and k_2 is the rate constant.

8. Cost Analysis of Adsorbents

Several authors indicated that the application of bio adsorbents derived from microorganisms and forest and agricultural waste is lower than the cost of traditional treatment methods. Nonetheless, none of these research works considered the cost analysis in their final assessment. For a cost effective system, the volume of adsorbent used, the simplicity of preparation or processing, green chemistry ideas, and the activation process used are the factors that need to be considered [260]. In contrast, another study emphasized the term "low-cost adsorbents" to refer to their initial costs, and their local availability, transportation, treatment process, recycling, and lifespan concerns, as well as regenerating and treatment methods [95]. Additionally, most of the previous research works on biomass-based adsorption were undertaken on a laboratory scale using simulated wastewater, thereby restricting the cost of the analysis to be undertaken.

The ability to remove Basic Red 09 dye from wastewater was investigated using coconut shell, groundnut shell and rice husk. The study revealed the cost of 1 g of adsorbent used to remove 4.54, 0.91, and 0.97g when operational expenses such as production, maintenance, feedstock, transportation, labor, and distribution costs are included [261]. Groundnut shell-based biochar showed the highest adsorption capacity (46.3 mg/g) and the lowest cost-per-unit in grams of Basic Red 09 dye removal (0.91). A phosphoric acid-functionalized locust bean pod adsorbent was produced for the removal of RhB dye, and the initial cost of this adsorbent was determined. They revealed that the activated carbon generated by these plant sources was roughly six times less costly than conventional activated carbon. The expense is mostly borne by phosphoric acid and deionized water [262].

9. Regeneration and Economic Challenges of Bio-Waste-Derived Adsorbents

The desorption process can induce the application of reused adsorbent, thus reducing waste and minimizing capital and operational costs [95]. Common desorption methods include thermal, acid (i.e., hydrochloric acid (HCl), H_2SO_4, phosphoric acid (H_3PO_4), and nitric acid (HNO_3)) NaOH, organic solvent (methanol), vacuum, and biological methods [92]. Solvent desorption through drying processes can vaporize and remove the dye with suitable a combination ratio between the adsorbent and the solvent. [95]. On top of reuse or regeneration of the adsorbent, the selection of an appropriate adsorbent, particularly at large scales, plays a vital role in terms of ensuring an efficient and economical treatment method. Powdered activated carbons were reported as being inappropriate for industrial applications due to their high costs and times, and complex recovery processes [263]. This can lead to high energy consumption due to inefficient processes. However, the post-treatment of adsorbent and effluent that contains contaminants is necessary after dye's removal from wastewater [264]. Immobilization and stabilization immobilization are two possible ways of securely disposing of the final effluent; for example, utilizing in concrete technology as a binder material [265,266].

Dahiru et al. [267] reported that the efficiency of banana peel adsorbent reduced to 64% after 5 uses. Despite there being numerous studies on the development of bio-waste-derived activated carbon, there were minimal efforts focused on the technoeconomic assessment and life cycle analysis of these applications. This signifies the need for the development of adsorbents that are more robust in order to maintain removal rates, optimize costs, and promote the sustainable regeneration of adsorbents.

10. Management of Post-Adsorption Materials

After usage, the adsorbent can be managed in a variety of ways, including regeneration, re-use, and safe disposal (Figure 7). Regeneration may be accomplished in a variety of ways, including with a chelating desorbing agent, an alkali desorbing agent, a salt desorbing agent, or via thermal regeneration [268]. In addition to the forementioned approaches, organic pollutants may be regenerated via ultrasonic regeneration, microbiological regeneration, microwave-assisted regeneration, thermal regeneration, chemical regeneration, ozonation, photo-assisted oxidation, and electrochemical oxidation [269]. After many adsorption–regeneration cycles, the adsorbent's efficacy decreases [270]. After many adsorption–regeneration cycles with the same pollutant, the method renders the adsorbent redundant. The used adsorbent may be disposed of in a landfill or burnt or recycled [271]. Prior to landfill disposal, used adsorbents containing hazardous elements can be stabilized/solidified [272], thus increasing the expense of the adsorbent's life cycle evaluation. Enhancing the adsorbent's sustainability may be accomplished by properly disposing and reusing it in other applications. The used adsorbent can be used in a variety of ways, including as a catalyst [273], in brick formulations [274], in road construction [275], or in cement clinkers [276]. The three major applications of wasted adsorbents are as follows: as a catalyzer, in the manufacturing of ceramics, and as a fertilizer.

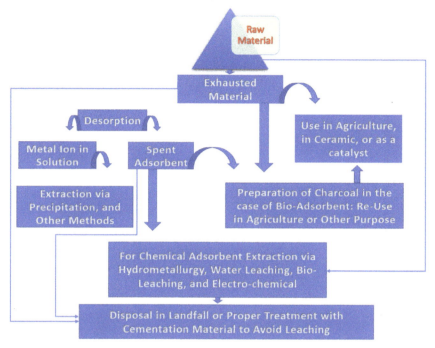

Figure 7. Adsorbent disposal management after adsorption.

10.1. Application as a Catalyst

Following the adsorption process, the used adsorbents can be employed as catalysts throughout the processes of photodegradation [277], nitrophenol's reduction to aminophenol [278], hydrocarbon oxidation [277], the conversion of xylose and xylan to furfural, and also the conversion of phenylacetylene to acetophenone [279]. Depending on the type of pollutant, the final product can be further analyzed using nuclear magnetic resonance (NMR) spectroscopy, high-performance liquid chromatography (HPLC) [279], gas chromatography [273], ultra-violet spectroscopy [278], and Fourier-transform infrared spectroscopy (FTIR) [277]. In some cases, the catalytic activity of the metal ion varies according to its position on the adsorbent, the conversion and selectivity inside the oxidation of cyclohexanol, as well as the increasing of ethyl benzene [280]. Despite the vast potential of expended adsorbents to induce catalysis, several issues must be addressed, with the most significant being the leachability of the pollutant or other materials from the adsorbent during their usage as catalysts. The majority of research employed either the California waste removal test or the Toxicity Characteristic Leaching Procedure (TCLP) for leaching measurement [281]. In many situations, wasted adsorbent contains dangerous elements, and environmental organizations (e.g., the USEPA in the United States, the CPCB in India, and DEFRA in the United Kingdom) enforce strict disposal rules. Consequently, this problem can be eliminated by increasing the use of nontoxic waste adsorbents [282].

10.2. Application in Ceramic Production

Used adsorbents can also be employed as ingredients in the manufacturing of ceramic materials, including as fillers in the cement industry. The issue of the adsorbent's hazardous nature may be mitigated by its application in the manufacturing of ceramics as well as in road building. The leaching of dangerous materials from the used adsorbent can be managed with the correct preparation conditions. Spent adsorbent (zeolite- and perlite-supported magnetite following molybdenum adsorption) was combined with sludge at a ratio of 3/97, which corresponded to the adsorption capacity of the loaded adsorbent [283]. Additionally, ceramic products may help in preventing the leaching of additional heavy metals (including Nickel, Chromium, Copper, Zinc, Arsenic and Cadmium) that spike during the application of the ceramic synthesis technique. This is advantageous in the treatment of polluted eluent generated during desorption operations. Additionally, the used adsorbent may be disposed of by immobilizing it inside the phosphoric glass matrix. It was also shown that around 20% of wasted adsorbent can be integrated during glass production [284].

10.3. Application as Fertilizer

The used adsorbent can be converted into a user-friendly material, including fertilizers. The properties required for fertilizer production include affinities for anions and cations and long-term stability in various environments. Charcoal is mainly used as a fertilizer. Calcium (Ca), Nitrogen (N), Potassium (K), and Phosphorus (P) are abundant in biomass. By applying this method, nutrients are returned to the soil, potentially improving soil fertility [285]. The use of biodegradable organic adsorbents as fertilizers is possible. It was reported that 20 days is required for carboxy methyl cellulose, a copper-removing chitosan, to break down [286].

The pyrolysis of discarded bio-sorbent, occurring as a result of the adsorption of contaminants in biochar or direct soil applications, yields charcoal, biochar, and a variety of products, each with a distinct economic value [287]. Toxic substances that are present in soil can be reduced by adding charcoal. The use of charcoal (15 g/kg) was found to lower chromium and cadmium contents in a plant by 33.50 and 28.73 percent, respectively [288]. Additionally, crops need nitrates and phosphates for their growth, and charcoal is inefficient in terms of serving these needs. Consequently, metal ions such as Ca, Mg, and Al may be added to charcoal [289]. In the case of phosphate, these components increased the formation of H bonds or precipitation, whereas in the case of nitrate, they increased the electrostatic attraction [289]. Meanwhile, the nonfuel fraction gases (carbon monoxide (CO), methane (CH_4), and other hydrocarbons) may be utilized to synthesize various chemical reagents in order to synthesize biofuels [290]. Additionally, the use of adsorbents as fertilizer can improve metal sequestration [291], improvement of soil's nutritive value [292], increased soil organic carbon (SOC) (due to the application of activated carbon) [282], and increased water-holding capacity of the soil [285]. The content of each heavy metal in charcoal has a specific threshold level. Lead concentrations in basic and premium biochar should be lower than 120 and 150 g/t. This includes the need for higher charcoal demands as compared to commercial fertilizer, and the regulated release of nutrients to prevent soil contamination as well as metal ion accumulation. This has influenced the initial capital cost of recovering all products from pyrolysis, such as heat and gases during biomass feedstocks [292].

11. Cost-Effectiveness: Desorption versus Disposal

Following adsorption process, adsorbents may be desorbed and renewed until the pollutant content in the effluent is maintained below the permitted level established by regulatory bodies. The used adsorbent may be repurposed for different applications such as catalyst synthesis, ceramic manufacture, and pollutant removal, or it can be discarded. The desorption of pollutants may be accomplished using an alkali or acid reagent, a chelating agent, or salt; or, for organic pollutants, chemical, thermal, microwave, or other processes can be used [293]. Alkali was reported as the most effective method for removing heavy metals from chemical-based adsorbents (Table 14). The employment of an acid, an alkali, a chelating molecule, or a chemical as a desorbing agent result in waste creation (secondary pollution) in contaminated eluent. As a result, this approach suffers from the same disposal issues as other approaches, such as wasted adsorbent, which have environmental and economic implications. Nonetheless, there are rare occasions when metals that are laced with other heavy metals can be recovered, such as chromium (Cr) being recovered from barium chloride (baCl), and mercury (Hg) from the ethylenediaminetetraacetic acid (EDTA)–Hg combination being recovered as mercury chloride (HgCl) [294].

Table 14. Desorbing agents for various adsorbents.

Adsorbents	Desorbing Agents	Agent	References
Chemical sorbents	Alkali	NaOH	[295]
Bio-adsorbents	Acid	HCl, H_2SO_4, HNO_3	[295]
Biomass (fungi, algae)	Complexing agents	EDTA	[295]

12. Limitations and Strategies

The primary disadvantage of the previously reported adsorption studies is that they are generally applied at the laboratory scale without any pilot study or commercial-scale column filtration system. On top of the limitations of the adsorbents used, the bulk of the research work employed batch mode experiments with simulated mono-pollutant solutions, with just a handful using genuine wastewater. Most investigations on bio-waste adsorption focused on removing a single contaminant from actual dye-containing effluent. To meet the needs of wastewater treatment, more research should be conducted in multi-pollutant systems with real textile wastewater. Additionally, the review demonstrates certain inherent limits of recent developments in the use of activated carbon in terms of operational efficiency, overall costs, energy consumption, and the potential to form harmful by-products, even when these approaches work well against a specific pollutant. Although most bio-wastes had high elimination efficiency up to 99%, various and different parameters were used as indicators in the previous research works, which limits the potential for comparative studies. Finally, most of the previous research works on biomass-based adsorption were undertaken at a laboratory scale using simulated wastewater, and thus, the undertaking of cost analyses was restricted.

13. Conclusions and Recommendations

Bio-waste is the richest economically available source of carbon synthesis and is often transformed into activated carbon. From 2012 to 2021, bio-waste has emerged as a low cost, effective, and renewable source of activated carbon for the removal of MB. Low-cost bio-waste-derived adsorbents can be characterized and defined in terms of their initial costs, local availability, stability, eco-friendliness, transportation, applied treatment processes, recycling, lifespan concerns, regeneration potential, and pore volume after deactivation. In terms of the parameters that influence performance, the most critical characteristic affecting the adsorption of cationic dyes is the pH level; high pH values are necessary to achieve maximum dye uptake. Additionally, the initial dye concentration, temperature, adsorbent dose, type, and contact duration are the parameters that determine the dye-adsorption ability.

The processing methods employed in the adsorption studies include activation by steam, carbon dioxide, and chemical methods. Steam activation is the most cost-effective approach, whereas chemical activation produces the highest porosity and surface area. In terms of regeneration processes, the available desorption methods include thermal acid and nitric acid, sodium hydroxide, organic solvent, vacuum, and biological approaches. For a cost-effective system, the volume of adsorbent used, the simplicity of preparation or processing, green chemistry ideas, and the activation process used are the factors that can be considered. Additionally, catalyzer, ceramic, and fertilizer applications all show potential in the management of post-adsorption material.

In a nutshell, the adsorbent's stability and affordability are other important characteristics that influence its applicability in terms of ensuring an efficient on-site treatment. Local availability, transportation, economic feasibility, potential for regeneration, and lifespan difficulties can also be investigated in future research works. Regeneration studies are also necessary to reduce process costs, recover adsorbed pollutants, and reduce waste generation.

Author Contributions: Conceptualization, H.N.H. and S.I.; methodology, H.N.H.; validation, S.I.; formal analysis, H.N.H.; investigation, H.N.H.; resources, S.I.; data curation, S.I.; writing—original draft preparation, H.N.H.; writing—review and editing, S.I.; visualization, S.I.; supervision, S.I. All authors have read and agreed to the published version of the manuscript.

Funding: This research was financially supported by Universiti Putra Malaysia through the Research Management Centre (Project Code: 9001103).

Institutional Review Board Statement: Not applicable.

Informed Consent Statement: Not applicable.

Data Availability Statement: Not applicable.

Acknowledgments: The authors would like to acknowledge the support received from Universiti Putra Malaysia (UPM). Additionally, all academicians in the Water Engineering Unit, and the assistant engineer, Department of Civil Engineering, Faculty of Engineering, UPM are thanked for their contributions to the success of this study.

Conflicts of Interest: The authors declare no conflict of interest.

References

1. Schwarzenbach, R.P.; Escher, B.I.; Fenner, K.; Hofstetter, T.B.; Johnson, C.A.; Von Gunten, U.; Wehrli, B. The challenge of micropollutants in aquatic systems. *Science* **2006**, *313*, 1072–1077. [CrossRef] [PubMed]
2. Inamuddin. Xanthan gum/titanium dioxide nanocomposite for photocatalytic degradation of methyl orange dye. *Int. J. Biol. Macromol.* **2018**, *121*, 1046–1053. [CrossRef] [PubMed]
3. Saratale, G.D.; Saratale, R.G.; Chang, J.S.; Govindwar, S.P. Fixed-bed decolorization of Reactive Blue 172 by Proteus vulgaris NCIM-2027 immobilized on Luffa cylindrica sponge. *Int. Biodeterior. Biodegrad.* **2011**, *65*, 494–503. [CrossRef]
4. Métivier-Pignon, H.; Faur-Brasquet, C.; Le Cloirec, P. Adsorption of dyes onto acti v ated carbon cloths: Approach of adsorption mechanisms and coupling of ACC with ultrafiltration to treat coloured wastewaters. *Sep. Purif. Technol.* **2003**, *31*, 3–11. [CrossRef]
5. Eshaq, G.; Elmetwally, A.E. Bmim[OAc]-Cu_2O/g-C_3N_4 as a multi-function catalyst for sonophotocatalytic degradation of methylene blue. *Ultrason Sonochem.* **2019**, *53*, 99–109. [CrossRef]
6. Mosbah, A.; Chouchane, H.; Abdelwahed, S.; Redissi, A.; Hamdi, M.; Kouidhi, S.; Neifar, M.; Masmoudi, A.S.; Cherif, A.; Mnif, W. PeptiDesalination Fixing Industrial Textile Dyes: A New Biochemical Method in Wastewater Treatment. *J. Chem.* **2020**, *2019*, 5081807.
7. Ihsanullah, I.; Jamal, A.; Ilyas, M.; Zubair, M.; Khan, G.; Atieh, M.A. Bioremediation of dyes: Current status and prospects. *J. Water Process Eng.* **2020**, *38*, 101680. [CrossRef]
8. Kumar, P.S.; Varjani, S.J.; Suganya, S. Treatment of dye wastewater using an ultrasonic aided nanoparticle stacked activated carbon: Kinetic and isotherm modelling. *Bioresour. Technol.* **2018**, *250*, 716–722. [CrossRef]
9. Katheresan, V.; Kansedo, J.; Lau, S.Y. Efficiency of Various Recent Wastewater Dye Removal Methods: A Review. *J. Environ. Chem. Eng.* **2018**, *6*, 4676–4697. [CrossRef]
10. Kishor, R.; Purchase, D.; Saratale, G.D.; Saratale, R.G.; Ferreira, L.F.R.; Bilal, M.; Chandra, R.; Bharagava, R.N. Ecotoxicological and health concerns of persistent coloring pollutants of textile industry wastewater and treatment approaches for environmental safety. *J. Environ. Chem. Eng.* **2021**, *9*, 105012. [CrossRef]
11. Ippolito, A.; Fait, G. PesticiDesalination in surface waters: From edge-of-field to global modelling. *Curr. Opin. Environ. Sustain.* **2019**, *36*, 78–84. [CrossRef]
12. Lim, J.Y.; Mubarak, N.; Abdullah, E.C.; Nizamuddin, S.; Khalid, M.; Inamuddin. Recent trends in the synthesis of graphene and graphene oxide based nanomaterials for removal of heavy metals—A review. *J. Ind. Eng. Chem.* **2018**, *66*, 29–44. [CrossRef]
13. Uddin, M.K.; Rehman, Z. *Application of Nanomaterials in the Remediation of Textile Effluents from Aqueous Solutions*; John Wiley & Sons, Inc.: Hoboken, NJ, USA, 2018. [CrossRef]
14. Isloor, A.M.; Nayak, M.C.; Inamuddin; Prabhu, B.; Ismail, N.; Ismail, A.; Asiri, A.M. Novel polyphenylsulfone (PPSU)/nano tin oxide (SnO_2) mixed matrix ultrafiltration hollow fiber membranes: Fabrication, characterization and toxic dyes removal from aqueous solutions. *React. Funct. Polym.* **2019**, *139*, 170–180. [CrossRef]

15. Ahmed, M.; Mashkoor, F.; Nasar, A. Development, characterization, and utilization of magnetized orange peel waste as a novel adsorbent for the confiscation of crystal violet dye from aqueous solution. *Groundw. Sustain. Dev.* **2020**, *10*, 100322. [CrossRef]
16. Oz, M.; Lorke, D.E.; Hasan, M.; Petroianu, G.A. Cellular and MolecularActions of Methylene Blue in the Nervous System. *Med. Res. Rev.* **2011**, *31*, 93–117. [CrossRef]
17. Brooks, M.M. Methylene blue as antidote for cyanide and carbon monoxide poisoning. *J. Am. Med. Assoc.* **1933**, *100*, 59. [CrossRef]
18. Kishor, R.; Saratale, G.D.; Saratale, R.G.; Ferreira, L.F.R.; Bilal, M.; Iqbal, H.M.; Bharagava, R.N. Efficient degradation and detoxification of methylene blue dye by a newly isolated ligninolytic enzyme producing bacterium Bacillus albus MW407057. *Colloids Surfaces B Biointerfaces* **2021**, *206*, 111947. [CrossRef] [PubMed]
19. Sun, Q.; Saratale, R.G.; Saratale, G.D.; Kim, D.S. Pristine and Modified Radix Angelicae Dahuricae (Baizhi) Residue for the Adsorption of Methylene Blue from Aqueous Solution: A Comparative Study. *J. Mol. Liq.* **2018**, *265*, 36–45. [CrossRef]
20. Alharby, N.F.; Almutairi, R.S.; Mohamed, N.A. Adsorption Behavior of Methylene Blue Dye by Novel Crosslinked O-CM-Chitosan Hydrogel in Aqueous Solution: Kinetics, Isotherm and Thermodynamics. *Polymers* **2021**, *13*, 3659. [CrossRef]
21. Owamah, H.I.; Chukwujindu, I.S.; Asiagwu, A.K. Biosorptive capacity of yam peels waste for the removal of dye from aqueous solutions. *Civ. Environ. Res.* **2013**, *3*, 36–48.
22. Muhamad Ng, S.N.; Idrus, S.; Ahsan, A.; Tuan Mohd Marzuki, T.N.; Mahat, S.B. Treatment of wastewater from a food and beverage industry using conventional wastewater treatment integrated with membrane bioreactor system: A pilot-scale case study. *Membranes* **2021**, *11*, 456. [CrossRef] [PubMed]
23. Rafatullah, M.; Sulaiman, O.; Hashim, R.; Ahmad, A. Adsorption of methylene blue on low-cost adsorbents: A review. *J. Hazard. Mater.* **2010**, *177*, 70–80. [CrossRef] [PubMed]
24. Erabee, I.K.; Ahsan, A.; Jose, B.; Aziz, M.M.A.; Ng, A.W.M.; Idrus, S.; Daud, N.N.N. Adsorptive Treatment of Landfill Leachate using Activated Carbon Modified with Three Different Methods. *KSCE J. Civ. Eng.* **2018**, *22*, 1083–1095. [CrossRef]
25. Cao, J.; Zhang, J.; Zhu, Y.; Wang, S.; Wang, X.; Lv, K. Novel polymer material for efficiently removing methylene blue, Cu (II) and emulsified oil droplets from water simultaneously. *Polymers* **2018**, *10*, 1393. [CrossRef]
26. Vakili, M.; Rafatullah, M.; Salamatinia, B.; Abdullah, A.Z.; Ibrahim, M.H.; Tan, K.B.; Gholami, Z.; Amouzgar, P. Application of chitosan and its derivatives as adsorbents for dye removal from water and wastewater: A review. *Carbohydr. Polym.* **2014**, *113*, 115–130. [CrossRef]
27. Salleh, M.A.M.; Mahmoud, D.K.; Karim, W.A.W.A.; Idris, A. Cationic and anionic dye adsorption by agricultural solid wastes: A comprehensive review. *Desalination* **2011**, *280*, 1–13. [CrossRef]
28. Din, M.I.; Khalid, R.; Najeeb, J.; Hussain, Z. Fundamentals and photocatalysis of methylene blue dye using various nanocatalytic assemblies—A critical review. *J. Clean. Prod.* **2021**, *298*, 126567. [CrossRef]
29. Mashkoor, F.; Nasar, A. Magsorbents: Potential candidates in wastewater treatment technology—A review on the removal of methylene blue dye. *J. Magn. Magn. Mater.* **2020**, *500*, 166408. [CrossRef]
30. Santoso, E.; Ediati, R.; Kusumawati, Y.; Bahruji, H.; Sulistiono, D.O.; Prasetyoko, D. Review on recent advances of carbon based adsorbent for methylene blue removal from waste water. *Mater. Today Chem.* **2020**, *16*, 100233. [CrossRef]
31. Zhang, W.; Zhang, L.Y.; Zhao, X.J.; Zhou, Z. Citrus pectin derived porous carbons as a superior adsorbent toward removal of methylene blue. *J. Solid State Chem.* **2016**, *243*, 101–105. [CrossRef]
32. Gao, Y.; Xu, S.; Yue, Q.; Wu, Y.; Gao, B. Chemical preparation of crab shell-based activated carbon with superior adsorption performance for dye removal from wastewater. *J. Taiwan Inst. Chem. Eng.* **2016**, *61*, 327–335. [CrossRef]
33. Ghaedi, M.; Ghazanfarkhani, M.D.; Khodadoust, S.; Sohrabi, N.; Oftade, M. Acceleration of methylene blue adsorption onto activated carbon prepared from dross licorice by ultrasonic: Equilibrium, kinetic and thermodynamic studies. *J. Ind. Eng. Chem.* **2014**, *20*, 2548–2560. [CrossRef]
34. Zhao, M.; Peng, L. Adsorption of methylene blue from aqueous solutions by modified expanded graphite powder. *Desalination* **2009**, *249*, 331–336. [CrossRef]
35. El-Halwany, M. Study of adsorption isotherms and kinetic models for Methylene Blue adsorption on activated carbon developed from Egyptian rice hull (Part II). *Desalination* **2010**, *250*, 208–213. [CrossRef]
36. Heidarinejad, Z.; Rahmanian, O.; Fazlzadeh, M.; Heidari, M. Enhancement of methylene blue adsorption onto activated carbon prepared from Date Press Cake by low frequency ultrasound. *J. Mol. Liq.* **2018**, *264*, 591–599. [CrossRef]
37. Danish, M.; Ahmad, T.; Hashim, R. Comparison of surface properties of wood biomass activated carbons and their application against rhodamine B and methylene blue dye. *Surf. Interfaces* **2018**, *11*, 1–13. [CrossRef]
38. Karagöz, S.; Tay, T.; Ucar, S.; Erdem, M. Activated carbons from waste biomass by sulfuric acid activation and their use on methylene blue adsorption. *Bioresour. Technol.* **2008**, *99*, 6214–6222. [CrossRef]
39. Foo, K.Y.; Hameed, B.H. Adsorption characteristics of industrial solid waste derived activated carbon prepared by microwave heating for methylene blue. *Fuel Process. Technol.* **2012**, *99*, 103–109. [CrossRef]
40. Chiu, K.L.; Ng, D.H. Synthesis and characterization of cotton-made activated carbon fiber and its adsorption of methylene blue in water treatment. *Biomass Bioenergy* **2012**, *46*, 102–110. [CrossRef]
41. Cherifi, H.; Fatiha, B.; Salah, H. Kinetic studies on the adsorption of methylene blue onto vegetal fiber activated carbons. *Appl. Surf. Sci.* **2013**, *282*, 52–59. [CrossRef]
42. Li, Z.; Wang, G.; Zhai, K.; He, C.; Li, Q.; Guo, P. Methylene Blue Adsorption from Aqueous Solution by Loofah Sponge-Based Porous Carbons. *Colloids Surfaces A Physicochem. Eng. Asp.* **2018**, *538*, 28–35. [CrossRef]

43. Zaidi, A.A.; Feng, R.; Malik, A.; Khan, S.Z.; Shi, Y.; Bhutta, A.J.; Shah, A.H. Combining Microwave Pretreatment with Iron Oxide Nanoparticles Enhanced Biogas and Hydrogen Yield from Green Algae. *Processes* **2019**, *7*, 24. [CrossRef]
44. Liu, S.; Li, J.; Xu, S.; Wang, M.; Zhang, Y.; Xue, X. A Modified Method for Enhancing Adsorption Capability of Banana Pseudostem Biochar towards Methylene Blue at Low Temperature. *Bioresour. Technol.* **2019**, *282*, 48–55. [CrossRef] [PubMed]
45. Meili, L.; Lins, P.V.; Zanta, C.L.P.S.; Soletti, J.I.; Ribeiro, L.M.O.; Dornelas, C.B.; Silva, T.L.; Vieira, M.G.A. MgAl-LDH/Biochar composites for methylene blue removal by adsorption. *Appl. Clay Sci.* **2019**, *168*, 11–20. [CrossRef]
46. Tong, D.S.; Wu, C.W.; Adebajo, M.O.; Jin, G.C.; Yu, W.H.; Ji, S.F.; Zhou, C.H. Adsorption of methylene blue from aqueous solution onto porous cellulose- derived carbon/montmorillonite nanocomposites. *Appl. Clay Sci.* **2018**, *161*, 256–264. [CrossRef]
47. Li, X. Preparation and Adsorption Properties of Biochar/g-C3N4 Composites for Methylene Blue in Aqueous Solution. *J. Nanomater.* **2019**, *2019*, 2394184. [CrossRef]
48. Gan, Q.; Shi, W.; Xing, Y.; Hou, Y. A Polyoxoniobate/g-C_3N_4 Nanoporous Material with High Adsorption Capacity of Methylene Blue from Aqueous Solution. *Front. Chem.* **2018**, *6*, 7. [CrossRef]
49. Li, Y.; Du, Q.; Liu, T.; Peng, X.; Wang, J.; Sun, J.; Wang, Y.; Wu, S.; Wang, Z.; Xia, Y.; et al. Comparative study of methylene blue dye adsorption onto activated carbon, graphene oxide, and carbon nanotubes. *Chem. Eng. Res. Des.* **2013**, *91*, 361–368. [CrossRef]
50. Jin, Q.; Li, Y.; Yang, D.; Cui, J. Chitosan-derived three-dimensional porous carbon for fast removal of methylene blue from wastewater. *RSC Adv.* **2018**, *8*, 1255–1264. [CrossRef]
51. Narvekar, A.A.; Fernandes, J.B.; Tilve, S.G. Adsorption behavior of methylene blue on glycerol based carbon materials. *J. Environ. Chem. Eng.* **2018**, *6*, 1714–1725. [CrossRef]
52. Zhou, Q.; Jiang, X.; Guo, Y.; Zhang, G.; Jiang, W. An ultra-high surface area mesoporous carbon prepared by a novel MnO-templated method for highly effective adsorption of methylene blue. *Chemosphere* **2018**, *201*, 519–529. [CrossRef] [PubMed]
53. Chen, B.; Yang, Z.; Ma, G.; Kong, D.; Xiong, W.; Wang, J.; Zhu, Y.; Xia, Y. Heteroatom-doped porous carbons with enhanced carbon dioxide uptake and excellent methylene blue adsorption capacities. *Microporous Mesoporous Mater.* **2018**, *257*, 1–8. [CrossRef]
54. Duman, O.; Tunç, S.; Polat, T.G.; Bozoğlan, B.K. Synthesis of magnetic oxidized multiwalled carbon application in cationic Methylene Blue dye adsorption. *Carbohydr. Polym.* **2016**, *147*, 79–88. [CrossRef]
55. Wang, B.; Gao, B.; Zimmerman, A.R.; Lee, X. Impregnation of multiwall carbon nanotubes in alginate beads dramatically enhances their adsorptive ability to aqueous methylene blue. *Chem. Eng. Res. Des.* **2018**, *133*, 235–242. [CrossRef]
56. Manilo, M.; Lebovka, N.; Barany, S. Mechanism of Methylene Blue adsorption on hybrid laponite-multi-walled carbon nanotube particles. *J. Environ. Sci.* **2016**, *42*, 134–141. [CrossRef]
57. Zhang, Z.; Xu, X. Wrapping carbon nanotubes with poly (sodium 4-styrenesulfonate) for enhanced adsorption of methylene blue and its mechanism. *Chem. Eng. J.* **2014**, *256*, 85–92. [CrossRef]
58. Robati, D.; Mirza, B.; Ghazisaeidi, R.; Rajabi, M.; Moradi, O.; Tyagi, I.; Agarwal, S.; Gupta, V.K. Adsorption behavior of methylene blue dye on nanocomposite multi-walled carbon nanotube functionalized thiol (MWCNT-SH) as new adsorbent. *J. Mol. Liq.* **2016**, *216*, 830–835. [CrossRef]
59. Gong, J.; Liu, J.; Jiang, Z.; Wen, X.; Mijowska, E.; Tang, T.; Chen, X. A facile approach to prepare porous cup-stacked carbon nanotube with high performance in adsorption of methylene blue. *J. Colloid Interface Sci.* **2015**, *445*, 195–204. [CrossRef]
60. Sonune, A.; Ghate, R. Developments in wastewater treatment methods. *Desalination* **2004**, *167*, 55–63. [CrossRef]
61. Stasinakis, A.S.; Thomaidis, N.S.; Arvaniti, O.S.; Asimakopoulos, A.G.; Samaras, V.G.; Ajibola, A.; Mamais, D.; Lekkas, T.D. Contribution of primary and secondary treatment on the removal of benzothiazoles, benzotriazoles, endocrine disruptors, pharmaceuticals and perfluorinated compounds in a sewage treatment plant. *Sci. Total Environ.* **2013**, *463–464*, 1067–1075. [CrossRef]
62. Ye, S.; Yan, M.; Tan, X.; Liang, J.; Zeng, G.; Wu, H.; Song, B.; Zhou, C.; Yang, Y.; Wang, H.P.T. Facile assembled biochar-based nanocomposite with improved graphitization for efficient photocatalytic activity driven by visible light. *Appl. Catal. B Environ.* **2019**, *250*, 78–88. [CrossRef]
63. Foo, K.Y.; Hameed, B.H. An overview of dye removal via activated carbon adsorption process. *Desalin. Water Treat.* **2010**, *9*, 255–274. [CrossRef]
64. Saratale, R.G.; Sun, Q.; Munagapati, V.S.; Saratale, G.D.; Park, J.; Kim, D.S. The use of eggshell membrane for the treatment of dye-containing wastewater: Batch, kinetics and reusability studies. *Chemosphere* **2021**, *281*, 130777. [CrossRef] [PubMed]
65. Taylor, P.; Singh, K.; Arora, S. Removal of Synthetic Textile Dyes From Wastewaters: A Critical Review on Present Treatment Technologies. *Crit. Rev. Environ. Sci. Technol.* **2011**, *41*, 807–878. [CrossRef]
66. Cai, Z.; Sun, Y.; Liu, W.; Pan, F.; Sun, P.; Fu, J. An overview of nanomaterials applied for removing dyes from wastewater. *Environ. Sci. Pollut. Res.* **2017**, *24*, 15882–15904. [CrossRef]
67. Tara, N.; Siddiqui, S.I.; Rathi, G.; Chaudhry, S.A.; Inamuddin; Asiri, A.M. Nano-Engineered Adsorbent for the Removal of Dyes from Water: A Review. *Curr. Anal. Chem.* **2020**, *16*, 14–40. [CrossRef]
68. Uddin, M.K.; Baig, U. Synthesis of Co3O4 nanoparticles and their performance towards methyl orange dye removal: Characterisation, adsorption and response surface methodology. *J. Clean. Prod.* **2018**, *211*, 1141–1153. [CrossRef]
69. Gupta, V.K. Suhas Application of low-cost adsorbents for dye removal—A review. *J. Environ. Manag.* **2009**, *90*, 2313–2342. [CrossRef]
70. Ghoreishi, S.M.; Haghighi, R. Chemical catalytic reaction and biological oxidation for treatment of non-biodegradable textile effluent. *Chem. Eng. J.* **2003**, *95*, 163–169. [CrossRef]

71. Eulalia, M.; Tenorio, P. Valorización De Residuos Agroindustriales Como Adsorbentes Para La Remoción De Fármacos De Uso Común De Aguas Contami-Nadas. Ph.D. Thesis, Universidad de Zaragoza, Zaragoza, Spain, 2021.
72. Unuabonah, E.I.; Adie, G.U.; Onah, L.O.; Adeyemi, O.G. Multistage optimization of the adsorption of methylene blue dye onto defatted Carica papaya seeds. *Chem. Eng. J.* **2009**, *155*, 567–579. [CrossRef]
73. Crini, G. Non-conventional low-cost adsorbents for dye removal: A review. *Bioresour. Technol.* **2006**, *97*, 1061–1085. [CrossRef] [PubMed]
74. Rauf, M.A.; Shehadeh, I.; Ahmed, A.; Al-zamly, A. Removal of Methylene Blue from Aqueous Solution by Using Gypsum as a Low Cost Adsorbent. *World Acad. Sci. Eng. Technol.* **2009**, *55*, 608–613.
75. Shi, B.; Li, G.; Wang, D.; Feng, C.; Tang, H. Removal of direct dyes by coagulation: The performance of preformed polymeric aluminum species. *J. Hazard. Mater.* **2007**, *143*, 567–574. [CrossRef] [PubMed]
76. Zhou, Y.; Liang, Z.; Wang, Y. Decolorization and COD removal of secondary yeast wastewater effluents by coagulation using aluminum sulfate. *Desalination* **2008**, *225*, 301–311. [CrossRef]
77. Mishra, A.; Bajpai, M. The flocculation performance of Tamarindus mucilage in relation to removal of vat and direct dyes. *Bioresour. Technol.* **2006**, *97*, 1055–1059. [CrossRef]
78. Yue, Q.Y.; Gao, B.Y.; Wang, Y.; Zhang, H.; Sun, X.; Wang, S.G.; Gu, R.R. Synthesis of polyamine flocculants and their potential use in treating dye wastewater. *J. Hazard. Mater.* **2008**, *152*, 221–227. [CrossRef]
79. Lee, J.-W.; Choi, S.-P.; Thiruvenkatachari, R.; Shim, W.-G.; Moon, H. Evaluation of the performance of adsorption and coagulation processes for the maximum removal of reactive dyes. *Dye. Pigment.* **2006**, *69*, 196–203. [CrossRef]
80. Marzuki, T.N.T.M.; Idrus, S.; Musa, M.A.; Wahab, A.M.A.; Jamali, N.S.; Man, H.C.; Ng, S.N.M. Enhancement of Bioreactor Performance Using Acclimatised Seed Sludge in Anaerobic Treatment of Chicken Slaughterhouse Wastewater: Laboratory Achievement, Energy Recovery, and Its Commercial-Scale Potential. *Animals* **2021**, *11*, 3313. [CrossRef]
81. Robinson, T.; Mcmullan, G.; Marchant, R.; Nigam, P. Remediation of dyes in textile e, uent: A critical review on current treatment technologies with a proposed alternative. *Biores. Technol.* **2001**, *77*, 247–255. [CrossRef]
82. Crini, G.; Lichtfouse, E. Advantages and disadvantages of techniques used for wastewater treatment. *Environ. Chem. Lett.* **2019**, *17*, 145–155. [CrossRef]
83. Holkar, C.; Jadhav, A.; Pinjari, D.V.; Mahamuni, N.M.; Pandit, A.B. A critical review on textile wastewater treatments: Possible approaches. *J. Environ. Manag.* **2016**, *182*, 351–366. [CrossRef]
84. Collivignarelli, M.C.; Abbà, A.; Miino, M.C.; Damiani, S. Treatments for color removal from wastewater: State of the art. *J. Environ. Manag.* **2019**, *236*, 727–745. [CrossRef] [PubMed]
85. Ahmed, M.B.; Zhou, J.L.; Ngo, H.H.; Guo, W.; Thomaidis, N.S.; Xu, J. Progress in the biological and chemical treatment technologies for emerging contaminant removal from wastewater: A critical review. *J. Hazard. Mater.* **2017**, *323*, 274–298. [CrossRef] [PubMed]
86. Koyuncu, I.; Güney, K. Membrane-Based Treatment of Textile Industry Wastewaters. *Encycl. Membr. Sci. Technol.* **2013**, 1–12. [CrossRef]
87. Hasan, M.; Ahmad, A.L.; Hameed, B.H. Adsorption of reactive dye onto cross-linked chitosan/oil palm ash composite beads. *Chem. Eng. J.* **2008**, *136*, 164–172. [CrossRef]
88. Jain, R.; Gupta, V.K.; Sikarwar, S. Adsorption and desorption studies on hazardous dye Naphthol Yellow S. *J. Hazard. Mater.* **2010**, *182*, 749–756. [CrossRef]
89. Gupta, V.K.; Kumar, R.; Nayak, A.; Saleh, T.A.; Barakat, M.A. Adsorptive removal of dyes from aqueous solution onto carbon nanotubes: A review. *Adv. Colloid Interface Sci.* **2013**, *193–194*, 24–34. [CrossRef] [PubMed]
90. Vital, R.K.; Saibaba, K.V.N.; Shaik, K.B. Dye Removal by Adsorption: A Review. *J. Bioremediation Biodegrad.* **2016**, *7*, 371. [CrossRef]
91. Piaskowski, K.; Świderska-Dąbrowska, R.; Zarzycki, P.K. Dye removal from water and wastewater using various physical, chemical, and biological processes. *J. AOAC Int.* **2018**, *101*, 1371–1384. [CrossRef]
92. Hassan, M.M.; Carr, C.M. Biomass-derived porous carbonaceous materials and their composites as adsorbents for cationic and anionic dyes: A review. *Chemosphere* **2021**, *265*, 129087. [CrossRef]
93. Shu, J.; Cheng, S.; Xia, H.; Zhang, L.; Peng, J.; Li, C.; Zhang, S. Copper loaded on activated carbon as an efficient adsorbent for removal of methylene blue. *RSC Adv.* **2017**, *7*, 14395–14405. [CrossRef]
94. Uğurlu, M.; Gürses, A.; Açıkyıldız, M. Comparison of textile dyeing effluent adsorption on commercial activated carbon and activated carbon prepared from olive stone by $ZnCl_2$ activation. *Microporous Mesoporous Mater.* **2008**, *111*, 228–235. [CrossRef]
95. Zhou, Y.; Lu, J.; Zhou, Y.; Liu, Y. Recent advances for dyes removal using novel adsorbents: A review. *Environ. Pollut.* **2019**, *252*, 352–365. [CrossRef] [PubMed]
96. Wang, Y.; Zhang, Y.; Li, S.; Zhong, W.; Wei, W. Enhanced methylene blue adsorption onto activated reed-derived biochar by tannic acid. *J. Mol. Liq.* **2018**, *268*, 658–666. [CrossRef]
97. El-Shafey, E.; Ali, S.N.; Al-Busafi, S.; Al Lawati, H. Preparation and characterization of surface functionalized activated carbons from date palm leaflets and application for methylene blue removal. *J. Environ. Chem. Eng.* **2016**, *4*, 2713–2724. [CrossRef]
98. Que, W.; Jiang, L.; Wang, C.; Liu, Y.; Zeng, Z.; Wang, X.; Ning, Q.; Liu, S.; Zhang, P.; Liu, S. Influence of sodium dodecyl sulfate coating on adsorption of methylene blue by biochar from aqueous solution. *J. Environ. Sci.* **2017**, *70*, 166–174. [CrossRef]
99. Karaer, H.; Kaya, I. Synthesis, characterization of magnetic chitosan/active charcoal composite and using at the adsorption of methylene blue and reactive blue4. *Microporous Mesoporous Mater.* **2016**, *232*, 26–38. [CrossRef]

100. Vishnu, D.; Dhandapani, B.; Panchamoorthy, G.K.; Vo, D.-V.N.; Ramakrishnan, S.R. Comparison of surface-engineered superparamagnetic nanosorbents with low-cost adsorbents of cellulose, zeolites and biochar for the removal of organic and inorganic pollutants: A review. *Environ. Chem. Lett.* **2021**, *19*, 3181–3208. [CrossRef]
101. Björklund, K.; Li, L.Y. Adsorption of organic stormwater pollutants onto activated carbon from sewage sludge. *J. Environ. Manag.* **2017**, *197*, 490–497. [CrossRef]
102. Al-Degs, Y.S.; El-Barghouthi, M.I.; El-Sheikh, A.H.; Walker, G.M. Effect of solution pH, ionic strength, and temperature on adsorption behavior of reactive dyes on activated carbon. *Dye. Pigment.* **2008**, *77*, 16–23. [CrossRef]
103. Wahab, A.A.; Nuh, A.; Rasid, Z.A.; Abu, A.; Tanasta, Z.; Hassan, M.Z.; Mahmud, J. Tensile behaviours of single-walled carbon nanotubes: Dehnungsverhalten einwandiger Kohlenstoffnanoröhren. *Mater. Werkst.* **2018**, *49*, 467–471. [CrossRef]
104. Wang, D.; Wu, G.; Zhao, Y.; Cui, L.; Shin, C.H.; Ryu, M.H.; Cai, J. Study on the copper (II)-doped MIL-101 (Cr) and its performance in VOCs adsorption. *Environ. Sci. Pollut. Res.* **2018**, *25*, 28109–28119. [CrossRef] [PubMed]
105. Shen, Y.; Zhang, N.; Fu, Y. Synthesis of high-performance hierarchically porous carbons from rice husk for sorption of phenol in the gas phase. *J. Environ. Manag.* **2019**, *241*, 53–58. [CrossRef]
106. Xie, Z.Z.; Wang, L.; Cheng, G.; Shi, L.; Zhang, Y.B. Adsorption properties of regenerative materials for removal of low concentration of toluene. *J. Air Waste Manag. Assoc.* **2016**, *66*, 1224–1236. [CrossRef] [PubMed]
107. Tahriri, M.; Del Monico, M.; Moghanian, A.; Yaraki, M.T.; Torres, R.; Yadegari, A.; Tayebi, L. Graphene and its derivatives: Opportunities and challenges in dentistry. *Mater. Sci. Eng. C* **2019**, *102*, 171–185. [CrossRef] [PubMed]
108. Na, C.J.; Yoo, M.J.; Tsang, D.C.W.; Kim, H.W.; Kim, K.H. High-performance materials for effective sorptive removal of formaldehyde in air. *J. Hazard. Mater.* **2019**, *366*, 452–465. [CrossRef]
109. Mekki, A.; Boukoussa, B. Structural, textural and toluene adsorption properties of microporous–mesoporous zeolite omega synthesized by different methods. *J. Mater. Sci.* **2019**, *54*, 8096–8107. [CrossRef]
110. Zhu, L.; Meng, L.; Shi, J.; Li, J.; Zhang, X.; Feng, M. Metal-organic frameworks/carbon-based materials for environmental remediation: A state-of-the-art mini-review. *J. Environ. Manag.* **2019**, *232*, 964–977. [CrossRef]
111. Liu, C.; Cai, W.; Liu, L. Hydrothermal carbonization synthesis of Al-pillared montmorillonite@carbon composites as high performing toluene adsorbents. *Appl. Clay Sci.* **2018**, *162*, 113–120. [CrossRef]
112. Huang, W.; Xu, J.; Tang, B.; Wang, H.; Tan, X.; Lv, A. Adsorption performance of hydrophobically modified silica gel for the vapors of n-hexane and water. *Adsorpt. Sci. Technol.* **2018**, *36*, 888–903. [CrossRef]
113. Wu, Q.; Huang, W.; Wang, H.-J.; Pan, L.-L.; Zhang, C.-L.; Liu, X.-K. Reversely swellable porphyrin-linked microporous polyimide networks with super-adsorption for volatile organic compounds. *Chin. J. Polym. Sci.* **2015**, *33*, 1125–1132. [CrossRef]
114. Mart, J.; Otero, M. Dye adsorption by sewage sludge-based activated carbons in batch and fixed-bed systems. *Bioresour. Technol.* **2003**, *87*, 221–230.
115. Inorg, O. Microporous activated carbons from a bituminous coal. *Fuel* **1996**, *75*, 966–970.
116. Illa, M.J. Activated Carbons from Spanish Coals. 2. Chemical Activation. *Energy Fuels* **1996**, *10*, 1108–1114.
117. Karaca, S.; Gürses, A.; Bayrak, R. Effect of some pre-treatments on the adsorption of methylene blue by Balkaya lignite. *Energy Convers. Manag.* **2004**, *45*, 1693–1704. [CrossRef]
118. Aadil, A.; Shahzad, M.; Kashif, S.; Muhammad, M.; Rabia, A.; Saba, A. Comparative study of adsorptive removal of congo red and brilliant green dyes from water using peanut shell. *Middle East J. Sci. Res.* **2012**, *11*, 828–832.
119. Ahmad, R.; Kumar, R. Adsorptive removal of congo red dye from aqueous solution using bael shell carbon. *Appl. Surf. Sci.* **2010**, *257*, 1628–1633. [CrossRef]
120. Dawood, S.; Sen, T.K. Removal of anionic dye Congo red from aqueous solution by raw pine and acid-treated pine cone powder as adsorbent: Equilibrium, thermodynamic, kinetics, mechanism and process design. *Water Res.* **2012**, *46*, 1933–1946. [CrossRef]
121. Vaishnav, V.; Chandra, S.; Daga, K. Adsorption Studies of Zn (II) Ions from wastewater using Calotropis procera as an adsorbent. *Res. J. Recent Sci.* **2012**, *1*, 160–165.
122. Gopalakrishnan, S.; Kannadasan, T.; Velmurugan, S.; Muthu, S.; Vinoth Kumar, P. Biosorption of Chromium (VI) from Industrial Effluent using Neem Leaf Adsorbent. *Res. J. Chem. Sci.* **2013**, *3*, 48–53.
123. Bernard, E.; Jimoh, A.; Odigure, J.O. Heavy metals removal from industrial wastewater by activated carbon prepared from coconut shell Heavy Metals Removal from Industrial Wastewater by Activated Carbon Prepared from Coconut Shell. *Res. J. Chem. Sci.* **2013**, *2231*, 606X.
124. Tiwari, A.; Kathane, P. Superparamagnetic PVA-Alginate Microspheres as Adsorbent for Cu^{2+} ions Removal from Aqueous Systems. *Int. Res. J. Environ. Sci.* **2013**, *2*, 44–53.
125. Gupta, G.; Khan, J.; Singh, N.K. Application and efficacy of low-cost adsorbents for metal removal from contaminated water: A review. *Mater. Today Proc.* **2021**, *43*, 2958–2964. [CrossRef]
126. Shichi, T.; Takagi, K. Clay minerals as photochemical reaction fields. *J. Photochem. Photobiol. C Photochem. Rev.* **2000**, *1*, 113–130. [CrossRef]
127. Doğan, M.; Abak, H.; Alkan, M. Adsorption of methylene blue onto hazelnut shell: Kinetics, mechanism and activation parameters. *J. Hazard. Mater.* **2009**, *164*, 172–181. [CrossRef]
128. Krysztafkiewicz, A.; Binkowski, S.; Jesionowski, T. Adsorption of dyes on a silica surface. *Appl. Surf. Sci.* **2002**, *199*, 31–39. [CrossRef]

129. Woolard, C.D.; Strong, P.J.; Erasmus, C.R. Evaluation of the use of modified coal ash as a potential sorbent for organic waste streams. *Appl. Geochem.* **2002**, *17*, 1159–1164. [CrossRef]
130. Ahmed, M.N.; Ram, R.N. Removal of basic dye from waste-water using silica as adsorbent. *Environ. Pollut.* **1992**, *77*, 79–86. [CrossRef]
131. Kumar, K.; Kirnaji, N.P.; Bagewadi, C.S. Decohering Environment And Coupled Quantum States And Internal Resonance In Coupled Spin Systems And The Conflict Between Quantum Gate Operation And Decoupling A Cormorant-Barnacle Model. *Adv. Phys. Theor. Appl.* **2002**, *6*, 24–31.
132. Ozdemir, O.; Armagan, B.; Turan, M.; Çelik, M.S. Comparison of the adsorption characteristics of azo-reactive dyes on mezoporous minerals. *Dye. Pigment.* **2004**, *62*, 49–60. [CrossRef]
133. Calzaferri, G.; Brühwiler, D.; Megelski, S.; Pfenniger, M.; Pauchard, M.; Hennessy, B.; Maas, H.; Devaux, A.; Graf, U. Playing with dye molecules at the inner and outer surface of zeolite L. *Solid State Sci.* **2000**, *2*, 421–447. [CrossRef]
134. Batool, A.; Valiyaveettil, S. Chemical transformation of soya waste into stable adsorbent for enhanced removal of methylene blue and neutral red from water. *J. Environ. Chem. Eng.* **2021**, *9*, 104902. [CrossRef]
135. Siddiqui, S.I.; Fatima, B.; Tara, N.; Rathi, G.; Chaudhry, S.A. Recent advances in remediation of synthetic dyes from wastewaters using sustainable and low-cost adsorbents. In *The Impact and Prospects of Green Chemistry for Textile Technology*; Elsevier: Amsterdam, The Netherlands, 2018; pp. 471–507.
136. Almeida, E.J.R.; Corso, C.R. Decolorization and removal of toxicity of textile azo dyes using fungal biomass pelletized. *Int. J. Environ. Sci. Technol.* **2019**, *16*, 1319–1328. [CrossRef]
137. Nasar, A.; Mashkoor, F. Application of polyaniline-based adsorbents for dye removal from water and wastewater—a review. *Environ. Sci. Pollut. Res.* **2019**, *26*, 5333–5356. [CrossRef]
138. Ni Law, X.; Cheah, W.Y.; Chew, K.W.; Ibrahim, M.F.; Park, Y.-K.; Ho, S.-H.; Show, P.L. Microalgal-based biochar in wastewater remediation: Its synthesis, characterization and applications. *Environ. Res.* **2022**, *204*, 111966. [CrossRef]
139. Karthik, V.; Saravanan, K.; Sivarajasekar, N.; Suriyanarayanan, N. Bioremediation of dye bearing effluents using microbial biomass. *Ecol. Environ. Conserv.* **2016**, *22*, S423–S434.
140. Roy, U.; Manna, S.; Sengupta, S.; Das, P.; Datta, S.; Mukhopadhyay, A.; Bhowal, A. Dye Removal Using Microbial Biosorbents. In *Green Adsorbents for Pollutant Removal*; Springer: Cham, Switzerland, 2018; pp. 253–280. [CrossRef]
141. Pearce, C.I.; Lloyd, J.R.; Guthrie, J.T. The removal of colour from textile wastewater using whole bacterial cells: A review. *Dye. Pigment.* **2003**, *58*, 179–196. [CrossRef]
142. Sarvajith, M.; Reddy, G.K.K.; Nancharaiah, Y.V. Textile dye biodecolourization and ammonium removal over nitrite in aerobic granular sludge sequencing batch reactors. *J. Hazard. Mater.* **2018**, *342*, 536–543. [CrossRef]
143. Singh, N.B.; Nagpal, G.; Agrawal, S. Rachna Water purification by using Adsorbents: A Review. *Environ. Technol. Innov.* **2018**, *11*, 187–240. [CrossRef]
144. Liu, Y.; Shao, Z.; Reng, X.; Zhou, J.; Qin, W. Dye-decolorization of a newly isolated strain Bacillus amyloliquefaciens W36. *World J. Microbiol. Biotechnol.* **2021**, *37*, 1–11. [CrossRef]
145. Kim, S.Y.; Jin, M.R.; Chung, C.H.; Yun, Y.S.; Jahng, K.Y.; Yu, K.Y. Biosorption of cationic basic dye and cadmium by the novel biosorbent Bacillus catenulatus JB-022 strain. *J. Biosci. Bioeng.* **2015**, *119*, 433–439. [CrossRef] [PubMed]
146. Kousha, M.; Tavakoli, S.; Daneshvar, E.; Vazirzadeh, A.; Bhatnagar, A. Central composite design optimization of Acid Blue 25 dye biosorption using shrimp shell biomass. *J. Mol. Liq.* **2015**, *207*, 266–273. [CrossRef]
147. Upendar, G.; Dutta, S.; Chakraborty, J.; Bhattacharyya, P. Removal of methylene blue dye using immobilized bacillus subtilis in batch & column reactor. *Mater. Today Proc.* **2016**, *3*, 3467–3472. [CrossRef]
148. Ahmed, H.A.B.; Ebrahim, S.E. Removal of methylene blue and congo red dyes by pretreated fungus biomass-equilibrium and kinetic studies. *J. Adv. Res. Fluid Mech. Therm. Sci.* **2020**, *66*, 84–100.
149. Argumedo-Delira, R.; Gómez-Martínez, M.J.; Uribe-Kaffure, R. Trichoderma biomass as an alternative for removal of congo red and malachite green industrial dyes. *Appl. Sci.* **2021**, *11*, 448. [CrossRef]
150. Nouri, H.; Azin, E.; Kamyabi, A.; Moghimi, H. Biosorption performance and cell surface properties of a fungal-based sorbent in azo dye removal coupled with textile wastewater. *Int. J. Environ. Sci. Technol.* **2021**, *18*, 2545–2558. [CrossRef]
151. Gül, Ü.D. Treatment of dyeing wastewater including reactive dyes (Reactive Red RB, Reactive Black B, Remazol Blue) and Methylene Blue by fungal biomass. *Water SA* **2013**, *39*, 593–598. [CrossRef]
152. Dai, Y.; Sun, Q.; Wang, W.; Lu, L.; Liu, M.; Li, J.; Yang, S.; Sun, Y.; Zhang, K.; Xu, J.; et al. Utilizations of agricultural waste as adsorbent for the removal of contaminants: A review. *Chemosphere* **2018**, *211*, 235–253. [CrossRef]
153. Azam, R.; Kothari, R.; Singh, H.M.; Ahmad, S.; Ashokkumar, V.; Tyagi, V. Production of algal biomass for its biochemical profile using slaughterhouse wastewater for treatment under axenic conditions. *Bioresour. Technol.* **2020**, *306*, 123116. [CrossRef]
154. Essekri, A.; Hsini, A.; Naciri, Y.; Laabd, M.; Ajmal, Z.; El Ouardi, M.; Addi, A.A.; Albourine, A. Novel citric acid-functionalized brown algae with a high removal efficiency of crystal violet dye from colored wastewaters: Insights into equilibrium, adsorption mechanism, and reusability. *Int. J. Phytoremed.* **2021**, *23*, 336–346. [CrossRef]
155. Angelova, R.; Baldikova, E.; Pospiskova, K.; Maderova, Z.; Safarikova, M.; Safarik, I. Magnetically modified Sargassum horneri biomass as an adsorbent for organic dye removal. *J. Clean. Prod.* **2016**, *137*, 189–194. [CrossRef]
156. Sen, S.K.; Raut, S.; Bandyopadhyay, P.; Raut, S. Fungal decolouration and degradation of azo dyes: A review. *Fungal Biol. Rev.* **2016**, *30*, 112–133. [CrossRef]

157. Singh, S.; Kumar, V.; Datta, S.; Dhanjal, D.S.; Sharma, K.; Samuel, J.; Singh, J. Current advancement and future prospect of biosorbents for bioremediation. *Sci. Total Environ.* **2020**, *709*, 135895. [CrossRef]
158. Al-Najar, J.A.; Lutfee, T.; Alwan, N.F. The action of yeast as an adsorbent in wastewater treatment: A Brief Review. In Proceedings of the Fifth International Scientific Conference on Environment and Sustainable Development, Baghdad, Iraq, 1–2 June 2021; Volume 779, p. 012054. [CrossRef]
159. Ruscasso, F.; Bezus, B.; Garmendia, G.; Vero, S.; Curutchet, G.; Cavello, I.; Cavalitto, S. Debaryomyces hansenii F39A as biosorbent for textile dye removal. *Rev. Argent. Microbiol.* **2021**, *53*, 257–265. [CrossRef]
160. Semião, M.A.; Haminiuk, C.W.I.; Maciel, G.M. Residual diatomaceous earth as a potential and cost effective biosorbent of the azo textile dye Reactive Blue 160. *J. Environ. Chem. Eng.* **2020**, *8*, 103617. [CrossRef]
161. Lin, H.-H.; Inbaraj, B.S.; Kao, T.-H. Removal Potential of Basic Dyes and Lead from Water by Brewer's Yeast Biomass. *J. Am. Soc. Brew. Chem.* **2019**, *77*, 30–39. [CrossRef]
162. Mo, J.; Yang, Q.; Zhang, N.; Zhang, W.; Zheng, Y.; Zhang, Z. A review on agro-industrial waste (AIW) derived adsorbents for water and wastewater treatment. *J. Environ. Manag.* **2018**, *227*, 395–405. [CrossRef]
163. Deniz, F.; Kepekci, R.A. Bioremoval of Malachite green from water sample by forestry waste mixture as potential biosorbent. *Microchem. J.* **2017**, *132*, 172–178. [CrossRef]
164. Boukhlifi, F.; Chraibi, S.; Alami, M. Evaluation of the adsorption kinetics and equilibrium. *J. Environ. Earth Sci.* **2013**, *3*, 181–190.
165. Ahmad, A.A.; Hameed, B.H.; Aziz, N. Adsorption of direct dye on palm ash: Kinetic and equilibrium modeling. *J. Hazard. Mater.* **2007**, *141*, 70–76. [CrossRef]
166. Acemioğlu, B. Adsorption of Congo red from aqueous solution onto calcium-rich fly ash. *J. Colloid Interface Sci.* **2004**, *274*, 371–379. [CrossRef]
167. Rastogi, K.; Sahu, J.N.; Meikap, B.C.; Biswas, M.N. Removal of methylene blue from wastewater using fly ash as an adsorbent by hydrocyclone. *J. Hazard. Mater.* **2008**, *158*, 531–540. [CrossRef] [PubMed]
168. Saha, P.; Datta, S. Assessment on thermodynamics and kinetics parameters on reduction of methylene blue dye using flyash. *Desalin. Water Treat.* **2009**, *12*, 219–228. [CrossRef]
169. Santos, S.C.R.; Vilar, V.J.P.; Boaventura, R.A.R. Waste metal hydroxide sludge as adsorbent for a reactive dye. *J. Hazard. Mater.* **2008**, *153*, 999–1008. [CrossRef]
170. Wang, S.; Boyjoo, Y.; Choueib, A.; Zhu, Z.H. Removal of dyes from aqueous solution using fly ash and red mud. *Water Res.* **2005**, *39*, 129–138. [CrossRef]
171. Tor, A.; Cengeloglu, Y. Removal of congo red from aqueous solution by adsorption onto acid activated red mud. *J. Hazard. Mater.* **2006**, *138*, 409–415. [CrossRef]
172. Gupta, V.K.; Ali, I.; Saini, V.K. Removal of chlorophenols from wastewater using red mud: An aluminum industry waste. *Environ. Sci. Technol.* **2004**, *38*, 4012–4018. [CrossRef]
173. Benabbas, K.; Zabat, N.; Hocini, I. Study of the chemical pretreatment of a nonconventional low-cost biosorbent (*Callitriche obtusangula*) for removing an anionic dye from aqueous solution. *Euro-Mediterr. J. Environ. Integr.* **2021**, *6*, 54. [CrossRef]
174. Kumar, A.; Singh, R.; Kumar, S.K.U.S.; Charaya, M.U. Biosorption: The Removal of Toxic Dyes From Industrial Effluent Using Phytobiomass- a Review. *Plant Arch.* **2021**, *21*, 1320–1325. [CrossRef]
175. Piccin, J.S.; Cadaval, T.R.S.A.; De Pinto, L.A.A.; Dotto, G.L. Adsorption isotherms in liquid phase: Experimental, modeling, and interpretations. In *Adsorption Processes for Water Treatment and Purification*; Bonilla-Petriciolet, A., Mendoza-Castillo, D., Reynel, Á., Vila, H., Eds.; Springer: Cham, Switzerland, 2017; ISBN 9783319581361.
176. Wang, H.; Shen, H.; Shen, C.; Li, Y.N.; Ying, Z.; Duan, Y. Kinetics and Mechanism Study of Mercury Adsorption by Activated Carbon in Wet Oxy-Fuel Conditions. *Energy Fuels* **2019**, *33*, 1344–1353. [CrossRef]
177. Liu, Z.; Yang, Z.; Chen, S.; Liu, Y.; Sheng, L.; Tian, Y.; Huang, D.; Xu, H. A smart reaction-based fluorescence probe for ratio detection of hydrazine and its application in living cells. *Microchem. J.* **2020**, *156*, 104809. [CrossRef]
178. Mohammed, M.A.; Shitu, A.; Ibrahim, A. Removal of methylene blue using low cost adsorbent: A review. *Res. J. Chem. Sci.* **2014**, *4*, 91–102.
179. Oyelude, E.O.; Appiah-takyi, F. Removal of methylene blue from aqueous solution using alkali-modified malted sorghum mash. *Turk. J. Eng. Environ. Sci.* **2012**, *36*, 161–169. [CrossRef]
180. Rehman, M.S.U.; Kim, I.; Han, J.-I. Adsorption of methylene blue dye from aqueous solution by sugar extracted spent rice biomass. *Carbohydr. Polym.* **2012**, *90*, 1314–1322. [CrossRef]
181. Soni, M.; Sharma, A.K.; Srivastava, J.K.; Yadav, J.S. Adsorptive removal of methylene blue dye from an aqueous solution using water hyacinth root powder as a low cost adsorbent. *Int. J. Chem. Sci. Appl.* **2012**, *3*, 338–345.
182. Ahmed, M.J.; Dhedan, S.K. Equilibrium isotherms and kinetics modeling of methylene blue adsorption on agricultural wastes-based activated carbons. *Fluid Phase Equilibr.* **2012**, *317*, 9–14. [CrossRef]
183. Foo, K.Y.; Hameed, B.H. Dynamic adsorption behavior of methylene blue onto oil palm shell granular activated carbon prepared by microwave heating. *Chem. Eng. J.* **2012**, *203*, 81–87. [CrossRef]
184. Feng, Y.; Zhou, H.; Liu, G.; Qiao, J.; Wang, J.; Lu, H.; Yang, L. Methylene blue adsorption onto swede rape straw (*Brassica napus* L.) modified by tartaric acid: Equilibrium, kinetic and adsorption mechanisms. *Bioresour. Technol.* **2012**, *125*, 138–144. [CrossRef]
185. Liu, Y.; Zhao, X.; Li, J.; Ma, D.; Han, R. Characterization of bio-char from pyrolysis of wheat straw and its evaluation on methylene blue adsorption. *Desalination Water Treat.* **2012**, *46*, 115–123. [CrossRef]

186. Al-Aoh, H.A.; Yahya, R.; Maah, M.J.; Bin Abas, M.R. Adsorption of methylene blue on activated carbon fiber prepared from coconut husk: Isotherm, kinetics and thermodynamics studies. *Desalin. Water Treat.* **2013**, *52*, 6720–6732. [CrossRef]
187. Krishni, R.; Foo, K.Y.; Hameed, B. Adsorption of methylene blue onto papaya leaves: Comparison of linear and nonlinear isotherm analysis. *Desalin. Water Treat.* **2013**, *52*, 6712–6719. [CrossRef]
188. Toumi, L.B.; Hamdi, L.; Salem, Z.; Allia, K. Batch adsorption of methylene blue from aqueous solutions by untreated Alfa grass. *Desalin. Water Treat.* **2013**, *53*, 806–817. [CrossRef]
189. Patel, H.; Vashi, R.T. A Comparison Study of Removal of Methylene Blue Dye by Adsorption on Neem Leaf Powder (Nlp) and Activated Nlp. *J. Environ. Eng. Landsc. Manag.* **2012**, *21*, 36–41. [CrossRef]
190. Khodaie, M.; Ghasemi, N.; Moradi, B.; Rahimi, M. Removal of Methylene Blue from Wastewater by Adsorption onto $ZnCl_2$ Activated Corn Husk Carbon Equilibrium Studies. *J. Chem.* **2013**, *2013*, 383985. [CrossRef]
191. Kini Srinivas, M.; Saidutta, M.B.; Murty, V.R.C.; Kadoli Sandip, V. Adsorption of basic Dye from Aqueous Solution using HCl Treated Saw Dust (Lagerstroemia microcarpa): Kinetic, Modeling of Equilibrium, thermodynamic. *Int. Res. J. Environ. Sci.* **2013**, *2*, 6–16.
192. Lakshmipathy, R.; Sarada, N. Adsorptive removal of basic cationic dyes from aqueous solution by chemically protonated watermelon (*Citrullus lanatus*) rind biomass. *Desalin. Water Treat.* **2014**, *52*, 6175–6184. [CrossRef]
193. Khoo, E.-C.; Ong, S.-T.; Hung, Y.-T.; Ha, S.-T. Removal of basic dyes from aqueous solution using sugarcane bagasse: Optimization by Plackett—Burman and Response Surface Methodology. *Desalin. Water Treat.* **2013**, *51*, 7109–7119. [CrossRef]
194. Cottet, L.; Almeida, C.A.P.; Naidek, N.; Viante, M.F.; Lopes, M.C.; Debacher, N.A. Adsorption characteristics of montmorillonite clay modified with iron oxide with respect to methylene blue in aqueous media. *Appl. Clay Sci.* **2014**, *95*, 25–31. [CrossRef]
195. Bayat, M.; Javanbakht, V.; Esmaili, J. Synthesis of zeolite/nickel ferrite/sodium alginate bionanocomposite via a co-precipitation technique for efficient removal of water-soluble methylene blue dye. *Int. J. Biol. Macromol.* **2018**, *116*, 607–619. [CrossRef]
196. Wang, P.; Cao, M.; Wang, C.; Ao, Y.; Hou, J.; Qian, J. Kinetics and thermodynamics of adsorption of methylene blue by a magnetic graphene-carbon nanotube composite. *Appl. Surf. Sci.* **2014**, *290*, 116–124. [CrossRef]
197. Mahamadi, C.; Mawere, E. High adsorption of dyes by water hyacinth fixed on alginate. *Environ. Chem. Lett.* **2013**, *12*, 313–320. [CrossRef]
198. Amuda, O.S.; Olayiwola, A.O.; Alade, A.O.; Farombi, A.G.; Adebisi, S.A. Adsorption of methylene blue from aqueous solution using steam-activated carbon produced from Lantana camara stem. *J. Environ. Prot.* **2014**, *5*, 1352–1363. [CrossRef]
199. Zhou, L.; Huang, J.; He, B.; Zhang, F.; Li, H. Peach gum for efficient removal of methylene blue and methyl violet dyes from aqueous solution. *Carbohydr. Polym.* **2014**, *101*, 574–581. [CrossRef] [PubMed]
200. Banerjee, S.; Sharma, G.C.; Chattopadhyaya, M.; Sharma, Y.C. Kinetic and equilibrium modeling for the adsorptive removal of methylene blue from aqueous solutions on of activated fly ash (AFSH). *J. Environ. Chem. Eng.* **2014**, *2*, 1870–1880. [CrossRef]
201. Mubarak, N.M.; Fo, Y.T.; Al-Salim, H.S.; Sahu, J.N.; Abdullah, E.C.; Nizamuddin, S.; Jayakumar, N.S.; Ganesan, P. Removal of Methylene Blue and Orange-G from Waste Water Using Magnetic Biochar. *Int. J. Nanosci.* **2015**, *14*, 1–13. [CrossRef]
202. Ma, H.; Li, J.; Liu, W.; Miao, M.; Cheng, B.; Zhu, S. Novel synthesis of a versatile magnetic adsorbent derived from corncob for dye removal. *Bioresour. Technol.* **2015**, *190*, 13–20. [CrossRef]
203. Lou, Z.; Zhou, Z.; Zhang, W.; Zhang, X.; Hu, X.; Liu, P.; Zhang, H. Magnetized bentonite by Fe_3O_4 nanoparticles treated as adsorbent for methylene blue removal from aqueous solution: Synthesis, characterization, mechanism, kinetics and regeneration. *J. Taiwan Inst. Chem. Eng.* **2015**, *49*, 199–205. [CrossRef]
204. Zeng, L.; Xie, M.; Zhang, Q.; Kang, Y.; Guo, X.; Xiao, H.; Peng, Y.; Luo, J. Chitosan/organic rectorite composite for the magnetic uptake of methylene blue and methyl orange. *Carbohydr. Polym.* **2015**, *123*, 89–98. [CrossRef]
205. Wang, W.; Ding, Z.; Cai, M.; Jian, H.; Zeng, Z.; Li, F.; Liu, J.P. Synthesis and high-efficiency methylene blue adsorption of magnetic $PAA/MnFe_2O_4$ nanocomposites. *Appl. Surf. Sci.* **2015**, *346*, 348–353. [CrossRef]
206. Sun, X.; Liu, B.; Jing, Z.; Wang, H. Preparation and adsorption property of xylan/poly (acrylic acid) magnetic nanocomposite hydrogel adsorbent. *Carbohydr. Polym.* **2015**, *118*, 16–23. [CrossRef]
207. Yu, B.; Zhang, X.; Xie, J.; Wu, R.; Liu, X.; Li, H.; Chen, F.; Yang, H.; Ming, Z.; Yang, S. Magnetic graphene sponge for the removal of methylene blue. *Appl. Surf. Sci.* **2015**, *351*, 765–771. [CrossRef]
208. Cui, L.; Guo, X.; Wei, Q.; Wang, Y.; Gao, L.; Yan, L.; Yan, T.; Du, B. Removal of mercury and methylene blue from aqueous solution by xanthate functionalized magnetic graphene oxide: Sorption kinetic and uptake mechanism. *J. Colloid Interface Sci.* **2015**, *439*, 112–120. [CrossRef] [PubMed]
209. Cui, L.; Wang, Y.; Hu, L.; Gao, L.; Du, B.; Wei, Q. Mechanism of Pb(ii) and methylene blue adsorption onto magnetic carbonate hydroxyapatite/graphene oxide. *RSC Adv.* **2015**, *5*, 9759–9770. [CrossRef]
210. Wong, K.T.; Eu, N.C.; Ibrahim, S.; Kim, H.; Yoon, Y.; Jang, M. Recyclable magnetite-loaded palm shell-waste based activated carbon for the effective removal of methylene blue from aqueous solution. *J. Clean. Prod.* **2016**, *115*, 337–342. [CrossRef]
211. Chang, J.; Ma, J.; Ma, Q.; Zhang, D.; Qiao, N.; Hu, M.; Ma, H. Adsorption of methylene blue onto Fe_3O_4/activated montmorillonite nanocomposite. *Appl. Clay Sci.* **2016**, *115*, 337–342. [CrossRef]
212. Mu, B.; Tang, J.; Zhang, L.; Wang, A. Preparation, characterization and application on dye adsorption of a well-defined two-dimensional superparamagnetic clay/polyaniline/Fe_3O_4 nanocomposite. *Appl. Clay Sci.* **2016**, *132–133*, 7–16. [CrossRef]
213. Shao, Y.; Zhou, L.; Bao, C.; Ma, J.; Liu, M.; Wang, F. *Magnetic Responsive Metal-Organic Frameworks Nanosphere with Core-Shell Structure for Highly Efficient Removal of Methylene Blue*; Elsevier B.V.: Amsterdam, The Netherlands, 2015; ISBN 8693189123.

214. Online, V.A.; Deng, J.; Zhou, X.; Bai, R. Removal of mercury (II) and methylene blue from a wastewater environment with magnetic graphene oxide: Adsorption kinetics, isotherms and mechanism. *Rsc Adv.* **2016**, *6*, 82523–82536. [CrossRef]
215. Ge, H.; Wang, C.; Liu, S.; Huang, Z. Synthesis of citric acid functionalized magnetic graphene oxide coated corn straw for methylene blue adsorption. *Bioresour. Technol.* **2016**, *221*, 419–429. [CrossRef]
216. Li, L.; Liu, F.; Duan, H.; Wang, X.; Li, J.; Wang, Y.; Luo, C. The preparation of novel adsorbent materials with efficient adsorption performance for both chromium and methylene blue. *Colloids Surfaces B Biointerfaces* **2016**, *141*, 253–259. [CrossRef]
217. Altıntıg, E.; Altundag, H.; Tuzen, M.; Sarı, A. Effective removal of methylene blue from aqueous solutions using magnetic loaded activated carbon as novel adsorbent. *Chem. Eng. Res. Des.* **2017**, *122*, 151–163. [CrossRef]
218. Abuzerr, S.; Darwish, M.; Mahvi, A.H. Simultaneous removal of cationic methylene blue and anionic reactive red 198 dyes using magnetic activated carbon nanoparticles: Equilibrium, and kinetics analysis. *Water Sci. Technol.* **2018**, *2017*, 534–545. [CrossRef] [PubMed]
219. Wan, X.; Zhan, Y.; Long, Z.; Zeng, G.; He, Y.; Zhan, Y.; Long, Z.; Zeng, G.; He, Y. Core @ double-shell structured magnetic halloysite nanotube nano-hybrid as efficient recyclable adsorbent for methylene blue removal. *Chem. Eng. J.* **2017**, *330*, 491–504. [CrossRef]
220. Mahdavinia, G.R.; Soleymani, M.; Sabzi, M.; Azimi, H.; Atlasi, Z. Novel magnetic polyvinyl alcohol/laponite RD nanocomposite hydrogels for efficient removal of methylene blue. *J. Environ. Chem. Eng.* **2017**, *5*, 2617–2630. [CrossRef]
221. Baruah, S.; Devi, A.; Bhattacharyya, K.G.; Sarma, A. Developing a biosorbent from Aegle Marmelos leaves for removal of methylene blue from water. *Int. J. Environ. Sci. Technol.* **2016**, *14*, 341–352. [CrossRef]
222. Kuppusamy, S.; Kadiyala, V.; Palanisami, T.; Yong, B.L.; Ravi, N.; Mallavarapu, M. Quercus robur acorn peel as a novel coagulating adsorbent for cationic dye removal from aquatic ecosystems. *Ecol. Eng.* **2017**, *101*, 3–8. [CrossRef]
223. Maingi, F.M.; Mbuvi, H.M.; Ng'ang'a, M.M.; Mwangi, H. Adsorption Kinetics and Isotherms of Methylene blue by Geopolymers Derived from Adsorption Kinetics and Isotherms of Methylene Blue by Geopolymers Derived from Common Clay and Rice Husk. *Phys. Chem.* **2017**, *7*, 87–97. [CrossRef]
224. Meili, L.; Da Silva, T.S.; Henrique, D.C.; Soletti, J.I.; de Carvalho, S.H.V.; Fonseca, E.J.D.S.; de Almeida, A.R.F.; Dotto, G.L. Ouricuri (*Syagrus coronata*) fiber: A novel biosorbent to remove methylene blue from aqueous solutions. *Water Sci. Technol.* **2017**, *75*, 106–114. [CrossRef]
225. Zirak, M.; Abdollahiyan, A.; Eftekhari-Sis, B.; Saraei, M. Carboxymethyl cellulose coated $Fe_3O_4@SiO_2$ core—Shell magnetic nanoparticles for methylene blue removal: Equilibrium, kinetic, and thermodynamic studies. *Cellulose* **2017**, *25*, 503–515. [CrossRef]
226. Alijani, H.; Beyki, M.H.; Kaveh, R.; Fazli, Y. Rapid biosorption of methylene blue by in situ cellulose-grafted poly 4-hydroxybenzoic acid magnetic nanohybrid: Multivariate optimization and isotherm study. *Polym. Bull.* **2017**, *75*, 2167–2180. [CrossRef]
227. Cojocaru, C.; Humelnicu, A.C.; Samoila, P.; Pascariu, P.; Harabagiu, V. Optimized formulation of $NiFe_2O_4$@Ca-alginate composite as a selective and magnetic adsorbent for cationic dyes: Experimental and modeling study. *React. Funct. Polym.* **2018**, *125*, 57–69. [CrossRef]
228. Talbot, D.; Queiros Campos, J.; Checa-Fernandez, B.L.; Marins, J.A.; Lomenech, C.; Hurel, C.; Godeau, G.D.; Raboisson-Michel, M.; Verger-Dubois, G.; Obeid, L.; et al. Adsorption of Organic Dyes on Magnetic Iron Oxide Nanoparticles. Part I: Mechanisms and Adsorption-Induced Nanoparticle Agglomeration. *ACS Omega* **2021**, *6*, 19086–19098. [CrossRef] [PubMed]
229. Pooresmaeil, M.; Mansoori, Y.; Mirzaeinejad, M.; Khodayari, A.L.I. Efficient Removal of Methylene Blue by Novel Magnetic Hydrogel Nanocomposites of Poly(acrylic acid). *Adv. Polym. Technol.* **2016**, *37*, 262–274. [CrossRef]
230. Othman, N.H.; Alias, N.H.; Shahruddin, M.Z.; Abu Bakar, N.F.; Him, N.R.N.; Lau, W.J. Adsorption kinetics of methylene blue dyes onto magnetic graphene oxide. *J. Environ. Chem. Eng.* **2018**, *6*, 2803–2811. [CrossRef]
231. Meili, L.; Lins, P.; Costa, M.; Almeida, R.; Abud, A.K.; Soletti, J.; Dotto, G.L.; Tanabe, E.; Sellaoui, L.; de Carvalho, S.H.V.; et al. Adsorption of methylene blue on agroindustrial wastes: Experimental investigation and phenomenological modelling. *Prog. Biophys. Mol. Biol.* **2018**, *141*, 60–71. [CrossRef] [PubMed]
232. Esmaeili, H.; Foroutan, R. Adsorptive Behavior of Methylene Blue onto Sawdust of Sour Lemon, Date Palm, and Eucalyptus as Agricultural Wastes. *J. Dispers. Sci. Technol.* **2018**, *40*, 990–999. [CrossRef]
233. Luo, L.; Wu, X.; Li, Z.; Zhou, Y.; Chen, T.; Fan, M.; Zhao, W. Synthesis of activated carbon from biowaste of fir bark for methylene blue removal. *R. Soc. Open Sci.* **2019**, *6*, 190523. [CrossRef]
234. Rashid, J.; Tehreem, F.; Rehman, A.; Kumar, R. Synthesis using natural functionalization of activated carbon from pumpkin peels for decolourization of aqueous methylene blue. *Sci. Total Environ.* **2019**, *671*, 369–376. [CrossRef]
235. Shrestha, L.; Thapa, M.; Shrestha, R.; Maji, S.; Pradhananga, R.; Ariga, K. Rice Husk-Derived High Surface Area Nanoporous Carbon Materials with Excellent Iodine and Methylene Blue Adsorption Properties. *C J. Carbon Res.* **2019**, *5*, 10. [CrossRef]
236. Gherbia, A.; Chergui, A.; Yeddou, A.R.; Selatnia Ammar, S.; Boubekeur, N. Removal of methylene blue using activated carbon prepared from date stones activated with NaOH. *Glob. Nest J.* **2019**, *21*, 374–380. [CrossRef]
237. Shen, X.; Huang, P.; Li, F.; Wang, X.; Yuan, T.; Sun, R. Compressive alginate sponge derived from seaweed biomass resources for methylene blue removal from wastewater. *Polymers* **2019**, *11*, 961. [CrossRef]
238. Sakr, F.; Alahiane, S.; Sennaoui, A.; Dinne, M.; Bakas, I.; Assabbane, A. Removal of cationic dye (Methylene Blue) from aqueous solution by adsorption on two type of biomaterial of South Morocco. *Mater. Today Proc.* **2020**, *22*, 93–96. [CrossRef]

239. Dos Santos, K.J.L.; de Souza dos Santos, G.E.; de Sá, Í.M.G.L.; de Carvalho, S.H.V.; Soletti, J.I.; Meili, L.; da Silva Duarte, J.L.; Bispo, M.D.; Dotto, G.L. Syagrus oleracea–activated carbon prepared by vacuum pyrolysis for methylene blue adsorption. *Environ. Sci. Pollut. Res.* **2019**, *26*, 16470–16481. [CrossRef] [PubMed]
240. Rawat, A.P.; Kumar, V.; Singh, D.P. A combined effect of adsorption and reduction potential of biochar derived from Mentha plant waste on removal of methylene blue dye from aqueous solution. *Sep. Sci. Technol.* **2020**, *55*, 907–921. [CrossRef]
241. Shafiq, M.; Alazba, A.A.; Amin, M.T. Synthesis, characterization, and application of date palm leaf waste-derived biochar to remove cadmium and hazardous cationic dyes from synthetic wastewater. *Arab. J. Geosci.* **2019**, *12*, 63. [CrossRef]
242. Sahu, S.; Pahi, S.; Sahu, J.K.; Sahu, U.K.; Patel, R.K. Kendu (Diospyros melanoxylon Roxb) fruit peel activated carbon—an efficient bioadsorbent for methylene blue dye: Equilibrium, kinetic, and thermodynamic study. *Environ. Sci. Pollut. Res.* **2020**, *27*, 22579–22592. [CrossRef]
243. Myneni, V.R.; Kanidarapu, N.R.; Vangalapati, M. Methylene blue adsorption by magnesium oxide nanoparticles immobilized with chitosan (CS-MgONP): Response surface methodology, isotherm, kinetics and thermodynamic studies. *Iran. J. Chem. Chem. Eng.* **2020**, *39*, 29–42.
244. Bayomie, O.S.; Kandeel, H.; Shoeib, T.; Yang, H.; Youssef, N.; El-Sayed, M.M.H. Novel approach for effective removal of methylene blue dye from water using fava bean peel waste. *Sci. Rep.* **2020**, *10*, 7824. [CrossRef] [PubMed]
245. Gago, D.; Chagas, R.; Ferreira, M.; Velizarov, S.; Coelhoso, I. A novel cellulose-based polymer for efficient removal of methylene blue. *Membranes* **2020**, *10*, 13. [CrossRef]
246. Othman, I.; Abu Haija, M.; Kannan, P.; Banat, F. Adsorptive Removal of Methylene Blue from Water Using High-Performance Alginate-Based Beads. *Water Air Soil Pollut.* **2020**, *231*, 1–16. [CrossRef]
247. Thabede, P.M.; Shooto, N.D.; Naidoo, E.B. Removal of methylene blue dye and lead ions from aqueous solution using activated carbon from black cumin seeds. *S. Afr. J. Chem. Eng.* **2020**, *33*, 39–50. [CrossRef]
248. Jawad, A.H.; Saud Abdulhameed, A.; Wilson, L.D.; Syed-Hassan, S.S.A.; ALOthman, Z.A.; Rizwan Khan, M. High surface area and mesoporous activated carbon from KOH-activated dragon fruit peels for methylene blue dye adsorption: Optimization and mechanism study. *Chin. J. Chem. Eng.* **2021**, *32*, 281–290. [CrossRef]
249. Dao, M.U.; Le, H.S.; Hoang, H.Y.; Tran, V.A.; Doan, V.D.; Le, T.T.N.; Sirotkin, A.; Le, V.T. Natural core-shell structure activated carbon beads derived from Litsea glutinosa seeds for removal of methylene blue: Facile preparation, characterization, and adsorption properties. *Environ. Res.* **2021**, *198*, 110481. [CrossRef] [PubMed]
250. Do, T.H.; Nguyen, V.T.; Dung, N.Q.; Chu, M.N.; Van Kiet, D.; Ngan, T.T.K.; Van Tan, L. Study on methylene blue adsorption of activated carbon made from Moringa oleifera leaf. *Mater. Today Proc.* **2020**, *38*, 3405–3413. [CrossRef]
251. Abdulhameed, A.S.; Firdaus Hum, N.N.M.; Rangabhashiyam, S.; Jawad, A.H.; Wilson, L.D.; Yaseen, Z.M.; Al-Kahtani, A.A.; Alothman, Z.A. Statistical modeling and mechanistic pathway for methylene blue dye removal by high surface area and mesoporous grass-based activated carbon using K2CO3activator. *J. Environ. Chem. Eng.* **2021**, *9*, 105530. [CrossRef]
252. Zhang, Z.; Xu, L.; Liu, Y.; Feng, R.; Zou, T.; Zhang, Y.; Kang, Y.; Zhou, P. Efficient removal of methylene blue using the mesoporous activated carbon obtained from mangosteen peel wastes: Kinetic, equilibrium, and thermodynamic studies. *Microporous Mesoporous Mater.* **2021**, *315*, 110904. [CrossRef]
253. Yağmur, H.K.; Kaya, İ. Synthesis and characterization of magnetic $ZnCl_2$-activated carbon produced from coconut shell for the adsorption of methylene blue. *J. Mol. Struct.* **2021**, *1232*, 130071. [CrossRef]
254. Zhou, Y.; Li, T.; Shen, J.; Meng, Y.; Tong, S.; Guan, Q.; Xia, X. Core-shell structured magnetic carboxymethyl cellulose-based hydrogel nanosorbents for effective adsorption of methylene blue from aqueous solution. *Polymers* **2021**, *13*, 3054. [CrossRef]
255. Misran, E.; Bani, O.; Situmeang, E.M.; Purba, A.S. Banana stem based activated carbon as a low-cost adsorbent for methylene blue removal: Isotherm, kinetics, and reusability. *Alex. Eng. J.* **2021**, *61*, 1946–1955. [CrossRef]
256. Alamin, N.U.; Khan, A.S.; Nasrullah, A.; Iqbal, J.; Ullah, Z.; Din, I.U.; Muhammad, N.; Khan, S.Z. Activated carbon-alginate beads impregnated with surfactant as sustainable adsorbent for efficient removal of methylene blue. *Int. J. Biol. Macromol.* **2021**, *176*, 233–243. [CrossRef]
257. El Nemr, A.; Shoaib, A.G.M.; El Sikaily, A.; Mohamed, A.E.D.A.; Hassan, A.F. Evaluation of Cationic Methylene Blue Dye Removal by High Surface Area Mesoporous Activated Carbon Derived from Ulva lactuca. *Environ. Process.* **2021**, *8*, 311–332. [CrossRef]
258. Sulaiman, N.S.; Amini, M.H.M.; Danish, M.; Sulaiman, O.; Hashim, R. Kinetics, Thermodynamics, and Isotherms of Methylene Blue Adsorption Study onto Cassava Stem Activated Carbon. *Water* **2021**, *13*, 2936. [CrossRef]
259. Sun, Z.; Qu, K.; Cheng, Y.; You, Y.; Huang, Z.; Umar, A.; Ibrahim, Y.S.A.; Algadi, H.; Castañeda, L.; Colorado, H.A.; et al. Corncob-derived Activated Carbon for Efficiently Adsorption Dye in Sewage. *ES Food Agrofor.* **2021**, *4*, 61–73. [CrossRef]
260. Bulgariu, L.; Escudero, L.B.; Bello, O.S.; Iqbal, M.; Nisar, J.; Adegoke, K.A.; Alakhras, F.; Kornaros, M.; Anastopoulos, I. The utilization of leaf-based adsorbents for dyes removal: A review. *J. Mol. Liq.* **2019**, *276*, 728–747. [CrossRef]
261. Praveen, S.; Gokulan, R.; Pushpa, T.B.; Jegan, J. Techno-economic feasibility of biochar as biosorbent for basic dye sequestration. *J. Indian Chem. Soc.* **2021**, *98*, 100107. [CrossRef]
262. Bello, O.S.; Adegoke, K.A.; Sarumi, O.O.; Lameed, O.S. Functionalized locust bean pod (Parkia biglobosa) activated carbon for Rhodamine B dye removal. *Heliyon* **2019**, *5*, e02323. [CrossRef]
263. Moosavi, S.; Lai, C.W.; Gan, S.; Zamiri, G.; Akbarzadeh Pivehzhani, O.; Johan, M.R. Application of efficient magnetic particles and activated carbon for dye removal from wastewater. *ACS Omega* **2020**, *5*, 20684–20697. [CrossRef]

264. Patel, H. Review on solvent desorption study from exhausted adsorbent. *J. Saudi Chem. Soc.* **2021**, *25*, 101302. [CrossRef]
265. Dey, M.D.; Das, S.; Kumar, R.; Doley, R.; Bhattacharya, S.S.; Mukhopadhyay, R. Vermiremoval of methylene blue using Eisenia fetida: A potential strategy for bioremediation of synthetic dye-containing effluents. *Ecol. Eng.* **2017**, *106*, 200–208. [CrossRef]
266. Saha, A.; Basak, B.B.; Ponnuchamy, M. Performance of activated carbon derived from Cymbopogon winterianus distillation waste for scavenging of aqueous toxic anionic dye Congo red: Comparison with commercial activated carbon. *Sep. Sci. Technol.* **2020**, *55*, 1970–1983. [CrossRef]
267. Dahiru, M.; Zango, Z.U.; Haruna, M.A. Cationic Dyes Removal Using Low-Cost Banana Peel Biosorbent. *Am. J. Mater. Sci.* **2018**, *8*, 32–38. [CrossRef]
268. Yang, X.; Debeli, D.K.; Shan, G.; Pan, P. Selective adsorption and high recovery of La3+ using graphene oxide/poly (N-isopropyl acrylamide-maleic acid) cryogel. *Chem. Eng. J.* **2020**, *379*, 122335. [CrossRef]
269. Naghizadeh, A.; Momeni, F.; Derakhshani, E. Efficiency of ultrasonic process in the regeneration of graphene nanoparticles saturated with humic acid. *Desalin. Water Treat.* **2017**, *70*, 290–293. [CrossRef]
270. Zhang, A.; Li, X.; Xing, J.; Xu, G. Adsorption of potentially toxic elements in water by modified biochar: A review. *J. Environ. Chem. Eng.* **2020**, *8*, 104196. [CrossRef]
271. Mohan, D.; Pittman, C.U., Jr. Arsenic removal from water/wastewater using adsorbents—A critical review. *J. Hazard. Mater.* **2007**, *142*, 1–53. [CrossRef] [PubMed]
272. Paudyal, H.; Ohto, K.; Kawakita, H.; Inoue, K. Recovery of fluoride from water through adsorption using orange-waste gel, followed by desorption using saturated lime water. *J. Mater. Cycles Waste Manag.* **2020**, *22*, 1484–1491. [CrossRef]
273. He, D.; Zhang, L.; Zhao, Y.; Mei, Y.; Chen, D.; He, S.; Luo, Y. Recycling Spent Cr Adsorbents as Catalyst for Eliminating Methylmercaptan. *Environ. Sci. Technol.* **2018**, *52*, 3669–3675. [CrossRef] [PubMed]
274. Avinash, A.; Murugesan, A. Judicious Recycling of Biobased Adsorbents for Biodiesel Purification: A Critical Review. *Environ. Prog. Sustain. Energy* **2019**, *38*, e13077. [CrossRef]
275. Mukherjee, S.; Halder, G. *A Review on the Sorptive Elimination of Fluoride from Contaminated Wastewater*; Elsevier: Amsterdam, The Netherlands, 2018; Volume 6, ISBN 3432754078.
276. Saikia, J.; Goswamee, R.L. Use of carbon coated ceramic barriers for adsorptive removal of fluoride and permanent immobilization of the spent adsorbent barriers. *SN Appl. Sci.* **2019**, *1*, 634. [CrossRef]
277. Kolinko, P.A.; Smirniotis, P.G.; Kozlov, D.V.; Vorontsov, A.V. Cr modified TiO_2-loaded MCM-41 catalysts for UV-light driven photodegradation of diethyl sulfide and ethanol. *J. Photochem. Photobiol. A Chem.* **2012**, *232*, 1–7. [CrossRef]
278. Meng, J.; Rao, F.; Changmei, S.; Rongjun, Q.; Ying, Z. Silica gel-based adsorbents prepared via homogeneous and heterogeneous routes: Adsorption properties and recycling as heterogeneous catalysts. *Polym. Int.* **2017**, *66*, 1913–1920. [CrossRef]
279. Fu, Y.; Jiang, J.; Chen, Z.; Ying, S.; Wang, J.; Hu, J. Rapid and selective removal of Hg(II) ions and high catalytic performance of the spent adsorbent based on functionalized mesoporous silica/poly(m-aminothiophenol) nanocomposite. *J. Mol. Liq.* **2019**, *286*, 110746. [CrossRef]
280. Dutta, D.; Roy, S.K.; Talukdar, A.K. Effective removal of Cr(VI) from aqueous solution by diamino-functionalised mesoporous MCM-48 and selective oxidation of cyclohexene and ethylbenzene over the Cr containing spent adsorbent. *J. Environ. Chem. Eng.* **2017**, *5*, 4707–4715. [CrossRef]
281. Mondal, M.K.; Garg, R. A comprehensive review on removal of arsenic using activated carbon prepared from easily available waste materials. *Environ. Sci. Pollut. Res.* **2017**, *24*, 13295–13306. [CrossRef] [PubMed]
282. Reddy, D.H.K.; Vijayaraghavan, K.; Kim, J.A.; Yun, Y.-S. Valorisation of post-sorption materials: Opportunities, strategies, and challenges. *Adv. Colloid Interface Sci.* **2017**, *242*, 35–58. [CrossRef] [PubMed]
283. Verbinnen, B.; Block, C.; Van Caneghem, J.; Vandecasteele, C. Recycling of spent adsorbents for oxyanions and heavy metal ions in the production of ceramics. *Waste Manag.* **2015**, *45*, 407–411. [CrossRef]
284. Majumder, A.; Ramrakhiani, L.; Mukherjee, D.; Mishra, U.; Halder, A.; Mandal, A.K.; Ghosh, S. Green synthesis of iron oxide nanoparticles for arsenic remediation in water and sludge utilization. *Clean Technol. Environ. Policy* **2019**, *21*, 795–813. [CrossRef]
285. Bădescu, I.S.; Bulgariu, D.; Ahmad, I.; Bulgariu, L. Valorisation possibilities of exhausted biosorbents loaded with metal ions—A review. *J. Environ. Manag.* **2018**, *224*, 288–297. [CrossRef]
286. Manzoor, K.; Ahmad, M.; Ahmad, S.; Ikram, S. Synthesis, Characterization, Kinetics, and Thermodynamics of EDTA-Modified Chitosan-Carboxymethyl Cellulose as Cu (II) Ion Adsorbent. *ACS Omega* **2019**, *4*, 17425–17437. [CrossRef]
287. Abdallah, M.M.; Ahmad, M.N.; Walker, G.; Leahy, J.J.; Kwapinski, W. Batch and Continuous Systems for Zn, Cu, and Pb Metal Ions Adsorption on Spent Mushroom Compost Biochar. *Ind. Eng. Chem. Res.* **2019**, *58*, 7296–7307. [CrossRef]
288. Bashir, S.; Hussain, Q.; Akmal, M.; Riaz, M.; Hu, H.; Ijaz, S.S.; Iqbal, M.; Abro, S.; Mehmood, S.; Ahmad, M. Sugarcane bagasse-derived biochar reduces the cadmium and chromium bioavailability to mash bean and enhances the microbial activity in contaminated soil. *J. Soils Sediments* **2018**, *18*, 874–886. [CrossRef]
289. Yin, Q.; Zhang, B.; Wang, R.; Zhao, Z. Biochar as an adsorbent for inorganic nitrogen and phosphorus removal from water: A review. *Environ. Sci. Pollut. Res.* **2017**, *24*, 26297–26309. [CrossRef] [PubMed]
290. Guedes, R.E.; Luna, A.S.; Torres, A.R. Operating parameters for bio-oil production in biomass pyrolysis: A review. *J. Anal. Appl. Pyrolysis* **2018**, *129*, 134–149. [CrossRef]
291. Kan, T.; Strezov, V.; Evans, T.J. Lignocellulosic biomass pyrolysis: A review of product properties and effects of pyrolysis parameters. *Renew. Sustain. Energy Rev.* **2016**, *57*, 1126–1140. [CrossRef]

292. Cole, A.J.; Paul, N.A.; de Nys, R.; Roberts, D.A. Good for sewage treatment and good for agriculture: Algal based compost and biochar. *J. Environ. Manag.* **2017**, *200*, 105–113. [CrossRef] [PubMed]
293. Vakili, M.; Deng, S.; Cagnetta, G.; Wang, W.; Meng, P.; Liu, D.; Yu, G. Regeneration of chitosan-based adsorbents used in heavy metal adsorption: A review. *Sep. Purif. Technol.* **2019**, *224*, 373–387. [CrossRef]
294. Zelmanov, G.; Semiat, R. Iron (Fe^{+3}) oxide/hydroxide nanoparticles-based agglomerates suspension as adsorbent for chromium (Cr^{+6}) removal from water and recovery. *Sep. Purif. Technol.* **2011**, *80*, 330–337. [CrossRef]
295. Lata, S.; Singh, P.K.; Samadder, S.R. Regeneration of adsorbents and recovery of heavy metals: A review. *Int. J. Environ. Sci. Technol.* **2015**, *12*, 1461–1478. [CrossRef]

MDPI
St. Alban-Anlage 66
4052 Basel
Switzerland
Tel. +41 61 683 77 34
Fax +41 61 302 89 18
www.mdpi.com

Polymers Editorial Office
E-mail: polymers@mdpi.com
www.mdpi.com/journal/polymers

CPSIA information can be obtained
at www.ICGtesting.com
Printed in the USA
BVHW011200030223
657818BV00010B/856

9 783036 562438